INDUSTRIAL PLASTICS:
Theory and Application
Second edition

Terry L. Richardson
Northern State College
Aberdeen, South Dakota

Delmar Publishers Inc.®

NOTICE TO THE READER

COVER PHOTO CREDITS:

Delmar Staff
Production Editor: Mary Ormsbee
Design Coordinator: Susan Mathews
Publications Coordinator: Karen Seebald

For information, address Delmar Publishers Inc.,
2 Computer Drive West, Box 15-015
Albany, New York 12212-5015

COPYRIGHT © 1989
BY DELMAR PUBLISHERS INC.

Printed in the United States of America
Published simultaneously in Canada
by Nelson Canada,
A Division of International Thomson Limited

10 9 8 7 6 5 4 3 2 1

Library of Congress Cataloging in Publication Data
Richardson, Terry L.
    Industrial plastics: theory and application/Terry L.
Richardson. — 2nd ed.
        p.   cm.
    Includes index.
    ISBN 0-8273-3392-7 (pbk.).   ISBN 0-8273-3393-5 (instructor's
guide
    1. Plastics.   I. Title.
TP1120.R49 1988
668.4 — dc19

                                                                88-3683
                                                                CIP

CONTENTS

FOREWORD

The Society of Plastics Engineers is pleased to sponsor and endorse the second edition of *INDUSTRIAL PLASTICS: Theory and Application,* by Dr. Terry Richardson. One of the leading educators in plastics technology, Dr. Richardson first wrote this invaluable basic introduction to plastics in 1982. It was published in early 1983 and has proven to be a best seller, in that a second edition was undertaken less than five years later.

The basic chapter headings remain the same. However, overall content has been substantially revised to keep pace with technological advances. For example, extensive material on polymer composites and composite technology has been added to several chapters. And Chapter 3 has been rewritten to reflect current issues and trends regarding health and safety in the plastics industry.

SPE, through its Technical Volumes Committee, has long sponsored books on various aspects of plastics and polymers. Its involvement has ranged from identifying needed volumes to recruiting authors. An everpresent ingredient has been review of the final manuscript to insure technical accuracy.

This technical competence pervades all SPE activities, not only in publication of books but in other areas such as technical conferences and educational programs. In addition, the Society publishes five periodicals—*Plastics Engineering, Polymer Engineering and Science, Polymer Processing and Rheology, Journal of Vinyl Technology,* and *Polymer Composites* — as well as conference proceedings and other selected publications, all of which are subject to similar rigorous technical review procedures.

The resource of some 25,000 practicing plastics engineers has made SPE the largest organization of its type in plastics worldwide. Further information is available from the Society at 14 Fairfield Drive, Brookfield Center, Connecticut 06805, U.S.A.

Robert D. Forger
Executive Director
Society of Plastics Engineers

Technical Volumes Committee
Thomas W. Haas, Chairman
Virginia Commonwealth University

PREFACE

INDUSTRIAL PLASTICS: Theory and Application is designed to meet the educational needs of the plastics industry of today. The purpose of this text is to introduce an important class of engineering materials, processes, and applications in the plastics industry. It should be of interest to all those involved in the industry, whether in learning, training, manufacturing, or designing.

Charles Kettering used to say "Chance favors the prepared mind." An increasing number of people, in a wide variety of industries, are finding they need to know about plastics. Each family of plastics has unique characteristics that are as different from metals, wood, and ceramics as they are different from each other.

Despite the existence of plastics technology for a number of years, new, innovative, creative ideas are yet to be realized from it.

The plastics industry has never been stagnant, but continues to expand. New materials, processes, and applications are introduced each year. Polymer composites continue to experience rapid growth. There are not enough qualified technical personnel to meet the needs of this ever expanding industry. This lack of knowledge has sometimes led to product failure, over-design, wasted resources, lower profits, and environmental and health concerns.

This text can be used at various levels, at the technician level or to upgrade the practicing technician. The information should be an invaluable reference, resource, or asset to all those who will need to select the correct material, process, design, or product application.

The confidential and proprietary details outlined or other information provided is intended only as a guide. It is not to be taken as a license under which to operate or as a recommendation to infringe on any patents.

ORGANIZATION

The plastics technology content in *INDUSTRIAL PLASTICS:* Theory and Application is covered in a practical rather than a theoretical manner.

The material is organized to enable the reader to reach a basic knowledge of chemistry, properties, testing, designing, processing, fabricating, tooling, and manufacturing plastics. Each chapter may be studied independently and not in the exact order presented. Seven appendices are included to reinforce the technical presentations.

FEATURES

Special effort has been made to make *INDUSTRIAL PLASTICS* an effective learning tool. Chapters are divided into short units of instruction. The text is readable. Language and style have been carefully monitored for easy comprehension. All technical terms are defined and used in context. Textual presentations are illustrated extensively. Within each chapter, there is a series of learning reinforcements and immediate feedback through a vocabulary development program and a "test your knowledge" activity.

INDUSTRIAL PLASTICS: Theory and Application is supported by an instructor's manual and key that contains a test bank and key to the end-of-chapter questions. To assist the instructor, a series of transparency/duplicating masters is provided in the manual. These may be used for a visual presentation on the overhead projector, duplicated for class handouts or for class assignments. This teaching/learning program will provide the reader with a basic understanding of the most versatile and useful materials on earth.

Chapter 1

Introduction to Plastics

In this chapter you will see how advances in chemistry and technology have made possible a new industry, and you will learn how that industry — plastics — has changed our lives.

Early humans probably learned about fire from nature. Their earliest tools were stone. Their earliest fuel was wood. Their earliest mode of transportation was on foot.

Early people probably found raw materials — copper, gold, and iron — in the earth. Later, they found methods of processing raw materials. As humans began to industrialize, they needed to make goods that were strong, malleable, durable, and practical. However, materials with all of these properties could not be found in natural form. Synthetic materials, which do not occur naturally, were needed. Over the period of time since humans began to find the materials they needed, many synthetic materials have been developed.

1-1 CHEMISTRY

In the Middle Ages, alchemists tried but failed to produce gold from common metals such as tin, iron, and lead. The modern words *chemistry* and *chemist* are derived from the medieval Latin word *alchimista*.

Chemistry is a science that deals with the composition of matter and how that composition changes. The two broad classes of chemistry are *inorganic* and *organic*. Inorganic chemistry studies matter that is mineral in origin. Organic chemistry deals with matter that contains the element *carbon*. The term organic was used originally to mean com-

pounds of plant or animal origin, but now it includes also many synthetic materials that have been developed through research. One such group of synthetic organic materials is called *plastics*.

1-2 THE AGE OF PLASTICS

What material may be as hard as stone, as transparent as glass, as elastic as rubber? What can be strong yet light, moisture and chemical resistant, and any of a rainbow of colors? The answer, of course, is plastics. According to the Society of the Plastics Industry, Inc. (SPI), production by volume of plastics will surpass that of steel and all other materials combined by the year 2000 (Figs. 1-1 and 1-2). Today may truly be called the *Plastics Age*.

The word *polymer* is derived from the Greek *poly,* meaning many, and *meros,* meaning unit or part. Polymers are also called resins (incorrectly), macromolecules (from the Greek *macros,* meaning long), elastomers, and plastics. We will discuss polymers further in Chapter 2.

Elastomer describes any polymer that can be stretched more than two hundred percent. The American Society for Testing and Materials (ASTM) defines an elastomer as "a polymeric material which at room temperature can be stretched at least twice its original length and upon immediate release of the stress will return quickly to approximately its original length."

The word *plastic* is an adjective meaning pliable and capable of being shaped by pressure. It is often incorrectly used as the generic word for the plastics industry and its products. A major reason

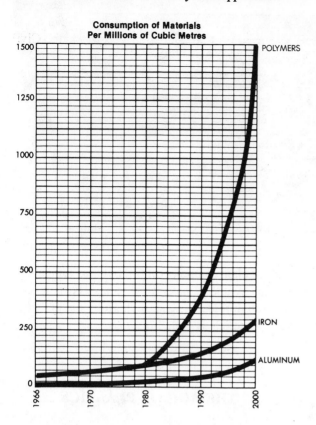

**Consumption of Materials
Per Millions of Cubic Metres**

Fig. 1-1. Projected annual world consumption of selected materials through the year 2000. *Based on presentations at the World Chemical Engineering Congress.* (SPI)

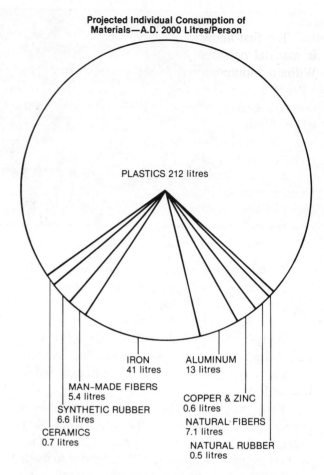

**Projected Individual Consumption of
Materials—A.D. 2000 Litres/Person**

PLASTICS 212 litres

IRON
41 litres

ALUMINUM
13 litres

MAN–MADE FIBERS
5.4 litres

COPPER & ZINC
0.6 litres

SYNTHETIC RUBBER
6.6 litres

NATURAL FIBERS
7.1 litres

CERAMICS
0.7 litres

NATURAL RUBBER
0.5 litres

Fig. 1-2. Projected annual consumption of materials, per person, by the year 2000. *Based on presentations at the World Chemical Engineering Congress.* (SPI)

for confusion is that the word may be used as a noun or as an adjective with both singular and plural meanings.

The word *plastics* comes from the Greek word *plastikos*. It means "to form or fit for molding." The Society of the Plastics Industry has defined plastics as follows:

> Any one of a large and varied group of materials consisting wholly or in part of combinations of carbon with oxygen, nitrogen, hydrogen, and other organic or inorganic elements which, while solid in the finished state, at some stage in its manufacture is made liquid, and thus capable of being formed into various shapes, most usually through the application, either singly or together, of heat and pressure.

Because plastics are closely related to resins, the two are often confused. Resins are gum-like solid or semisolid substances used in making such products as paints, varnishes, and plastics.

A resin is not a plastics unless and until the resin has become a "solid in the finished state." Plastics products are often made from resins that are processed and made solid. Both natural and synthetic resin materials are composed of a series of molecules bonded together with some resins having several thousand bonded molecules.

1-3 HISTORY OF PLASTICS

Observations by the Swedish chemist J. J. Berzelius led, in 1833, to the production of one of the first synthetic compounds containing several thousand molecules.

The term *polymer* was first used by the chemist H. V. Regnault, when, in 1835, he synthesized the plastics vinyl chloride. A polymer is a substance that has large molecules made by joining many small molecules of one or several substances.

The first known commercial use of polymeric material occurred in 1843, by Dr. George IV William Montgomerie, a Malayan surgeon. He noted that the Malayans used a natural polymer material, *gutta percha,* to make knife handles and whips. Resins collected from the gutta percha trees were sent to England to interested scientists and manufacturers.

Michael Faraday, a pioneer in electricity, found that gutta percha was a good electrical insulator in water and so it was used as insulation for the first transatlantic cable. Modern undersea cables still use plastics for bases, housings, shields, and cores. Figure 1-3 shows how modern technologies depend upon the dielectric and other electrical insulating properties of polymers.

In 1862, Alexander Parks of Birmingham, England, displayed a new plastics at the International Exhibition in London. Called *Parkesine,* this plastics was made from nitrocellulose containing less than 12 percent nitrogen. When the nitrogen content of nitrocellulose, a combination of nitric acid and cellulose, is more than 13 percent, the material is known as *guncotton,* an explosive used in the Civil War and World War I.

John W. Hyatt, a New York printer, began synthetic plastics production in the United States. In 1868, Hyatt responded to an advertisement to find a substitute for gutta percha and for the ivory used for billiard balls. There is no record that Hyatt ever received the reward, but he did succeed in producing a new plastics. Figure 1-4 shows the *Celluloid* billiard ball produced by John Hyatt and his brother, Isaiah.

Hyatt accidentally mixed camphor with some *pyroxylin* (a nitrocellulose with low nitrogen content), and Celluloid resulted. This new, easily molded material was less explosive than previous nitrocellulose plastics.

At the end of the Civil War, there were huge quantities of surplus nitrocellulose. Hyatt and his brother bought a large amount of this surplus to use as raw material and they were granted over 75 United States patents for the production of plastics. An elaborately molded comb and brush produced in the early years of the plastics industry are shown in Figure 1-5.

In 1897, W. Krische found that protein from milk could be used to make a new plastic material called *casein.* A Bavarian, Adolf Spittler, found that treating the compressed protein sheets with formaldehyde improved their water-resistance. Casein plastics found little favor in the United States, except as adhesives.

Based on the earlier work of a German chemist, Adolf von Baeyer, Leo Hendrik Baekeland of

Fig. 1–3. Modern technologies depend on the dielectric and other electrical insulating properties of polymers. (B. F. Goodrich)

Fig. 1-4. The Hyatt billiard ball, an early Celluloid product. (Celanese Plastic Materials Co.)

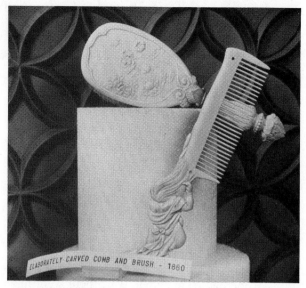

Fig. 1-5. A comb and brush produced from Celluloid in 1880. (Celanese Plastic Materials Co.)

Table 1-1. Chronology of Plastics

Date	Material	Example
1868	Cellulose nitrate	Eyeglass frames
1909	Phenol-formaldehyde	Telephone handset
1909	Cold molded	Knobs and handles
1919	Casein	Knitting needles
1926	Alkyd	Electrical bases
1926	Analine-formaldehyde	Terminal boards
1927	Cellulose acetate	Toothbrushes, packaging
1927	Polyvinyl chloride	Raincoats
1929	Urea-formaldehyde	Lighting fixtures
1935	Ethyl cellulose	Flashlight cases
1936	Acrylic	Brush backs, displays
1936	Polyvinyl acetate	Flash bulb lining
1938	Cellulose acetate butyrate	Irrigation pipe
1938	Polystyrene or styrene	Kitchen housewares
1938	Nylon (polyamide)	Gears
1938	Polyvinyl acetal	Safety glass interlayer
1939	Polyvinylidene chloride	Auto seat covers
1939	Melamine-formaldehyde	Tableware
1942	Polyester	Boat hulls
1942	Polyethylene	Squeezable bottles
1943	Fluorocarbon	Industrial gaskets
1943	Silicone	Motor insulation
1945	Cellulose propionate	Automatic pens and pencils
1947	Epoxy	Tools and jigs
1948	Acrylonitrile-butadiene-styrene	Luggage
1949	Allylic	Electrical connectors
1954	Polyurethane or urethane	Foam cushions
1956	Acetal	Automotive parts
1957	Polypropylene	Safety helmets
1957	Polycarbonate	Appliance parts
1959	Chlorinated polyether	Valves and fittings
1962	Phenoxy	Bottles
1962	Polyallomer	Typewriter cases
1964	Ionomer	Skin packages
1964	Polyphenylene oxide	Battery cases
1964	Polyimide	Bearings
1964	Ethylene-vinyl acetate	Heavy-gauge flexible sheeting
1965	Parylene	Insulating coatings
1965	Polysulfone	Electrical and electronic parts
1965	Polymethylpentene	Food bags
1970	Poly(amide-imide)	Films
1970	Thermoplastic polyester	Electrical and electronic parts
1972	Thermoplastic polyimides	Valve seats
1972	Perfluoroalkoxy	Coatings
1972	Polyaryl ether	Recreation helmets
1973	Polyethersulfone	Oven windows
1974	Aromatic polyesters	Circuit boards
1974	Polybutylene	Pipes
1975	Nitrile barrier resins	Packaging
1976	Polyphenylsulfone	Aerospace components
1978	Bismaleimide	Circuit boards
1982	Polyetherimide	Ovenable containers
1983	Polyetheretherketone	Wire coating

the United States produced a new resin, phenol-formaldehyde, that advanced the commercial use and acceptance of plastics. In 1909, Baekeland secured patents to and began producing plastics products using the trade name *Bakelite*.

Plastics technology developed as new polymers were discovered and new methods of forming and shaping them were devised. New steels for use in molding plastics had to be made. Machinery builders, engineers, educators, salespeople, and others had to learn about this new technology. This education process continues today (Table 1-1).

It might be said that the plastics industry was in the bronze age of plastics development during the 1970s. Most development during this time was directed toward discovering new monomers or homopolymers.

Today, polymer chemists and engineers are combining (or, in a sense, alloying) and assembling many synthetic polymers to improve properties, improve processing, or lower costs. The idea of combining or assembling two or more different materials is much older than modifying the polymer chain. Many of the demands made upon modern materials are so severe that individual materials no longer perform or have the desired properties.

1-4 CLASSIFICATION OF POLYMERS

Polymers may be classified by six different systems: *source, light penetration, heat reaction,*

Table 1-1. Chronology of Plastics (Continued)

Date	Material	Example
1983	Interpenetrating Networks (IPN)	Shower stalls
1983	Polyarylsulfone	Lamp housings
1984	Polyimidesulfone	Convey links
1985	Polyketone	Automotive engine parts
1985	Polyether sulfonamide	Cams
1985	Liquid-crystal polymers	Electronic components

Table 1-2. Principal Polymer Sources

Category	Source	Origin
Natural	Amber Asphalt Copal Lignin Shellac Tar	Naturally occuring deposits; tree gum, sap, and wood cell binder; insect byproduct
Modified Natural	Casein Cellulose Gelatin Protein Rubber	Animal protein, esters, and regenerated animal protein; Plant and animal protein; natural latex
Synthetic	Agricultural products	Fatty acids, alcohols, formaldehydes, and esters from plants; Glycerin, ureas, alcohols, and formaldehyde from animals
	Coal products	Benzene, toluene, coke, xylene, and naphthalene
	Natural gas products	Acetylene and methane
	Petroleum products	Acetylene, ethylene propylene, butene, benzene, toluene, xylene, and naphthalene

polymerization reactions, molecular structure, and *crystal structure.*

Source

When classifying polymers by source, there are three principal categories. These are natural, modified natural, and synthetic polymers (Table 1-2).

Natural Sources of resins include animals, vegetables, and minerals.

Naturally occurring polymers are easily shaped but have little strength. Some common examples are the following:

- *Rosin* is a byproduct of turpentine distilling. You may have seen rosin oozing out of pine tree stumps or lumber. It was once widely used in making linoleum and for electrical insulating compounds.
- *Asphalt,* sometimes called pitch, is found in a natural state formed from the decaying remains of plants and animals. Today, most asphalt is a by-product of the petroleum industry. Asphalt was once used to mold battery cases and electrical insulators.
- *Tar* is obtained by distilling organic materials. Wood, waste fats, petroleum, coal, and peat are sources. Tar is still used on road surfaces, roofs, and in making some color dyes.
- *Amber* is a fossilized resin that formed from the oily sap of ancient coniferous trees. It was once used to mold knife handles and other objects.
- *Copal* is a resin derived from tropical trees. Copal varies from white to brown in color and its resins were once used in paints, linoleum, and varnishes.
- *Lignin* is a resinous binder that surrounds each wood cell. The lignin content of wood varies from 35 to 90 percent depending on the species. Today, most lignin is used as filler for plastics, or as an adhesive to bind wood chips together under pressure.

- *Shellac* was the first natural resin to be molded. At one time, large quantities of shellac were molded into phonograph records but today, it is primarily used as a filler, wood finish, and electrical insulation.

Modified Natural Sources of resins include cellulose and protein.

Modifying natural sources of resins was a natural step toward improving polymeric materials and developing new polymers and product applications. *Cellulose* is a major part of all plants, thus, it is readily available as a raw material for plastics.

One of the purest forms of cellulose is cotton linters. These are the short fibers that remain on the cotton seeds after the longer fibers have been separated. Polymers made from linters do not yellow and are very transparent. Another major source of cellulose is wood pulp.

Other cellulose sources are plant wastes and residues, such as straw, cornstalks, corncobs, grass, and weeds. All of these materials have proved useful, but are difficult to collect and process.

Cellulose is a very complex material, and many types of plastics are made from it. There are more than ten basic resins in the cellulose family. The most familiar resins are cellulose acetate, cellulose nitrate, cellulose acetate butyrate, and cellulose propionate.

Another modified natural resin source is *protein*, which comes from milk, soybeans, peanuts, coffee beans, and corn. Casein made from skim milk is the only protein-derived plastics that has had some commercial success.

There are other protein sources including hair, feathers, bones, and similar wastes. There has been little interest in these sources because collection is not easy and there are not many uses for protein-derived plastics. Most are used for adhesives and coatings. Protein plastics absorb moisture and swell, therefore items made from them may warp and change shape.

Gelatin is a protein mixture made from bones, hoofs, and animal skins. It is used as an odorless, tasteless, and transparent filler in candies, meats, ice cream, and drugs.

A very important polymer, *rubber*, is made from a milky juice called *latex*. When people talk about natural rubber, they are usually referring to elastomers made from the latex of the *Hevea* tree. A mature rubber tree will yield about 1.8 kilograms (kg) or 4 pounds (lb) of rubber per year. A polymer that is chemically like rubber is gutta percha. Gutta percha trees that yield the milky latex for this polymer are found in Malaya, Borneo, and Sumatra.

Although there is still a large market for natural rubber products, synthetic rubbers have largely replaced natural resins for most uses.

Synthetic sources are the most important sources of resins for plastics. Agriculture, petroleum, and coal are the three main sources of chemicals for synthetic plastic materials. The most important of these three is petroleum. As it comes from the earth, petroleum is mainly a mixture of solid, liquid, and gaseous *hydrocarbons* that are almost useless. In refined forms, petroleum is the largest volume product, other than water, dispensed to people in the U. S.

Distilling and refining petroleum by a process called *fractional distillation* yields various crude oil components. These include heavy asphalt and tar, lighter oils, and a gaseous portion including propane, butane, and other hydrocarbons (Fig. 1-6).

Residues collected near the *bottom* of the refining tower include asphalt and thick oils. Lighter fuels (kerosene, diesel fuel, gasoline, and benzene) are extracted near the top of the distilling tower. Light gases (methane, ethylene, propane, and propylene) are byproducts of further refining.

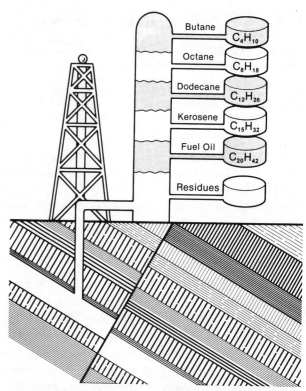

Fig. 1-6. Petroleum is found in deposits deep underground. In a fractionating tower, the material is vaporized by heating. Different products are drawn off at various points.

By varying the proportions of hydrocarbons, salt, chlorine, formaldehyde, nitrogen, and other chemicals, chemists create a great variety of polymers. Hydrocarbons, of course, are the building blocks of polymers. Hydrocarbons also may come from agricultural products, such as cottonseed, linseed, soybean, lard, and safflower. Methane and other gases may come from sewage sludge and crushed coal.

Grain (starch, protein and oil products) has been made into biodegradable plastics, semipermeable membranes, and protective films. A wide range of plastics are produced from agricultural products. Cornstarch-based plastics film is used as a mulch and protective covering that keeps in moisture and heat. Polyethylene-based resins may be made both photodegradable and biodegradable by incorporating cornstarch into the molecular chain. (See Chapter 3.) More than 125 million pounds of petroleum-based plastics film were used as mulch in 1985.

The human race is draining the world's reserves of oil and coal. Soon we will have to use other hydrocarbon sources to produce polymers; as a result, we will need to use more farm and human waste products than we do today.

A study of global prospects for energy up to the year 2000 concludes that "oil is essential for just two major uses: transportation and petrochemical feedstocks." (See "Chemical Emergency Preparedness Program Interim Guidance," in Appendix G, Bibliography.)

Plastics technologists are also working on the secrets of *photosynthesis*. Photosynthesis, the process by which plants convert light and carbon dioxide into sugar, may lead to new ways of producing simple organic chemicals. Bacteria or small plants may ferment sugar to produce alcohol, a raw substance for making many plastics.

Light Penetration

Because many plastics possess unique optical properties they are often classified relative to light penetration. The following optical properties summarize this classification.

- *Opaque* — Light will not pass through. Cannot be seen through.
- *Transparent* — Light will pass through. Can be seen through.
- *Translucent* — Light will pass through. Cannot be seen through.
- *Luminescent* — (a) *fluorescent*: emits light only when electrons are being excited, usually transparent; (b) *phosphorescent*: gives off light energy more slowly than it takes on light, translucent.

See the further discussion of optical properties in Chapter 4.

Heat Reaction

All plastics fall into two broad categories with regard to their reaction to heating. They are either thermoplastic or thermosetting.

Thermoplastic materials become soft when heated and solid when cooled to room temperature. This softening and setting may be repeated many times. It is something like melting and cooling wax. When cooled, the plastics become firm.

The most useful members of the thermoplastic group are acrylics, cellulosics, polyamide, polystyrene, polyethylene, fluoroplastics, polyvinyls, polycarbonate, and polysulfone. See Chapter 7 for a complete discussion of thermoplastics materials.

Thermosetting materials may not be reheated and softened again. Once the structural framework is set, these plastics cannot be reformed. A simplified analogy is that hardening a thermosetting plastics is like baking a cake or boiling an egg.

Useful members of the thermosetting group include aminos, casein, epoxies, phenolics, polyesters, silicones, and polyurethanes. See Chapter 8 for a complete discussion of thermosetting plastics materials.

Polymerization Reaction

Molecules are formed or produced into polymers by two types of reactions: 1) addition or 2) condensation. These two methods of joining many (*poly*) units (*meros*) will be explained in Chapter 2.

Molecular Structure

Polymers may be composed of different repeating units with different structures. All properties of the polymer are affected by the arrangement of the repeating units. The three major molecular structural categories, 1) linear, 2) branched, and 3) cross-linked or network, are described in Chapter 2.

Crystal Structure

Crystallinity is determined primarily by molecular structure. Polymers may be classified as crystalline (actually semicrystalline) and amorphous. Many polymers possess a significant degree of crystallinity. The ordered, three-dimensional arrangement of molecules results in crystallinity. The skeleton or fibrous aggregates of polymer chains (lamellae) generally define the principal direction of the structure geometry. These directional features result in materials that are generally anisotropic. It should be noted that polymers invariably contain a proportion of amorphous material. Polymers have varying percentages or degrees of crystallinity. Polyethylene, polypropylene, polyamide, and polytetrafluoroethylene are classed as crystalline or semicrystalline polymers. Only during the melt stage is a crystalline polymer amorphous.

Increased degree of crystallinity raises the melting point, increases density, and generally improves mechanical properties. Increased crystallinity also generally lowers impact strength, solubility, and optical clarity.

Amorphous polymers in contrast lack any internal skeleton or structure. They are generally described as isotropic materials. The random molecular structures of amorphous polymers do not contain crystalline regions. Polystyrene, poly-methyl methacrylate, polyester polyurethane, urea-formaldehyde, and epoxy are familiar amorphous polymers.

Further information on crystalline structures is provided in Chapter 2. Also see polyolefins in Chapter 7.

1-5 ADVANTAGES AND DISADVANTAGES OF PLASTICS

A list of ten general advantages and disadvantages of plastics follows. It is included to show positive and negative aspects of plastics in general. For example, not all plastics are difficult to repair, nor do all deteriorate easily. Cost factors may be considered an advantage in many plastics products.

Advantages of plastics

1. Corrosion and chemical resistance
2. Good thermal and electrical insulating properties
3. May be made isotropic or anisotropic
4. Good strength-to-mass ratio
5. Light (mass) weight
6. Ease of processing
7. Available in a variety of forms
8. Capable of being foamed or made flexible
9. Available as transparent, translucent, or opaque
10. Available in wide range of colors

Disadvantages of plastics

1. Dimensional instability
2. Limited useful thermal range
3. Fragility (may break, crack, or scratch easily)
4. Flammability (many burn easily)
5. Absorb moisture
6. Non-degradability (some do not decompose)
7. Subject to attack by chemicals (deteriorate)
8. Odors or chemical fumes in processing
9. Difficulty of repair (thermosets)
10. Cost (vary by family)

1-6 MODERN INDUSTRIAL PLASTICS

The first commercial plastics, *Celluloid,* was developed about 100 years ago, but in recent decades, explosive growth and diversification have occurred in the industry. Since World War II, the plastics industry has been growing at a rate double that of other industries. Compare the modern plas-

Fig. 1-7. A modern plant for manufacturing plastics. (Chemplex Company)

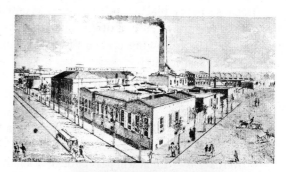

Fig. 1-8. The first plastics plant in the United States, with some of the personnel. (Celanese Plastic Materials Co.)

tics plant shown in Figure 1-7 with the first plastics plant, shown in Figure 1-8.

Plastics are widely used in everyday living — at home, on the job, and even at the frontiers of space. From housewares to exotic aerospace applications, plastics are replacing more traditional materials (Fig. 1-9). The volume of plastics consumed in the United States has exceeded the volume of metals consumed in 1983 (Fig. 1-10).

Although most people think of plastics as housewares or toys, their biggest growth areas are elsewhere (Fig. 1-11). Construction is the number one market for plastics, despite limited acceptance by labor unions and the public. Electronics now uses about 1 000 000 tons of plastics, worth more than three billion dollars, in the manufacture of

(A) Autos with bodies and trim of plastics. (Borg-Warner Chemicals)

(B) Solid vinyl siding doesn't rot or peel, and requires no painting. (Bird & Sons, Inc.)

(C) Composite materials and other plastics parts make up more than 30% of this B-1B bomber. (Rockwell International)

Fig. 1-9. Diverse uses of plastics.

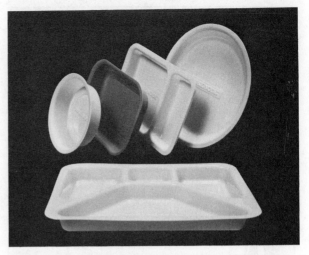

(D) It is hard to imagine modern life without the convenience of plastics packaging. These CPET trays are dual ovenable. (Signode® All-OvenWare™ Tray Division)

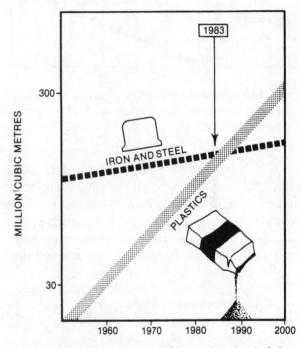

Fig. 1-10. World consumption of iron and steel and of plastics, through the year 2000. From the pamphlet *The Need For Plastics Education*. (SPI/SPE)

components. The growing use of plastics in all modes of transportation is most evident in the automobile industry. The packaging industry has long been a high volume user of plastics, and that trend continues to grow.

1-7 DESIGNING WITH PLASTICS

Plastics have helped produce new designs in automobiles, trucks, buses, rapid transit vehicles,

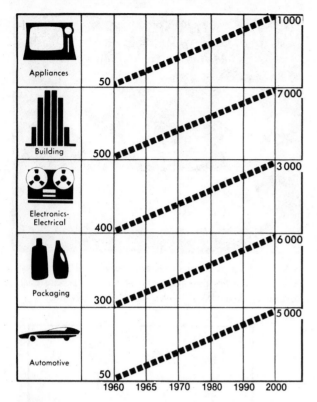

Fig. 1-11. Estimated consumption of plastics by selected industries through the year 2000, in thousands of metric tonnes.

matic advances in aerospace technology. New world-records are being set with the aid of composite bicycles, sleds, canoes, tennis rackets, and other athletic equipment.

Many plastics are stronger and lighter than the metallic parts they replace.

It is frequently necessary to combine several materials into a composite. A composite is formed when two or more materials are combined, usually with the intention of achieving better results than can be obtained with a single, homogenous material.

Some thermoplastic polymers are combinations of more than one polymeric resin system. Alloying and blending techniques are used to improve physical properties, lower cost, or improve processing. These alloys and blends generally have a synergistic effect. The resulting combination may produce a polymer with attributes better than those of either component.

With increasing demands for emission control and fuel economy, plastics are an important automotive design element. Reducing mass is one of the most efficient ways of increasing fuel economy.

Researchers have estimated that every 100 pounds reduced from a 2 500 pound automobile produces fuel savings of 0.3 miles per gallon. The energy resource requirements per pound of most plastics are only half that of aluminum or steel.

Because plastics materials are easily processed, they have become the perfect substitutes for other materials. Often better results may be gained by using plastics as the original design material rather

aircraft, boats, trains, musical instruments, furniture, appliance housings, biocompatible implants, and other applications. Aerospace and military research in composite designs have changed our concept of design and traditional construction techniques. Lightweight composites have allowed dra-

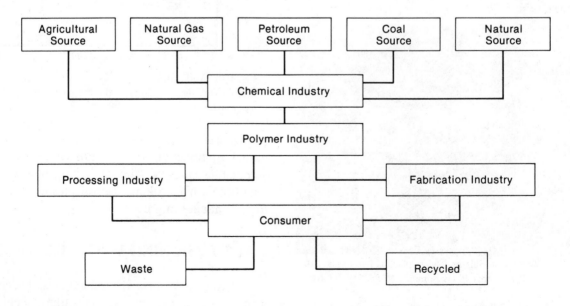

Fig. 1-12. Flow chart of the polymer industry.

than as a substitute material. Some of the early uses of plastics were failures because plastics products were designed with little regard for the properties and limitations of the material. A lack of structural design knowledge is a problem even today, because plastics are not sufficiently understood by some designers, engineers, and technicians.

1-8 EXTENT OF THE PLASTICS INDUSTRY

What is the extent of the plastics industry? With each passing day, it becomes increasingly difficult to define plastics as a discrete industry. Plastics are products of companies in practically every industry including steel, paper, chemicals, petroleum, and electronics, to name only a few.

Plastics are known by many names, some of which are hard to pronounce. For this reason, manufacturers use trade names for their brands of resin, knowing these names are easier for consumers to pronounce and remember. For example, *Acrylite*, *Lucite*, and *Plexiglas* are trade names for polymethyl methacrylate, plastics manufactured by three different companies. *Acrylic* is used as the common name. (See Appendixes B and C.)

The plastics industry may be divided into three large categories that sometimes overlap:

1. The material manufacturer who produces the basic plastics resins from chemical sources
2. The processor who converts the basic plastics into solid shapes
3. The fabricator and the finisher who further fashions and decorates the plastics.

In Figure 1-12, a simple flow chart of the polymer industry is shown. There may be additional connecting lines between the polymer industry, processing industry, and fabrication industry. Also, a connecting line could be drawn to the waste and recycle stage. There are waste products in all stages. Plastics and other chemicals are recycled in each stage. Recycling is more economical by saving energy and material resources.

Odor, noise, and, in processing, radiation pollution are among problems that affect the plastics industry. Pollutants have been given increased attention because of their effect on people and the environment. These effects can involve a remarkable range of problems including chemical, physical, social, physiological, and even psychological damage. The plastics industry must apply the knowledge and technology at hand to solve these problems.

VOCABULARY

The following vocabulary words are found in this chapter. Look up the definition of any of these words you do not understand in the glossary, Appendix A.

Anisotropic
Celluloid
Composites
Elastomer
Feedstocks
Gutta percha
Homopolymers
Hydrocarbons
Inert gases
Isotropic
Lac
Latex
Monomers
Nitrocellulose
Photosynthesis
Plastics
Polymerization reaction
Polymers
Resin
Shellac
Strength-to-mass ratio
Synthetic
Technology
Thermoplastic
Thermosetting
Trade name

1-1. The two broad branches of chemistry are ___?___ and ___?___ .

1-2. The branch of chemistry that deals with plastics is ___?___ chemistry.

1-3. Plastics products must be plastic while they are being ___?___ .

1-4. Plastics products must be ___?___ in the finished state, to meet the definition.

1-5. The usual method of forming plastics is by applying ___?___ and/or ___?___ .

1-6. True or false. Resins and plastics are two names for the same thing.

1-7. The first commercial plastics in the U.S. was ___?___ .

1-8. The first commercial thermosetting plastics was ___?___ .

1-9. The historical importance of ___?___ is that it was the first polymer used commercially.

1-10. Nitrocellulose becomes explosive if the nitrogen content exceeds ___?___ .

1-11. Three systems for classifying polymers are by ___?___ , ___?___ , and ___?___ .

1-12. When classifying polymers by source, the three major categories are ___?___ , ___?___ , and ___?___ polymers.

1-13. Classify the following materials by source:
Copal _____
Casein _____
Benzene _____
Acetylene _____
Gelatin _____
Lignin _____

1-14. The major source of raw material for synthetic plastics is ___?___ .

1-15. If fossil fuels could not be used for plastics production, a good source of raw materials would be ___?___ .

1-16. Most of the cellulose used in plastics comes from ___?___ and ___?___ .

1-17. The only plastics in commercial production that is made from protein is ___?___ .

1-18. The two kinds of luminescence are ___?___ , and ___?___ .

1-19. The two categories of plastics with regard to heat softening are ___?___ and ___?___ .

1-20. With regard to heat reaction, an egg is analogous to a ___?___ resin.

1-21 Classify the following as thermoplastics or thermosetting:
Cellulosics _____
Polystyrene _____
Polyethylene _____
Polysulfone _____

1-22. Waste from processing a ___?___ polymer can usually be recycled.

1-23. Classify the following as thermoplastic or thermosetting:
Silicones _____
Polyurethanes _____
Phenolics _____

1-24. A disadvantage of plastics is dimensional ___?___ .

1-25. For their mass (weight), most plastics have ___?___ strength.

1-26. Since World War II, the rate of growth of the plastics industry has been ___?___ as fast as the rate of growth of other industries.

1-27. Three major areas that consume large amounts of plastics are the ___?___ , ___?___ and ___?___ .

1-28. Three trade names of polymethyl methacrylate are ___?___ , ___?___ , and ___?___ , while its common name is ___?___ .

1-29. The three major categories within the plastics industry are the ___?___ , ___?___ , and ___?___ .

1-30. A resin molecule may be made up of ___?___ bonded molecules.

1-31. Plastics products can/cannot usually be reshaped at room temperature.

1-32. In general, plastics are good thermal and electrical ___?___ .

1-33. A resin, also known as pitch is named ___?___ .

1-34. The resinous binder that surrounds each wood cell is called ___?___ .

1-35. Name a material that is capable of being repeatedly softened by heat and hardened by cooling.

1-36. The building blocks for most plastics are ___?___ .

1-37. Name the four classifications of plastics relative to light penetration.

1-38. The most important natural polymer available today is ___?___ .

1-39. Name five major advantages of plastics.

1-40. Name five major disadvantages of plastics.

Basic Polymer Chemistry

In this chapter, you will learn the importance of chemistry in the development of polymers. Polymer development has created new materials, products, processes, and careers.

To understand the materials, properties, and processing of plastics, you must have a general understanding of their chemical structure. An understanding of the basic principles of polymer chemistry will provide this knowledge. This chemical knowledge is easy to acquire, but do not attempt to memorize this chapter. Use this chapter as a reference to the many possible variations of polymers.

There are continuing developments in chemical and process technology, bringing improved understanding of polymers. A knowledge of structure and property relationships is important in producing tailor-made or "engineered polymers."

The terms "engineered plastics," "advanced composites," or "engineered composites" should be avoided. This implies that other materials are less advanced or not engineered.

These terms are widely used in literature. *Engineered plastics* commonly refers to polymers used or tailored to end-product needs. Acetal, ABS (acrylonitrile-butadiene-styrene), polyamide, polyamide-imide, polycarbonate, polysulfone, polyethersulfone, and polyarylsulfone are familar examples. The term is also used for any polymer with physical properties good enough for use as structural materials. *Advanced composites* or *engineered composites* refer to the group of materials usually associated with aerospace structures. The word *advanced* infers that new, engineered reinforcements or matrices are being used rather than traditional reinforcements of glass with a matrix of polyester. Kevlar, carbon, and boron fibers in an epoxy ma-

trix are used to produce aircraft structures, athletic equipment, and automobile components. All these terms denote an improvement in materials for producing high-performance products.

2-1 THE ATOM

Nearly 2,000 years ago, the Greek philosopher-scientist Democritus stated that everything can be broken down into small particles. Only in relatively recent times have other scientists given atomic theory serious thought. It was Democritus who gave us the word *atom*. He used the Greek word *atomos,* meaning "something that cannot be cut, or something that is indivisible", to describe the smallest particles of matter.

Aristotle

The Greek philosopher Aristotle told his students that atoms did not exist. He contended that there were only four basic qualities: hot, cold, moist, and dry. This basic philosophy of Aristotle's was disarmingly simple, and his ideas governed Western thought for many centuries.

Scientists like Galileo Galilei, Pierre Gassendi, and Sir Isaac Newton disputed the accepted teachings of Aristotle. But experimental proof of the atom's existence still lay a long time in the future.

Boyle

In about 1650, Irish researcher Robert Boyle observed that alchemists had searched for many

15

years for ways of making gold from common materials. Boyle concluded that gold must be one of a limited number of simple substances that cannot be broken into simpler materials. He called these simple substances *elements*. There were very few elements known in Boyle's time, as shown in Table 2-1.

British scientists discovered the elements hydrogen and oxygen. A Frenchman, Lavoisier, discovered nitrogen and a Swede, Scheele, discovered chlorine.

Dalton

In 1808, John Dalton, a British schoolmaster, physicist, and chemist, noted with the aid of a microscope that crystals of the known elements appeared different from one another. Crystals of copper always looked alike, as did crystals of gold; however, gold and copper crystals did not resemble each other. Dalton concluded that atoms of these substances must also be different, and he even went so far as to draw pictures of these atoms while theorizing that an unknown force held the atoms together.

Although Dalton is associated with modern atomic theory, he made a basic mistake. He assumed that one gram of hydrogen and eight grams of oxygen contain the same number of atoms, because these amounts combine to produce nine grams of water. In chemical symbols, he said:

$$H + O \rightarrow HO$$
$$\text{1 gram} \quad \text{8 grams} \quad \text{9 grams}$$

Table 2-1. Elements Known in Boyle's Time.

Elements Known A. D. 100	Elements Added by A. D. 1600	Elements Added by A. D. 1700
Gold	Zinc	Phosphorous
Silver	Antimony	Oxygen
Tin	Bismuth	Hydrogen
Lead	Arsenic	
Copper	Reference made in	
Mercury	1557 to "unmeltable	
Iron	metal"—Platinum	
Carbon	(not then named)	
Sulphur		

Avogadro

Three years after Dalton announced his atomic theory, the Italian physicist Amadeo Avogadro discovered the law that has been linked to his name: Equal volumes of different gases, measured under the same conditions of temperature. and pressure, contain the same number of gas particles (molecules).

Avogadro solved the problem that faced Dalton and corrected his water formula by applying *Boyle's Law*. Boyle's law states that an amount of gas enclosed in a container will double in pressure if squeezed into half the space.

Avogadro had only to weigh a litre of hydrogen and a litre of oxygen to prove that oxygen weighs sixteen times as much as hydrogen. Because oxygen and hydrogen are gases, there are the same number of atoms in each litre container. Thus, there are the same number of atoms in one-half gram of hydrogen as in eight grams of oxygen. Avogadro reasoned that two hydrogen atoms and one oxygen atom make up water (H_2O), and in so doing corrected Dalton's mistake.

When atoms compounded or combined, Avogadro called them *molecules,* meaning little masses. According to Avogadro, chemical *elements* are composed of single atoms or molecules with like atoms, while chemical *compounds* are molecules composed of two or more different atoms.

2-2 THE ELEMENTS

Current theory explains that the structure of matter is composed of atoms of elements, which form molecules of compounds and elements.

Matter is anything that has mass (weight) and occupies space, and may exist in the form of a solid, liquid, or gas.

There are 92 natural and 13 artificial elements that make up all matter (Table 2-2). Discovery of element 104 is claimed by both the Soviet Union and the United States, but neither claim has been verified. Element 105 also has not yet been confirmed.

When two or more elements combine, they produce a new material called a compound

Table 2-2. The Elements

Natural Elements

Atomic Number	Name	Symbol	Atomic Number	Name	Symbol
1	Hydrogen	H	47	Silver	Ag
2	Helium	He	48	Cadmium	Cd
3	Lithium	Li	49	Indium	In
4	Beryllium	Be	50	Tin	Sn
5	Boron	B	51	Antimony	Sb
6	Carbon	C	52	Tellurium	Te
7	Nitrogen	N	53	Iodine	I
8	Oxygen	O	54	Xenon	Xe
9	Fluorine	F	55	Cesium	Cs
10	Neon	Ne	56	Barium	Ba
11	Sodium	Na	57	Lanthanum	La
12	Magnesium	Mg	58	Cerium	Ce
13	Aluminum	Al	59	Praseodymium	Pr
14	Silicon	Si	60	Neodymium	Nd
15	Phosphorus	P	61	Promethium	Pm
16	Sulfur	S	62	Samarium	Sm
17	Chlorine	Cl	63	Europium	Eu
18	Argon	A	64	Gadolinium	Gd
19	Potassium	K	65	Terbium	Tb
20	Calcium	Ca	66	Dysprosium	Dy
21	Scandium	Sc	67	Holmium	Ho
22	Titanium	Ti	68	Erbium	Er
23	Vanadium	V	69	Thulium	Tm
24	Chromium	Cr	70	Ytterbium	Yb
25	Manganese	Mn	71	Lutetium	Lu
26	Iron	Fe	72	Hafnium	Hf
27	Cobalt	Co	73	Tantalum	Ta
28	Nickel	Ni	74	Tungsten	W
29	Copper	Cu	75	Rhenium	Re
30	Zinc	Zn	76	Osmium	Os
31	Gallium	Ga	77	Iridium	Ir
32	Germanium	Ge	78	Platinum	Pt
33	Arsenic	As	79	Gold	Au
34	Selenium	Se	80	Mercury	Hg
35	Bromine	Br	81	Thallium	Tl
36	Krypton	Kr	82	Lead	Pb
37	Rubidium	Rb	83	Bismuth	Bi
38	Strontium	Sr	84	Polonium	Po
39	Yttrium	Y	85	Astatine	At
40	Zirconium	Zr	86	Radon	Rn
41	Niobium (Columbium)	Nb	87	Francium	Fr
			88	Radium	Ra
42	Molybdenum	Mo	89	Actinium	Ac
43	Technetium	Tc	90	Thorium	Th
44	Ruthenium	Ru	91	Protactinium	Pa
45	Rhodium	Rh	92	Uranium	U
46	Palladium	Pd			

Artificial Elements

Atomic Number	Name	Symbol	Atomic Number	Name	Symbol
93	Neptunium	Np	100	Fermium	Fm
94	Plutonium	Pu	101	Mendelevium	Mv
95	Americium	Am	102	Nobelium	No
96	Curium	Cm	103	Lawrencium	Lw
97	Berkelium	Bk	104	Not Named	
98	Californium	Cf	105	Not Named	
99	Einsteinium	E			

(Fig. 2-1). As noted earlier, Avogadro's formula defines water as a compound containing one oxygen and two hydrogen atoms (H_2O). He also pointed out that a *molecule* is the smallest particle to which a compound can be reduced before it breaks down into its elements.

The atom is the smallest particle to which an element can be reduced while still keeping the properties of the element. The atom may be broken down still further into subatomic particles.

2-3 SUBATOMIC PARTICLES

There are three basic subatomic particles: neutrons, protons, and electrons. The atoms of each element differ in the number of these particles they contain. Thus, no two elements have identical atoms.

The *neutron* is a particle with no electrical charge. The *proton* is a positively charged (+) particle. Both occupy a dense, compact central area of the atom called the *nucleus*. Protons are very small, very dense particles.

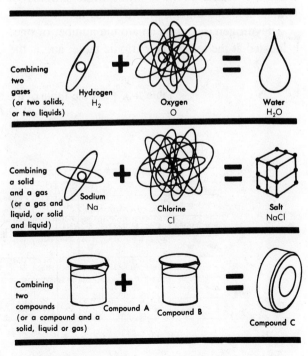

Fig. 2-1. Ways that compounds may be formed.

Electrons are negatively charged (−) particles. They are located in orbits or shells around the nucleus. Electrons are nearly three times as large in diameter as protons, and they are about 1 840 times lighter. This is one factor that makes protons harder to move than electrons.

The simplest atom is that of hydrogen. It has one proton (as its nucleus), one electron, and no neutrons (Fig. 2-2).

For all elements, the atom itself is electrically neutral. The negative charge of the electrons moving around the nucleus is equal to the positive charge of the protons in the nucleus. Neutrons, as noted, are electrically neutral. If an atom gains an electron, it becomes *electronegative* (a negative ion). If it loses an electron, it becomes *electropositive* (a positive ion). An *ion* is an atom that has either gained or lost one or more electrons.

2-4 THE PERIODIC TABLE

Elements are classified in the periodic table according to their *atomic number,* which is the number of protons in the nucleus of the atom. (See Table 2-3.) For example, the atomic number of carbon, six, is the number of protons in its nucleus, and also the number of electrons around the nucleus (Fig. 2-3). The atom is three-dimensional, of course, not two-dimensional as shown.

Hydrogen (H), with an atomic number of one, is located at the top of the periodic table, and is the

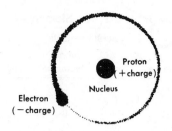

Fig. 2-2. Diagram of a hydrogen atom.

first element in Group IA. It is followed (reading downward) by Li, Na, K, Rb, Cs, and Fr. All of the elements in Group IA have one electron in their outer shell and have similar chemical properties. The elements in Group I are highly chemically active, while elements in Group VIII are inert elements and will not readily combine chemically.

Carbon (C), chlorine (Cl), fluorine (F), oxygen (O), hydrogen (H), nitrogen (N), and silicon (Si) are the elements that are of primary concern in the study of plastics. Note the positions of these elements in the periodic table.

The *atomic mass* (weight) of an element is the mass of an atom of that element compared with the mass of an atom of carbon. A system of relative atomic masses of the elements is used, setting carbon at 12 g [0.42 oz]. The relative atomic mass of an element in grams is called a *gram-atom,* thus one gram-atom of carbon is 12 g. It is known that one gram-atom of any element contains 6.02×10^{23} atoms. The atomic mass of an element in grams

Table 2-3. Periodic Table of the Elements. Note Position of Elements C, N, O, F, Si, and Cl.

Periods Principal Quantum Number n					IA	IIA	Transition Elements						VIII			IB	IIB	IIIB
n = 1 to He				O Rare Gases	1.00797 H 1													
n = 2 to Ne	IVA	VA	VIA	VIIA	4.0026 He 2	6.939 Li 3	9.0122 Be 4											10.811 B 5
n = 3 to Ar	12.01115 C 6	14.0067 N 7	15.9994 O 8	18.9984 F 9	20.183 Ne 10	22.9898 Na 11	24.312 Mg 12	IIIA	IVB	VB	VIB	VIIB						26.9815 Al 13
n = 4 to Kr	28.086 Si 14	30.9738 P 15	32.064 S 16	35.453 Cl 17	39.948 Ar 18	39.102 K 19	40.08 Ca 20	44.956 Sc 21	47.90 Ti 22	50.942 V 23	51.996 Cr 24	54.9380 Mn 25	55.847 Fe 26	58.9332 Co 27	58.71 Ni 28	63.54 Cu 29	65.37 Zn 30	69.72 Ga 31
n = 5 to Xe	72.59 Ge 32	74.9216 As 33	78.96 Se 34	79.909 Br 35	83.80 Kr 36	85.47 Rb 37	87.62 Sr 38	88.905 Y 39	91.22 Zr 40	92.906 Nb 41	95.94 Mo 42	(97) Tc 43	101.07 Ru 44	102.905 Rh 45	106.4 Pd 46	107.870 Ag 47	112.40 Cd 48	114.82 In 49
n = 6 to Rn	118.69 Sn 50	121.75 Sb 51	127.60 Te 52	126.9044 I 53	131.30 Xe 54	132.905 Cs 55	137.34 Ba 56	138.91 La 57	178.49 Hf 72	180.948 Ta 73	183.85 W 74	186.2 Re 75	190.2 Os 76	192.2 Ir 77	195.09 Pt 78	196.967 Au 79	200.59 Hg 80	204.37 Tl 81
n = 7 to Lw	207.19 Pb 82	208.980 Bi 83	(209) Po 84	(210) At 85	(222) Rn 86	(223) Fr 87	(226.05) Ra 88	(227) Ac 89										

Lanthanide Series

140.12 Ce 58	140.907 Pr 59	144.24 Nd 60	(145) Pm 61	150.35 Sm 62	151.96 Eu 63	157.25 Gd 64	158.924 Tb 65	162.50 Dy 66	164.930 Ho 67
167.26 Er 68	168.934 Tm 69	173.04 Yb 70	174.97 Lu 71						

Actinide Series

232.038 Th 90	(231) Pa 91	238.03 U 92	(237) Np 93	(244) Pu 94	(243) Am 95	(247) Cm 96	(247) Bk 97	(251) Cf 98	(254) Es 99
(257) Fm 100	(256) Md 101	(255) No 102	(257) Lw 103	(260) 104					

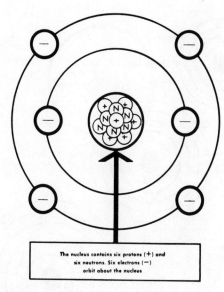

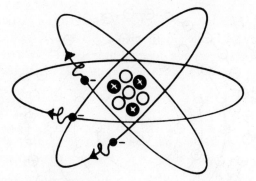

Fig. 2–4. Spinning electrons in shell orbit around the nucleus of a lithium atom.

Fig. 2-3. The carbon atom, in a two-dimensional representation.

(one gram-atom) divided by the number of atoms per gram-atom gives the mass of an atom in grams:

$$\text{mass of atom in grams} = \frac{\text{atomic mass in grams}}{6.02 \times 10^{23}}$$

For carbon, then,

$$\text{mass of carbon atom in grams} = \frac{12 \text{ g}}{6.02 \times 10^{23}}$$
$$= 2 \times 10^{-23} \text{ g}$$

The number 6.02×10^{23} is referred to as *Avogadro's number*.

2-5 ELECTRONS AND ELECTRON SHELLS

The arrangement of the electrons about the nucleus of an atom greatly influences the chemical and physical properties of an element. Electrons may arrange themselves in a number of possible shells, or energy levels, around the nucleus. An electron spins like a top and vibrates as it orbits the nucleus in its shell, or energy level.

Figure 2-4 represents a lithium atom showing three spinning electrons in two shells. The spirals indicate electron spin. The lithium nucleus contains three protons and four neutrons. A molecular model of this type has a basic weakness. It does not account for the geometry of the molecule, which is very important for polymers.

Atoms may have up to seven shells or orbital paths for electrons. The first, or lowest, energy level of an atom (the shell nearest the nucleus) is labeled the *K shell*. The K shell can contain no more than two electrons. Electrons seek the lowest energy level possible, filling the lower shells first. The remaining electrons assume positions at progressively higher energy levels. There are a few exceptions where a higher energy-level shell will fill with electrons before a lower one. The elements in the scandium-to-nickel series and the yttrium-to-palladium series are examples.

The common names for the shells, in order of increasing electron energy, are the *K, L, M,* and *N* shells. In place of letter designations, energy levels may also be identified as 1, 2, 3, and 4 energy levels. Atoms lose electrons more readily from their outer shells. More energy would be needed to move electrons from atoms with L-level electrons than those with only K-level electrons.

The formula $2n^2$, where n is the energy level number, can be used to determine the maximum number of electrons possible in any energy level. As shown in Figure 2-5, the greatest number of electrons in each energy level is:

First Shell (K)	2
Second Shell (L)	$2 + 6 = 8$
Third Shell (M)	$2 + 6 + 10 = 18$
Fourth Shell (N)	$2 + 6 + 10 + 14 = 32$

The numbers 2, 8, 18, and 32 are identical to the lengths of the various periods in the periodic table. (See Table 2-3.) Period I ($n = 1$) with two elements begins with hydrogen (H) and ends with helium (He). Period II ($n = 2$) begins with lithium (Li) and ends with neon (Ne), for exactly eight elements. Period III has eight, period IV eighteen,

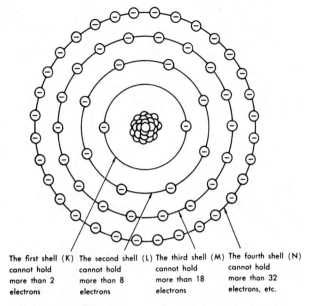

The first shell (K) cannot hold more than 2 electrons

The second shell (L) cannot hold more than 8 electrons

The third shell (M) cannot hold more than 18 electrons

The fourth shell (N) cannot hold more than 32 electrons, etc.

Fig. 2-5. Each energy shell around the nucleus can hold a given maximum number of electrons.

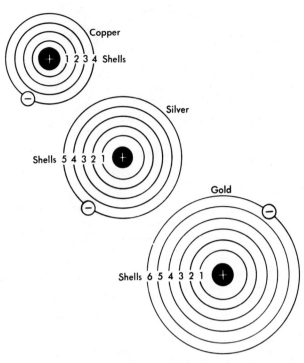

Fig. 2-6. Atoms with only one valence electron are good conductors. (Only valence electrons are shown here.)

period V eighteen, period VI thirty-two, and period VII thirty-two elements.

The electrons in the highest energy levels of an atom are known as *valence electrons*. These electrons are the primary determiners of an atom's chemical properties. Nonvalence electrons are closely associated with the nucleus of the atom. They do not enter into reactions with other atoms.

The *outer* shell of any atom will have no more than eight electrons. Atoms are stable when they have their outer valence shell filled. There are never more than two electrons in the first (K) level and never more than eight electrons in the outermost shell when filled.

If an atom has only one electron in its outer shell, less energy is needed to attract or repel the electron. Elements made up of these atoms are conductors of electricity. Figure 2-6 shows atoms of three elements that are good conductors.

If the outer shell of an atom is more than half filled, but contains less than eight electrons, then that element will be a good electrical insulator. The atom attempts to become stable by filling its valence shell. It is hard to free an electron from the valence shell in such a situation.

When the outer shell is completely filled with eight electrons, the atom is said to be *stable*. Atoms will attract or repel electrons from their valence shell in an effort to have eight electrons and become stable. It is this property that determines how atoms combine with one another to form molecules.

2-6 BONDING IN PLASTICS

Most elements are combinations of two or more atoms bonded together to form molecules. Molecules in crystals assume specific geometric patterns. Attractive forces form invisible bonds between atoms and influence the properties of all materials. The fundamental principles of bonding must be studied and understood.

Bonds (ionic bonds, covalent bonds, and van der Waals' forces) are all-important considerations in the study of plastics. These bonds determine many of the physical characteristics of the plastics product.

All bonds between atoms and between molecules are electrical in nature. Individual atoms are held together by the electrical forces of their individual particles (protons and electrons). It is these forces, either negative or positive, that bond atoms into molecules as well. The valence electrons provide the connecting link or bond that creates a molecule.

Most elements constantly try to reach a stable state. They become stable by: (a) receiving extra electrons, (b) releasing electrons, or (c) sharing electrons.

2-7 PRIMARY BONDS

Ionic and covalent bonding forces are known as *primary* bonds. The simplest bond to understand is the *ionic* bond. In ionic bonding, there is an actual lending of one or more electrons from one atom to another (Fig. 2-7). For example sodium has only one electron in its outermost shell. Chlorine has only seven electrons in its outer shell, but tends to become stable with eight. Since less energy is required to lend one electron than seven, the sodium electron jumps to the outer shell of the chlorine atom. This converts these two previously neutral atoms to oppositely charged ions. A strong electrical attraction between the ions provides the bond that holds together the sodium atom and a chlorine atom in a compound as sodium chloride (NaCl) molecule. Ionic bonding can occur only between elements on opposite sides of the periodic table, for example, between an element in group IA and an element in group VIIA, as occurs to form the NaCl molecule.

The major requirement in an ionically bonded material is that the number of positive charges equal the number of negative charges.

Another primary force of attraction is the *covalent* bond. In covalent bonding, electrons are shared and the valence orbitals between adjacent

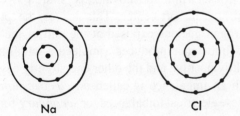

(A) Sodium valence electron is easily lost to chlorine atom.

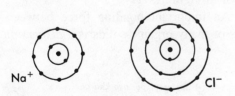

(B) Sodium atom is now positively charged, and the chlorine atom is negatively charged.

(C) Electrical or ionic attraction occurs to form the NaCl molecule.

Fig. 2-7. Atoms of sodium and chlorine form sodium chloride (NaCl) by ionic bonding.

atoms overlap (Fig. 2-8). Covalent bonding makes it possible for elements adjacent to each other on the periodic table or two atoms of the same element to bond together.

The simplest example of shared electrons is found in the hydrogen molecule, H_2. The element hydrogen has only one electron in its K shell. Hydrogen rarely occurs this way in nature because the K shell level seeks completion. As noted, only two electrons are required in the K level.

When two hydrogen atoms combine, each shares an electron, as shown in Figure 2-8, and a single covalent bond is formed. Unlike sodium and chlorine, which are oppositely charged when combined, hydrogen atoms remain electrically neutral after joining forces to make a molecule. The atoms draw closely enough together so that their electrons begin orbiting around both nuclei. The strong bond resulting from this sharing of electrons is what holds most giant molecules together. When two electrons from each atom are shared, a double bond is formed. An even closer bond is produced when more electrons are shared.

For example, a carbon atom may be surrounded by four hydrogen atoms sharing electrons to form a methane molecule (Fig. 2-9). The schematic arrangement of electrons in diatomic (two-atom) molecules is shown in Figure 2-10.

Because carbon has four valence bonds, many organic compounds can be made. Bonds may be formed by sharing one, two, or three pairs of electrons. In the conventional symbols below, a set of two dots or a dash between carbon atoms both indicate a single covalent bond formed by one pair of electrons.

C:C	C::C	C⋮⋮C
or	or	or
C—C	C=C	C≡C
Single bonds	*Double bonds*	*Triple bonds*

Some of the more common elements used in the building of plastics are shown in Table 2-4.

The subjects of bonding lengths, angles, and energy levels are beyond the scope of this text, as is the third type of primary bonding, *metallic bonding*.

2-8 SECONDARY BONDS

Secondary bonding forces, or van der Waals' forces, are much weaker than the primary bonds.

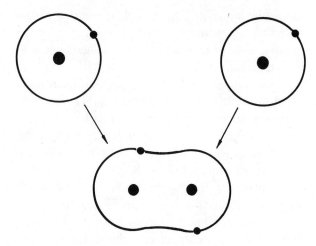

Fig. 2-8. Hydrogen atoms share a pair of electrons in covalent bonding.

Table 2-4. Elements Used in Building Plastics

Element	Symbol	Valence Bonds	Covalent Bonding Symbol
Carbon	C	4	$-\overset{\vert}{\underset{\vert}{C}}-$
Chlorine	Cl	1	Cl—
Fluorine	F	1	F—
Hydrogen	H	1	H—
Nitrogen	N	3	$-\overset{\vert}{N}-$
Oxygen	O	2	$-O-$
Silicon	Si	4	$-\overset{\vert}{\underset{\vert}{Si}}-$
Sulfur	S	2	$-\overset{\vert}{\underset{\vert}{S}}-$

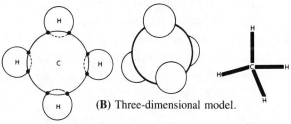

(B) Three-dimensional model.

(A) Two-dimensional representation.

(C) Covalent bonds of atoms.

Fig. 2-9. Models of methane (CH_4)

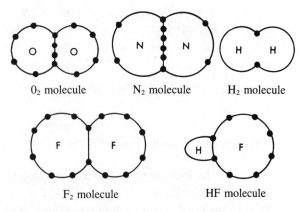

O_2 molecule N_2 molecule H_2 molecule

F_2 molecule   HF molecule

Fig. 2-10. Examples of diatomic molecules.

These are the forces that bind the atoms of inert gases together to form a liquid or solid.

Secondary bonds that hold molecules together are related to the electrical forces that hold atoms together. Ionic and covalent bonds occur between atoms to make molecules. Van der Waals' forces are the bonds between the molecules or between the

atoms in different molecules. Most van der Waals' forces of attraction arise from three sources:

1. dipole bonds
2. dispersion bonds
3. hydrogen bonds

The dipole-bonding effect may be seen in Figure 2-11. When hydrogen and fluorine atoms combine, they are left with an electrical imbalance. This occurs even though the K shell of the hydrogen atom and the L shell of the fluorine atom are filled. The fluorine nucleus is completely surrounded by electrons, while the covalently shared hydrogen electrons are off-center. This leaves the positive hydrogen nucleus exposed at the end of the bond. This imbalance produces a molecule with one end slightly positive and the other end slightly negative. Such an imbalance is called an *electric dipole*. It is these electrical imbalances, or *secondary* forces of attraction, that cause many molecules to join, as shown in Figure 2-11.

An important bonding force between molecules of many organic solids is a *dispersion bond*.

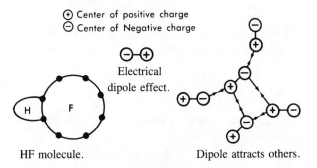

⊕ Center of positive charge
⊖ Center of Negative charge

Electrical dipole effect.

HF molecule.   Dipole attracts others.

Fig. 2-11. The dipole bonding effect.

As the electrons orbit around the nucleus of each atom, a momentary dipole effect (polarization) is established. With an atom of hydrogen, it is easy to see how one side of the atom can become more positive for a moment. Figure 2-12 shows the attraction that results between the electron of a one hydrogen atom and the nucleus of another hydrogen atom.

A weak secondary van der Waals' force, similar to the dipole bond, is the *hydrogen bond*. This force is produced by the exposed nucleus (proton) of the hydrogen atom attracting the unshared electrons of nearby molecules (Fig. 2-13).

Because all van der Waals' forces form relatively weak bonds, plastics and other materials involving these forces are comparatively soft and have low melting points. Table 2-5 is a summary of bonding types.

Table 2-5. Summary of Bonding Types

Form of Bonding	Associated Properties	Example
Bonds between atoms:		
Ionic	Crystalline, high melting point	NaCl
Covalent		
Polar	Unequal sharing; strong, stable bonds	HCl
Nonpolar	Equal sharing; strong, stable bonds	Cl_2
Metallic	Lustrous, ductile; conductor	Cu
Bonds between molecules:		
Van der Waals	Low melting points	$C_6H_{12}O_6$
Hydrogen	Higher melting points	H_2O

2-9 POLYMERIC MOLECULES

Molecules of organic materials are composed of carbon as a base element, most commonly joined to hydrogen by covalent bonds. These molecules are referred to as *hydrocarbons*. As the name suggests, hydrocarbons are composed of hydrogen and carbon. Because of the ability of carbon atoms to link up with other atoms, forming chains, rings, and other complex molecules, thousands of hydrocarbons are known to the organic chemist.

At one time, all organic compounds came from plants or animals. Today, coal, oil, and natural gas are good sources of carbon for the chemicals used in making plastics.

Many organic hydrocarbons are obtained directly from plant or animal sources. Oils such as cottonseed, linseed, and soybean oils, lard, and similar products all yield hydrocarbons for plastics production. Corncobs and oathulls are used as well. New techniques in the collection of methane and other gases from sewage sludge may yield additional sources for the chemist.

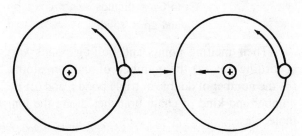

Fig. 2-12. Dispersion bonding of hydrogen atoms.

2-10 ORGANIC NOMENCLATURE

In the study of hydrocarbons, there are two major organic groups to consider: (a) aliphatic or straight-chain molecules, and (b) cyclic or ringshaped molecules, which do not have a general formula.

The aliphatic hydrocarbons may be further reduced into three hydrocarbon series: (a) alkanes or paraffins, (b) alkenes or olefins, and (c) alkynes or acetylenes.

The aliphatic alkanes are linked together by single covalent bonds. Common, shared electrons are usually depicted by a single line. Molecules of this type possess strong intramolecular covalent bonds and the weaker intermolecular van der Waals' bonds.

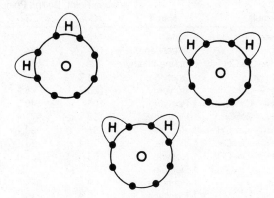

Fig. 2-13. Hydrogen bonding between molecules.

The alkane series has the general formula C_nH_{2n+2}. In this formula, n is the number of carbon atoms in the molecule. Methane gas, colorless and odorless, is the simplest of the alkane series (Table 2-6).

The name octane is used with gasoline. Octane contains 8 carbon atoms and 18 hydrogen atoms. Alkanes are *saturated* hydrocarbons because all their carbon-carbon bonds are single bonds. When a molecule has a multiple carbon-carbon bond, it is considered to be an *unsaturated hydrocarbon.*

The alkenes (Table 2-7) have the formula C_nH_{2n} and these each possess one double covalent bond.

Table 2-6. Alkane Series with Melting and Boiling Points

Formula	Name	Melting Point, °C	Boiling Point, °C
CH_4	Methane	−182.5	−161.5
C_2H_6	Ethane	−183.3	−88.6
C_3H_8	Propane	−187.7	−42.1
C_4H_{10}	Butane	−138.4	−0.5
C_5H_{12}	Pentane	−129.7	+36.1
C_6H_{14}	Hexane	−95.3	68.7
C_7H_{16}	Heptane	−90.6	98.4
C_8H_{18}	Octane	−56.8	125.7
C_9H_{20}	Nonane	−53.5	150.8
$C_{10}H_{22}$	Decane	−30	174
$C_{11}H_{24}$	Undecane	−26	196
$C_{12}H_{26}$	Dodecane	−10	216
$C_{15}H_{32}$	Pentadecane	+10	270
$C_{20}H_{42}$	Eicosane	36	345
$C_{30}H_{62}$	Triacontane	66	distilled at
$C_{40}H_{82}$	Tetracontane	81	reduced
$C_{50}H_{102}$	Pentacontane	92	pressure to
$C_{60}H_{122}$	Hexacontane	99	avoid
$C_{70}H_{142}$	Heptacontane	105	decomposition

Table 2-7. Alkene Series with Melting and Boiling Points

Formula	Name	Melting Point, °C	Boiling Point, °C
$CH_2{=}CH_2$	Ethene (Ethylene)	−169	−103.7
$CH_3{-}CH{=}CH_2$	Propene (Propylene)	−185	−47.7
$C_2H_5{-}CH{=}CH_2$	1-Butene	−185	−6.3
$C_3H_7{-}CH{=}CH_2$	1-Pentene	−165	+30.0
$C_4H_9{-}CH{=}CH_2$	1-Hexene	−140	64.6
$C_5H_{11}{-}CH{=}CH_2$	1-Heptene	−119	93.6
$C_6H_{13}{-}CH{=}CH_2$	1-Octene	−102	121.3
$C_7H_{15}{-}CH{=}CH_2$	1-Nonene	−81	146.9
$C_{10}H_{20}$	1-Decene	−66	170.5
$C_{15}H_{30}$	1-Pentadecene	−3.3	268.6
$C_{20}H_{40}$	1-Eicosene	+30.1	344

If a hydrocarbon molecule has two double covalent bonds, it is a diolefin, with the formula C_nH_{2n-2}. The name of the material will end with *diene*.

Butadiene

As noted, *alkanes* have only one kind of bond, a single covalent bond. The molecule has all the hydrogen atoms it can hold.

Members of the *alkynes* (Table 2-8) are known by a triple bond between two carbon atoms. They are also *unsaturated* hydrocarbons. Alkynes also have the general formula C_nH_{2n-2}.

The alkanes, alkenes, and alkynes are very much alike. All are colorless compounds with low molecular mass (weight). With chain length of only one to four carbon atoms, these compounds are gases. With five to ten carbon atoms, they are liquids. Compounds in these series are solids at room temperature only when the length of the carbon chain is above ten:

C_1 to C_4 = gases
C_5 to C_{10} = liquids
C_{10} and up = solids

Their melting points and boiling points vary according to: (a) the number of carbon atoms, (b) the number of double or triple bonds, and (c) the number and kind of chain branches along the long molecules.

Table 2-8. Alkyne Series with Melting and Boiling Points

Formula	Name	Melting Point, °C	Boiling Point, °C
$CH{\equiv}CH$	Ethyne (Acetylene)	−81	−84
$CH_3{-}C{\equiv}CH$	Propyne (Methylacetylene)	−102.7	−23.2
$C_2H_5{-}C{\equiv}CH$	1-Butyne	−125.7	+8.1
$C_3H_7{-}C{\equiv}CH$	1-Pentyne	−106	40.2
$C_4H_9{-}C{\equiv}CH$	1-Hexyne	−132	71.4
$C_5H_{11}{-}C{\equiv}CH$	1-Heptyne	−81	99.7
$C_6H_{13}{-}C{\equiv}CH$	1-Octyne	−79.4	126.2
$C_7H_{15}{-}C{\equiv}CH$	1-Nonyne	−58	151.0
$C_{10}H_{18}$	1-Decyne	−40	174.2
$C_{15}H_{28}$	1-Pentadecyne	+9	270.7
$C_{20}H_{38}$	1-Eicosyne	39	345

Compounds that have the same molecular formula but different structural properties are called *isomers*. For example, here are two different compounds with the same molecular formula, C_4H_{10}:

$$H-\underset{\underset{H}{|}}{\overset{\overset{H}{|}}{C}}-\underset{\underset{H}{|}}{\overset{\overset{H}{|}}{C}}-\underset{\underset{H}{|}}{\overset{\overset{H}{|}}{C}}-\underset{\underset{H}{|}}{\overset{\overset{H}{|}}{C}}-H \quad \text{and}$$

Butane

$$H-\underset{\underset{H}{|}}{\overset{\overset{H}{|}}{C}}-\underset{\underset{\underset{\underset{\underset{H}{|}}{H}}{|}}{\overset{\overset{H}{|}}{C}}}{}-\underset{\underset{H}{|}}{\overset{\overset{H}{|}}{C}}-H$$

Methyl propane

Hydrogen atoms may be replaced by atoms from the very active nonmetallic chemical elements: fluorine, chlorine, bromine, iodine, and astatine.

Ethylene (C_2H_4) may be formed into vinyl chloride when one atom of hydrogen is replaced by one of chlorine.

$C=C$	$C=C$	$C=C$	$C=C$
Ethylene	Vinyl chloride	Vinylidene chloride	Tetrafluoro ethylene

As a result of the double bonding, long chains of molecules may link together and if many of these unsaturated hydrocarbons link together, they form a polymer. This linking together is called polymerization. As an example, the vinyl chloride polymer becomes polyvinyl chloride.

2-11 CYCLIC HYDROCARBONS

There is nothing to prevent the ends of molecules from joining, if the bonding site is acceptable. Propane, from the alkane series, may join carbon atoms and produce a molecule called cyclopropane. The carbon atoms of the propane molecule join ends, forming a circle.

In general, compounds containing rings of atoms are called *cyclic* or *aromatic* hydrocarbons. The most important cyclic carbon ring is the molecule of benzene. Do not confuse benzene with the common solvent, benzine. Benzene has the molecular formula C_6H_6, with alternating double and single bonds.

Toulene: an aromatic hydrocarbon

It would appear that the benzene ring should be highly unsaturated and easily undergo additional reactions. This hexagonal ring has an unusual bond

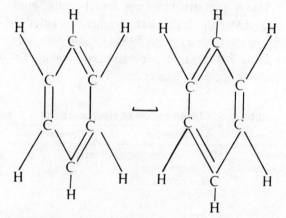

The symbol ⟷ *means oscillation between two structures.*

which is a cross between single and double bonding. One authority theorizes that the electrons are shared by all the carbon atoms in the ring. Another suggests that the double bonds oscillate quickly between the positions.

Because of the number of possible bonding sites and elements, there are many possible cyclic compounds. The compound chlorobenzene, with the formula C_6H_5Cl, may be shown several ways. A plain hexagon figure is often used for simplicity.

Benzene

Chlorobenzene (there is no difference between these symbols)

A dichlorobenzene atom with the formula $C_6H_4Cl_2$ may be shown several ways, depending on the position of the bonding site:

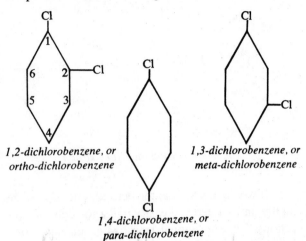

1,2-dichlorobenzene, or
ortho-dichlorobenzene

1,3-dichlorobenzene, or
meta-dichlorobenzene

1,4-dichlorobenzene, or
para-dichlorobenzene

These ring structures are found in the compounds that help form such synthetic plastics as epoxy, polystyrene, polyurethane, and polyester.

Table 2-9 classifies aromatic and aliphatic hydrocarbons for easy reference.

Table 2-9. Classification of Hydrocarbons

Name	Description	Examples
Aromatic hydrocarbons	Have structures and properties related to the compound benzene	Benzene
		Naphthalene
Aliphatic hydrocarbons	Structures not related to benzene	
Saturated	Have only single bonds	Ethane
Alkanes (paraffins)	A general name for saturated hydrocarbons	
Unsaturated Alkenes (olefins)	Have multiple bonds Contain at least one —C=C—	Ethene (ethylene)
Alkynes (acetylenes)	Contain at least one —C≡C—	Ethyne (acetylene) H—C≡C—H

There are a few plastics not based upon hydrocarbons. (See Cellulosics in Chapter 7 and Silicones in Chapter 8.)

The active chemical bonds of carbon can also be satisfied by attaching other elements or groups of elements. These groups are called functional or *radical* groups; they replace one or more of the hydrogen atoms.

A compound that contains the OH group is called an *alcohol*.

H—C—C—OH   or CH_3CH_2OH or ethyl alcohol
(ethanol)

A compound that contains a CHO group is called an *aldehyde*.

H—C—C=O or CH_3CHO or acetaldehyde (ethanal)

The group C—C—C called a *ketone* is shown as:

H—C—C—C—H or $CH_3 C CH_3$ or acetone
(propanone)

The compound will be an *acid* if it contains a COOH group.

H—C—C—OH or CH_3COOH or acetic acid

A compound containing a NH_2 group is called an *amine*.

H—C—N or CH_3NH_2 or methyl amine

The most common group in plastics is the CH_3 o. *methyl* group.

H—C=C   or CH_3CHCH_3 or propylene

2-12 POLYMERS

When a large number of molecules bond together in a regular pattern, a polymer is formed. *Poly* means many and *mer* means unit. A polymer is made of many units repeated in a pattern.

This repeating unit (*mer*) should be considered bifunctional or difunctional (two reactive bonding sites).

In the example below, many units are repeated to form a long molecule of a single-bonded polymer. These mers cannot stand alone because of the requirement of the carbon's four bonding sites. This mer of the linear polymer is not a molecule. It is only a segment of the molecular chain.

A mer unit is shown in the example of polyvinyl chloride (PVC). The number of times this mer unit is repeated in the polymer is called degree of polymerization (DP). A subscript *n* is used to denote the number of like units,

$$\left[\begin{array}{cc} H & Cl \\ | & | \\ -C - C - \\ | & | \\ H & H \end{array} \right]_n$$

where H = hydrogen atom
  Cl = chloride atom
  C = carbon atom
  n = DP, or degree of polymerization
  [ ] = mer
  — or | = bond

In Figure 2-14, the growing chain length is shown from a molecule of methane gas to a solid polyethylene plastics.

Ethylene, an important monomer, is a gas. Six mers of ethylene form a liquid; 36, a grease; 140, a wax; and 500, polyethylene plastics.

A *monomer* is the initial molecule used in forming a polymer. It may occur alone because it is a molecule with a single mer. Most monomers are liquids.

The mechanism or chemical reaction that joins monomers together is called *polymerization* (meaning many mers joined). If the polymer consists of similar repeating units it is known as a *homopolymer* (from *hom* or *homo,* meaning same). It is the repeating unit that distinguishes one polymer from another. When two or three different monomers are polymerized, the polymer is known as a *copolymer* or a *terpolymer.* (See plastics alloy and polyblend.)

Polymerization falls into two general categories: *addition* and *condensation.* These two general types of polymerization (and some modifications of each), cause small molecules to link together in a variety of patterns.

MONOFUNCTIONAL MONOMERS AND THEIR POLYMERS

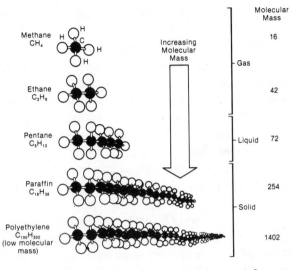

Fig. 2-14. Growing chain length (molecular mass) from a molecule of gas to a solid plastics.

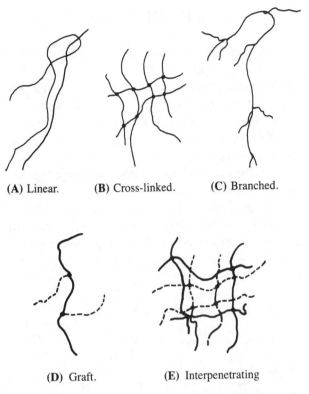

(A) Linear.   **(B)** Cross-linked.   **(C)** Branched.

(D) Graft.   **(E)** Interpenetrating

Fig. 2-15. Various molecular structure arrangements of monomers.

2-13 MOLECULAR STRUCTURE

Monomers form five basic molecular structures: 1) linear, 2) cross-linked, 3) branched, 4) graft, and 5) interpenetrating.

Linear structures are long chains of molecules without appendages, something like rope (Figure 2-15A). The primary forces that hold the monomeric units together within polymers are covalent bonds. These are relatively strong bonds in the order of 600 kJ/mol for carbon-to-carbon bonds. Collectively the secondary bonds which provide the attraction between molecules are known as van der Waals' forces. These molecular or intermolecular secondary forces include dispersion, induced permanent, dipole, and hydrogen bonding. The dispersion force between each pair of adjacent $-CH_2-$ mers on two polyethylene chains is about 200 times weaker than the primary force within each $C-H$ bond. These secondary bonds allow the molecular chains to slide by each other and be deformed when heated.

Linear polymers have long chains (with no branching) of varying lengths, which tend to crowd close together.

Not all linear molecules are homopolymers. We may want to attach other monomers or molecules along the linear chains.

One of the most important aspects of polymer growth is the ability to modify the existing polymers and obtain new properties. When two differ-

ent monomers are joined in a regular sequence, it is called *copolymerization*. A copolymer plastics is the result.

There are three linear copolymer classes: 1) alternating, 2) random, and 3) block.

An *alternating* copolymer has a definite ordered alternation of the A and B monomer units:

ABABABABABABABABABAB

In *random* copolymers, units are situated along the chains in no definite order:

ABBBAABAABABABABBBBBBB

The term *block* copolymer refers to long sequences or blocks of repeating monomers. For example, when forming a polymer from two monomers, A and B, a block pattern is formed:

AAAABBBBBBBAAAAAABBB

A random terpolymer (*ter* meaning three) may be represented as:

ABCBACBCBABCBCBABCABCABA

A common terpolymer of styrene, acrylonitrile, and butadiene rubber forms the plastics ABS. Through combinations of this type, polymers can be custom-made to fit specific technological requirements.

In addition to linear chain polymers there are materials with structures consisting of two- or three-dimensional structures or networks.

Crosslinked or *network* structures involve primary, covalent bonds between the molecular chains (Figure 2-15B). For crosslinking to be very extensive, there must be a number of unsaturated carbon atoms. Crosslinking restricts movement between adjacent chains. This type of polymer is commonly referred to as a thermosetting material.

Branched structures represent three-dimensional branching of the linear chain (Figure 2-15C). Polymers having pendant groups, such as the methyl groups in polypropylene, are not considered branched polymers. Low-density polyethylene (LDPE) is a branched polymer because branches of polyethylene are irregularly spaced along the chain.

Graft polymers are similar to the crosslinked and branch-structured copolymers. The term *graft copolymer* has been used to denote a controlled, planned, or engineered addition to the main polymer chain. These side chains permit less movement. As a result, deformation is more difficult—or impossible. There may be some branching, which occurs naturally during polymerization and molding.

In the formation of graft polymers, a backbone of one monomer has branches of another:

```
AAAAA A AAAAAAAAAAAAAAAAAAAAAA A AAAA
      |                        |
      B                        B
      B                        B
      B                        B
      B                        B
```

The main difference between block and graft polymers is the point of attachment of the two monomers. With block polymers, the attachment point is at the end of the A structure. In graft polymers, it occurs at intervals along A's length.

There are several other important types of copolymer systems.

An *interpenetrating* polymer network (IPN) is an entangled combination of two cross-linked polymers that are not bonded to each other. The techniques of combining dissimilar polymers to form interpenetrating networks results in a plastics alloy. (See Plastics Alloy.) The two types of molecules that are intertwined by permanent entanglement have discrete physical characteristics, but

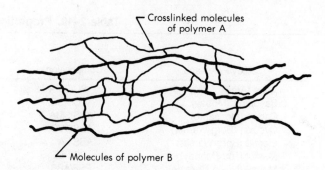
Crosslinked molecules of polymer A

Molecules of polymer B

there is often a synergistic effect in processing or property improvements. The two polymer molecules do not need to be chemically similar or compatible. A semi-interpenetrating polymer network is a combination of two polymers, one crosslinked and one linear, at least one of which was synthesized and/or cross-linked in the immediate presence of the other. (See Polyesters and Polyimides in Chapter 8.)

Properties of a polymer may be uniquely influenced by periodically adding other molecules along the chain (See Polymer Structures). By replacing some of the hydrogen atoms in polyethylene with chlorine (Cl), methyl group (CH_3), benzene (C_6H_6), or fluorine (F), the polymers PVC, PP, PS and PTFE will be formed. (See Appendix B for abbreviations of materials.)

Compatible families of polymers may be alloyed together. Alloys of ABS/PVC, PVC/PMMA, PMMA/PS, and ABS/PC are common.

A *plastics alloy* is formed when two or more different polymers are physically mixed together in the melt. Although there may be some bond attachment between chains, these should not be considered copolymers (See Table 2-10).

The term *polyblend* refers to a plastics which has been modified by the addition of an elastomer. Elastomer modified PS, PVC, PMMA, and ABS are most common. A plastics alloy or polyblend may appear as:

```
BBBBBBBBBBBBBBBBBBBB   AAAAAAAAAAA
     AAAAAAAAAAAAA   BBBBBBBBBB
```

These alloys and polyblends are composed of two or more polymers with no primary bonds between the dissimilar polymer chains.

2-14 MOLECULAR MASS (WEIGHT)

The mass of a molecule is determined by adding together the atomic masses of the various

Table 2–10. Properties of Alloy/Polyblends.

Alloy/Polyblend	Composition*	Flex. Mod. (10⁵ psi)	Heat-Defl. Temperature @264 psi (°F)	Impact Str. Izod (ft-lb/in.)
Arloy (Arco)	PC/SMA	3.2	250	12
Bayblend (Mobay)	PC/ABS	3.7	246	10
Cadon (Monsanto)	ABS/SMA	3.0	175	5
Cycovin (Borg-Warner)	ABS/PVC	3.2	147	12.5
Elemid (Borg-Warner)	ABS/PA	—	—	—
Makroblend (Mobay)	PC/PET	3.0	190	18
Mindel (Union Carbide)	PSF/ABS	3.5	302	9.5
Noryl (General Electric)	PPO/HIPS	3.5	225	6.5
GX-200 (General Electric)	ASA/PC	3.5	200	2.0
Prevex (Borg-Warner)	PPO/PS	3.6	265	5.0
Proloy (Borg-Warner)	ABS/PC	3.3	—	10
Xenoy (General Electric)	PC/PBT	3.0	210	15
Zytel (DuPont)	PA/PO	2.5	160	17

*See Appendix B.

atoms making up the molecule. The atomic mass of hydrogen is about 1. The atomic mass of oxygen is about 16. Thus, the molecular mass of a single molecule of water (H_2O), is $1 + 1 + 16$, or 18.

The side chains in branched structures permit greater interlocking of the structure and a wide range of *molecular mass* (weight).

Molecular weight (MW) is an obsolete term. Molecular mass should be used when referring to the relative molecular mass (defined as the sum of the relative atomic masses of the constituent atoms) of polymers. Methane would have a molecular mass of $12 + (4 \times 1.008) = 16.03$ g/mol.

The length (molecular mass) of the polymer chain has a profound effect on processing (moldability) and properties of the plastics. Increasing chain length may increase toughness, melt temperature, melt viscosity, creep, and stress resistance. It may also increase (with higher melt viscosity) difficulty in processing, material costs, and possible polymer degradation.

Because the molecular structure may vary in shape and length, molecular mass average ($\overline{M}$) is used. Molecular mass average is the product of the average number of repeating units or mers in the polymer. It would be logical to expect polymers to contain molecular chains with a wide distribution of lengths (some short, some long). If all molecules have the same mass, the polymer is said to have a narrow molecular mass distribution. (See Polyolefin.) Polymers that do not have sufficient molecular mass (many long molecules) to cause some chain entanglement are generally weak, have lower melting points, and are readily attacked by appropriate reactants. Some place the minimum

molecular mass required for good physical and mechanical properties at about 20 000 g/mol.

The number-average chain length $\overline{N}_n$ is obtained by determining the total length of the polymer chain and dividing this by the total number of molecules. A number-average molecular mass $\overline{M}_n$ is calculated by dividing the sum of the individual molecular mass values by the number of molecules. Most thermodynamic properties are related to the number of particles present and thus are dependent on $\overline{M}_n$.

The molecular mass of most polymers falls in the 10 000 g/mol to 1 000 000 g/mol range. The molecular mass of polymers has a great effect on properties such as melt viscosity.

There is a significant change in properties with increasing molecular mass in the example of going from methane gas to polyethylene plastics in Figure 2-14. Molecular mass (weight) averages for selected polymers are shown in Table 2-11.

Sometimes conflicting requirements include using high molecular mass to obtain the desired physical properties and low enough molecular mass to permit reasonable processing (molding) conditions. Commercial PVC ranges from 30 000 g/mol to about 90 000 g/mol.

In commercial plastics production, a polymer normally falls in the range of 75 to 750 mers per molecule. Polyvinyl chloride may contain 1 000 carbon, 1 500 hydrogen, and 500 chlorine atoms per 500 mers. It may have a molecular mass of more than 31 000.

To calculate the average molecular mass of a polymer, the degree of polymerization (DP) must be known. The DP is the number of mers in a poly-

Table 2–11. Average Molecular Mass of Selected Polymers.

Polymer	Molecular mass
HDPE	25 000
LDPE	20 000
PA	20 000
Paraffin	800
PETP	25 000
PP	40 000
PVC	40 000
UHDPE	40 000

mer molecule. In chemical laboratories, this DP number is measured by several methods: melt index, viscosity, osmotic pressure, and light scattering.

For example, consider a typical sample of polyvinyl chloride. A DP of 500 might be found by a polymer chemist, using one of the above methods.

Then for one gram-atom of the mer, C_2H_3Cl, mer molecular mass equals

$$24 + 3 + 35.5 = 62.5$$

$$\text{Since DP} = \frac{\text{polymer molecular mass}}{\text{mer molecular mass}} \text{ or } \frac{\text{mers}}{\text{mole}}$$

$$= \frac{31\,256}{62.5}$$

$$= 500$$

If there were 500 mers of vinyl chloride per polymer, the molecular mass would be:

$$
\begin{array}{cc}
H & Cl \\
| & | \\
C & = C \\
| & | \\
H & H
\end{array}
$$

$$\text{Grams/mer} = \frac{\text{g/mer}}{\text{mers/mer mass}}$$

$$= \frac{(2)(12) + (3)(1) + (35.5)}{6.02 \times 10^{23}}$$

Molecular mass of a polymer chain equals

(g/mer mass) (mers/mole)

$$= (62.5)(500)$$

$$= 3.125 \times 10^4 \text{ g/molecular mass}$$

This means that Avogadro's number of these molecules will have a mass of 31 250 grams, or roughly 48 pounds.

Only a few polymers occur in nature. People must cause monomers to join in various combinations and arrangements to meet their needs. A *catalyst* is used to bring about the union of the monomers to form the plastics polymer. A catalyst is a substance that can change the speed of reaction between two or more materials, without being itself permanently changed or used up. This changing or joining of the monomers will occur without a catalyst. But the reactions are hardly perceptible, because they are so slow. A catalyst speeds them up enormously.

The mechanisms of catalysts are complicated and based on speculation. Two theories are commonly expressed. One speculates that a catalyst *grabs* molecules in a way that distorts their structure. The strained molecules will then react when hit by other reactants. The second theory states that the catalyst actually forms an intermediate compound with one of the reactants, which is then vulnerable when it collides with the other reactant.

Catalysts may react with other chemical agents sometimes called *promoters*. Promoters are weak solutions, or reactive agents, which help in the polymerization process. See additives in Chapter 6. Any energy-causing agent could be classified as a catalyst in joining monomers. Heat energy is used to cause polymerization in both natural and synthetic polymers. Ionic or electrical energy can also be used.

2-15 ADDITION POLYMERIZATION

Most of the polymers in current use are produced by the addition method. In *addition* polymerization, unsaturated hydrocarbons are caused to form large molecules. The mechanism of the reaction involves three basic phases: (a) initiation, (b) propagation, and (c) termination.

Initiation is a phase in which the monomers are caused to join by the aid of a high-energy source, such as a catalyst or heat. A typical example is that of the gas, ethylene (C_2H_4). Since a double bond is present, two bonding sites (R) can be made available:

$$
\begin{array}{l}
\quad\ H\ \ H \qquad\qquad\qquad H\ \ H\ \ H\ \ H\ \ H \\
\quad\ |\ \ \ | \qquad\qquad\qquad |\ \ \ |\ \ \ |\ \ \ |\ \ \ | \\
R-C-C-R \rightarrow \cdots -C-C-C-C-C- \cdots \\
\quad\ |\ \ \ | \qquad\qquad\qquad |\ \ \ |\ \ \ |\ \ \ |\ \ \ | \\
\quad\ H\ \ H \qquad\qquad\qquad H\ \ H\ \ H\ \ H\ \ H \\
\ \ \ \text{Initiation} \qquad\qquad\quad \text{Propagation}
\end{array}
$$

Energy must be supplied to break the double bonds. In the *propagation phase* the addition process continues. The final product is the polymer. In the example above, it would be polyethylene.

The *termination phase* is exactly what the name implies. The polymerization process is completed or caused to stop. Termination of growth can occur in several ways. In one method, the final molecular mass depends on the catalytic reaction causing many or few mers to form the molecular chain growth. A special ionic catalyst may be used to bond with the free ends of the polymer chain completely stabilizing its growth. The catalyst terminates the bonding of individual mers into a repeating molecular pattern.

In theory, there is no limit to the length of a linear addition polymer. However, length seldom exceeds several thousand bonded atoms. If each chain could be positioned so that a bonding site was available for the next monomer chain end, an endless molecular chain would form. Unfortunately, it is not possible for this to take place. The monomers are constantly vibrating and moving about causing the chains to become entangled like a plate of spaghetti. This long chain growth is often referred to as a *macromolecule*. *Macro* means large or, in this case, long.

The term *macromolecule* may be misleading because it does not convey the concept of a repeating unit.

Some polymers may be polymerized by several methods. Polyamide may be produced using addition or condensation methods. Because there are several addition polymerization techniques, only three will be briefly summarized.

In *bulk* or *mass* polymerization, one or more monomers are joined by adding catalysts and initiators. The high yielding reaction produces the highest purity polymer with excellent optical prop-

erties. Cast epoxy and methyl methacrylate may be produced by this process.

In *solution* polymerization, the monomer, solvents, initiators, and catalysts are heated together. Throughout the polymerization process, the solvent is removed. There is added expense in recovery and handling of solvents. The optical properties do not equal those of bulk polymerization because the polymer contains small quantities of solvent. Many of these polymers are used in coating and impregnating applications where the solvent is allowed to evaporate.

Water rather than organic solvent is used to distribute energy, emulsifier, and initiators in *emulsion* polymerization. High molecular weight polymers are produced. Water based polymer adhesives and coatings are familiar examples.

Suspension, *bead*, or *pearl* polymerization are terms used to describe the polymerization technique of suspending monomer droplets in water. Polystyrene can be made by any of the four addition polymerization methods. Addition polymerization methods are shown in Table 2-12.

2-16 CONDENSATION POLYMERIZATION

The second general process of polymerization is called *condensation* polymerization. A polymer different from those produced by the addition method and having varied properties, may be produced by condensation polymerization. The condensation process is a chemical reaction in which a small molecule, often water, is eliminated from the polymer. For this reason, it is known as a condensa-

Table 2–12. Addition Polymerization Methods

Polymer	Addition Methods
Chloroprene (CR)	Emulsion
Low Density Polyethylene (LDPE)	Bulk
High Density Polyethylene (HDPE)	Solution
Polyacetals	Solution
Polyamide (PA)	Bulk, suspension
Polycarbonate (PC)	Bulk
Polyethylene terephthalate (PET)	Bulk
Polyisoprene (IR)	Solution
Polymethyl methacrylate (PMMA)	Bulk, suspension, solution
Polypropylene (PP)	Solution
Polystyrene (PS)	Emulsion, suspension, bulk, solution
Polysulphides	Suspension
Polytetrafluoroethylene (PTFE)	Suspension
Polyvinyl chloride (PVC)	Emulsion, suspension
Polyvinyl acetate (PVA)	Emulsion
Stryrene-Butadiene (SBR)	Emulsion

tion action (See Index for A, B, and C stages of phenolic plastics.)

Probably one of the best examples of condensation polymerization is phenol-formaldehyde, also called Bakelite. When polymerization is in progress between formaldehyde (CH_2O) and phenol (C_6H_5OH), a bridge occurs between the benzene rings and a water molecule is left as *condensation*.

In *addition* polymers, unsaturated monomers are completely used up in the final product. In *condensation* polymers, a by-product is produced during polymerization. Dacron (PET), Melmac (MF), Mylar (PET), and Nylon (PA) are a few of the many plastics that are polymerized by condensation. There are other conditions for the polymerization process. Simply placing monomers close together will not ensure automatic polymerization. The application of heat, light, or pressure energy or a chemical catalyst may be needed.

It is important to note that condensation polymerization can lead to either thermoplastic or thermosetting polymers. Some condensation polymers are shown in Table 2-13.

2-17 POLYMER STRUCTURES

The properties of plastics are dramatically affected by the arrangement of the atoms and molecules. The atomic arrangement of polymers can be classified as crystalline, amorphous, or molecular.

Crystallinity is the three-dimensional arrangement of atoms, ions, or molecules in a regular pattern. Diamonds and table salt are two common materials with easily observed crystalline structures. Because of the great length of polymer molecules, there is less crystallization. Only weak van der Waals' forces are available to align the molecules, and a great number of atoms must be maneuvered to produce any degree of crystallinity in the polymer.

Table 2–13. Condensation Polymers.

Phenol-formaldehyde
Polyacetal
Polyamide
Polyamide-imide
Polybutylene terephthalate
Polycarbonate
Polyester
Polyether
Polyimides
Polysiloxane (silicones)
Polysulfides
Polyurea
Polyurethane
Urea-formaldehyde

Because the molecular chains are only partly ordered, most crystalline plastics are not transparent in the solid state. Partially crystalline (semi-crystalline) polymers, such as the linear polyolefins, polyacetals, and polyamides, are as a rule translucent to opaque.

The crystalline content of some polymers may exceed 95%, while the remainder is amorphous. Crystallinity of LDPE is about 65% while HDPE may exceed 95%.

The crystallization of bulk polymers is characterized by the forming of large crystalline aggregates called *spherulites*. The folded-chain molecule alignment is called a *lamellar* (plate-like) structure. In ordered polymers, chains are folded back on themselves to produce parallel chains perpendicular to the face of the crystal.

These crystalline structures within the polymers act to scatter light. They give polyethylene and polypropylene a milky appearance. With the use of an electron microscope, X-ray defraction, or even the naked eye, it is possible to see these spherulite crystals (Fig. 2-16). With normal processing, a crystalline polymer may begin to form spherulites as it cools. Processing may help align or orient the molecular chains.

The benzene side groups in PS (atactic) normally prevent chain alignment of significant crystal formation. By forcing PS through or around narrow restrictions in molds, some crystalline structures will form. These areas will have a cloudy or milky appearance. Oriented alignment of molecules on one axis may be desirable because this produces greater crystallinity and higher tensile strength (Fig. 2-17). There are few hard and fast rules when it comes to crystallinity in polymers. Controlling the

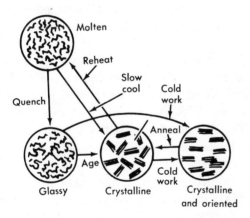

Fig. 2–16. At a magnification of 10 000 diameters, fine fibers can be seen linking the radial arms of a single spherulite of semicrystalline polyethylene. (Bell Telephone Laboratories)

 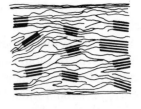

(A) Amorphous. **(B)** Oriented crystalline.

(C) Mixed amorphous crystalline.

Fig. 2–18. Ordered semicrystalline and disordered amorphous polymers.

Fig. 2–17. Deformation of linear polymers from glassy, molten, amorphous materials to a solid, cooled, semicrystalline state.

crystallization process is a basic consideration in determining the physical properties of the polymer.

Materials that are *amorphous*, meaning "without form," have atoms or molecules that are noncrystalline and without long-range order (Fig. 2-18). One analogy for an amorphous polymer is a mass of cooked spaghetti. Polymers are amorphous in a molten state. A rapid quenching may preserve most of this amorphous state.

Rapid quenching may prevent a large number of the chains from realigning to form crystals. This technique is sometimes used in the production of PET and PE to improve transparency.

Amorphous polymers are usually less rigid than crystalline ones. Polymers are transparent if most crystallinity can be prevented. Low-density

polyethylene is more transparent if crystallinity is reduced. High-density polyethylene is semicrystalline. It has a higher melting point.

Polystyrene is amorphous. It will be clear but more easily attacked by solvents.

The effect of the *molecular* chain structure may be grasped by viewing the examples of polymer shapes in Figure 2-19.

The physical size and bonding angle have a significant role in polymer properties. The hydrogen atom is about the same size as carbon.

(A) Polyethylene.

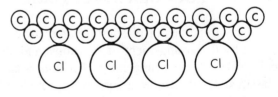

(B) Polyvinyl chloride.

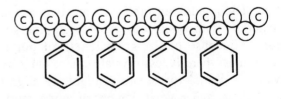

(C) Polystyrene.

Fig. 2-19. Examples of molecular chains.

The polyethylene chain shown in Fig. 2-19A has only carbon atoms with simple, uniform structure. The second polymer chain, polyvinyl chloride, has chlorine atoms attached (Fig. 2-19B). The third polymer chain is polystyrene, and it has a benzene structure attached to its molecular chain (Fig. 2-19C). These last two polymers are not as flexible as polyethylene. The van der Waals' forces are greater and the additional chlorine or benzene atoms make molecular movement more difficult.

2-18 STEREOISOMERISM

There are several molecular chain arrangements that permit close packing of the molecules, which may lead to high crystallinity.

An *isotactic* polypropylene chain is shown in Figure 2-20A. It has an ordered placement of the CH_3 groups. This type of arrangement may lead to close molecular-chain packing and crystallinity. By proper polymerization control, an *atactic* polypropylene polymer is formed that has a random placement of the CH_3 groups (Fig. 2-20B). This random placement prevents chain packing, and a noncrystalline polymer is formed. This polymer is soft, weak, and less cloudy than the more crystalline isotactic and syndiotactic forms.

A *syndiotactic* arrangement of the same CH_3 group, regularly alternating the CH_3 molecule on either side of the backbone of the chain, will lead toward a tougher, more dense crystalline structure (Fig. 2-20C).

The dense packing of the molecules bears a direct relationship to the strength of forces between molecules, and thus affects mechanical properties. There is little commercial use for atactic PP because it has few useful properties. Commercial PS is atactic and PMMA is largely syndiotactic.

2-19 RESINS

There are a number of ingredients of plastics, but resins are the basic organic materials from which plastics are formed. Fillers, solvents, plasticizers, stabilizers, and colorants influence many of the characteristics of plastics. A resin is the polymeric material that helps impart many of the physical characteristics of the solid plastics. It is the molecular arrangement of the resin that

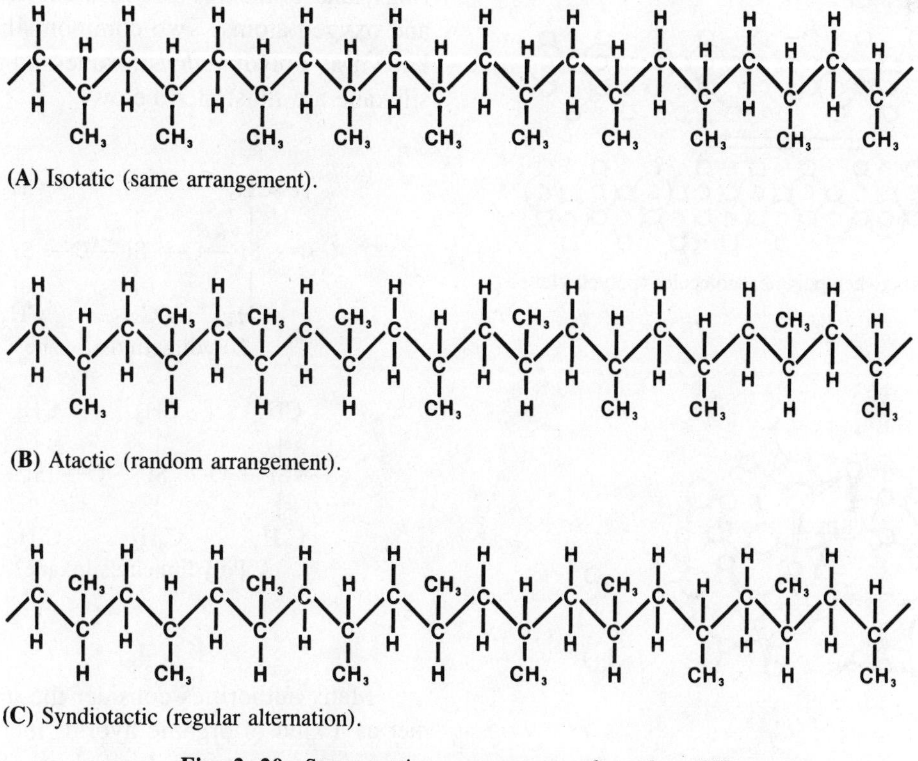

(A) Isotatic (same arrangement).

(B) Atactic (random arrangement).

(C) Syndiotactic (regular alternation).

Fig. 2–20. Stereotactic arrangements of pendent CH_3 groups in polypropylene.

determines whether a plastics is *thermosetting* or *thermoplastic*.

Thermoplastic materials increase in plasticity with temperature. They become soft when heated and solid when cooled to room temperature. One analogy is a dish of butter. When heated, the butter becomes soft; when cooled to room temperature, it solidifies. Thermoplastics are easily formed into useful products because weak van der Waals' forces allow slippage between the molecules (Fig. 2-21). The molecules themselves are held by the stronger covalent bonds.

There is a practical limit to the number of times a thermoplastic material may be formed because repeated processing may cause some of the additives to be lost.

Thermosetting plastics are polymeric materials with structural frameworks that do not allow deformation or slip to occur between molecular chains. They are composed of strong, primary covalent bonds and may be thought of as one large molecule (Fig. 2-22).

In thermoplastics, only secondary van der Waals forces, dipoles, and hydrogen bonds exist between chains. In thermosetting materials, heat is commonly used to cause a chemical reaction (polymerization) resulting in crosslinks between chains. While in a low molecular mass state, heat and pressure are commonly used to cause the thermosetting material to flow into a mold cavity.

Once solidified, these materials may not be reshaped or formed by applying heat. These plastics have a permanent *set* once they have been polymerized.

Thermosetting plastics are stronger than thermoplastics, and have a higher product-service temperature. Thermoplastics may hold many process advantages over thermosets, however. A major advantage is that thermoplastics may be ground and recycled into other useful products.

2-20 INORGANIC POLYMERS

Expanded research in the field of inorganic polymers has led chemists to begin working with many inorganic materials. Polymers may be based on silicon, nitrogen, phosphorus, boron, tin or germanium. One polymer material with covalent bonding capacity comparable to that of carbon is silicon, which forms silicone plastics. Silicon itself constitutes 28 percent of the earth's crust.

The silicone plastics may be based on chains, rings, and branching networks of alternate silicon and oxygen atoms. Two common silicon polymers known as polydi*methyl*siloxane and polydi*phenyl*siloxane are illustrated below:

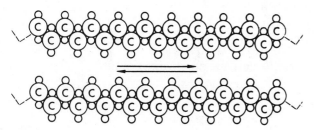

Fig. 2–21. Slip between polymer molecules (polyethylene) in melt.

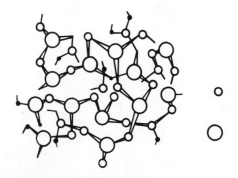

Fig. 2–22. Because of the structural framework of thermosetting plastics, slip is impossible.

$$
\begin{array}{ccccc}
 & CH_3 & & CH_3 & & CH_3 \\
 & | & & | & & | \\
- & Si & -O- & Si & -O- & Si & - \\
 & | & & | & & | \\
 & CH_3 & & CH_3 & & CH_3
\end{array}
$$

Polydimethylsiloxane

$$
\begin{array}{ccccc}
 & CH_3 & & CH_3 & & CH_3 \\
 & | & & | & & | \\
- & Si & -O- & Si & -O- & Si & -O- \\
 & | & & | & & | \\
 & C_6H_5 & & C_6H_5 & & C_6H_5
\end{array}
$$

Polydiphenylsiloxane

Many authorities consider the inorganic polymer as a kind of organic hybrid. Increased interest along with newer technologies have given rise to the study of inorganic polymers.

VOCABULARY

The following vocabulary words are found in this chapter. Look up the definition of any of these words you do not understand as they apply to plastics in the glossary, Appendix A.

Addition polymerization
Aliphatic molecules
Alkanes
Alkenes
Alkynes
Amorphous
Aromatic hydrocarbons
Atomic mass
Atomic number
Atoms
Avogadro's number
Benzene
Bifunctional
Block polymer
Branching
Bulk polymerization
Catalyst
Compound
Condensation polymerization
Copolymerization
Covalent bonding
Cross-linking
Crystallization
Cyclic hydrocarbons
Degree of polymerization (DP)
Difunctional
Electrical dipole
Electron
Emulsion polymerization
Graft copolymer
Hydrocarbons

Hydrocarbon plastics
Inert (rare) gases
Ion
Ionic bonding
Initiation phase
Intermolecular forces
Interpenetrating polymer network
Isomers
Lamellar
Linear
Macromolecules
Mass and weight
Mer
Molecule
Molecular mass
Monomer
Periodic table
Polyblend
Plastics alloy
Polymerization
Primary bonds
Promoter
Propagation phase
Proton
Secondary bonds
Shell
Solution polymerization
Spherulite
Stereoisomerism
   Atactic
   Isotactic
   Syndiotactic
Suspension polymerization
Termination phase
Van der Waals' forces
Valence electrons

2-1. Atoms with fewer than ___?___ valence electrons in their outer shell are good conductors.

2-2. Carbon atoms have ___?___ electrons in their valence shell.

2-3. The total number of covalent bonds in the compound CH_4 is ___?___ .

2-4. A bond between two carbon atoms is a ___?___ bond.

2-5. A single carbon atom may have ___?___ hydrogen atoms attached to it.

2-6. A CH_3 group is called a ___?___ radical.

2-7. A compound containing an OH group is called an ___?___ .

2-8. A compound containing an NH_2 group is called an ___?___ .

2-9. A compound containing a ___?___ group is called an aldehyde.

2-10. A dash between carbon atoms indicates a ___?___ bond.

2-11. The formula below that represents isopropyl alcohol is ___?___ .

```
      H   H   H              H       H
      |   |   |              |       |
  H—C — C —C—C        H—C—C—C—H
      |   |   |              |   |   |
      H  OH   H              H   O   H
          A         or           B
```

2-12. Hydrocarbons containing one triple carbon-to-carbon bond are called the ___?___ series.

2-13. Hydrocarbons that are straight or branched chains with single bonds between carbon atoms are called the ___?___ series.

2-14. Hydrocarbons that are straight or branched chains with one or more double bonds between carbon atoms are called ___?___ .

2-15. Most thermoplastic plastics are produced by the ___?___ polymerization chemical reaction.

2-16. Most thermosetting plastics are produced by the ___?___ polymerization chemical reaction.

2-17. The gas that is an important monomer for polyethylene is called ___?___ .

2-18. Benzene has the molecular formula of ___?___ , with alternating double and single bonds.

2-19. The element silicon has ___?___ valence bonds.

2-20. The small repeating units that make up a plastics molecule are called ___?___ .

2-21. Weak secondary forces holding some plastics molecules together are called ___?___ forces.

2-22. Cloudiness in plastics is caused by ___?___ .

2-23. What does poly mean?

2-24. The smallest particle that a compound can be reduced to before it breaks down into its elements is ___?___ .

2-25. The heavy center of an atom is called the ___?___ .

2-26. The maximum number of electrons in the fourth or N shell of an atom is ___?___ .

2-27. What name is given to the relative mass of an atom of any element as compared with that of one atom of carbon?

2-28. Name seven elements primarily used in the production of plastics.

2-29. Name the law that states equal volumes of gases measured under the same condition of temperatures and pressure, contain the ___?___ number of gas particles.

2-30. What element was assigned the atomic mass of 12?

2-31. The three kinds of atomic bonding methods used in plastics are ___?___ , ___?___ and ___?___ bonds.

2-32. The elements hydrogen, fluorine and chlorine each have ___?___ bonding site.

2-33. Mechancial and other properties of a plastics depend upon ___?___ bonding.

2-34. Name three bonds that arise from van der Waals' forces.

2-35. Because ___?___ do not have all bonding sites filled, they may not exist alone.

2-36. Because not all molecules in a polymer are the same length or shape, the molecules mass ___?___ is used.

2-37. The three basic arrangements that monomers form in the molecular structure of plastics are ___?___ , ___?___ and ___?___ .

2-38. The number of structural units per average molecular mass (weight) is known as ___?___

2-39. A ___?___ is an additive used to cause monomer resins to combine into long chains.

2-40. Name four forms of energy used to cause polymerization:

2-41. Polymer systems used as mechanisms for branched or repeating copolymerization are called ___?___ and ___?___ .

2-42. In ___?___ reactions a by-product is given off during polymerization.

2-43. Crystalline plastics are often more rigid and not as transparent as ___?___ plastics.

2-44. Name four properties affected by molecular chain arrangements.

2-45. The polymeric material that gives a plastics its physical characteristics is called a ___?___ .

2-46. A ___?___ material may be softened repeatedly when heated and hardened when cooled.

2-47. A term used to describe the tying together of adjacent polymer chains is ___?___ .

2-48. The backbone element used in the manufacture of the inorganic polymer silicone is ___?___ .

2-49. If two different mers make up the composition of a polymer, it is called a ___?___ .

2-50. Molecules with the same composition but different structures are called ___?___ .

Most industries present their workers with potential health and safety hazards. Many of the compounds and processes used in the plastics industry are potentially dangerous, therefore each reputable company carries on its own safety education program to protect its workers and improve their working environment. This chapter discusses the health and safety hazards and their correction and prevention, along with the ecological problems and solutions unique to the plastics industry.

The toxicology of polymers, the burning characteristics of polymers, waste disposal, and the use of personal protective equipment are the safety considerations most often discussed in professional journals or at professional meetings.

3-1 HAZARDS IN PLASTICS PROCESSING

As raw chemicals are made into useful products, hazardous methods and materials must be used. Hazards may be broken down into three groups: (a) chemical, (b) physical, and (c) biological or biomechanical.

Responsible designers and managers must consider the safety of the public and of the personnel operating processing equipment. A safety program for all personnel may help avert hazards and provide factual information. A safety education program may consider access to equipment, fire protection, storage and labeling of toxic chemical hazards, waste disposal, sewer problems, product liability or other potentially hazardous scenarios. The greatest emphasis in safety training should be on inhalation, since nearly 90 percent of toxicity cases are attributed to lung absorption.

Chemical Hazards

Chemical hazards are found in the form of liquids, mists, vapors, gases, and dusts. Some chemicals are corrosive, many can cause skin irritation, while others produce a narcotic action. Many are considered carcinogenic (cancer causing) if a sufficient dosage is ingested, inhaled, or absorbed.

Chemical hazards are related to the physical and chemical properties of the substances present in the process stream, which may contain flammable, toxic, corrosive, explosive, or highly reactive materials. Some of the chemical reactions in manufacturing and processing produce hazardous by-products or intermediates. Processing may also yield potentially hazardous wastes in the form of solids, sludges, gases, or liquids. (See Waste Laws and Pollution.)

Physical Hazards

Besides the obvious hazards in operation of machinery, noise and vibration may produce hearing and nerve damage. Physical hazards exist whenever machines are operated improperly, without guards, or with other safety devices ignored. Hydraulic, steam, air, or other pressure systems may also present danger if faulty or operated improperly. In addition, workers may be exposed to ionizing, ultraviolet, microwave, or thermal radiation.

Equipment hazards are related to such factors as construction or installation of equipment, age, maintenance, reliability of instrumentation and controls, and human error.

Biological or Biomechanical Hazards

Biological or biomechanical hazards may develop because of poor working conditions. Repetitive motions, improper tool and machine design, poor visual conditions, or lack of ventilation may cause fatigue and result in accidents.

Human engineering (ergonomics) may reduce the physiological and psychological problems associated with biological or biomechanical hazards.

According to the U.S. Center for Disease Control (CDC), parent agency of the National Institute for Occupational Safety and Health (NIOSH), "there is increasing evidence that an unsatisfactory work environment may contribute to psychological disorders including affective disturbances (anxiety and irritability), behavioral problems (substance abuse, sleeping difficulties), psychiatric disorders (neuroses), and somatic complaints (headaches and gastrointestinal symptoms)." The report links job stress to mass psychogenic illnesses, saying the smell of strong odors may trigger an outbreak of illness throughout the plant with complaints of headaches and nausea.

3-2 TOXICITY OF MATERIAL

Although safety hazards related to the physical characteristics of a chemical can be objectively defined in terms of testing requirements (e.g. flammability), health hazard definitions are less precise and more subjective. Some may cause measurable changes in the body, such as decreased pulmonary function. Others are non-measurable, subjective feelings such as shortness of breath or a tired feeling.

Although any material can be used safely if you know how to handle it properly, it is important to know about the toxicity of chemicals. In this chapter, it is impossible to describe all chemical compounds used in the plastics industry and their specific toxic characteristics, therefore only general information is given here. Overexposure to a toxic material is usually the result of carelessness or ignorance.

Almost all plastics, when completely reacted, are inert and nontoxic. Some, in fact, are used as surgical implants in the human body. Others are used to store and package food.

Packaging of food or pharmaceuticals must use FDA approved polymers and additives. Poor selection of additives and polymers may lead to poisoning of consumers by ingestion.

In 1987, ethylene vinyl alcohol (EVOH) copolymer for multilayer packaging of retorted foods has met all the Environmental Protection Agency (EPA) requirements as set forth in Title 21, CFR. It has also ammended Title 21, CFR 177.1315 regulation to allow the safe use of nonoriented CHDM modified PET (1 to 34 mole percent CHDM) in contact with foods containing 25 percent aqueous alcohol. Most toxicity hazards occur in the use or manufacture of monomers.

Butadiene, chloroprene, acrylonitrile, styrene, vinyl acetate, vinyl chloride, and acrylic are commonly used, potentially hazardous monomers. Furfural, formaldehyde, phenol, and epichlorhydrin are also used in the polymerization process of some polymers. Each of these materials may expose workers to potential health hazards (See Table 3-1).

Some health hazards are associated with the plasticizers, lubricants, stabilizers, colors, fillers, and other modifiers found in some plastics. Some of these chemical additives may leach out and come in contact with the skin or be absorbed into food.

Many chemicals and solvents can cause severe irritation to the skin, therefore, every effort should be made to avoid contact with chemicals and solvents.

Some monomers and solvents may be absorbed (cutaneous absorption) by healthy skin and give rise to the same symptoms of poisoning as that due to inhalation.

Organic solvents such as chlorinated hydrocarbons, trichloroethane, methyl alcohol, methyl chloride, homologues of benzene, chlorobenzene, and hydrogenated cyclic hydrocarbons are considered highly toxic, and must be used with caution.

If skin color changes or irritation develops, seek medical help at once.

Chemicals such as triethanolamine, osmium compounds, and ammonia can impair eyesight or cause blindness. β-naphthylamine crysene and selected polynuclear aromatic compounds in sufficient quantity may produce various forms of cancer.

Carbon tetrachloride, a chlorinated hydrocarbon, may cause permanent damage to lungs. Other compounds can cause chronic poisoning of the liver, kidneys, and nervous system (Table 3-1).

Toxicity refers to the potential for harm arising from the action of specific chemicals or other agents on living tissue. Unfortunately, there is little factual information on the hazards to humans from chemical exposures. Most data assumes that humans may be as sensitive to a chemical sub-

Table 3-1. Toxicity and Health Hazards of Selected Materials

Hazardous Waste Number	Material	Toxicity	Flash Point °C	TLV ppm · mg/m³	Health Hazard
V002	Acetone (dimethyl ketone)	Narcosis	−18	750 · 1780	Skin irritation, moderate narcosis from inhalation
V003	Acrylonitrile (vinyl cyanide)	Chronic	5	2 · 4.5	Absorbed through skin, inhalation carcinogenic
	Amyl acetate (banana oil)	Discomfort	25	100 · 530	Irritation to nose and throat from inhalation
	Asbestos	Chronic	—	2 fibers/cc	Respiratory (inhalation) diseases, carcinogenic
V019	Benzene (benzol)	Chronic	11	10 · 30	Poisoning by inhalation, carcinogenic skin irritation and burns
V021	Bisphenol A	Discomfort	—	—	Skin and nasal irritation
	Boron fibers	Discomfort	—	—	Irritant, respiratory discomfort
	Carbon dioxide	Chronic	gas	5000 · 9000	Asphyxiation possible, chronic (inhalation) poisoning in small amounts
	Carbon fibers	Discomfort	—	—	Irritant, respiratory discomfort
	Carbon monoxide	Acute	gas	50 · 55	Asphyxiation
V211	Carbon tetrachloride	Acute	—	5 · 30	Inhaled and absorbed, chronic poisoning in small amounts, carcinogenic
	Chlorine	Acute	—	1 · 3	Bronchial distress, poisoning and chronic affects
V037	Chlorobenzene (phenyl chloride)	Narcosis	29	75 · 450	Absorbed and inhaled, paralyzing in acute poisoning
	Cobalt	Acute	—	0.1	Possible (inhalation) pneumoconiosis and dermatitis
V056	Cyclohexane	Chronic	−20	300 · 1050	Liver and kidney damage, inhalation
	Cyclohexanol	Discomfort	66	50 · 200	Inhaled and absorbed, possible organ damage
V057	Cyclohexanone	Discomfort	44	25 · 100	Inhaled and absorbed, possible organ damage
	Diatomaceous earth	Chronic	—	5	Produces silicosis
V070	Dichlorobenzene	Chronic	66	50 · 300	Possible liver damage, inhalation, percutaneous
V077	1,2-Dichloroethane (ethylene dichloride)	Narcosis	13	10 · 40	Anesthetic and narcotic, inhalaion, possible nerve damage
	Epichlorohydrin	Acute	95	2 · 10	Highly irritating to eyes, percutaneous and respiratory tract, carcinogenic
V173	Ethanol (ethyl alcohol)	Acute	13	1000 · 1900	Possible liver damage, narcotic effects
	Fluorine	Chronic	—	—	Respiratory distress, acute in high concentrations
V122	Formaldehyde	Acute	—	1 · 1.5	Skin and bronchial irritations, inhalation, carcinogenic
	Glass fibers	Discomfort	—	10	Irritant, respiratory discomfort
	Isopropyl alcohol	Discomfort	14	400 · 980	Narcotic, irritation to respiratory tract, dermatitis
V154	Methanol (methyl alcohol)	Chronic	11	200 · 260	Chronic if inhaled, poisoning may cause blindness
	Methyl acrylates	Acute	3	200 · 610	Inhaled and absorbed, liver, kidney, and intestinal damage
V080	Methylene chloride (dichloromethane)	Narcosis	gas	50 · 105	Possible liver damage, narcotic, severe skin irritation, moderate narcosis through inhalation, ingestion, carcinogenic
V159	Methyl ethyl ketone	Chronic	−6	200 · 590	Slightly toxic, effects disappearing after 48 h
	Mica	Chronic	—	20	Pneumoconiosis, respiratory discomfort

Table 3-1. Toxicity and Health Hazards of Selected Materials (continued)

Hazardous Waste Number	Material	Toxicity	Flash Point °C	TLV ppm · mg/m³	Health Hazard
	Nickel	Acute	—	1	Chronic eczema, carcinogenic
V188	Phenol (carbolic acid)	Acute	80	5 · 19	Inhaled, absorbed through skin, narcotic, tissue damage, skin irritation
	Phosgene	Acute	—	0.1 · 0.4	Lung damage
V196	Pyridine	Acute	20	5 · 15	Liver and kidney damage
	Silane	Acute	—	—	Inhalation, organ damage
	Silica	Chronic	—	30	Respiratory discomfort, silicosis, potential carcinogen
V220	Toluene	Acute	4	100 · 375	Similar to benzene, possible liver damage
	Vinyl acetate	Discomfort	8	10 · 30	Inhalation, irritant
V043	Vinyl chloride	Acute	—	5 · 10	Carcinogenic

stance as test animals. Individuals, however, may react differently to harmful materials.

Lethal Dose fifty, designated as LD_{50}, is a calculated dose capable of killing 50 percent of a population of experimental animals exposed through a route other than respiration. Dose concentrations are expressed in milligrams per kilogram of body weight (mg/kg) or as parts of vapor or gas per million parts of air by volume (ppm) at 25 °C and 760 mm Hg pressure. The mineral dusts are expressed as millions of particles per cubic foot (mppcf).

Threshold Limit Value (TLV) is another toxicity rating for exposure during a normal workday without harmful effects. The TLV may be considered to be more applicable than the LD_{50} for restrictions. The TLVs listed in Table 3-1 are those currently recommended by The American Congress of Governmental Industrial Hygienists (ACGIH), but it should be noted that federal and state agencies as well as foreign governments often issue different standards. These standards are also frequently altered, as a result of new evidence from epidemiological and experimental research.

If a substance is exactly as toxic to man as to a guinea pig and the LD_{50} is 20 mg/kg, the LD_{50} for an 80 kg man is $20 \times 80 = 1.60$ g.

New studies on the effects of methylene chloride are the focus of an Occupational Safety and Health Administration (OSHA) study to determine the propriety of lowering the allowable average exposure limit established at 100 ppm in 1986. Several groups have called for lower standards. Recent legislation could cut the use of methylene

chloride and limit it to less than 10 ppm. Methylene chloride is an important ingredient used to manufacture cellular plastics.

It is impossible to list or discuss the toxic properties of all chemical compounds used by the polymer industry. The EPA has a list of known toxic chemcials.

The EPA has developed terminology for decision models that can be used in screening and classifying hazardous compounds and wastes.

Hazardous materials have been defined in the Code of Federal Regulations (CFR) as substances that have been found capable of posing an unreasonable risk to health, safety, and property. The CFR is kept current by daily publication (Monday through Friday) of the *Federal Register*.

Toxicity is determined by the rate of intake, onset of symptoms, and duration of symptoms. *Acute* toxicity is typically characterized by rapid absorption and exposure to the substance. *Chronic* toxicity implies absorption of a harmful material over a long period of time. *Narcosis* may produce a gradual paralysis of the central nervous system. Some of the effects may not be permanent but may be addictive. *Discomfort* refers to short exposure times with little exposure to danger. Many toxins at this level cause skin irritation or dermatitis. (See Table 3-1.)

3-3 FIRE TOXICITY

Fire toxicity of plastics is associated with the fumes, smoke, and gases produced from combustion. Both natural and synthetic materials give off toxic gases when burned.

Heat (direct burns), oxygen deficiency, and panic are other contributing factors which may result in death or incapacitation.

Carbon monoxide, by far the greatest danger, is a colorless, odorless gas that can produce unconsciousness in less than three minutes. In a study conducted by the Fire Safety Center of the University of San Francisco, test animals (mice) were exposed to pyrolysis gases of selected materials. Table 3-2 gives the relative fire toxicity of selected polymers and fibers.

Care must be taken in the extrapolation of toxic fire data. Laboratory tests cannot duplicate all combustion or biological factors. The actual plastics product should be tested to estimate the toxic potential of burning.

Table 3-2. Relative Fire Toxicity of Selected Polymers and Fibers

Material	Approximate Time to Death, min	Approximate Time to Incapacitate, min
Acrylonitrile-butadiene-styrene	12	11
Bisphenol A polycarbonate	20	15
Chlorinated polyethylene	26	9
Cotton fiber, 100%	13	8
Polyamide	14	12
Polyaryl sulfone	13	10
Polyester fiber, 100%	11	8
Polyether sulfone	12	11
Polyethylene	17	11
Polyisocyanurate rigid foam	22	19
Polymethyl methacrylate	16	13
Polyphenyl sulfone	15	13
Polyphenylene oxide	20	9
Polyphenylene sulfide	13	11
Polystyrene	23	17
Polyurethane flexible foam	14	10
Polyurethane rigid foam	15	12
Polyvinyl chloride	17	9
Polyvinyl fluoride	21	17
Polyvinylidene fluoride	16	7
Silk fiber, 100%	9	7
Wood	14	10
Wool fiber, 100%	8	5

Note: Information from Fire Safety Center of the University of San Francisco with the support of the National Aeronautics and Space Administration. All data modified to show approximate values.

Wool and silk are the most toxic materials, with wool rated as more toxic than many polymers. Although they are highly toxic, it is not likely that the use of such generally accepted materials as wool, silk, and wood will be restricted.

The National Fire Protection Association (NFPA) has designated an integer numbering system from zero (0), indicating no special hazard and an inflammable material, to four (4), indicating a severe hazard and very flammable substance. This system for materials is used in three categories: health, flammability, and reactivity. The American Institute of Chemical Engineers also provides a guide (Dow) to assist in selecting, designing, and providing preventive protective safety features for new plants. The guide divides material and process hazards into six ranges: mild, light, moderate, moderately heavy, heavy, and extreme. The larger the fire and explosion index, the greater the hazard.

Although many particles are only a nuisance, some may cause accumulated damage to the lungs. Prolonged exposure to asbestos, cotton, and silica dust may result in asbestosis, pneumoconiosis, and silicosis respectively. There is evidence that asbestos may be carcinogenic. The best way to prevent respiratory problems is to reduce the amount of airborne particles in the work environment.

All motors and electric lights should have combustion covers, and a mechanical-draft duct system should be used to remove dusts or corrosive vapors. A static charge that builds-up on printing equipment in many plastics-processing techniques can pose vapor- or dust-ignition hazards.

When heated, polyvinyl chloride and ethylene sulfide give off hydrogen chloride and sulfur dioxide gases, respectively. These gases corrode metals and electrical equipment. Overheated polytetrafluoroethylene may cause a symptom similar to that produced by inhaling zinc fumes. Proper ventilation and protective clothing can prevent irritation to eyes, nose, and lungs.

As stated by the Factory Mutual Engineering Corporation, 23 percent of reported fires were from electrical causes, 18 percent from smoking, and only 8 percent from overheated materials.

As defined by the Occupational Safety and Health Administration (OSHA) and the National Fire Prevention Association (NFPA) a *flammable* liquid is any material having a flash point below 38 °C [100 °F]. *Combustible* liquids are those with flash points at or above 38 °C. Flash point is the lowest temperature at which enough vapors are given off to form an ignitable mixture of vapor and air. Flash points of some materials are listed in Table 3-1.

Fire Hazards in Manufacture, Processing, and Fabrication

The greatest hazard of fire comes from the many solvents, diluents, plasticizers, and other ingredients essential for producing and processing resins and plastics. They must be used with care and proper precaution. (See Appendix D.)

Cellulose nitrate (nitrocelluose or pyroxylin) is unique in the plastics field. It has the most vigorous burning characteristics of all plastics and must be handled with special precautions. Little cellulose nitrate is used today.

Most plastics can be made self-extinguishing or fire retardant.

All thermosets are self-extinguishing. Polyamide, polyphenylene oxide, polysulfone, polyarylsulfone, polycarbonate, polyvinyl, chlorinated polyether, chlorotrifluoroethylene, vinylidene fluoride and others may be made self-extinguishing. Glass and other inorganic reinforcement may reduce flammability. Most flame-retardant additives act by interfering chemically with the flame reactions. The plastics labeled "slow-burning" or "self-extinguishing" often burn vigorously under actual fire conditions. They may give off quantities of smoke or toxic fumes. Once the source of the flame is removed, the plastics will extinguish themselves.

Flammability tests for polymers include tests of ignition (ASTM-D-1929-77), rate of burning (ASTM-D-635-74), measurement of smoke density (ASTM-D-2843-70), and the oxygen index (ASTM-D-2863-74). The latter has been widely used in comparing the fire properties of polymers. In Table 3-3, the oxygen index values are the minimum concentration of oxygen in an oxygen-nitrogen mixture that will support combustion.

The National Bureau of Standards, National Fire Prevention Association, Underwriter's Laboratories, and the American Society for Testing and Materials have developed tests for the flammability of plastics. (See Chapter 4.)

Many processes involve solvent-recovery systems and the use of heating elements. Others emit explosive or flammable gases. Static charges may be generated on fast-moving plastics films during processing. Flammable blowing agents and solvents that burn rapidly make it imperative that sources

Table 3-3. Health Hazards Associated with Processing and Fabrication

Material	Health Hazard
Acrylics	Avoid skin contact with monomers and catalysts; may cause dermatitis. Normal processing temperatures cause no problems.
Acrylonitrile	Avoid skin contact with monomers. Causes dermatitis and respiratory irritation. No particular problems at normal processing temperatures.
Alkyds	Dermatitis may develop.
Aminos	Urea and melamine-formaldehyde resins cause dermatitis and respiratory discomfort. Dust may produce dermatitis and discomfort.
Cellulose acetate	No particular problems at processing temperatures.
Cellulose acetate butyrate	No particular problems at processing temperatures.
Epoxies	Dermatitis from resin and hardener.
Fluorocarbons	When overheated, cause polymer-fume fever with flu-like symptoms.
Phenolics	Phenol may cause burns and may be absorbed through the skin. Formaldehyde is irritating. Phenolics cause dermatitis and respiratory hazards.
Polyamides	Contact dermatitis has been reported, but rare.
Polyester	Monomers, promoters, and catalysts cause irritation and respiratory problems. Dermatitis may be severe.
Polyethylene	No problems at processing temperatures.
Polystyrene	Monomers are dangerous, but plastics cause no particular problems at processing temperatures.
Polyurethanes	Isocyanates are irritants to skin and respiratory tract. Avoid contact with catalysts and blowing agents.
Silicones	No problems reported. Avoid contact with hardeners with RTV rubbers.
Vinyls	Few problems at normal molding temperatures. Greatest danger from additives. Hydrogen chlorine gas may cause discomfort. Vinyl chloride is a carcinogen.

of electric sparks and flame be avoided. "No hazardous areas. All vapors should be quickly recovered or removed. Flammable liquids should be carefully stored.

Note the flash points of the materials listed in Table 3-1. Flash point is determined by ASTM-D-56 standard tag closed cup test method. It is the lowest temperature at which a sufficient quantity of vapors are given off to form an ignitable mixture of vapor and air immediately above the surface of the liquid.

NFPA Classes of Fires

The NFPA lists four classes of fires. Fire extinguishers are labeled with the class of fire for which they are of primary value. For extinguishing plastics-related fires, a Class B-C extinguisher should be used.

Class A fires involve ordinary combustible materials (such as wood, cloth, paper, rubber, and many plastics) requiring the heat-absorbing (cooling) effects of water or water solutions, or the coating effects of certain dry chemicals that retard combustion.

Class B fires involve flammable or combustible liquids, flammable gases, greases, and similar materials where extinguishment is most readily secured by excluding air (oxygen), inhibiting the release of combustible vapors, or interrupting the combustion chain reaction.

Class C fires involve energized electrical equipment where safety to the operator requires the use of electrically nonconductive extinguishing agents. (Note: When electrical equipment is de-energized, the use of Class A or B extinguishers may be indicated.)

Class D fires involve certain combustible metals, such as magnesium, titanium, zirconium, sodium, or potassium, requiring a heat-absorbing extinguishing medium not reactive with the burning metals.

3-4 EQUIPMENT SAFETY

Much of the potential danger of cutting blades, moving shafts, shearing blades, or hot surfaces can be overcome if safety covers, screens, or other locking devices are installed on machinery and equipment. Two-hand controls may be installed to ensure that the operator's hands are not in the moving mechanism. Equipment and machinery manufacturers do help the efforts of the plastics industry to make processing safer.

Accidents are often caused while maintenance or setup operations are in progress. Machine operators often circumvent safety devices by such means as taping down limit switches. Guards and safety devices are of little value if the operator does not use them or is unaware of their purpose.

High pressure steam, air, or hydraulic systems may rupture and present a danger to the operator.

The petroleum fluids used in hydraulic systems may present fire hazards if the equipment is ruptured or leaking. Fire-resistant emulsions or liquids are often substituted in hydraulic pressure systems as a preventive measure.

Good housekeeping, knowledge of the materials and processes, adequate maintenance, and planned safety can prevent accidents. Many industrial associations offer safety-prevention help in the form of literature, organized programs, and authoritative consulting.

All materials should be kept in properly designed and labeled containers. Some are designed with safety *pop-off* valves. Chemicals should be stored at temperatures below 30 °C (85 °F), and kept in a storage area away from other products. The use of fireproof chests and separate rooms or buildings is ideal for storage.

Personal Protective Wear

Devices such as goggles, masks, helmets, hearing protectors, air respirators, and protective clothing such as disposable polyethylene gloves should be used as needed. They may prevent skin reaction, respiratory problems, and vision or hearing damage.

The ACGIH has published "Guidelines for the Selection of Chemical Protective Clothing," and the American National Standards Institute (ANSI) and OSHA standards address eye and face protection. Protective barrier cream systems may be used to protect workers from minor skin irritants and can curb the incidence of dermatitis.

Allergic-reactive people should be warned of possible bronchial or skin irritants. Any irritation to the skin, eyes, nose, or throat should be dealt with promptly.

3-5 OSHA

The federally enacted William-Steiger Bill, better known as the Occupational Safety and Health Act (OSHA) has special significance for the plastics industry. The act is applicable to all businesses engaged in interstate commerce, and people in schools should also be aware of this important law.

The importance of OSHA cannot be overemphasized. Many states are requiring that all institutions comply with OSHA standards, and are enforcing them at the state level.

Chemical manufacturers were required to develop a Material Safety Data Sheet (MSDS) for all hazardous chemicals, clearly place product labels on all containers, and make MSDS information available to the consumers. In 1986, employers (Standard Industrial Classifications Code 20-39) must train employees, label all in-plant containers, and have a chemical list, MSDS file, and a complete written program.

The Department of Transportation (DOT) has been made responsible for regulating the transportation of hazardous materials by rail, air, water, highway, and pipelines, as part of the Transportation Safety Act (1974) and as described in Title 49, CFR. The EPA has created the Interagency Regulatory Liaison Group (IRLG), the Consumer Product Safety Commission (CPSC), the Food and Drug Administration (FDA), and OSHA. These four departments play major roles in protecting human health and the environment from the adverse effects of toxic materials.

The Toxic Substances Control Act of 1976 authorized the EPA to gather certain kinds of basic information on chemical risks from those who manufacture and process chemicals. The Act may require manufacturers to test for toxic effects, and it may limit or ban those chemicals which exhibit unreasonable risks to human health or the environment.

The EPA requires processors to provide notification to the state emergency response commission if they have extremely hazardous substance in amounts exceeding the threshold planning quantities. Among the 400 substances included are: acrylonitrile, butadiene, cresols, epichlorohydrin, ethylene oxide, formaldehyde, phenol, phosgene, propylene oxide, and vinyl acetate monomer.

In 1976 and 1984, the Resource Conservation and Recovery Act (RCRA) established a system to track hazardous wastes. The RCRA has given the EPA the authority to manage, identify, and set standards for the storage, generation, handling, and disposal of hazardous wastes.

The Comprehensive Environmental Response, Compensation, and Liability Act of 1980 (CERCLA) gave the government authority to clean up hazardous waste sites and respond to hazardous spills. Guidelines and procedures for implementing the CERCLA are listed in a regulation known as the National Contingency Plan (NCP). This plan allows the govenment to respond to cleanups by removal or long term remedial actions. The Trust Fund for Federal and State governments (superfund) is financed by taxes on the manufacture of some chemicals and from the general revenues. Although the EPA is responsible for managing cleanup projects financed through (superfund) Trust Funds, it is making a strong effort to hold individual companies liable for cleanup. The EPA has also identified over 500 hazardous waste sites targeted for long-term action under the "superfund" law. These sites are identified on the National Priorities List (NPL).

For more familiarity with the wide-ranging provisions of the act, various "hazard categories" have been selected from Part 1910 of Title 29 of the Code of Federal Regulations and reprinted in the following paragraphs. These selections are to be used only as a guide, to highlight some of the hazard categories.

3-6 ECOLOGY

People have become increasingly aware of the need to take proper care of our air, water, and land.

Most people are aware also of the need to conserve energy. The plastics industry is often criticized for wasting energy, but a study done for the EPA by Midwest Research Institute found that the plastics industry is not using energy resources inefficiently. Petrochemical production consumes about 7 percent of U.S. petroleum and natural gas resources. These petrochemicals are the basis for synthetic fibers, rubbers, plastics, agri-chemicals, pharmaceuticals, solvents, and other important products. The plastics industry consumes less than 4 percent of the total.

A Massachusetts Institute of Technology study of global prospects for energy up to the year 2000 concludes that "oil is essential for just two major uses: transportation and petrochemical feedstocks."

We must develop alternative energy sources such as solar, wind, nuclear, fusion, geothermal, tidal, and chemical energy to allow conservation of oil supplies.

Waste Laws

The hazardous waste laws are a network of legal requirements specified under three different statutes. These include the RCRA, as amended by the Hazardous and Solid Waste Amendments of 1984 (HSWA); the Toxic Substances Control Act of 1976 (TSCA); and CERCLA, commonly known as Superfund. Each of these laws deals with different subject matter and has its own unique provisions and special terminology. Each is accompanied by

Selected Rules and Regulations of OSHA

Paragraph	Hazard Regulations

Walking-Working Surfaces
1910.22-a All places of employment, passageways, storerooms, and service rooms shall be kept clean and orderly and in a sanitary condition.

1910.36-a This subpart contains general fundamental requirements essential to providing a safe means of egress from fire and like emergencies.

Occupational Health and Environmental Control
1910.94 This section lists numerous acceptable ceiling concentrations for possible air contaminants. When ceiling limits or controls are not feasible to achieve to full compliance, protective equipment or other protective measures shall be used to keep the exposure of employees to air contaminants within the limits prescribed.

1910.94-b Permissible exposure to airborne concentrations of asbestos fibers shall not exceed five fibers, longer than 5 micrometres, per cubic centimetre of air.

1910.94-3a In reference to abrasive blasting operations: all air inlets and access openings shall be baffled or so arranged that by the combination of inward air flow and baffling the escape of abrasive or dust particles into an adjacent work area shall be minimized.

Grinding, Polishing, and Buffing Operations
1910.94-8b All power-driven rotatable wheels composed all or in part of textile fabrics, wood, felt, leather, and paper, may be coated with abrasives on the periphery of the wheel for purposes of polishing, buffing, and light grinding. Grinding wheels or discs for horizontal single or double spindle grinders shall be hooded to collect the dust generated by the grinding operation...

1910.94 All operations involving the immersion of materials in liquids, or the vapors of such liquids, for the purpose of cleaning, the toxic, flammable, or explosive nature of the vapor, gas, or mist shall be determined and held below accepted limits for workers.

Occupational Noise Exposure
1910.95-a Protection against the effects of noise exposure shall be provided when the sound levels exceed those shown below.
8 hours duration... 90 dBA sound level
4 hours duration.... 95 dBA sound level
1 hour duration.... 105 dBA sound level
Impact noise not to exceed 140 dB peak sound pressure level.

Ionizing Radiation
1910.96-a Radiation includes alpha, beta, gamma, x-rays, neutrons, high-speed electrons, protons, and other atomic particles. The body should not be exposed to more than 3 Rems during any calendar quarter.

Selected Rules and Regulations of OSHA (Continued)

Paragraph	Hazard Regulations

All radioactive areas, materials or sources should be labeled with the prescribed radiation caution colors.
Every employer shall supply appropriate personnel monitoring equipment, such as film badges, pocket chambers, dosimeters, or film rings for the purpose of measuring the radiation dose received.

Nonionizing radiation
1910.97 This section applies to radiation in the radiofrequency region including microwave and radar frequencies.
For normal environmental conditions and for incident energy frequencies from 10 MHz to 100 GHz, the radiation protection guide is 10 mW/cm^2 (milliwatt per square centimetre) as averaged over any possible 0.1 h period. This applies whether the radiation is continuous or intermittent.

Hazardous Materials
1910.101-a Each employer shall determine that compressed gas cylinders are in safe condition by visual and other inspections.

1910.102 Applies to installation of gaseous acetylene systems.

1910.103 Applies to the installation of gaseous hydrogen systems.

1910.104 Applies to the installation of oxygen systems.

1910.106 Discusses the storage, ventilation, and fire control of flammable and combustible liquids. Storage of flammable or combustible liquids in drums or other containers including aerosols not exceeding 272 L [60 gal] individual capacity and portable tanks not exceeding 3 000 L [660 gal] individual capacity is permitted if placed in proper containers.
Not more than 272 L [60 gal] of flammable or 545 L [120 gal] of combustible liquids may be stored in a storage cabinet.
Every inside storage room shall be provided with either a gravity or mechanical exhaust ventilation system.
A portable fire extinguisher having a rating of not less than 12-B units shall be located outside of, but not more than 3 M [10 f] from, the storage room door.

1910.107 Spray finishing using flammable and combustible materials, including aerated solid powders such as fluidized powder bed, dip, and spray areas.
All spraying areas shall be provided with mechanical ventilation to remove flammable vapors, mists, or powders.
"No smoking" signs shall be conspicuously posted at all spraying areas.
Electrostatic apparatus shall be equipped with automatic controls which operate without time delay to disconnect the power supply. Electrostatically charged hand-

Selected Rules and Regulations of OSHA
(Continued)

Paragraph	Hazard Regulations
	held spraying equipment shall be designed so as not to produce a spark of sufficient intensity to ignite any vapor-air mixtures or result in appreciable shock hazard upon coming in contact with a grounded object.
1910.108	Dip tanks containing flammable or combustible liquids in which articles are immersed for the purpose of coating, finishing, treating, or similar processing shall have ventilating systems. No open flames, spark-producing devices, or heated surfaces having sufficient temperature to ignite vapors shall be used. Adequate arrangements shall be made to prevent sparks from static electricity.

Personal Protective Equipment

Paragraph	Hazard Regulations
1910.132-a	Protective equipment, including personal protective equipment for eyes, face, head, and extremities, protective clothing, respiratory devices, and protective shields and barriers shall be provided, used, and maintained in a sanitary and reliable condition wherever it is necessary by reason of hazards of processes or environment. Irritants may cause injury or impair the function of any part of the body through absorption, inhalation, or physical contact.
1910.133	Eye and face protection shall be required where there is a reasonable probability of injury. Suitable eye protectors shall be provided where machines or operations present the hazard of flying objects, glare, or liquids injurious radiation, or a combination of these hazards.
1910.134-a2	Respirators shall be provided by the employer when necessary to protect the health of the employee.

General Environmental Controls

Paragraph	Hazard Regulations
1910.141	All places of employment, passageways, storerooms, and service rooms shall be kept clean and orderly and in a sanitary condition. Potable water shall be provided for drinking.... Adequate toilet facilities which are separate for each sex shall be provided.

Selected Rules and Regulations of OSHA
(Continued)

Paragraph	Hazard Regulations
1910.144	Safety color codes for marking physical hazards shall be used. Red: for fire protection equipment and apparatus. Orange: for designating dangerous parts of machines. Yellow: for marking physical hazards. Green: for designating "Safety" and first aid equipment. Blue: for warning against starting or moving equipment under repair. Purple: for designating radiation hazards. Black/white: designating traffic.

Machinery and Machine Guarding

Paragraph	Hazard Regulations
1910.216-4bi	A safety trip control shall be provided in front and in back of each mill having a 46-inch roll height or over. These bars shall operate readily by pressure of the mill operator's body.
1910.216-3ci	A safety triprod, cable or wire center cord shall be provided across each pair of in-running rolls extending the length of the face of the rolls. It shall be readily accessible and operate whether pushed or pulled. The safety tripping devices shall be located within reach of the operator and the bite.

Toxic and Hazardous Substances

Paragraph	Hazard Regulations
1900.1000	In this section the acceptable work environment concentrations of various materials and maximum air contaminants to the employee are listed.
1900.1200-a1	The purpose of this section is to ensure that the hazards of all chemicals produced or imported by chemical manufacturers or importers are evaluated, and that information concerning their hazards is transmitted to affected employers and employees within the manufacturing sector. This transmittal of information is to be accomplished by means of comprehensive hazard communication programs, which are to include container labeling and other forms of warning, material safety data sheets and employee training.

long and complicated regulations detailing the general requirements of the statute, including specific duties and liabilities.

New regulations and amendments to regulations are published in the *Federal Register,* available from the Superintendent of Documents. (See Appendix G.)

Bills proposing bans, taxes, or deposits on plastics products are being introduced at local, state, and national levels. Most bills are intended to encourage using recyclable or degradable products and to discourage litter and incineration of plastics. Many of these measures appear to be biased against plastics products.

One bill would prohibit dumping of plastics on land or sea and require that ring connectors for can six-packs be biodegradable. Another state bill would place a disposal tax on nonrecyclable consumer goods. Several propose a mandatory deposit on beverage containers.

Sources of Plastics Waste

The Society of the Plastics Industry, Inc. (SPI) estimated in a published report that the U.S.

would use more than 42 billion pounds of plastics in 1987, with about 35 billion pounds contributing to waste. The use of plastics amounts to nearly double the combined output of steel, aluminum, and copper. According to one source, plastics waste consists of about 83% thermoplastics, 13% thermosets, and 4% polyurethane foams. In 1970 less than 3% of the solid waste was plastics. In 1984 about 7% was plastics. Plastics are expected to grow to almost 10% of U.S. municipal solid waste by the year 2000.

In one respect, plastics waste may be classified as follows: 1) *waste plastics,* which consists of plastics resins or products that must be reprocessed or disposed of, 2) *industrial plastics waste,* consisting of waste generated by various industrial sectors including the resin manufacturer, fabricator, compounder, reprocessor, assembler, and distributor, 3) *postconsumer plastics waste,* consisting of discarded consumer products and accounting for more than 90% of the total plastics waste, 4) *nuisance plastics,* or those waste plastics that we lack technology to reprocess under existing technoeconomic conditions, and 5) *scrap plastics,* or waste components capable of being reprocessed into useful products.

Pollution Problems

A special report, *Plastics and Solid Waste Disposal,* by the Phillips Petroleum Company showed that pollution can be grouped as: (a) air pollution, (b) water pollution and (c) solid waste.

Air Pollution. Vehicle exhaust fumes, industrial smoke, and open-dump burning are only a few examples of air pollution. Plastics, specifically polyvinyl chloride and polystyrene, have received strong criticism with respect to air pollution since they both emit smoke. Polyvinyl chloride releases hydrogen chloride gas when improperly burned in open dumps. When these gases combine with moisture in the air, hydrochloric acid (HCl) is formed. However, other household and municipal wastes, such as common table salt from food, paper, grass clippings, wood, and leather are also sources of chlorine. Wool, when improperly burned, produces deadly hydrogen cyanide gas (HCN). Other nonplastics components are responsible for sulfur oxides and other corrosive and polluting agents.

When plastics are burned in open pits or dumps the pollution is indistinguishable from that produced by other sources. In the nation, over 80 percent of solid waste is disposed of by burning in open dumps. The chief advantage of burning is the reduction of mass (weight) and volume of the refuse by as much as 90 percent during proper incineration. Modern incinerator technology may be one way to reduce the volume of waste and still prevent pollution of the air.

A study by DeBell & Richardson, Inc. (an independent research firm), based on experiments in several nations, concludes that when plastics are properly incinerated, carbon dioxide (CO_2) and water (H_2O) are the principal products. It is when wastes are improperly incinerated (in open dumps) that carbon monoxide (CO), soot, and black smoke result. This report shows that properly designed incinerators could handle domestic, municipal, or industrial wastes containing ten times the present levels of polyvinyl chloride (PVC). These incinerators would still be within established safety limits for hydrogen chloride emission.

The charge that plastics clog and burn out incinerator grates is unfounded. Plastics are desirable fuels in modern incinerators because of their high energy or heat content. They have three to four times the heat content of other combustible wastes. Plastics may have a heating value of 40 000 kilojoules (kJ) [37 883 British thermal units] per kg [2.2 lbs.], as compared to 12 000 Btu/lb. for coal.

Professor Elmer R. Kaiser of New York University, an expert on incinerator processes, reports: "When grates get clogged or burned out under ordinary conditions, it is not the fault of plastics". Plastics volatilize (turn to vapor) at [600 °F] 315 °C. The volatiles burn *above* the grate. They do not touch the grate, and no incinerator grate will burn out at 600 °F.

The high energy content of plastics in waste has been used for years in heat recovery incinerators in Europe. In the United States, Chicago has an incinerator system that disposes of waste with an enormous heat energy that has the potential of generating electric power. Several other American cities are exploring the idea of building efficient, nonpolluting incinerators that could serve as power plants.

According to the Bureau of Solid Waste Management, 75 percent of the nation's 300 municipal incinerators are substandard; that is, they do not burn refuse fully or properly.

Modern pollution control equipment and properly controlled combustion will help solve many air pollution problems.

Water Pollution. The waste materials emitted from industries, municipal sewage-disposal plants, commercial shipping, agriculture, and other sources are often water pollutants. Many of these waste products are of value, and modern sewage plant operators are exploring methods of collecting methane and other gases for possible polymer use. Many companies and industries are installing adequate disposal systems rather than dumping wastes into sewer lines or rivers. All litter in water or soil is unattractive and potentially hazardous. Most plastics do not decompose, release explosive gases, or contaminate water.

Solid Waste. Automobile bodies, grass trimmings, packaging materials, and agricultural wastes are examples of solid waste pollution. The four most-used methods for disposal of solid waste are: (a) open dumping (often involving burning), (b) sanitary landfill, (c) incineration, and (d) composting (shredding or pulverizing wastes into the soil as conditioners).

Open dumping is a health hazard and an eyesore. Many communities have banned this method of refuse disposal. 140 million tons of solid waste is generated each year in the United States. Rubbish, trash, and garbage are now recognized as a major problem, but only 8 percent of this total refuse is composed of plastics.

Unfortunately, open dumping is still the most common method of disposing of refuse. These dumps are breeding ground for vermin and insects. They threaten disease, cause fires, emit odors, pollute ground water, and blight the land.

Today, about 7 percent of municipal solid waste goes into *sanitary landfills*. The most desirable material for sanitary landfills is dirt, broken concrete bricks, and other dense materials. Sanitation authorities welcome plastics in sanitary landfills because they do not break down or produce polluting odors, gases, or liquids. Other organic materials in landfills leach out chemicals, which may pollute ground water supplies.

Degradable landfills are often overrated and may create many secondary problems as they rot and burn. Paper, for example, may take more than 60 years to fully degrade.

Incineration of municipal waste (with its potential use in generating electric power) can significantly reduce air pollution and refuse volume.

Incineration may be the only practical method of disposal for many cities. It may not be practical to separate composites for recycling into new plastics composites or products, because so many of them are thermosetting with glass, paper, or metallic layers or fillers. Nevertheless, they still contain BTUs of energy, which are released when the plastics are recycled for incineration and production of electricity.

Plastics are good as bulk factors and mulch, however, most plastics do not break down. *Composting* of plastics is desirable because this inert substance acts as a bulk factor, retarding soil compaction. This, in turn, allows air to circulate and thus speeds the breakdown of other organic materials.

There are more biodegradable plastics on the market each year, however, they represent only a small proportion of the total used in packaging. Polyethylene and polystyrene packaging films degrade much faster than either wood or paper.

The use of edible and water-soluble plastics films in packaging is growing. Examples include casings for meats, drug capsules, and coatings for fruits. Cellulose films are based on alpha-cellulose and have been modified to form hydroxypropyl methyl cellulose, which is soluble both in organic solvents and in water. This film is a thermoplastic resin and may be extruded, injection-molded, and blow-molded.

A nondigestible material used as a protective coating on poultry and other food products is based upon 30 to 40 percent cellulose acetate butyrate and 60 to 70 percent acetylated monoglycerides. The cellulose acetate butyrate component is not digestible.

Water-soluble, nonedible films may include polyvinyl alcohol and polyethylene oxide. Polyvinyl alcohol films are used as protective coatings and for packaging products that are ultimately dissolved in water. Polyethylene oxide films are used for soluble laundry bags and for packaging a variety of industrial and household products. Both water-soluble films are made so that they will go into complete solution at temperatures as low as 5 °C [40 °F] without any trace of residue (Fig. 3-1).

As of 1987, no degradable FDA-sanctioned food packages exist. It is expected that hydroxyhatyratevelerate (PHBV) aliphatic polyester copolymer and biodegradable starch additives may be approved in food packaging films and bottles. Photodegradable products are proven safe as agricultural mulch film.

Starch is used to fill ethylene-acrylic acid or PE, after which microorganisms attach to the natural polymer causing the package to degrade. These additives may impede recycling, because small volumes may introduce degradability into other plastic products.

Although Figure 3-2 shows that 8% of solid waste is attributable to packaging and other plastics refuse, many cities estimate that plastics may compose more than 15 percent of the total waste volume (lightweight materials are more voluminous).

(**A**) Closeup of plastics film strip.

(**B**) Note width and layout of strips.

(**C**) Film around plant.

Fig. 3-1. Rows of vegetables, planted through a degradable plastics film, grow vigorously. The film eliminates weeding, conserves ground moisture, and warms the soil by absorbing heat energy from the sun. (Princeton Chemical Research, Inc.)

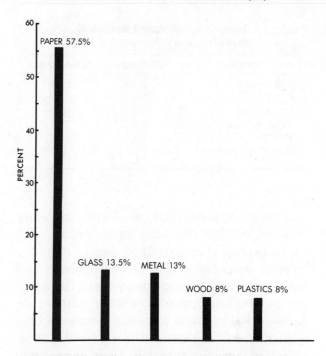

Fig. 3–2. Percent of solid waste attributable to disposable packaging and single-service containers. (Updated from *The Role of Packaging in Solid-Waste Management*, Public Health Service Bulletin No. 1855)

3-7 RECYCLING

The ultimate solution to pollution and solid waste problems is reclamation or recycling. The largest single obstacle to reclamation of plastics is the need for a practical, low-cost method of separating plastics from other wastes.

Plastics recycling is an infant industry. Metals and papers have long-established reclamation technologies, industries, and associations interested in recovery. The Center for Plastics Recycling Research (CPRR), SPI, the Society of Plastics Engineers (SPE), and other agencies and industries recognize the importance of reclaiming and reusing materials. See Table 3-4.

There are four general classifications for recycling plastics waste. 1) *Primary recycling* is the processing of scrap plastics into the same product or one similar to that from which it was generated, using standard plastics processing methods. Only thermoplastics waste can be directly recycled and reprocessed, and the material must be clean. Primary recycling causes general reduction of mechanical properties through a general thermoexodative degradation or, in some cases, branching and crosslinking. The aspect ratio of reinforcements, chain length, and matrix interface are generally lower. Up to 100% reprocessed material may be

**Table 3-4. Estimated Municipal
Waste Recovery Percentage 1988**

Material	% Waste	% Receovery
Aluminum	2.5	28.5
Glass	13.0	7.0
Iron & Steel	10.5	2.5
Paper & Paperboard	62.5	18.0
Plastics	8.5	1.0
Rubber & Leather	3.0	2.0

used for less critical applications. An industry source estimates that 75% of plastics waste generated during fabrication is reused. 2) *Secondary recycling* involves reprocessing plastics into products with less demanding properties and quality. The waste may be obtained from postconsumer waste or industrial plastics waste sources. There may be a mixture of different types of plastics and a small amount of nonplastic materials. Drain pipes, cargo skids, flower pots, posts, and cable reels are familiar products formed by intrusion (flow) molding, compression molding, and extrusion. 3) *Tertiary recycling* involves a pyrolysis process in which simple chemical compounds are made of mixtures of waste. The process does not cause air pollution, reduces the volume of waste by 90%, and results in valuable chemicals (tar, gases, light oils, and other organic compounds). 4) *Quaternary recycling* is the recovery of energy from waste plastics by incineration, hydrogenation, or anaerobic digestion.

The litter and waste disposal problem may be helped if more materials could be economically collected and reprocessed. New York State Supreme Court Justice, Saul S. Streit, pointed out in a 1971 decision that ". . . there is not one shred of evidence presented herein which demonstrates that any form of container, glass, metal, or paperboard, is any more recyclable than plastics containers. . . . It thus appears that the discrimination against plastics containers does not rest on any reasonable basis in relation to the objective of promoting the recycling of containers, plastics or otherwise."

Litter is a social problem for which an estimated $2 billion or more per year is spent for collection from along streets and highways and in parks. In spite of the cost of cleanup, our society seems unwilling to limit the variety of goods and the packaging conveniences provided by the use of plastics.

VOCABULARY

The following vocabulary words are found in this chapter. Look up the definition of any of these words you do not understand as they apply to plastics in the glossary, Appendix A.

Acute toxicity
Air pollution
Carcinogen
CERCLA
CFR
Chronic toxicity
Composting
EPA
FDA
Fire resistant
Flash point
Incineration
Industrial plastics waste
Inhalation
LD
NFPA
Nuisance plastics
Open dumping
OSHA
Oxygen index
Postconsumer plastics waste
Primary recycling
Pyrolysis
Quarternary recycling
Recycling
Sanitary landfill
Scrap plastics
Secondary recycling
Self-extinguishing
Silicosis
Solid waste
Tertiary recycling
Toxic substance
TVL
Waste plastics
Water pollution

3-1. Breathing vapors of some solvents may cause ___?___ effects.

3-2. The ultimate solution to pollution and solid waste is ___?___ .

3-3 A highly toxic, colorless, odorless gas is named ___?___ .

3-4. Name three broad types of pollution.

3-5. The name ___?___ indicates that a material does not easily burn.

3-6. The name ___?___ is a term used to describe the ability of a substance to stop burning when the flame is removed.

3-7. About 8 percent of our municipal solid waste goes into ___?___ .

3-8. The production of petrochemicals consumes about ___?___ percent of our petroleum and natural gas.

3-9. Combustible liquids are those with flash points at or above ___?___ °C.

3-10. Name three natural materials that may be more toxic than plastics.

3-11 Most toxicity hazards occur in the use or manufacture of ___?___ .

3-12. Two major reasons for overexposure to toxic materials are ___?___ and ___?___ .

3-13. Name three broad categories of hazards in working with and processing chemicals.

3-14. Name four materials that are classified as inducing acute toxicity.

3-15. Two of the most toxic materials in fire tests may be ___?___ and ___?___ .

3-16. The overheating of ___?___ plastics may cause polymer-fume fever with flu-like symptoms.

3-17. Liquids that have a flash point below 38 °C are called ___?___ .

3-18. According to OSHA, the most noise exposure allowed during an 8 hour period is ___?___ .

3-19. What percent of solid waste may be attributed to plastics?

3-20. What is the ignition temperature of cellulose acetate plastics dust? (See Appendix D)

3-21. Numerous additives or chemicals, and ___?___ of many plastics may cause cancerous ulcers on the body.

3-22. Name four compounds that may cause chronic poisoning of the liver, kidneys and nervous system.

3-23. Name four materials that may cause dermatitis or be absorbed through the skin.

3-24. Chemicals and plastics should be stored at temperatures below ___?___ and in properly designed containers.

3-25. Many plastics degrade when overheated. Name one that when heated releases a form of chlorine gas that causes discomfort.

3-26. The ___?___ of 1971 provides regulations for safety and standards for equipment manufacturers.

3-27. Plastics are ___?___ in modern incinerators because they aid in burning and increase energy output.

3-28. Name four sources of air pollution:

3-29. A class ___?___ fire extinguisher should be chosen for plastics related fires.

3-30. A process in which waste materials are heated in the absence of oxygen, usually under pressure, to produce grease like compounds is called ___?___ .

3-31. The difference between ___?___ and sanitary land fills is that waste is not covered with earth in a controlled manner.

3-32. Plastics ___?___ adds bulk and allows air to circulate through the soil of a compost pile.

3-33. Name three products where edible or water soluble plastics are used.

3-34. Name several methods for disposing of waste plastics.

3-35. Plastics are desirable as landfill because most do not ___?___ or produce polluting odors, gases or liquids.

3-36. Any unwanted, undesirable condition of the air, water and land is called ___?___ .

3-37. Where and how should chemicals and plastics be stored?

3-38. How can dust, steam, air and hydraulic pressure be dangerous?

3-39. Name some chemicals that may be toxic when they are inhaled.

3-40. Name several elements and compounds that are produced from the process called pyrolysis.

Chapter

Properties and Tests of Selected Plastics

In this chapter, you will become familiar with methods of identifying plastics and testing plastics. Both will help you select the proper plastics and product design to meet the requirements of consumer use.

The product designer must have a good understanding of polymer properties and be able to interpret information from data sheets and specifications. See Chapter 21 for further information on design considerations and specifications. You will also begin to become familiar with the SI metric system of measurement as used in the plastics industry.

4-1 METRICS AND PLASTICS

The metric system of measurement is becoming a reality for everyone, especially in the plastics industry. Those who teach plastics or those who design plastics parts and components must know how to express industrial and physical-property values for plastics materials in metric terms.

To prevent misunderstanding and to stop the proliferation and misuse of obsolete customary or metric units, only *Le Système International d'Unités* (SI) metric system units should be used. Units such as Btu, centigrade, Fahrenheit, dyne, erg, foot-pound, horsepower, inch, gallon, psi, therm, ton, yard, and the like are slowly becoming obsolete.

The SI metric system consists of the seven *base units* shown in Table 4-1, along with two supplementary angular units. Quantities measured in SI base units may vary greatly; therefore, to avoid ex-

Table 4-1. SI Base Units

Quantity	Unit	Symbol
Length	metre	m
Mass	kilogram	kg
Time	second	s
Thermodynamic temperature	kelvin	K
Electric current	ampere	A
Luminous intensity	candela	cd
Amount of substance	mole	mol

tremely large or small numbers, prefixes are used (Table 4-2). Symbols of base units used with prefixes save time and help reduce errors. Selected *derived units* useful in the plastics industry are shown in Table 4-3. All derived units are expressed in terms of SI base units. More complete lists of derived units are available from many sources.

The example following shows the use of SI metric units as they apply to the properties and testing of polyvinyl chloride. Typical values for polyvinyl chloride are chosen. Although the example of metric use may not be especially meaningful on first reading, it will be useful as you study this chapter. It is placed at the beginning of the chapter so that you will easily find it for future reference.

Relative Density

Density is the mass per unit volume of a plastics. It is obtained by dividing the mass of plastics by volume. The proper SI derived unit of density is kilograms per cubic metre, although it is commonly expressed by its submultiple, grams per cubic centimetre.

$$\text{Example:} \quad \text{Density} = \frac{\text{mass (kg)}}{\text{volume (m}^3)}$$

Table 4-2. Prefix and Numerical Expression

Symbol	Prefix	Phonic	Decimal Equivalent	Factor	Prefix Origin	Original Meaning
E	exa	*x'a*	1 000 000 000 000 000 000	10^{18}	Greek	colossal
P	peta	*pet'a*	1 000 000 000 000 000	10^{15}	Greek	enormous
T	tera	*ter'a*	1 000 000 000 000	10^{12}	Greek	monstrous
G	giga	*ji'ga*	1 000 000 000	10^{9}	Greek	gigantic
M	mega	*meg'a*	1 000 000	10^{6}	Greek	great
k	kilo	*kil'o*	1 000	10^{3}	Greek	thousand
h	hecto	*hek'to*	100	10^{2}	Greek	hundred
da	deka	*dek'a*	10	10^{1}	Greek	ten
d	deci	*des'i*	0.1	10^{-1}	Latin	tenth
c	centi	*cen'ti*	0.01	10^{-2}	Latin	hundredth
m	milli	*mil'i*	0.001	10^{-3}	Latin	thousandth
μ	micro	*mi'kro*	0.000 001	10^{-6}	Greek	small
n	nano	*nan'o*	0.000 000 001	10^{-9}	Greek	very small
p	pico	*pe'ko*	0.000 000 000 001	10^{-12}	Spanish	extremely small
f	femto	*fem'to*	0.000 000 000 000 001	10^{-15}	Danish	fifteen
a	atto	*at'to*	0.000 000 000 000 000 001	10^{-18}	Danish	eighteen

Table 4-3. Selected Derived SI Units

Quantity	Unit	Symbol	Formula
absorbed dose	gray	Gy	J/kg
area	square metre	m^2	
volume	cubic metre	m^3	
frequency	hertz	Hz	s^{-1}
mass density (density)	kilogram per cubic metre	kg/m^3	
speed, velocity	metre per second	m/s	
acceleration	metre per second squared	m/s^2	
force	newton	N	$kg \cdot m/s^2$
pressure (mechanical stress)	pascal	Pa	N/m^2
kinematic viscosity	square metre per second	m^2/s	
dynamic viscosity	pascal second	Pa · s	$N \cdot s/m^2$
work, energy, quantity of heat	joule	J	N · m
power	watt	W	J/s
quantity of electricity	coulomb	C	A · s
electric potential, potential difference, electromotive force	volt	V	W/A
electric field strength	volt per metre	V/m	
electric resistance	ohm	Ω	V/A

For polyvinyl chloride:

$$\text{Density} = \frac{1\ 300\ \text{kg}}{1\ \text{m}^3}$$

$$= 1\ 300\ \text{kg/m}^3 \text{ or } 1.3\ \text{g/cm}^3$$

The relative density of a plastics is the density of a given substance divided by the density of water. Relative density is a constant number under controlled conditions of air pressure and temperature. It is a dimensionless quantity; consequently, the ratio or relative density will be the same in any measurement system.

Example:

$$\text{relative density} = \frac{\text{density of polyvinyl chloride}}{\text{density of water}}$$

$$= \frac{1\ 300\ \text{kg/m}^3}{1\ 000\ \text{kg/m}^3}$$

$$= 1.3$$

Tensile Strength

In order to comprehend the unit of measurement that describes the property of stress we must understand the metric unit of force. The newton is the SI derived unit of *force*. A kilogram is a unit of *mass*. The standard value for the acceleration caused by gravity on earth is 9.806 65 metres per second squared. The force of gravity acting on one kilogram is thus 9.806 65 newtons.

Example:

$$Force (N) = mass (kg) \times acceleration (m/s^2)$$
$$= 1 \text{ kg} \times 9.806 \text{ } 65 \text{ m/s}^2$$
$$= 9.806 \text{ } 65 \text{ N}$$

If you were to hold 100 g of PVC in your hand, what force is required to keep it from falling?

Example: Force = mass × acceleration
$$= 100 \text{ g} \times 9.8 \text{ m/s}^2$$
$$= 0.100 \text{ kg} \times 9.8 \text{ m/s}^2$$
$$= 0.98 \text{ N}$$

The metric unit used to express *pressure* or force applied over an area is the pascal (Pa). One pascal is equal to a force of one newton exerted on an area of one square metre.

Example: Pressure (Pa) = force (N)/area (m²)
$$1 \text{ Pa} = 1 \text{ N/m}^2$$

All kinds of stress (pressure) including tensile strength, modulus of elasticity, shear strength, compression strength, flexural strength, and adhesion are measured in pascals. Air, vacuum, and fluidic pressures are expressed in pascals. Do not use kg/cm². Stress is calculated by dividing force by area.

Example: stress (pressure) = force/area
$$Pa = N/m^2$$

It is easy to calculate the force and pressure exerted on a surface by a 2-kg piece of PVC measuring 50 mm by 100 mm at its base.

Example: area = length × width
$$= 50 \text{ mm} \times 100 \text{ mm}$$
$$= 5 \text{ } 000 \text{ mm}^2 \text{ or } 0.005 \text{ m}^2$$
force = mass × acceleration
$$= 2 \text{ kg} \times 9.8 \text{ m/s}^2$$
$$= 19.6 \text{ N}$$
pressure = force/area
$$= 19.6 \text{ N}/0.005 \text{ m}^2$$
$$= 3 \text{ } 920 \text{ Pa or } 3.92 \text{ kPa}$$

Tensile strength may be calculated by dividing the maximum load (force) by the original cross-sectional area.

Example:
$$\text{tensile strength (Pa)} = \frac{\text{pulling force (N)}}{\text{cross-sectional area (m}^2)}$$

Since a pascal is a very small pressure (force/area), megapascal or gigapascal may be used.

Impact Resistance

The Izod and Charpy impact resistance tests are measures of the energy required to break a specimen. All forms of *energy* (work) are measured in joules (J).

Example: energy (J) = force (N) × distance (m)
$$1 \text{ J} = 1 \text{ N} \cdot \text{m}$$

In order for a man or machine to lift a 50 kg bag of PVC to a height of 1 m, 490 J of energy are required.

Example: force = mass × acceleration
$$= 50 \text{ kg} \times 9.8 \text{ m/s}^2$$
$$= 490 \text{ N}$$
work (energy) = force × distance
$$= 490 \text{ N} \times 1 \text{ m}$$
$$= 490 \text{ N} \cdot \text{m or } 490 \text{ J}$$

The SI derived unit for a twisting or rotating motion is the newton metre (N · m). It is often called a *moment of force* or *torque*.

Impact resistance must consider the thickness of the materials for energy comparisons. The values are expressed in joules per metre (J/m).

In one test (drop-impact), a mass piece called a *tup* is dropped onto the test specimen. A 1-kg tup may be dropped from a height of 2 m. If the specimen can withstand this impact without cracking or breaking, its impact resistance is 19.6 J.

Example: force = mass × acceleration
$$= 1 \text{ kg} \times 9.8 \text{ m/s}^2$$
$$= 9.8 \text{ N}$$
energy (J) = force (N) × distance (m)
$$= 9.8 \text{ N} \times 2 \text{ m}$$
$$= 19.6 \text{ J}$$

Hardness

Hardness is an arbitrary unit; consequently, the numerical value will not change. Rockwell 110-M and Shore 75-A are typical hardness values for unfilled rigid polyvinyl plastics. The Durometer instrument registers the amount of indentation caused by a small, pointed indentor. Readings are taken immediately after the indentor is forced into the plastics. The scale range is 0 to 100.

Thermal Expansion

The coefficient of expansion is used to determine thermal expansion in length, area, or volume per unit of temperature rise. It is expressed as a ratio per degree Celsius.

If a PVC rod 2 m in length is heated from -20 °C to 50 °C, it will change 7 mm in length.

Example:

change in length = coefficient of
                linear expansion $\times$ original
                length $\times$ change in temperature
           = 0.000 050/°C $\times$ 2 m $\times$ 70 °C
           = 0.007 m or 7 mm

Since area is a product of two lengths, value of the coefficient must double. Similarly, we must triple the coefficient value to obtain thermal expansion for volume.

Resistance to Heat

Kelvin is the preferred thermodynamic scale; however, the Celsius scale is permitted and is in general use. Add 273.15 to the Celsius reading to obtain Kelvin value. Heat resistance for PVC is from 65 to 80 °C.

Dielectric Strength

This is a measure of electrical voltage required to break down or arc through a plastics material. The units are reported as volts per millimetre of thickness. PVC has a dielectric strength of 15 000 to 19 500 V/mm.

Mold Shrinkage

The mold (linear) shrinkage is the ratio of the decrease in length to the original length of a molded test specimen; consequently, the numerical values do not change with the units used. The units are usually expressed as millimetres per millimetre, for clarity.

Many plastics companies are committed to hard metric conversion. The practice of expressing measurements in both customary and equivalent metric units is called soft conversion. Soft conversion serves little purpose, is confusing and more costly, and errors are more easily made.

4-2 IDENTIFYING PLASTICS

Plastics are complex materials. Because they can contain several polymers and other ingredients, including fillers, identification is not easy. The insolubility of some plastics adds to the problem. Correct identification requires complex tools and techniques.

A student or a consumer may have to identify a plastics so that repairs or new parts can be made. Laboratory tests can help to identify ingredients of the unknown material. The methods shown in this chapter are meant for easy identification of basic polymer types. Identification methods that involve complex instruments are not covered.

More sophisticated methods include infrared spectroscopy analysis, which is the only accurate method of obtaining the quantitative identification of unknown polymers. Another highly complex and costly method is X-ray diffraction, which is used for identification of solid crystalline compounds.

The elements nitrogen, sulfur, chlorine, and fluorine are often found in plastics. Testing for these elements serves to identify many plastics. For example, casein, cellulose nitrate, polymide, melamine-formaldehyde, and polyurethane will show the presence of the element nitrogen.

The Beilstein test is a simple method of determining the presence of a halogen (chlorine, fluorine, bromine, and iodine). For this test, heat a clean copper wire in a Bunsen flame until it glows. Quickly touch the hot wire to the test sample, then return the wire to the flame. A green flame shows the presence of a halogen.

Plastics containing chlorine are polychlorotrifluoroethylene, polyvinyl chloride, polyvinylidene chloride, and others. They give positive results to the halogen test. If the halogen test is negative, the polymer may be composed of only carbon, hydrogen, oxygen, or silicon.

For further chemical analysis, the Lassaigne procedure of sodium fusion may be used. *WARNING: While very useful, this test is dangerous because sodium is highly reactive. It should be performed with great care.*

To conduct the Lassaigne test, place five grams of the sample in an ignition tube with 0.1 gram of sodium. Heat the tube until the sample decomposes, keeping the open end of the tube pointing away from you. While the tube is still red-hot, place it in distilled water and grind it with a mortar and pestle device. Filter out the carbon and glass fragments while the mixture is still hot. Divide the resulting filtrate into four equal portions. Use these portions to perform standard tests for nitrogen, chlorine, fluorine, and sulphur.

4-3 IDENTIFICATION METHODS

There are five broad identification methods for plastics:

1. trade name
2. appearance
3. effects of heat
4. effects of solvents
5. relative density

Trade Names

Numerous trade names are in use today. Many are used to identify the product of a fabricator, a manufacturer, or a processor. The trade name may be associated with the product or with the plastics material. In either case, trade names can serve as a guide to identification of the plastics. See Appendix C for trade names of plastics materials on the commercial market.

If the trade name is known, the supplier or manufacturer may be the most reliable source of information about the kind of plastics, ingredients, additives, or physical properties. Batch and lot numbers may vary, but most of the essential information will be known by the supplier. Like gasoline brands, additives may vary in each family of plastics manufactured.

Appearance

Many physical or visual clues can be used to help identify plastics materials. Plastics in the raw, uncompounded, or pellet stage are harder to identify than finished products. Thermoplastics are generally produced in powder, granular, or pellet forms. Thermosetting materials are usually made in the form of powders, preforms, or resins.

The method of fabrication and the product application are good clues to identity of plastics. Thermoplastic materials are commonly extruded, injection-formed, calendered, blow-molded, and vacuum-molded. Thermosetting plastics are usually compression-molded, transfer-molded, cast, or laminated. Polyethylene, polystyrene, and cellulosics are used extensively in the container and packaging industry. Harsh chemicals and solvents are likely to be stored in polyethylene containers. Polyethylene, polytetrafluoroethylene, polyacetals, and polyamides have a waxy feel not found in most polymers. Some plastics are not available in transparent colors. Others are generally reinforced or heavily filled with inexpensive filler material. Some identifiable characteristics of plastics are given in Table 4-4.

Effects of Heat

When plastics specimens are heated in test tubes, distinct odors of specific plastics may be identified. The actual burning of the sample in an open flame may provide further clues. Polystyrene and its copolymers burn with black (carbon) smoke. Polyethylene burns with a clear, blue flame and drips when molten (see Table 4-5).

The actual melting point may provide further clues to identification. Thermosetting materials don't melt. Several thermoplastics melt at less than 203 °F [95 °C]. An electric soldering gun can be pressed on the surface of the plastics. If the material softens and the hot tip sinks into it, the material is a thermoplastic. If the material stays hard and merely chars, it is a thermoset.

The melting or softening point may be observed by placing a small piece of the unknown thermoplastic on an electrically heated platen or in an oven. The temperature must be carefully controlled and recorded. When the specimen is within a few degrees of the suspected melting point, the temperature should be increased at a rate of 1 °C/min.

A standard method of testing polymers is given in ASTM D-2117 (a procedure established by the American Society for Testing and Materials). For polymers that have no definite melting point (such as polyethylene, polystyrene, acrylics, and cellulosics) or for those that melt with a broad transition temperature, the *Vicat softening point* may be of some aid in identification. The Vicat softening point test is described later in this chapter. Melting points of selected plastics are shown in Table 4-5.

Effects of Solvents

Tests for the solubility or insolubility of plastics are easily performed identification methods. Except for polyolefins, acetals, polyamides, and fluoroplastics, thermoplastic materials can be considered soluble at room temperature. Thermosetting plastics may be considered solvent resistant. ***WARNING:*** *When making solubility tests,*

Table 4-4.   Identification of Selected Plastics

Plastics	Appearance	Applications
ABS	Styrenelike, tough, metal-like ring when struck, translucent	Appliance and tool housings, instrument panels, luggage, packing crates, sporting goods.
Acetal	Tough, hard, metal-like ring when struck, waxy feel, translucent	Aerosol stem valves, cigarette lighters, conveyor belts, plumbing, zippers
Acrylics	Brittle, hard, transparent	Models, glazing, wax
Alkyds	Hard, tough, brittle, usually bulk-filled, opaque	Electrical, paints
Allyl	Hard, filled, reinforced, transparent to opaque	Electrical, sealers
Aminos	Hard, brittle, opaque with some translucence	Appliance knobs, bottle caps, dials, handles
Cellullosics	Varies; tough, transparent	Explosives, fabrics, packaging, pharmaceuticals, handles, toys
Chlorinated Polyesters	Tough, translucent or opaque	Electrical, laboratory equipment, plumbing
Epoxies	Hard, mostly filled, reinforced, transparent	Adhesives, casting, finishes
Fluoroplastics	Tough, waxy feel, translucent	Anti-stick coatings, bearings, gaskets, seals, electrical
Ionomers	Tough, impact resistant, transparent	Containers, paper coatings, safety glasses, shields, toys
Nitrile barrier plastics	Tough, transparent, impact resistant	Packaging
Phenolics	Hard, brittle, filled, reinforced, transparent	Adhesives, billiard balls, handles, molding powders
Phenylene oxide	Tough, hard, often filled, reinforced, opaque	Appliance housings, consoles, electrical, respirators
Polyamides	Tough, waxy feel, translucent	Combs, door catches, castors, gears, valve seats
Polyaryl ether	Impact resistant, polycarbonatelike, translucent to opaque	Appliances, automobile paint, electrical
Polyaryl sulfone	Tough, stiff, opaque, polycarbonatelike	High-temperature uses in aerospace, industry, and consumer goods
Polycarbonate	Styrenelike, tough, metal-like ring when struck, translucent	Beverage dispensers, films, lenses, light fixtures, small appliances, windshields
Polyester, aromatic	Stiff, tough, opaque	Coatings, insulation, transistors
Polyester, thermoplastic	Hard, tough, opaque	Beverage bottles, packaging, photography, tapes, labels
Polyester, unsaturated	Hard, brittle, filled, transparent	Furniture, radar domes, sports equipment, tanks, trays
Polyolefins	Waxy feel, tough, soft, translucent	Carpeting, chairs, dishes, medical syringes, toys
Polyphenylene sulfide	Stiff, hard, opaque	Bearings, gears, coatings
Polystyrene	Brittle, white bend marks, metal-like ring when struck, transparent	Blister packages, bottle caps, dishes, lenses, transparent display boxes
Polysulfone	Rigid, polycarbonatelike, transparent to opaque	Aerospace, distributor caps, hospital equipment, shower heads
Silicones	Tough, hard, filled, reinforced, some flexible, opaque	Artificial organs, grease, inks, molds, polishes, Silly Putty, waterproofing
Urethanes	Tough castings, mostly foams, flexible, opaque	Bumpers, cushions, elastic thread, insulation, sponges, tires
Vinyls	Tough, some flexible, transparent	Balls, dolls, floor coverings, garden hoses, rainwear, wall tiles, wallpaper

Table 4-5. Identification Test for Selected Plastics

Thermoplastic	Flame-Burn, Smoke & Flame Danger	Odor and Respiratory Danger	Melting Point, °C
ABS	Yellow flame, black smoke, drips, continues to burn	Rubber, sharp, biting	100
Acetal	Blue flame, no smoke, drips, melts, may burn, continues to burn	Formaldehyde	181
Acrylic	Blue flame, yellow top, white ash, black smoke, popping sound, spurts, continues to burn	Fruit, floral	105
Cellulose acetate	Yellow or yellow-orange-to-green flame, melts, drips, continues to burn, black smoke	Burnt sugar, acetic acid, burning paper	230

Table 4-5. Identification Test for Selected Plastics (Continued)

Thermoplastic	Flame-Burn, Smoke & Flame Danger	Odor and Respiratory Danger	Melting Point, °C
Cellulose acetate butyrate	Blue flame, yellow top, sparks, melts, drips, dripping may burn, continues to burn	Camphor, rancid butter	140
Ethyl cellulose	Pale yellow to blue-green flame with blue edge, melts, drips, drips burn	Burnt wood, burnt sugar	135
Fluoronated ethylene propylene	Melts, decomposes, poison gases formed	Slight acid or burned hair. DO NOT INHALE	275
Ionomer	Yellow flame with blue edge, continues to burn, some black smoke, melts, bubbles, drips burn	Paraffin	110
Phenoxy	Burns, no drips	Acid	93
Polyallomer	Yellow or yellow-orange flame, with blue edge, continues to burn, black smoke, melts clear, spurts, drips burn	Paraffin	120
Polyamides 6, 6	Blue flame, yellow tip, melts and drips, self-extinguishing, froths	Burned wool or hair	265
Polycarbonate	Decomposes, chars, self-extinguishing, dense black smoke, spurts orange flame	Characteristic, sweet, compare known sample	150
Polychlorotrifluoro-ethylene	Yellow flame, won't support combustion	Slight, acidic fumes. DO NOT INHALE	220
Polyethylene	Blue flame, yellow top, drippings may burn, transparent hot area, burns rapidly, continues to burn	Paraffin	110
Polyimides	Chars, brittle, blue flame		300
Polyphyenylene oxide	Yellow to yellow-orange flame, no drips, spurts, difficult to ignite, thick black smoke, decomposes	Paraffin, phenol	105
Polypropylene	Blue flame, drips, transparent hot area, burns slowly, trace of white smoke, melts, swells	Heavy, sweet, paraffin, burning asphalt	176
Polysulfones	Yellow or orange flame, black smoke, drips, self-extinguishing, sparkles, decomposes	Acid	200
Polystyrene	Yellow flame, dense smoke, clumps of carbon in air, drips, continues to burn, bubbles	Illuminating gas, sweet, marigold, floral	100
Polytetrafluoro-ethylene	Yellow flame, slightly green near base, won't support combustion, self-extinguishing, turns clear	None. DO NOT INHALE	327
Polyvinyl acetate	Yellow flame, black smoke, spurts, continues to burn, some soot, green on copper wire test	Vinegar, acetic acid	60
Polyvinyl alcohol	Yellow, smoky	Unpleasant, sweet	105
Polyvinyl chloride	Yellow flame, green at edges, black or gray smoke, chars, self-extinguishing, leaves an ash	Hydrochloric acid	75
Polyvinyl fluoride	Pale yellow	Acid	230
Polyvinylidene chloride	Yellow with green base, spurts green smoke, smoky, self-extinguishing	Pungent	210

Thermosetting	Flame-Burn, Smoke & Flame Danger	Odor, Respiratory Danger
Casein	Yellow flame, burns with flame contact, chars	Burnt milk
Diallyl phthalate	Yellow flame, green-blue edge, self-extinguishing	Acid
Epoxy	Yellow flame, some soot, spits black smoke, chars, continues to burn	Phenolic phenol, acid
Melamine formaldehyde	Difficult to burn, self-extinguishing, swells, cracks, yellow flame, blue-green base, turns white	Fishlike, formaldehyde
Phenolic	Cracks, deforms, difficult to burn, self extinguishing, yellow flame, little black smoke	Phenolic phenol
Polyester	Yellow flame, blue edge, burns, ash and black beads, continues to burn, dense black smoke, no drips	Sweet, bitter-sweet, burning coal
Polyurethane	Yellow with blue base, thick black smoke, spurts, may melt and drip, continues to burn	Acid
Silicone	Low, bright yellow-white flame, white smoke, white ash, continues to burn	None
Urea formaldehyde	Yellow flame with green-blue edge, self-extinguishing	Pancakes

remember that solutions may be flammable, may give off toxic fumes, may be absorbed through the skin, or all three. Appropriate safety precautions should be taken.

Before a solubility test can be made, a chemical solvent must be selected. To help identify polymers and solvents that may react molecularly with each other, solubility parameter numbers have been assigned to selected polymers and solvents (Table 4-6). In principle, a polymer will dissolve in a solvent with a similar or lower solubility parameter. But, because of high-energy hydrogen bonding, crystallization, or other forms of molecular interaction, some polymers may dissolve in solvent having different solubility parameters.

When making solubility tests, use a ratio of one volume of plastics sample to twenty volumes of boiling or room temperature solvent. A water-cooled reflux condenser may be used to collect solvent or minimize solvent loss during heating. Table 4-7 shows selected solvents with selected plastics.

Relative Density

The presence of fillers or other additives and the degree of polymerization (DP) can make identification of plastics by relative density tests difficult. The presence of fillers and additives can cause the relative density to differ greatly from that of the plastics itself. Polyolefins, ionomers, and low-density polystyrene will float in water (which has a relative density of 1.00). For comparison of relative densities of various selected materials, see Table 4-8, page 65.

The degree of polymerization and the density of selected plastics may be measured in solutions of known relative density. These solutions are contained in individual cylinders, as shown in

Table 4-6. Solubility Parameters of Selected Solvents and Plastics

Solvent	Solubility Parameter
Water	23.4
Methyl alcohol	14.5
Ethyl alcohol	12.7
Isopropyl alcohol	11.5
Phenol	14.5
n-Butyl alcohol	11.4
Ethyl acetate	9.1
Chloroform	9.3
Trichloroethylene	9.3
Methylene chloride	9.7
Ethylene dichloride	9.8
Cyclohexanone	9.9
Acetone	10.0
Isopropyl acetate	8.4
Carbon tetrachloride	8.6
Toluene	9.0
Xylene	8.9
Methyl isopropyl ketone	8.4
Cyclohexane	8.2
Turpentine	8.1
Methyl amyl acetate	8.0
Methyl cyclohexane	7.8
Heptane	7.5

Plastics	Solubility Parameter
Polytetrafluoroethylene	6.2
Polyethylene	7.9–8.1
Polypropylene	7.9
Polystyrene	8.5–9.7
Polyvinyl acetate	9.4
Polymethyl methacrylate	9.0–9.5
Polyvinyl chloride	9.38–9.5
Bisphenol A polycarbonate	9.5
Polyvinylidene chloride	9.8
Polyethylene terephthalate	10.7
Cellulose nitrate	10.56–10.48
Cellulose acetate	11.35
Epoxide	11.0
Polyacetal	11.1
Polyamide 6, 6	13.6
Coumarone indene	8.0–10.6
Alkyd	7.0–11.2

Table 4-7. Identification of Selected Plastics by Solvent Test Method

Plastics	Acetone	Benzene	Furfuryl Alcohol	Toluene	Special Solvents
ABS	Insoluble	Partially soluble	Insoluble	Soluble	Ethylene dichloride
Acrylic	Soluble	Soluble	Partially soluble	Soluble	Ethylene dichloride
Cellulose acetate	Soluble	Partially soluble	Soluble	Partially soluble	Acetic acid
Cellulose acetate butyrate	Soluble	Partially soluble	Soluble	Partially soluble	Ethyl acetate
Fluorocarbon	Insoluble (most)	Insoluble	Insoluble	Insoluble	Dimethyacetamide (not **FEP-TFE**)
Polyamide	Insoluble	Insoluble	Insoluble	Insoluble	Hot aqueous ethanol
Polycarbonate	Partially soluble	Partially soluble	Insoluble	Partially soluble	Hot benzene-toluene
Polyethylene	Insoluble	Insoluble	Insoluble	Insoluble	Hot benzene-toluene
Polypropylene	Insoluble	Insoluble	Insoluble	Insoluble	Hot benzene-toluene
Polystyrene	Soluble	Soluble	Partially soluble	Soluble	Methylene dichloride
Vinyl acetate	Soluble	Soluble	Insoluble	Soluble	Cyclohexanol
Vinyl chloride	Soluble	Insoluble	Insoluble	Partially soluble	Cyclohexanol

Table 4-8. Relative Densities of Selected Materials

Substance	Relative Density
Gases (based on air)	
Air	1.00
Acetylene	0.898
Carbon dioxide	1.529
Ethylene	0.967
Hydrogen	0.069
Nitrogen	0.97
Oxygen	1.105
Woods (based on water)	
Ash	0.73
Birch	0.65
Fir	0.57
Hemlock	0.39
Red oak	0.74
Walnut	0.63
Liquids	
Acid, muriatic	1.20
Acid, nitric	1.217
Benzine	0.71
Kerosene	0.80
Turpentine	0.87
Water, 4 °C	1.00
Metals	
Aluminum	2.67
Brass	8.5
Copper	8.85
Iron, cast	7.20
Iron, wrought	7.7
Steel	7.85
Plastics	
ABS	1.02–1.25
Acetal	1.40–1.45
Acrylic	1.17–1.20
Allyl	1.30–1.40
Aminos	1.47–1.65
Casein	1.35
Celullosics	1.15–1.40
Chlorinated polyesters	1.4
Epoxies	1.11–1.8
Fluoroplastics	2.12–2.2
Ionomers	0.93–0.96
Phenolic	1.25–1.55
Phenylene oxide	1.06–1.10
Polyamides	1.09–1.14
Polycarbonate	1.2–1.52
Polyester	1.01–1.46
Polyolefins	0.91–0.97
Polysytrene	0.98–1.1
Polysulfone	1.24
Silicones	1.05–1.23
Urethanes	1.15–1.20
Vinyls	1.2–1.55

RELATIVE DENSITY 1.20

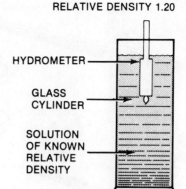

Fig. 4-1. Arrangement for measuring density and DP of plastics.

isopropyl alcohol may be used. Full-strength isopropyl alcohol has a relative density of 0.92. By adding small amounts of distilled water, this value may be raised.

Remember that the density of a plastics may vary depending on DP, fillers, reinforcements, plasticizers, and other additives present in the sample. It is best to test only known samples. Saw marks, dirt, or grease should be removed in order to avoid entrapping air when immersing the specimen.

Use of the relative density of solutions is a handy, rapid method of finding the density of plastics.

If a plastics floats in a solution with a relative density of 0.94, it may be a medium or low-density polyethylene plastics. If the sample floats in a solution of 0.92, it must be either low-density polyethylene or polypropylene. If the sample sinks in all solutions below a relative density of 2.00, the sample is a fluorocarbon plastics.

All of the solutions may be stored in clean containers and reused. The density of solutions should be checked during the testing procedure, however. Factors such as temperature and evaporation may radically change the relative density value.

Another method for determining relative density is to weigh the sample in air and in water (ASTM D-792). A fine wire may be used to suspend the plastics sample in the water from a laboratory balance, as shown in Figure 4-2. You may calculate the relative density by the following formula:

$$D = \frac{a - b}{a - b + c - d}$$

D = density at 68 °F [20 °C]

a = mass of specimen and wire in air

b = mass of wire in air

Figure 4-1. In industry, a density-gradient column (ASTM D-1505) is used to measure the densities of small plastics samples.

Solutions of distilled water and calcium nitrate are mixed and measured with technical-grade hydrometers until a desired relative density is obtained. For densities less than that of water (1.00),

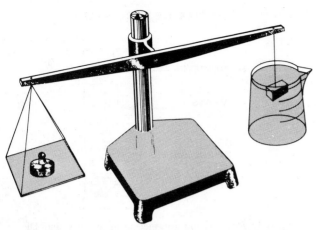

Fig. 4-2. An analytical balance is used as shown to determine the relative density of plastics samples.

c = mass of wire with end immersed
in water

d = mass of wire and specimen
immersed in water

4-4 TESTING OF PLASTICS

Tests allow us to describe plastic materials and to understand their characteristics. It is important to learn the nomenclature and testing methods used by the plastics industry. Test data are valuable for identification, for ensuring uniformity, and for indicating likely performance of plastics products.

The manufacturer, who produces the basic plastics from chemicals, conducts tests to help describe the product. This testing provides reference points in quality control during manufacture. The processor, who converts the basic plastics into solid shapes, performs standard tests to measure factors that affect handling and processability. Testing allows the fabricator and finisher, who further fashion and decorate the plastics forms, to select the best plastics material to suit a particular use.

Most tests are performed on specified samples under controlled laboratory conditions. Tests should be used only as indicators for product service and design, and testing limitations should be kept in mind when making judgments about plastics.

Testing is especially important because the properties and concomitant performance can vary widely with the materials, processing, and design parameters. Materials can vary from batch to batch and between suppliers. Improper processing during mass production can significantly alter properties.

Poor product design may dramatically influence the results of testing.

It is important to remember that polymers have a wide spectrum of properties depending upon environment. To ensure product success, the service requirements, design, and properties of the plastics product must be carefully considered. The true test is use of the plastics product in actual service conditions.

Testing Agencies

There are several agencies that conduct and publish testing specifications for plastics. The American National Standards Institute, the United States military services, and the American Society for Testing and Materials are currently involved in testing of plastics.

The ASTM is an international, nonprofit technical society devoted to "... the promotion of knowledge of the materials of engineering, and the standardization of specifications and methods of testing." The standardization and procedures for testing plastics are under the jurisdiction of the ASTM Committee D on Plastics. The *Book of ASTM Standards* is published annually. Most of the testing methods referred to in this unit were developed or approved by the ASTM. Table 4-19 at the end of this unit shows a summary of the ASTM testing methods for plastics. See Chapter 21 and standards.

Destructive Testing

Destructive testing procedures damage or destroy the sample. Selected production pieces are tested in this way as one method of quality control. Not all quality control procedures require destructive mechanical tests.

All testing adds to the production cost of goods, however, testing is vital. It helps to prevent production loss and to ensure customer satisfaction.

Nondestructive Testing

Nondestructive tests are designed to obtain data without damaging the plastics part. The tests involve more than external visual inspection (Fig. 4-3). The following tests may be used in nondestructive testing: electrical conductivity, electromagnetic induction, magnetic field, electric field, heat conduction, penetrating radiation, mechanical vibration, luminous energy, and static electricity, and the chemical spot test.

(A) Ultrasonics are used to measure the thickness. (Panametrics)

(B) Polarized filters were used to locate points and lines at stress in this injected molded spoon.

Fig. 4-3. Two kinds of nondestructive tests.

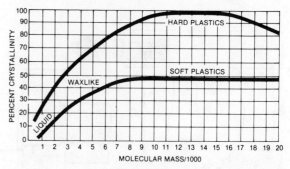

Fig. 4-4. The mechanical properties of selected plastics show the relationship of crystallinity to molecular mass.

4-5 MECHANICAL PROPERTIES

Mechanical properties are associated with the reaction that results when a force or load is applied to a material. These properties are sometimes referred to as physical properties; however, the term *mechanical properties* is preferred.

Mechanical properties include compressive strength, tensile strength, modulus of elasticity, impact strength, flexural strength, shear strength, fatigue and flexing, hardness, indention, and friction.

Figure 4-4 is a graph showing the relationship of certain mechanical properties to degree of crystallinity and molecular mass of selected plastics.

The crystalline axes of crystalline polymers generally produce a slightly anisotropic material. The crystallinity of a material provides relatively rigid regions in the molecular structure, something similar (not the same) to crosslinks. When stretched, a crystalline polymer tends to orient itself so that its crystals are stretched in the direction of the force. Mechanical drawing and fabrication may also align molecular segments in the direction of the applied stress, which may be desirable in fiber production or some product designs.

In linear polymers, the chains tend to pack closely, increasing density with the percentage of crystallinity. The bulky side groups in branched polymers keep the polymer chains farther apart, thus decreasing the tendency to crystallize.

Amorphous materials have the long-range disorder of liquids and are isotropic. Remember that a portion of all polymers have amorphous regions. As the molecular mass and crystallinity increase, there is generally an increase in softening temperature, strength, stiffness, hardness, creep resistance, and impermeability to gas and liquids. There is less space between molecular chains in crystalline materials for moisture and gases to permeate. Short molecular chains are more easily attached by reactants. It is also important to have a relatively narrow distribution of molecular chain lengths for best properties. Short molecular mass chains are more easily attacked and may act as plasticizers, softening the material. Increasing molecular mass reduces creep because the long chain length tends to entangle, thus resisting slippage. As you know, melt viscosity increases rapidly as the molecular mass increases. It takes much more energy to break the large molecular chains.

Thermosetting materials begin as low molecular mass materials but increase dramatically in molecular mass during processing (molding) and curing. There are three direct kinds of stress (force) which may be applied to materials: *compression, tension,* and *shear* (Fig. 4-5). The rate (time) at which all three types are applied is speci-

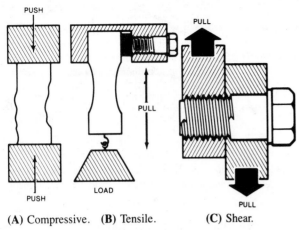

(A) Compressive.   (B) Tensile.   (C) Shear.

Fig. 4-5. Three types of stress.

fied in standard procedures. Forces applied too rapidly or slowly during testing yield false values. Most tensile and compressive testing equipment operates at constant crosshead speed.

Polymers are considered to be viscoelastic. They behave something like elastic metals and viscous oils. For some tests, the rate (time) at which the polymer is stressed may be measured in microseconds, minutes, days, or longer periods. Polymers subject to a load (stress) for a period of time tend to deform (strain). The degree of deformation depends on the load duration. Impact strength measures the polymer's resistance to a quick blow and constitutes a much faster experiment. Most stress-strain studies are relatively slow, in the order of millimetres per second or inches per minute. There is viscoelastic deformation of the polymer chains in stress-strain and impact tests. (See Dimensional Stability.)

Compressive Strength (ASTM D-695)

Compressive strength is a value that shows how much force is needed to rupture or crush a material. The values are expressed in multiples of the pascal, such as kPa, MPa and GPa. Compressive strength is calculated by dividing the maximum load (force) in newtons by the area of the specimen in square metres.

$$\text{Compressive strength (Pa)} = \frac{\text{force (N)}}{\text{cross-sectional area (m}^2\text{)}}$$

If 50 kg is required to rupture a 1.0 mm^2 plastics bar:

$$\text{Force (N)} = 50 \text{ kg} \times 9.8 \text{ m/s}^2$$
where 9.8 m/s^2 is the gravity constant

$$\text{Compressive strength (Pa)} = \frac{(50 \times a\ 9.8)\,\text{N}}{1 \text{ mm}^2}$$

$$= \frac{490 \text{ N}}{1 \text{ mm}^2}$$

$$= \frac{490 \text{ N}}{0.000\ 001 \text{ m}^2}$$

$$= 490 \text{ MPa or } 490\ 000 \text{ kPa}$$
[71 076 psi]

Compressive strength values may be useful in distinguishing between grades of plastics and in comparing plastics to other materials. Compressive strength is especially significant in testing cellular or foamed plastics.

Tensile Strength (ASTM D-638)

If stress is applied to a material by pulling it until it breaks, *tensile strength* may be calculated. Tensile strength is calculated by dividing the maximum load (force) by the original cross-sectional area:

$$\text{Tensile strength (Pa)} = \frac{\text{pulling force (N)}}{\text{cross-sectional area (m}^2\text{)}}$$

Strain. Pulling stress usually causes material to deform by thinning and stretching in length. The change in length, in relation to the original length of the material, is called *strain* (Fig. 4-6). Strain is measured in millimetres per metre, inch per inch of length and in percent of elongation:

$$\text{(Linear) Strain} = \frac{\text{deformation (mm)}}{\text{original dimension (mm)}}$$

(Elongation) Strain (%)
$$= \frac{\text{final length} - \text{original length}}{\text{original length}} \times 100$$

Strain is evident in testing most plastics. Plastics that are not reinforced deform in cross-sectional area before breaking (Fig. 4-7). The breaking strength may be less than the greatest tensile strength, which occurs when strain has reduced the cross-sectional area of the specimen (Fig. 4-7B).

Stress-strain diagrams (Fig. 4-8) are convenient means of expressing and plotting the strength of plastics.

A certain amount of strain (recoverable) is called *elastic strain*. As the name indicates, the molecules return to their original position once the stress is removed. *Plastic strain,* however, occurs

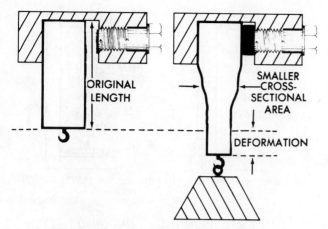

Fig. 4-6. *Strain* is deformation due to pulling stress.

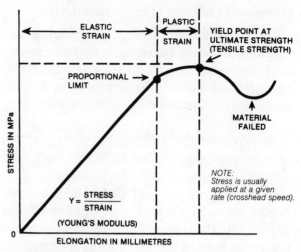

Fig. 4-9. Elastic strain, plastic strain, and the proportional limit.

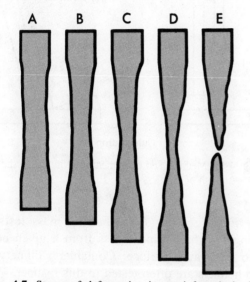

Fig. 4-7. Stages of deformation in unreinforced plastics.

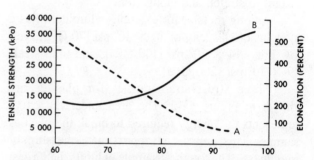

Fig. 4–10. The effect of crystallinity on the tensile strength and elongation of polyethylene. Because intermolecular bonding forces are a function of crystallinity, strength increases and ductility (elongation) decreases with degree of crystallinity. (See Polyethylene)

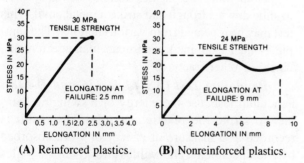

(A) Reinforced plastics.   **(B)** Nonreinforced plastics.

Fig. 4-8. Typical stress-strain curves.

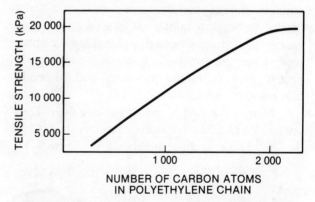

Fig. 4–11. The length of the carbon chain determines the tensile strength of polyethylene. Remember — the long chain length of PE allows for entanglement and is long enough to connect individual stems together within a lamellar crystallite by chain folding. Increases in molecular mass (chain length) improve strength.

when the molecules begin to slip and slide past each other in permanent (irreversible) deformation. The point where elastic strain ends and plastic strain begins is referred to as the *propotional limit* of the material (Fig. 4-9). Stress and strain are no longer proportional!

In Figures 4-10 and 4-11, the effects of crystallinity and carbon chain length on tensile strength are shown for polyethylene.

Modulus of Elasticity (Tensile Modulus)

The modulus of elasticity, also called tensile modulus and Young's modulus, is the ratio between the stress applied and the strain, within the elastic range. It is calculated by dividing the stress (load) in pascals by the strain in millimetres per millimetre.

$$\text{Young's modulus of elasticity} = \frac{\text{stress (Pa)}}{\text{strain (mm/mm)}}$$

Young's modulus has no meaning at stress above the proportional limit. Plastic strain and permanent deformation cannot be allowed in the design of products. The ratio of tensile force to elongation is useful in learning how far a plastics specimen will stretch under a given load. A large tensile modulus indicates that the plastic is rigid; it would be resistant to stretch and elongation.

Among metals, for example, aluminum has a modulus of elasticity of 10×10^6 psi [70 GPa], yellow brass 14×10^6 psi [100 GPa], and mild steel, 29×10^6 psi [200 GPa].

Three stress-strain curves for plastics are shown in Figure 4-12. The area under the curve represents the energy required to break the plastics sample. This area is an approximate measure of *toughness*. The toughest sample exhibits the largest area under the stress-strain curve (Fig. 4-12C.)

Impact Strength

Although the term *toughness* has not been clearly defined, it is used to measure *impact strength*. Impact strength is not a measure of stress needed to break a sample. It is a measure of the *energy* needed, or absorbed, in breaking the specimen. There are two basic methods for testing impact strength: (a) falling mass tests, and (b) pendulum tests.

Many variations of impact tests are used. They are needed to adjust for various product shapes and designs. Some of these are shown in Figure 4-13.

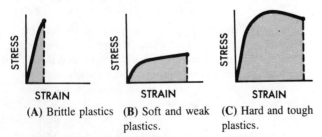

(A) Brittle plastics   **(B)** Soft and weak plastics.   **(C)** Hard and tough plastics.

Fig. 4–12. Toughness is a measure of the amount of energy needed to break a material. It is often defined as the total area under the stress-strain curve.

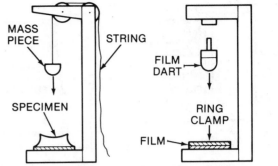

(A) Falling mass method.   **(B)** Film dart test (*ASTM D-1709*).

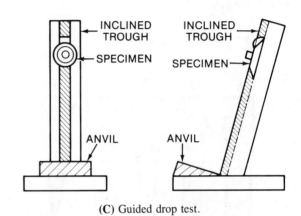

(C) Guided drop test.

Fig. 4-13. Methods and equipment used in testing the impact strength of plastics.

Falling Mass Test. Falling mass tests are dropping a ball-shaped mass from a given height onto the plastic surface. Containers, dinnerware, and helmets are often tested in this manner.

A blunt dart is commonly used for testing films up to 0.01 in [0.254 mm] thick.

Sometimes the plastics sample itself is allowed to slide down a trough and strike a metal anvil. This test may be repeated from various heights. The sample is then examined for impact damage (cracks, chips, or other fractures).

Pendulum Test (ASTM D-256 and D-618). Pendulum tests use the energy of a swinging hammer to strike the plastics sample. The result is recorded as a measure of energy or work absorbed by the specimen. The values are given in joules (foot-pounds). The hammers of most plastics testing machines have a kinetic energy of 2 to 16 ft-lb [2.7 to 22 J]. Figure 4-14 shows two impact testing machines.

In the *Charpy* (simple beam) method, the test piece is supported at both ends without being held down. The sample is struck in the center by the hammer (Figs. 4-14A and 4-14B). In the *Izod* (cantilever beam) method the test specimen is

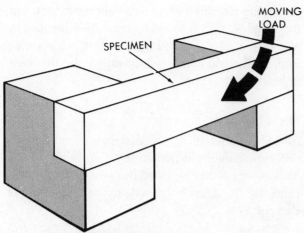

(A) Charpy pendulum method.

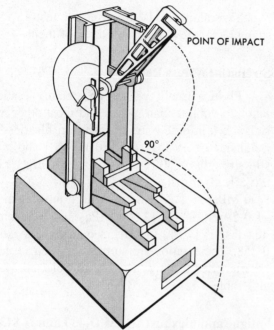

(B) Charpy simple beam impact machine. (Tinius Olsen Testing Machine Co., Inc.)

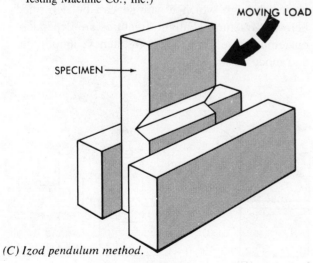

(C) Izod pendulum method.

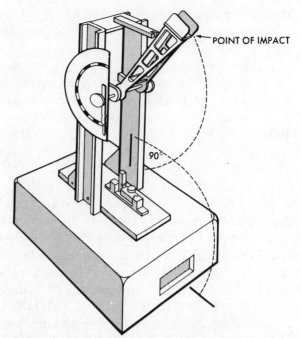

(D) Izod cantilever beam impact machine. (Tinius Olsen Testing Machine Co., Inc.)

(E) Impact tester for Charpy and Izod testing. (Tinius Olsen Testing Machine Co., Inc.)

Fig. 4-14. Charpy and Izod method testing equipment.

rigidly supported at one end and struck by the hammer. Either notched or unnotched samples may be used in Izod testing.

If the specimen is notched in the Charpy test, the notch is on the side away from the striker. In the Izod test, the notch is placed on the same side as the striker. Many polymers are notch sensitive. In tests with blunt notches (r = 2 mm), PVC has a higher impact strength than ABS. If the notch radius in these two specimens is sharp (r = 0.25 mm), ABS has a higher impact strength. Acetals, HDPE, PP, PVC, PET, and dry PA are notch brittle. Temperature and moisture are important considerations in the performance of polymers. All specimens must be properly conditioned before testing. Polyamides are hygroscopic (absorb moisture). Unconditioned, wet PA has an impact strength greater than 200 ft-lbs [20 kJ/m^2]. If dried to a conditioned state, PA is notch brittle with an impact strength of about 50 ft-lbs [5 kJ/m^2]. [5 kJ/m^2].

In either Charpy or Izod tests, impact values are usually expressed in terms of the energy needed to break the test sample (foot-pounds per inch of notch or J/m^2).

The ability to resist sharp blows is an important property of many plastics. Some possess very high impact strength, even surpassing the toughness of steel. The word *stretch* is a misnomer. It usually has the meaning of a critical stress but usage is firmly established in the literature.

Flexural Strength (ASTM D-790 and D-747)

Flexural strength is a measure of how much stress (load) can be applied to a material before it breaks. The sample is supported on test blocks 4 in. [100 mm] apart. The load is applied in the center (Fig. 4-15).

Because most plastics do not break when deflected, the flexural strength at fracture cannot be calculated easily. Most thermoplastics and elastomers are measured when 5 percent strain occurs in the samples. This is found by measuring the load in pascals that causes the sample to stretch 5 percent. Both tensile and compressive stresses are involved in bending the sample.

Shear Strength

Shear strength is the maximum load (stress) needed to produce a fracture by a shearing action. Shear strength may be found by dividing the applied force by the cross-sectional area of the sample sheared:

$$\text{Shear strength (Pa)} = \frac{\text{force (N)}}{\text{cross-sectional area (m}^2)}$$

Several methods are used in testing shear strength. Figure 4-16 shows three of them.

Strength-to-Mass Ratio

Plastics usually weigh considerably less in relation to volume than metals and most other materials. A reinforced structural foam plastics with a density of 35 lbs/ft^3 [550 kg/m^3] and a tensile strength of 100 × 10^6 psi [700 MPa] would have a strength-to-mass ratio of 550 kg/m^3 [700 MPa] = 1.272. Steel, with a tensile strength of 0.29 × 10^6 psi [2 000 MPa] and a density of 484 lbs/ft^3 [7 750 kg/m^3], would have a ratio of 0.258. These ratios are sometimes used in design criteria. (See Relative Density of Selected Materials, Table 4-8.)

Fatigue and Flexing (ASTM D-430 and D-813)

Fatigue is a term used to express the number of cycles through which a sample can be deformed before it fractures. Fatigue fractures are dependent on temperature, stress, and frequency, amplitude, and mode of stressing.

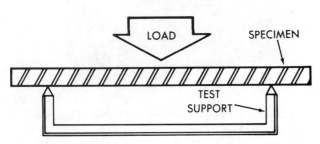

Fig. 4-15. The method used to test flexural strength (flexural modulus).

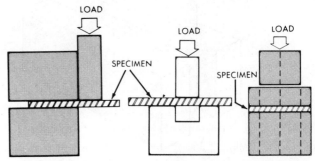

Fig. 4-16. Various methods used to test shear strength of plastics.

If the load (stress) does not exceed the elastic limit, many plastics may be stressed for a great many cycles without failure. In producing integral hinges and one-piece box-and-lid containers, the fatigue characteristics of the plastics must be considered (Fig. 4-17).

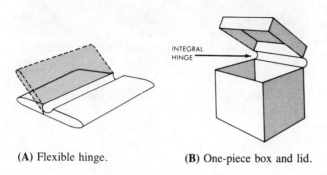

(A) Flexible hinge.   **(B)** One-piece box and lid.

(C) A folding endurance tester, which records on a dial the number of flexings that take place before a plastics sample breaks. (Tinius Olsen Testing Machine Co., Inc.)

Fig. 4-17. Fatigue characteristics of plastics articles are important.

Damping. Plastics can absorb or dissipate vibrations. This property is called *damping*. On an average, plastics have ten times more damping capacity than steel. Gears, bearing, appliance housings, and architectural applications of plastics make effective use of the vibration-reducing property.

Hardness (ASTM D-785, D-1706, D-618, and D-2240)

The term *hardness* does not describe a definite or single mechanical property of plastics. Scratch, mar, and abrasion resistance as well as strength are closely related to hardness. Surface wearing of floor tile and marring of optical lenses are affected by these properties.

A widely accepted definition of hardness, as applied to plastics, is resistance to penetration or indentation by another body. Stated in another way, the hardness of a plastics is the resistance of the plastics material to compression, indentation, and scratching.

There are several types of instruments used to measure hardness. The type of instruments or scale used is indicated when hardness values are given (Fig. 4-18A).

The Moh scale of hardness is used by geologists and mineralogists. It has been used to test the hardness of plastics samples. This scale is based on the fact that harder materials scratch softer ones.

Another simple test is based upon the hardness ranges of sharpened pencils. The hardness index is the first numbered pencil to scratch the test surface.

A nondestructive test may be performed with a *scleroscope* (Fig. 4-19). The scleroscope measures the rebound height of a free-falling hammer called a *tup*. The rebound method is of limited use in testing plastics.

Indention Hardness (ASTM D-2240)

Indentation test instruments are used for more sophisticated quantitative measurements. Rockwell, Wilson, Barcol, Brinell, and Shore are well-known testing tools. Table 4-9 shows the basic differences in hardness tests and scales. In these tests, either the depth or area of indentation is the measure of hardness.

The *Brinell test* relates hardness to the area of indention. Typical Brinell numbers for selected plastics are acrylic, 20; polystyrene, 25; polyvinyl chloride, 20; and polyethylene, 2. Figure 4-20 shows a Brinell hardness tester.

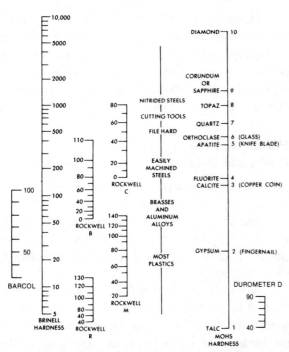

(A) Comparison of various hardness scales.

(C) Molded bar of ABS, positioned under the indentor of a Rockwell hardness tester. (Wilson Instrument Division of AACO)

Fig. 4-18. Hardness scales and tests used to measure hardness.

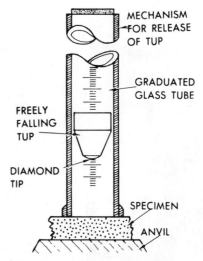

Fig. 4-19. Scleroscope for making hardness tests.

(B) Recording results of a Rockwell hardness test on an ABS bar sample. (Wilson Instrument Division of AACO)

The *Rockwell test* relates hardness to the difference in the depth of penetration of two different loads. The minor load (usually 10 k) and the major load (from 60 to 150 k) are applied to a ball-shaped indentor (Fig. 4-21). Typical Rockwell numbers for certain plastics are acrylic, M 100; polystyrene, M 75; polyvinyl chloride, M 115; and polyethylene, R 15.

The *Barcol tester* uses a sharp-pointed indentor (Fig. 4-22).

For soft or flexible plastics, the *Shore durometer* instrument may be used. There are two ranges of durometer hardness. Type A uses a blunt rod-shaped indentor to test soft plastics. Type D uses a pointed rod-shaped indentor to measure harder materials. The reading or value is taken after one or ten seconds of applying pressure by hand. The scale range is 0 to 100.

Fillers and reinforcements in a sample may cause local variations in hardness values. Many readings should be taken and expressed as an arithmetic mean (average).

Table 4-9. Comparison of Selected Hardness Tests

Instrument	Indentor	Load	Comments
Brinnell	Ball, 10 mm diameter	500 kg 3 000 kg	Averages out hardness differences in material. Load applied for 15–30 seconds. View through Brinnell microscope shows and measures diameter (value, of impression. Not for materials with high creep factors
Barcol	Sharp-point, rod 26° 0.157 mm flat tip	Spring loaded. Push against specimen with hand 5–7 kg	Portable. Readings taken after 1 or 10 s
Rockwell C	Diamond cone	Minor 10 kg Major 150 kg	Hardest materials, steel. Table model
Rockwell B	Ball 1.58 mm (¹⁄₁₆ in)	Minor 10 kg Major 100 kg	Soft metals and filled plastics
Rockwell R	Ball 12.7 mm (½ in)	Minor 10 kg Major 60 kg	Within 10 seconds after applying minor load, apply major load. Remove major load 15 s after application. Read hardness scale 15 s after removing major load or Apply minor load and zero within 10 s. Apply major load immediately after zero adjust. Read number of divisions pointer passed during 15 s of major load
Rockwell L	Ball 6.35 mm (¼ in)	Minor 10 kg Major 60 kg	
Rockwell M	Ball 6.35 mm (¼ in)	Minor 10 kg Major 100 kg	
Rockwell E	Ball 3.175 mm (⅛ in)	Minor 10 kg Major 100 kg	
Shore A	Rod, 1.40 mm diameter, sharpened to 35° 0.79 mm	Spring loaded. Push against specimen with hand pressure	Portable. Readings taken in soft plastics after 1 or 10 s
Shore D	Rod, 1.40 mm diameter. Sharpened to 30° point with 0.100 mm radius	As above	As above

Fig. 4-20. This Brinell hardness tester is air-operated. (Tinius Olsen Testing Machine Co., Inc.)

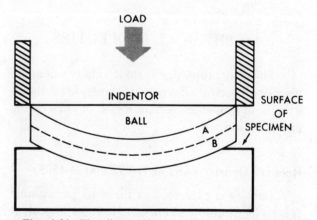

Fig. 4-21. The distance between Line *A* (minor load) and Line *B* (major load) is the basis for Rockwell hardness readings.

Abrasion Resistance (ASTM D-1044)

Abrasion is a process of wearing away the surface of a material by friction. The Williams, Lambourn and Tabor abraders are used. In each test, an abrader is rubbed against the sample. The amount of material loss (mass or volume) is then calculated.

$$\text{abrasion resistance} = \frac{\text{original mass} - \text{final mass}}{\text{relative density}}$$

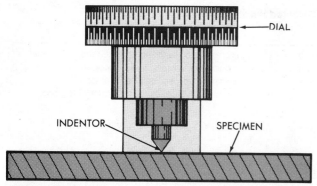

Fig. 4-22. A sharp-pointed indentor is used on the Barcol instrument (*ASTM D-2583*).

Brittleness (ASTM D-746)

Crystallinity will increase the brittleness of a plastics at low temperatures. The brittleness test determines the temperature at which 50 percent of the samples fail. The test is usually performed with an impact apparatus.

Friction

Some plastics are chosen for certain uses because of their frictional properties. Polyethylene, polytetrafluoroethylene (Teflon), polyacetal (Delrin), and polyamide (Nylon) all have low coefficients of friction.

4-6 PHYSICAL PROPERTIES

Physical properties include relative density, viscosity, specific volume, and dimensional stability. These properties pertain to the physics of the plastics materials.

Relative Density (ASTM D-792 and D-1505)

Density is expressed as the mass per volume (grams per cubic centimetre). It is an important physical property. Relative density is the ratio of the mass of a given volume of material to the mass of an equal volume of water at 73 °F [23 °C]. Relative density may be measured by placing samples in a density-gradient column. This column is composed of liquid layers decreasing in density from bottom to top. The layer to which the sample sinks shows its density.

A simple method of measuring relative density is to measure the mass of the plastics in air and in water. This method was described earlier in this chapter.

Relative density is an important factor in comparing material prices per unit volume. It is used to control raw material production and quality of molded parts.

Some typical relative densities of plastics are acrylic, 1.17 to 1.20; polystyrene, 1.04 to 1.09; polyvinyl chloride, 1.30 to 1.45; and polyethylene, 0.91 to 0.965.

Viscosity

The property of a liquid that describes its internal resistance to flow is called viscosity. The more sluggish the liquid, the greater its viscosity. Viscosity is measured in pascal-seconds (Pa·s) or units called *poises* (Table 4-10).

Viscosity is an important factor in transporting resins, injecting plastics in a liquid state, and obtaining critical dimensions of extruded shapes. Fillers, solvents, plasticizers, thixotropic agents (materials that are gel-like until shaken), degree of polymerization, and density all may affect viscosity. The viscosity of a resin such as polyester ranges from 1 to 10 Pa·s [1 000 to 10 000 centipoises]. One centipoise equals 0.01 poise. In the metric system, one centipoise equals 0.001 pascal-second. For a complete definition of poise, see any standard physics text or other reference work in which viscosity is described.

Specific Volume

Specific volume is the number of cubic metres or cubic centimetres of plastics materials with a mass of one kilogram.

Dimensional Stability

The term dimensional stability refers to a broad number of properties affecting dimensional changes

Table 4-10. Viscosity of Selected Materials

Material	Viscosity, Pa·s	Viscosity, centipoises
Water	0.001	1
Kerosene	0.01	10
Motor oil	0.01–1	10–100
Glycerine	1	1000
Corn syrup	10	10 000
Molasses	100	100 000
Resins	<0.1 to >10^3	<100 to >10^6
Plastics (hot, viscoelastic state)	<10^2 to >10^7	<10^5 to >10^{10}

in plastics products. Changes in dimensions may be caused by cold flow and creep, swelling or shrinking, internal stresses, temperature changes, and weathering and aging.

When a mass suspended from a test sample causes the sample to change shape over a given period of time, the strain is called *creep*. When creep occurs at room temperature, it is called *cold flow*. Creep is the strain in millimetres that occurs over a period of time (Fig. 4-23). Measurements of creep and cold flow must take into account temperature, humidity, and time. Fillers and other additives are used to help prevent this unstable state.

Creep must be considered as a factor in parts designed to rotate rapidly, such as rotors and blades. Creep and cold flow are very important properties to consider in the design of pressure vessels, pipes, and beams, where a constant load (pressure or stress) may cause deformation. These properties are also vital factors in designing parts where a constant pressure or stress may cause dimensional changes. Figure 4-24 shows a section of pipe being tested for burst strength. It is important to know how much stress and strain the pipe will sustain. Tests of rupture or burst strength are often carried out under water. Pressure is increased in the test sample until rupture occurs. This sample ruptured while under 848 psi [5 850 kPa] of pressure.

Plastics materials may be processed in a variety of physical states, but internal stresses may be locked into the finished product. This locked-in stress may cause warping, crazing, or cracking under varying environmental conditions. Solvents are also more likely to craze or crack parts held under stress. (See Chapters 6 and 15.)

Fig. 4-24. Testing the bursting strength of pipe. (Schloemann-Fellows)

Most cracks begin to form when the polymer molecules begin to shear-flow past one another. They usually begin at some flaw and propagate rapidly in the direction of stress, resulting in possible failure. Stress concentrations may be natural flaws, may result from incorrect design of the part, or may be formed when making holes or notches. This means that designs must consider the effects of sharp corners, threads, and other stress-concentrating notches. A craze is not an open crack. It is composed of highly oriented polymer chains (fibrils) which may extend from both surfaces of the polymer. A craze runs ahead of the crack tip. Craze lines are easily observed in PS and are formed at right angles to the stress. They appear as a frosty area in PS as it is bent or pulled (stressed).

4-7 THERMAL PROPERTIES

The important thermal properties of plastics are thermal conductivity, specific heat, coefficient of thermal expansion, heat deflection, resistance to cold, burning rate, flammability, melt index, glass transition point, and softening point (Fig. 4-25).

As thermoplastics are heated, molecules and atoms within the material begin to oscillate more rapidly. This causes the molecular chains to lengthen. More heat may cause slippage between molecules held by the weaker van der Waals' forces. The material may become a viscous liquid. In thermosetting plastics, bonds are not easily freed. They must be broken or decomposed.

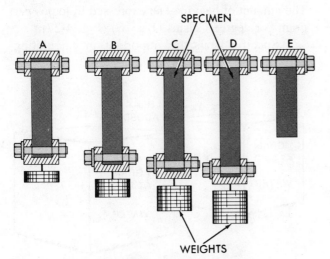

Fig. 4-23. Stages of creep and cold flow.

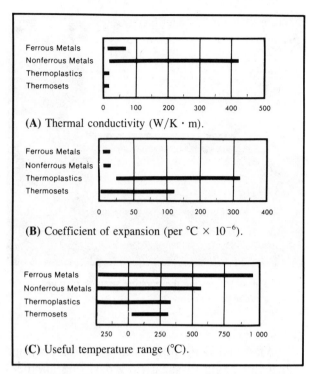

(A) Thermal conductivity (W/K · m).

(B) Coefficient of expansion (per °C × 10⁻⁶).

(C) Useful temperature range (°C).

Fig. 4-25. Three selected thermal properties of metals and plastics.

Fig. 4-26. The high thermal conductivity of the metal makes the ice melt rapidly.

Thermal Conductivity (ASTM C-177)

Thermal conductivity is the rate of transmission of heat energy from one molecule to another (Fig. 4-26). For the same molecular reasons that plastics are electrical insulators, they are also thermal insulators.

Thermal conductivity is expressed as a coefficient. Thermal conductivity is called the *k* factor. It should not be confused with the symbol *K* that indicates Kelvin temperature scale. Aluminum has a *k* factor of 122 W/K · m. Some foamed or cellular plastics have *k* values of less than 0.01 W/K · m (Table 4-11). The *k* values for most plastics show that they conduct heat less well than does an equal amount of metal.

Table 4-11. Thermal Conductivity of Selected Materials

Material	Thermal Conductivity (k-factor), W/K · m	Thermal Resistivity (R-factor)*, K · m/W	Thermal Conductivity, Btu · in/hr · ft² · °F
Acrylic	0.18	5.55	1.3
Aluminum (alloyed)	122	0.008	840
Copper (beryllium)	115	0.008	800
Iron	47	0.021	325
Polyamide	0.25	4.00	1.7
Polycarbonate	0.20	5.00	1.4
Steel	44	0.022	310
Window glass	0.86	1.17	6.0
Wood	0.17	5.88	1.2

*R-factor is the reciprocal of the *k* factor.

The flow of heat energy should be measured in watts, not calories per hour or Btu per hour. The watt (W) is the same as the joule per second (J/s). It is best to remember that the joule is a unit of energy. The watt is a unit of power.

Specific Heat (Heat Capacity)

Specific heat is the amount of heat required to raise the temperature of a unit of mass by one kelvin, or one degree Celsius. It should be expressed in joules per kilogram per kelvin (J/kg · K). The specific heat at room temperature for ABS is 104; for polystyrene, 125; and for polyethylene, 209 J/kg · K. This indicates that it will take more thermal energy to soften the crystalline plastics polyethylene than to soften ABS. The values for most plastics indicate that they require a greater amount of heat energy to raise their temperature than does water since the specific heat of water is 1. The amount of heat may be expressed in joules per gram per degree Celsius (J/g · °C) (Fig. 4-27).

Specific heats are important in calculating costs for processing and forming.

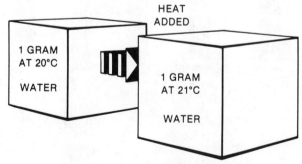

Fig. 4-27. How much heat was added?

Thermal Expansion (ASTM D-696 and D-864)

The *coefficient of thermal expansion* of plastics without fillers is relatively large. This number is used to represent expansion in length, area, or volume per unit of temperature rise.

Table 4-12 gives the coefficient of thermal expansion for several materials. From these values, a practical estimate of changes in length may be obtained. For example, a steel rod 20 m in length is heated from −30 °C to 40 °C. It will expand in length as follows:

$$\text{Expansion} = 70 \times 0.000\ 010\ 8 \times 20 \text{ m}$$
$$= 0.015\ 12 \text{ m or } 15.12 \text{ mm } [0.595 \text{ in.}]$$

High thermal expansion prevents use of plastics in some applications. Metals with low coefficients of expansion are used where close tolerances of parts are required. Fillers may be used in plastics for added stability.

Deflection Temperature (ASTM D-648)

In this test, a specimen (3.175 mm by 140 mm) is placed on supports 100 mm apart, then a force of 455 to 1 820 kPa is pressed on the sample. The temperature is raised 2 °C per minute. The temperature at which the sample will deflect 0.25 mm is reported as deflection temperature.

Deflection temperature (formerly called heat distortion) is the highest continuous operating temperature that the material will withstand. The product may be tested in an oven. The temperature is raised until the product chars, blisters, distorts, or loses appreciable strength. The boiling point of water is an important temperature level. Deflection temperature may be thought of as the highest useful temperature of the plastics product.

Under stress and high heat, some thermoplastics stand up better than others. As shown in Figure 4-28A, test bars of glass-reinforced polycarbonate, polysulfone, and thermoplastic polyester were locked in a laboratory vise. Equal 175 g loads were applied. After only 1 min under 155 °C [311 °F] heat from an infrared radiant heater, the polycarbonate bar began to deflect. One minute later, the polysulfone bar followed suit (Fig. 4-28B).

Table 4-12. Thermal Expansion of Selected Materials

Substance	Coefficient of Linear Expansion $\times 10^{-6}$, mm/mm · °C
Nonplastics	
Aluminum	23.5
Brass	18.8
Brick	5.5
Concrete	14.0
Copper	16.7
Glass	9.3
Granite	8.2
Iron, cast	10.5
Marble	7.2
Steel	10.8
Wood, pine	5.5
Plastics	
Diallyl phthalate	50–80
Epoxy	40–100
Melamine-formaldehyde	20–57
Phenol-formaldehyde	30–45
Polyamide	90–108
Polyethylene	110–250
Polystyrene	60–80
Polyvinylidene chloride	190–200
Polytetrafluoroethylene	50–100
Poly methyl methacrylate	54–110
Silicones	8–50

(A) Before heat was applied. Celanese Plastic Materials Co.) **(B)** After two minutes of heating. (Celanese Plastic Materials Co.)

(C) Deflection temperature tester determines heat distortion of up to five plastics samples at one time. (Tinius Olsen Testing Machine Co., Inc.)

Fig. 4-28. Testing method for heat distortion.

The thermoplastic polyester bar was still not bent after 6 min at 365 °F [185 °C].

Plastics are not often used where high heat resistance is needed. However, some phenolics have been subjected to temperatures as high as 4 032 °F [2 760 °C].

Ablative Plastics

Ablative plastics have been used for spacecraft and missiles. On re-entry, the temperature of the outer surface of a heat shield may be greater than 23 800 °F [13 000 °C], while the inner surface is no more than 203 °F [95 °C]. These ablative plastics may be composed of a phenolic or epoxy resin and graphite, asbestos, or silica matrixes.

In ablative materials, heat is absorbed through a process known as *pyrolysis*. This takes place in the near-surface layer exposed to heat energy. Much of the plastics is consumed and sloughed off as large amounts of heat energy are absorbed.

Resistance to Cold

As a rule, plastics have good resistance to cold. Polyethylene plastics are used to a large extent in the food-packaging industry at temperatures of −60 °F [−51 °C]. Some plastics can withstand the extreme low temperature of −319 °F [−196 °C] with little loss of physical properties.

Flammability (ASTM D-635, D-568, and E-84)

Flammability or *flame resistance* are loosely used terms indicating a measure of the ability of a material to support combustion. In one test, a plastics strip is ignited and the heat source (flame) removed. The time and amount of material consumed are measured, and the result is expressed in mm/min (Fig. 4-29). The more combustible plastics, such as cellulose nitrate, will have high values.

A rather loosely used term related to flammability is *self-extinguishing*. This indicates that the material will not continue to burn once the flame is removed. Nearly all plastics may be made self-extinguishing.

Note that plastics will combust when exposed to direct flame. There is no accepted universal test for fire retardancy (Fig. 4-30).

Table 4-13 gives further data on the flammability of various materials.

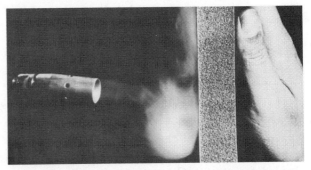

(A) This cellular polyurethane demonstrates the thermal insulating ability and flame resistance of special plastics formulations.

(B) When the flame is removed from a self-extinguishing plastics, burning ceases. (Henkel Corp.)

Fig. 4-29. Testing for flammability of plastics.

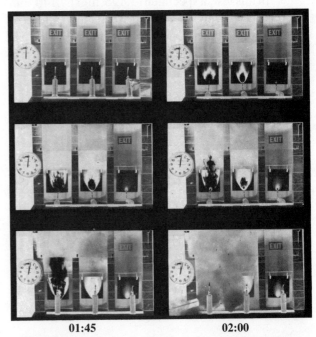

| 01:45 | 02:00 |

Fig. 4-30. These plastics panels were tested for ignition, flame spread, and smoke generation. The reinforced panel at right showed low flame spread and produced little smoke (Kemlite Corp.)

Melt Index (ASTM D-1238)

Viscosity and flow properties are major factors in processing plastics and in mold design. Melt index is a measure of the amount of material

Table 4-13. Ignition Temperatures and Flammability of Various Materials

Material	Flash-Ignition Temperature, °C	Self-Ignition Temperature, °C	Burning Ratio, mm/min
Cotton	230–266	254	SB
Paper, newsprint	230	230	SB
Douglas fir	260		SB
Wool	200		SB
Polyethylene	341	349	7.62–30.48
Polypropylene, fiber		570	17.78–40.64
Polytetrofluoro-ethylene		530	NB
Polyvinyl chloride	391	454	SE
Polyvinylidene chloride	532	532	SE
Polystyrene	345–360	488–496	12.70–63.5
Polymethyl methacrylate	280–300	450–462	15.24–40.64
Acrylic, fiber		560	SB
Cellulose nitrate	141	141	Rapid
Cellulose acetate	305	475	12.70–50.80
Cellulose triacetate, fiber		540	SE
Ethyl cellulose	291	296	27.94
Polyamide (Nylon)	421	424	SE
Nylon 6,6, fiber		532	SE
Phenolic, glass fiber laminate	520–540	571–580	SE–NB
Melamine, glass fiber laminate	475–500	623–645	SE
Polyester, glass fiber laminate	346–399	483–488	SE
Polyurethane, polyether, rigid foam	310	416	SE
Silicone, glass fiber laminate	490–527	550–564	SE

NB — Nonburning
SE — Self-extinguishing
SB — Slow burning

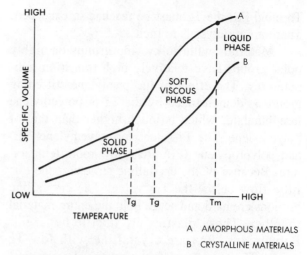

Fig. 4-31. Specific volume versus temperature for an amorphous and a crystalline plastics.

The melting point temperature (Tm) is higher than Tg. It is important to study the glass transition temperature of plastics to determine which plastics are more likely to be stable at room temperatures. Cold flow and processing techniques must be considered in polymer selection.

In Table 4-14 the Tg values and Tm values of selected plastics are shown. The amorphous polymers, polystyrene, polymethyl methacrylate, polycarbonate and polyvinyl chloride, have Tg values above room temperatures.

Glass transition has been found in semicrystalline and noncrystalline polymers. Plastics with little or no crystallinity must depend on the rigidity of their polymer chains for their strength above the Tg. A plastics with a low Tg and high crystallinity is strong, flexible, and tough. Amorphous plastics perform best below Tg. Atactic PS is a rigid, brittle, amorphous polymer at room temperature.

Table 4-14. Glass Transition and Melting Points of Plastics

Plastics	Tg, °C	Tm, °C
ABS	110	
PA	50	265
PAN	105	
PC	150	
PE	−90/−120	
PE (high density)		137
PE (linear, low density)		124
PET	65	265
PMMA	105	
PP	−10/−18	165
PS	100	
PVC	85	

that is extruded from a small orifice in 10 min at 43.5 psi [300 kPa]. Temperatures of 374 °F [190 °C] and 446 °F [230 °C] are specified for polyethylene and polypropylene respectively. A high melt index indicates a low-viscosity material. The melt index is calculated and given as grams per 10 min. High molecular mass materials are more resistant to flow.

Glass Transition Temperature

The temperature at which a plastics changes from a solid to a soft, rubbery state is called the glass transition temperature (Tg). Below Tg, the materials become hard, stiff, and solid. The molecular chain action is said to be frozen (Fig. 4-31).

To mold PS, the Tg must be reached or exceeded. There are no crystals to melt.

Materials with bulky side groups or highly polar groups have relatively high transition temperatures. The effect of the phenylene stiffening groups is demonstrated by the Tg of polyethylene terephthalate, which is much higher than that of polyethylene. The Tg values for polyethylene (PE) and polypropylene (PP) are below room temperature. Because of the crystalline structure, there is little flow below the Tm value. The crystalline portions are hard and act to hold the entire material together. These plastics are ductile but rigid. Polyamides (PA) are crystalline with low Tg points. They have excellent mechanical properties. Crystallinity in polymers depends on the regularity of structure; consequently, isotactic and syndiotactic polymers usually crystallize. Nonregularity of structure, such as copolymerization, random copolymers, and blends of isotactic and atactic polymers, reduce crystallinity.

Softening Point (ASTM D-1525)

In the Vicat softening point test, a sample is heated at a rate of 120 °F [50 °C] per hour. The temperature at which a needle penetrates the sample 0.039 in. [1 mm] is the Vicat softening point.

4-8 ENVIRONMENTAL PROPERTIES

Plastics are found in nearly every environment. They are used to contain chemicals, to store food in the freezer, and even inside the human body. Before a product is designed, plastics must be tested for endurance under the expected environmental extremes. The environmental properties of plastics include chemical resistance, weathering, ultraviolet resistance, permeability, water absorption, biochemical resistance, and stress cracking. (See Chapter 20, Radiation Processing.)

Chemical Properties

During the polymerization process, there is a breaking and recombining of chemical bonds. The types of bonds, the distance between bonds, and the energy needed to break these bonds are important in considering the chemical reactivity and solubility of plastics. Chemical deterioration is usually undesirable in plastics products.

The chemical resistance of the polyolefins and fluorocarbons is due to the C—C and C—F bonds.

These bonds are very stable. As a result these plastics are exceptionally inert, non-reactive, and resistant to chemical attack.

The hydroxyl groups (—OH) attached to the carbon backbone of cellulose and polyvinyl alcohol are very reactive. Water or other chemicals may break these bonds.

The statement that "most plastics resist weak acids, alkalies, moisture, and household chemicals" must be used only as a broad rule. Any statement about the response of plastics to chemical environments must be only a generalization. It is best to test each plastics to determine how it can be specifically applied and what chemicals each is expected to resist.

Temperature, fillers, plasticizers, stabilizers, colorants, and catalysts can affect the chemical resistance of plastics. In Table 4-15, chemical resistance of a number of plastics is listed. A concise statement of solubility parameters of many plastics is found in Table 4-6. Chemical environments may speed up cracking in plastics that are under stress.

In Figure 4-32, plastics test pieces are being exposed to a stress cracking test. A glass-reinforced

Table 4-15. Chemical Resistance of Selected Plastics at Room Temperature

Plastics	Strong Acids	Strong Alkalies	Organic Solvents
Acetal	Attacked	Resistant	Resistant
Acrylic	Attacked	Slight	Attacked
Cellulose acetate	Affected	Affected	Attacked
Epoxy	Slight	Slight	Slight
Ionomer	Slight	Resistant	Resistant
Melamine	Slight	Slight	Resistant
Phenolic	Resistant	Attacked	Affected
Phenoxy	Resistant	Resistant	Attacked
Pollallomer	Resistant	Resistant	Resistant
Polyamide	Attacked	Slight	Resistant
Polycarbonate	Resistant	Attacked	Attacked
Polychlorotrifluoroethylene	Resistant	Resistant	Resistant
Polyester	Slight	Affected	Affected
Polyethylene	Resistant	Resistant	Affected
Polyimide	Affected	Attacked	Resistant
Polyphenylene oxide	Resistant	Resistant	Slight
Polypropylene	Resistant	Resistant	Resistant
Polysulfone	Resistant	Resistant	Affected
Polystyrene	Affected	Resistant	Affected
Polytetrafluoroethylene	Resistant	Resistant	Resistant
Polyurethane	Resistant	Affected	Slight
Polyvinyl chloride	Resistant	Resistant	Affected
Silicone	Slight	Affected	Slight

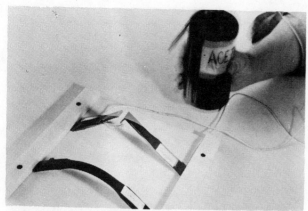

Fig. 4-32. An acetone spray causes a glass-reinforced polysulfone test bar to break. (Celanese Plastic Materials Co.)

polysulfone test bar at the rear cracked apart violently under a spray of acetone. This reaction broke an electrical connection, triggering the camera to take this picture.

The acetone did not affect the similarly stressed test bar of thermoplastic polyester in the foreground. Thermoplastic polyester withstands even higher stresses in atmospheres of carbon tetrachloride, methyl ethyl ketone, and other aromatic chemicals. Chlorinated hydrocarbons may cause cracking in polysulfone, polycarbonate, and other high-strength engineering thermoplastics.

Weathering (ASTM D-1435 and G-23)

Radiation, gases, and atmospheric exposures are used to test weathering resistance. Samples are placed at an angle of 45 degrees, facing south. Exposed and unexposed samples are then compared.

Accelerated weathering of a sample is described in ASTM test G-23. In both the normal and accelerated tests, the general weatherability of the plastics sample may be determined. The accelerated test greatly reduces the time needed for results.

The combined effects of temperature, light radiation, moisture, gases, and other chemical environments may cause dimensional and physical changes in plastics.

Ultraviolet Resistance (ASTM G-23 and D-2565)

Linked with weatherability is the resistance of plastics to the effects of direct sunlight or artificial weathering devices. Ultraviolet radiation (combined with water and other environmental oxidizing conditions) may cause color fading, pitting, crumbling, surface cracking, crazing, and brittleness. The Altas 18FR Fade-Ometer is used to check color stability. For artificial weathering, the Water-

Cooled Zenon-Arc Light and Water Exposure Apparatus is commonly utilized.

Permeability (ASTM D-1434 and E-96)

The transfer of water vapor and other gases is important in packaging. Permeability may be described as the volume or mass of gas or vapor penetrating an area of film in 24 h.

Permeability is an important concept in the food packaging industry. Polyvinyl chloride film is sometimes used to allow the passage of oxygen. This keeps meats and vegetables looking fresh.

Barrier films or layers are used to selectively prevent gases, moisture, and other agents from contaminating the package contents. (See Barrier Plastics.)

Water Absorption (ASTM D-570)

Some plastics, especially polyamides, tend to absorb moisture. They are said to be *hygroscopic*. After a sample is dried and weighed, it is immersed in water for a period of time. The sample is then reweighed and the percentage of additional mass is calculated.

Some plastics swell and lose optical and electrical properties when moisture is absorbed. Table 4-16 gives water absorption characteristics of various plastics.

Water absorption by epoxy and polyimide polymers results in decreases in short-term elevated-temperature strength. The moisture acts as a plasticizer, lowering the glass transition temperature. The swelling must be considered in designs with close tolerance dimensions and in composite designs. Moisture lowers the matrix and reinforcement bond. During processing and molding, moisture will vaporize into gas bubbles in the melt. Appearance and mechanical properties may be greatly affected by these bubbles in the molded parts.

Table 4-16. Water Absorption

Material	Water Absorbed, percent (24-Hour-Immersion)
Polychlorotrifluoroethylene	0.00
Polyethylene	0.01
Polystyrene	0.04
Epoxy	0.10
Polycarbonate	0.30
Polyamide	1.50
Cellulose acetate	3.80

(A) Plug in hot plate and calibrate it to a surface temperature of 518 °F (270 °C ±10°). Be sure surface is clean; place two glass slides on surface for 1-2 min.

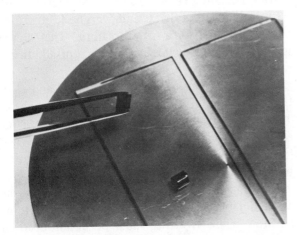

(B) When the glass surface temperature reaches 446-500 °F (230-260 °C) place four or five pellets on one glass slide using tweezers.

(C) Place the second hot slide over the pellets to form a sandwich.

(D) Press on the top slide with a tongue depressor until the pellets flatten to about 10 mm diameter.

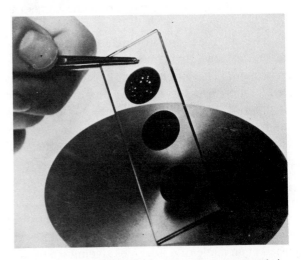

(E) Remove sandwich and allow to cool. Amount and size of bubbles indicate percentage of moisture.

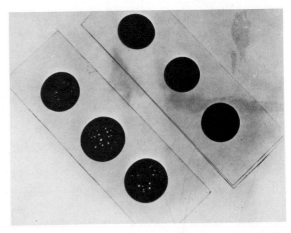

(F) Typical results. Slide at right indicates dry material; slide at left moisture-laden material. One or two bubbles may be only trapped air.

Fig. 4-33. The six simple steps of the Resin Moisture Test.

Two simple, low-cost methods of checking moisture content are the Tomasetti's Volatile Indicator (TVI) and the Test Tube/Hot Block (TTHB) techniques. The following procedure should be followed for the TVI technique:

1. Place two glass slides on a hot plate and heat from 1 to 2 min at 526 ±59 °F [275 ±15 °C] (Fig. 4-33).
2. Place four pellets or granular plastics samples on one of the glass slides.
3. Place the second hot slide on top of the sample and press the sandwiched pellets to about 0.393 in. [10 mm] diameter.
4. Remove the slides from the hot-plate platen and allow then to cool.
5. The number and size of bubbles seen in the plastics samples indicate percentage of moisture absorbed. Some bubbles may be the result of trapped air, but numerous bubbles indicate moisture-laden material. There will be a direct correlation between the number of bubbles and moisture content.

The TTHB procedure is:

1. Heat a hot block with test tube holes to 500 ±50 °F [260 ±10 °C].
2. Place 5.0 g of plastics in a 20 by 150 mm Pyrex test tube.
3. Place a stopper in the test tube, then carefully place the test tube in the hot block.
4. Allow the material to melt (about 7 min).
5. Remove the tube and sample from the hot block and allow it to cool for ten minutes.
6. Observe and record the correlation of moisture content with the surface area of the condensation in the test tube (Fig. 4-34).

Biochemical Resistance (ASTM G-21 and G-22)

Most plastics are resistant to bacteria and fungi, however, some plastics and additives are not. They may not be approved by the U. S. Food and Drug Administration (FDA) for use in packaging and as containers for foods or drugs. Various preservatives or antimicrobial agents may be added to plastics to make them resistant. (See Chapter 6– Ingredients of Plastics.)

Stress Cracking (ASTM D-1693)

Environmental stress cracking of plastics may be caused by solvents, radiation, or constant strain. There are several tests that expose the sample to a surface agent. One such test is shown in Figure 4-35.

4-9 OPTICAL PROPERTIES

Optical properties are closely linked with molecular structuring, such as chemical bonding and crystallinity. Thus, the electrical, thermal, and optical properties of the plastics are interrelated.

Gloss, luster, haze, transparency, color, clarity, and refractive index are only a few of the many optical properties of importance to plastics.

Luminous Transmittance (ASTM D-1003)

The light-reflective quality of a plastics surface is referred to as *gloss* or *luster*. A cloudy or milky appearance in plastics is known as *haze*. A plastics termed *transparent* is one that absorbs very little light in the visible spectrum. The selective absorption of light results in a material being *colored*. In general, materials with free electrons are opaque because free electrons absorb light energy. It is not likely, with present technology, that a transparent material can be a good electrical conductor. *Clarity* is a measure of distortion seen when viewing an object through a transparent plastics. Fillers, chemical bonds, or crystalline planes all distort or interfere with the passage of light.

Index of Refraction (ASTM D-542)

When light enters a transparent material, part of that light is reflected and part is refracted (Fig. 4-36). The index of refraction n may be expressed in terms of the angle of incidence i and the angle of refraction r

$$n = \frac{\sin i}{\sin r}$$

where i and r are taken relative to the perpendicular to the surface at the point of contact. The index of refraction for most transparent plastics is about 1.5. This is not greatly different from most window glass. For selected plastics, it may be 1.35. Table 4-17 gives indexes of refraction for some plastics.

4-10 ELECTRICAL PROPERTIES

The five basic properties that describe the electrical behavior of plastics are arc resistance, insulation resistance, dielectric strength, dielectric constant, and dissipation (power) factor. The predominantly covalent bonds of polymers limit their electrical conductivity thus most plastics may be treated as electrical insulators. With the addition of

(A) Plastics samples being heated to drive off moisture.

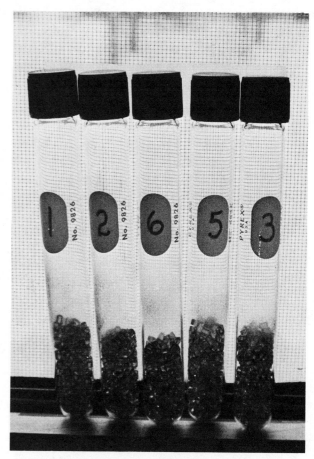

(B) Moisture condensed within test tubes.

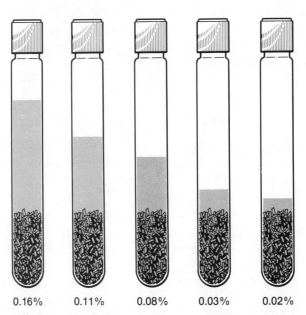

0.16%   0.11%   0.08%   0.03%   0.02%

(C) Area of condensation on test tube surface shows the percentage of moisture in plastics.

Fig. 4-34. The area of condensation on each test tube surface shows the percentage of moisture in each plastics sample. (General Electric Co.)

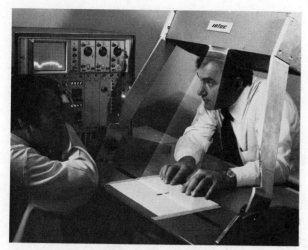

Fig. 4-35. Laser light from the scanner passes through the material and is collected by a receiver. The "spike" shown on the oscilloscope in the background indicates a defect in the material. (Intec Corporation)

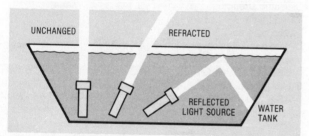

Fig. 4-36. Reflection and refraction of light.

Table 4-17. Optical Properties of Plastics

Material	Refractive Index	Light Transmission, %
Methyl methacrylate	<1.49	94
Cellulose acetate	1.49	87
Polyvinyl chloride acetate	1.52	83
Polycarbonate	1.59	90
Polystyrene	1.60	90

fillers such as graphite or metals, plastics can be made conductive or semiconductive.

Arc Resistance (ASTM D-495)

Arc resistance is a measure of the time needed for a given electrical current to render the surface of a plastics conductive because of carbonization. The measurement is reported in seconds. The higher the value, the more resistive the plastics is to arcing. This breakdown may be a result of corrosive chemicals. Ozone, nitric oxides, or a build-up of moisture or dust may be the cause.

Resistivity (ASTM D-257)

Insulation resistance is the resistance between two conductors of a circuit or between a conductor and ground when they are separated by an insulator. The insulation resistance is equal to the product of the resistivity of the plastics and the quotient of its length divided by its area:

$$\text{insulation resistance} = \text{resistivity} \times \frac{\text{length}}{\text{area}}$$

Resistivity is expressed in ohm-centimetres. Table 4-18 gives resistivities for certain plastics.

Dielectric Strength (ASTM D-149)

Dielectric strength is a measurement of electrical voltage required to break down or *arc through* a plastics material. The units are reported as volts per millimetre of thickness (V/mm). This electrical property gives an indication of the ability of a plastics to act as an electrical insulator. See Figure 4-37 and Table 4-18.

Dielectric Constant (ASTM D-150)

The dielectric constant of a plastics is a measure of the ability of the plastics to store elec-

Table 4-18. Electrical Properties of Selected Plastics

Plastics	Resistivity, ohm-cm	Dielectric Strength, V/mm	Dielectric Constant At 60 Hz	At 10^6 Hz	Dissipation (Power) Factor At 60 Hz	At 10^6 Hz
Acrylic	10^{16}	15 500–19 500	3.0–4.0	2.2–3.2	0.04–0.06	0.02–0.03
Cellulosic	10^{15}	8 000–23 500	3.0–7.5	2.8–7.0	0.005–0.12	0.01–0.10
Fluoroplastics	10^{18}	10 000–23 500	2.1–8.4	2.1–6.43	0.000 2–0.04	0.000 3–0.17
Polyamides	10^{15}	12 000–33 000	3.7–5.5	3.2–4.7	0.020–0.014	0.02–0.04
Polycarbonate	10^{16}	13 500–19 500	2.97–3.17	2.96	0.000 6–0.000 9	0.009–0.010
Polyethylene	10^{16}	17 500–39 000	2.25–4.88	2.25–2.35	<0.000 5	<0.000 5
Polystyrene	10^{16}	12 000–23 500	2.45–2.75	2.4–3.8	0.000 1–0.003	0.000 1–0.003
Silicones	10^{15}	8 000–21 500	2.75–3.05	2.6–2.7	0.007–0.001	0.001–0.002

trical energy. Plastics are used as dielectrics in the production of capacitors, which are used in radios and other electronic equipment. The dielectric constant is based upon air, which has a value of 1.0. Plastics with a dielectric constant of 5 will have five times the electricity-storing capacity of air or of a vacuum (Fig. 4-38).

Nearly all electrical properties of plastics will vary with time, temperature, or frequency. For example, the values may vary as frequency is increased. (See Table 4-18, for dielectric constant and dissipation factor.)

Dissipation Factor (ASTM D-150)

Dissipation (power) factor or *loss tangent* also varies with frequency. It is a measure of the power

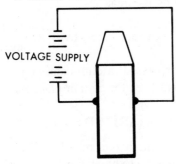

HOW GREAT CAN THE VOLTAGE BE BEFORE IT BREAKS THROUGH THE MATERIAL?

VOLTAGE SUPPLY

Fig. 4-37. Testing dielectric strength, an important characteristic of plastics for insulating applications.

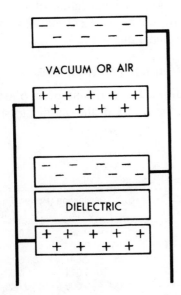

VACUUM OR AIR

DIELECTRIC

Fig. 4-38. The dielectric constant is the amount of electricity stored across an insulating material, divided by the amount of electricity stored across air or a vacuum.

Table 4-19. Summary of ASTM Testing Methods

Property	ASTM Test Method*	SI Units
Apparent density	D-1895	g/cm³
Free flowing		
Non-pouring		
Arc resistance	D-495	s
High voltage		
Low current		
Brittleness temperature	D-746	°C at 50%
Bulk factor	D-1895	Dimensionless
Chemical resistance	D-543	Changes recorded
Compression set	D-395	Pa
Compressive strength	D-695	Pa
Conditioning procedure	D-618	Metric units
Creep	D-2990	Pa
Creep rupture	D-2990	Pa
Deflection temperature	D-648	°C at 18.5 MPa
Density	D-1505	g/cm³
Dielectric constant	D-150	Dimensionless
Dissipation factor at 60 Hz, 1 KHz, 1 MHz		
Dielectric strength	D-149	V/mm
Short time		
Step by step		
Dynamic mechanical properties	D-2236	Dimensionless
Logarithmic decrement		
Elastic shear modulus		
Elasticity modulus		
Compressive	D-695	Pa
Tangent, flexural	D-790	Pa
Tensile	D-638	Pa
Elongation	D-638	%
Fatigue Strength	D-671	Number of cycles
Flammability	D-635	cm/min (burning), cm/s
Flexural strength	D-790	Pa
Flexural stiffness	D-747	Pa
Flow temperature	D-569	°C
Rossi-Peakes		
Gel time & peak exothermic temperature	D-2471	
Hardness		
Durometer	D-2240	Read dial
Rockwell	D-785	Read dial
Haze	D-1003	%
Impact resistance		
Dart	D-1709	Pa @ 50% failure
Izod	D-256	J/m
Indention Hardness	D-2583	Read dial
Barcol Impressor		
Linear coefficient of thermal expansion	D-696	mm/mm/°C
Load deformation	D-621	%
Luminous transmittance	D-1003	%
Melt-flow rate, thermoplastics	D-1238	g/10 min.
Melting point	D-2117	°C
Mold shrinkage	D-955	mm/mm
Molding index	D-731	Pa
Notch sensitivity	D-256	J/m

Table 4-19. Summary of ASTM Testing Methods (Continued)

Property	ASTM Test Method*	SI Units
Oxygen index	D-2863	%
Particle size	D-1921	Micrometre
Physical property changes	D-759	Changes recorded
Subnormal temperature		
Supernormal temperature		
Refraction index	D-542	Dimensionless
Relative density	D-792	Dimensionless
Shear strength	D-732	Pa
Solvent swell	D-471	J
Surface abrasion resistance	D-1044	Changes recorded
Tear resistance	D-624	Pa
Tensile strength	D-638	Pa
Thermal conductivity	C-177	W/K · m
Vicat softening point	D-1525	ohm-cm
Volume resistivity	D-257	%
1 min. at 500V		
Water absorption	D-570	%
24 hour immersion		
Long-term immersion		
Water vapor	E-96	g/24 h
Permeability		
Weathering	E-42	
	D-1435	Changes

*Use the latest revision of any ASTM method referenced.

(watts) lost in the plastics insulator. A test like those used in determining the value of the dielectric constant is used to measure this power loss. As a rule, measurements are made at one million hertz. They indicate the percent of alternating current lost as heat within the dielectric material. Plastics with low dissipation factors waste little energy and do not become overheated. For some plastics, this is a disadvantage since they cannot be preheated or heat-sealed by high-frequency methods of heating. (See Table 4-18 for various dissipation factors.)

The relationship between heat, current, and resistance is shown in the power equation:

$$P = I^2R$$

The power (P) used to perform wasted work is lost or dissipated power. In this formula, the amount of power can be decreased by lowering either the current (I) or the resistance (R). In electrical appliances designed to produce heat, a low dissipation factor is not considered desirable.

VOCABULARY

The following vocabulary words are found in this chapter. Look up the definition of any of these words you do not understand as they apply to plastics in the glossary, Appendix A.

Brittleness temperature
Centipoise
Cold flow
Compressive strength
Creep
Damping
Density
Density gradient column
Dimensional stability
Fatigue
Flexural modulus
Flexural strength
Glass transition temperature
Halogens
Hardness
Haze
Impact strength
Index of refraction
Ionomers
Parameters
Plastic strain
Proportional limit
Relative density
Scleroscope
Solvent resistance
Stiffness
Strain
Stress
Thixotropy
Toughness
Vicat softening point
Viscosity

4-1. The seven base units of the SI metric system are ___?___ , ___?___ , ___?___ , ___?___ , ___?___ , ___?___ , and ___?___ .

4-2. A gigahertz is equal to ___?___ Hz.

4-3. The SI metric unit for force is the ___?___ , which has the formula ___?___ .

4-4. Tensile strength, modulus of elasticity, and air pressure are all measured in ___?___ .

4-5. In the SI metric system, temperatures are measured in ___?___ .

4-6. Two sophisticated and costly indentification methods for plastics are ___?___ and ___?___ .

4-7. In the Lassaigne test, the specimen is heated with the metal ___?___ .

4-8. The five broad methods for plastics identification are ___?___ , ___?___ , ___?___ , ___?___ , and ___?___ .

4-9. Thermosetting plastics are usually formed by ___?___ molding, ___?___ molding, or ___?___ .

4-10. The plastics family with a waxy feel, used to produce carpeting, packaging, and toys is ___?___ .

4-11. A plastics with good weathering characteristics transparency, used for window glazing is ___?___ .

4-12. The thermoplastic material with the highest melting point (unfilled) is ___?___ .

4-13. A thermoplastics is entirely insoluble in all solvents except hot benzene-toluene. It is probably ___?___ or ___?___ .

4-14. The thermoplastics in Question 13 burns slowly, with a blue flame and a trace of white smoke. It is probably ___?___ .

4-15. The family of plastics with the greatest density is ___?___ .

4-16. Marlex is a trade name for ___?___ .

4-17. Toxic fumes are emitted when ___?___ are burned or overheated.

4-18. Since all tests have limitation, the truest test of a plastics product is ___?___ .

4-19. The international, nonprofit, technical society that develops standards and specifications for testing of plastics is ___?___ .

4-20. True or False. In testing mechanical properties, it is generally important to apply force at a specified rate.

4-21. Young's modulus is the ratio of ___?___ to ___?___ .

4-22. To choose a tougher plastics, choose one having a ___?___ area under the stress-strain curve.

4-23. The pendulum test measures ___?___ .

4-24. The Strength-to-mass ratio of plastics is ___?___ than that of most metals.

4-25. The important property for a plastics hinge would be ___?___ .

4-26. Resistance to transmitting vibration is called ___?___ .

4-27. Crystallinity of plastics ___?___ brittleness at low temperature.

4-28. The plastics materials with the lowest relative density are ___?___ and ___?___ .

4-29. A viscosity of 1 pascal-second is equal to ___?___ poise.

4-30. Viscosity is defined as a measure of the ___?___ of a fluid.

4-31. The elongation that occurs when a plastics sample has a force acting on it for a period of time called ___?___ or ___?___ .

4-32. Van der Waals' forces are ___?___ likely to be affected by heating than covalent bonds.

4-33. Handles of pots and pans are often made of plastics because of the property of low ___?___ .

4-34. If Plastics A has a heat deflection temperature below 65 °C while Plastics B has a heat deflection temperature above 150 °C, you would select ___?___ for use in boiling water.

4-35. Plastics for space-craft heat shields are chosen for their ___?___ properties.

4-36. The plastics with the lowest self-ignition temperature is ___?___ .

4-37. The plastics with higher melt index will have ___?___ viscosity.

4-38. Below the glass transition temperature, a plastics becomes ___?___ .

4-39. Water can break the bonds of ___?___ groups attached to the carbon backbone of polyvinyl alcohol.

4-40. Stresscracking tests combine physical stress and ___?___ stress.

4-41. Polychlorotrifluoroethylene has ___?___ resistance to acids, bases, and solvents.

4-42. A plastics that is hygroscopic is ___?___ .

4-43. The plastics with best light-transmission properties is ___?___ .

4-44. Fillers used to make plastics conduct electricity are ___?___ or ___?___ .

4-45. In the arc resistance test, the surface of a plastics becomes conductive because of ___?___ .

4-46. If the resistivity of a plastics is high, insulation resistance will be ___?___ .

4-47. Dielectric strength indicates suitability of a plastics for service as ___?___ .

4-48. Plastics used in electrical capacitors must have a high ___?___ .

4-49. For heat-sealing a plastics by high-frequency methods, the ___?___ must not be low.

4-50. What properties would you consider in making the following plastics products:
ice chest
freezer food container
gas tank
counter top
fishing rod
eye-glass lens
washing machine agitator

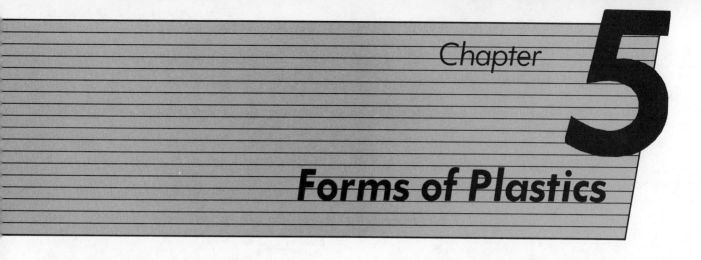

Chapter 5

Forms of Plastics

There are many ways to arrange chemicals into plastics. In this chapter, you will become familiar with how plastics can be made into nine basic forms.

Most people are aware that metals can be purchased in a variety of forms. Plastics, like metals, may be obtained as structural shapes, thin fibers, powders, and several other forms. However, in order to be a *plastics*, the material must be solid in the finished state. Not all the plastics forms described in this chapter are solid materials. Thus the names of some available forms of plastics are really a description of their use or product application. These forms may begin as monomer resins, partially polymerized resins, or fully polymerized solid plastics.

It is easy to visualize cutting, sawing, and welding I-beams, tubes, channel shapes, or flat sheets of plastic into various contours of a finished product. Craft classes have been cutting, sanding, buffing, and forming plastics for years. Perhaps this fluff-and-buff method of working with plastics has been your only contact with plastics forms up to this time. Industry does not typically work with the plastics forms found in craft classes.

In industrial plastics the resins and polymers are often formed in a die or mold to produce the finished product, thus some of the available forms of plastics are never seen by the consumer. Some, such as casting resins, are intermediate to the production of other plastics forms. Rods, sheets, coatings, and adhesives may begin in resinous form.

Combinations of plastics forms may be used in the production of some composites. For example, the composites may be composed of sheets of reinforced plastics bonded to a core of cellular plastics.

5-1 MOLDING COMPOUNDS

Many of the available forms of plastics begin as molding compounds. Large volumes of this form of plastics are molded into other familiar forms such as profile shapes, fibers, and films.

Molding compounds are available in different solid forms to facilitate molding. For example, *powdered* forms are used in rotational molding and in various coating applications. The powder grain diameter seldom exceeds 0.0005 in. [0.10 mm].

Thermoplastic pellets are easily dispensed and used in a wide variety of molding processes.

Thermosetting powders and granules are often made from B-stage resins and other additives. During this intermediate state, resins can be heated and caused to flow to the desired shape.

Preforms

Many polymer powders and granules are pressed into convenient *preform* shapes to be used in manufacturing the final shape of the plastics product. These preforms are often referred to as pills, tablets, biscuits, or premolds. Special machines weigh and compact plastics powder into the desired preform shape. Although many preforms are tablet or pill shaped, they may be made into any desired shape or size to fit the needs of the molding process. Preforms are widely used in the compression molding of thermosetting plastics. (See Chapter 10, Molding Processes.)

To make mold construction simple and shorten the molding cycle, the preform is usually preheated. This preheating helps to remove moisture and shortens the molding time for both thermosetting and thermoplastic compounds.

Preheating is achieved by several methods. Hot plates, ovens, infrared radiation, and high-frequency heating are used. The high-frequency method is in wide use, because it very rapidly and uniformly heats the preform (Fig. 5-1).

Pellets. Pellets are easily dispensed to equipment and molding machines. Granular pellets sel-

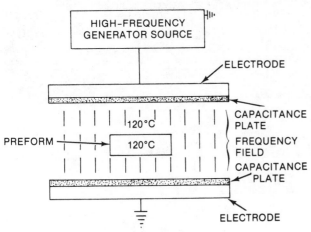

Fig. 5-1. High-frequency energy may be used to heat a preform before molding.

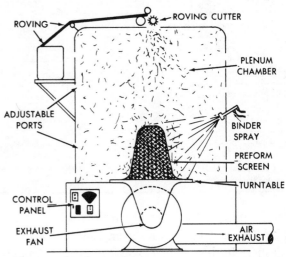

(A) Plenum chamber preform process.

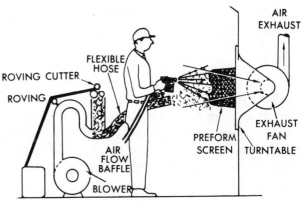

(B) Directed fiber preform process.

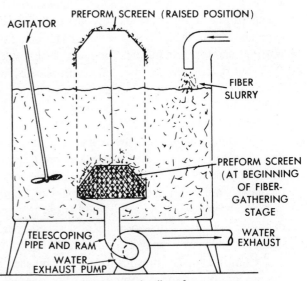

(C) Water slurry or "Aqua-glass" preform process.

Fig. 5–3. Three ways in which a coated, fibrous glass preform may be produced. (Owens-Corning Fiberglas Corp.)

dom exceed 0.25 in. [6 mm] in diameter (Fig. 5-2). Pellets are widely used in the injection and extrusion molding processes. These molding processes are discussed in Chapter 10.

Coated Preforms. Three methods of producing coated fibrous glass preform are shown in Figure 5-3. These preforms are easily handled, produce little waste, and result in short molding cycles. They have a greater resin-to-reinforcement ratio than many reinforced forms.

Large products, such as automobile bodies and boat hulls, may be produced by coated preforms because matched, molded preforms produce finished surfaces on both sides of the product.

Bulk Molding Compounds

Bulk molding compounds (BMC), formulated in a putty form, are referred to as a *premix*. This putty form is sometimes referred to as *gunk, putty, dough,* or *slurry molding.* Bulk molding compounds are often formed into log or rope shapes to aid molding and handling operations. A BMC is usually a mixture of monomer resin, short fibrous

Fig. 5-2. Various pellet preforms.

reinforcements, fillers, and other additives. When the reinforcing material is combined with a full complement of resin before molding, the mixture is called a **prepreg**. Some processes require additional resin to be added during the molding operation. The prepreg is generally in a continuous sheet or tape form and is composed of cloth, mat, or paper-impregnated B-stage resin and other additives. (See Composites, Reinforcements, BMC, SMC, TMC, or XMC)

The BMC premix may be molded by several processes. They are also used for filling cracks and voids in certain materials. Some are polymerized by adding a catalyst or heat. Others are colloidal suspensions that harden when the solvents evaporate.

Polyester or epoxy resins are typically mixed with the following percentages:

Premix Putty

Resin	38%
Chopped fibers	18%
China clay (filler)	44%
	100%

Sheet Molding Compounds

Sheet molding compounds (SMC) are produced as impregnated preforms. They are sometimes called the flow mat or mold mat.

The resins, fillers, and fibers are formed into final shape by the molding operation. Both thermosetting and thermoplastic materials are used to produce SMC. An upper and lower carrier film (usually polyethylene) is used with resins. This film allows the SMC to be stored neatly at hand and to be easily handled during processing. The carrier films are removed prior to the molding operation. (See Fig. 5-4.)

A typical SMC incoporates about 30 percent random 1 in. [25 mm] chopped glass fibers, 25 percent resin and 45 percent inorganic filler. SMC-25 indicates a 25 percent glass content. There are other SMC suited for compression, injection, pultrusion, transfer, and other molding techniques. (See Composites.) Unidirectional molding compound (UMC) usually has about 30% of its continuous reinforcing fibers aligned in one direction. This provides greater tensile strength in the direction of the fibers. Thick molding compounds (TMC) are produced up to two inches thick. This thicker sheet allows greater variation in part wall thickness and a wider choice of reinforcements. TMC are highly filled. Low-pressure molding compounds (LMC) are SMC formulated to allow low-pressure molding techniques. High-strength molding compound (HMC) may contain

more than 70% reinforcement for added strength and dimensional stability. Directionally reinforced molding compound (XMC) is a sheet containing about 75% directionally oriented continuous reinforcements. (See Reinforcements, Hybrid, Fibers, and Fillers)

As you can see, SMC are formulated to meet a variety of design parameters, economical considerations, and processing techniques.

5-2 ADHESIVES

Animal glue and some natural plastics have been used as adhesives for centuries. Wax and shellac were once widely used to seal letters and important documents. Many ancient civilizations used pitch to seal cracks in boats and rafts. Archeologists have evidence that more than 30 centruies ago the Egyptians used adhesives to attach gold leaf to wooden coffins and crypts.

Adhesive bonding is growing rapidly as a fastening method. Today, plastics adhesives have replaced most natural adhesives because they have proved to be strong, durable bonding materials (Fig. 5-5). Nailing, bolting, riveting, soldering, welding, and other assembly methods are giving way to adhesive bonding.

Because adhesives are used for so many different purposes and in so many different ways, the term is difficult to precisely define. Adhesives are a broad class of substances that adhere materials together by a surface bond. At one time, the word *glue* referred to an adhesive obtained from hides, cartilage, bones, and other animal materials. Today, this term is synonymous with adhesive. The term *glue* is probably used more often when referring to the bonding of wood.

Cement and liquid-glue adhesives are general terms used interchangeably. The term *cement* is applied to a liquid plastics adhesive with a synthetic or resin solvent base. The bond is chemical in nature and classed as cohesive bonding. Building blocks are held together with a mortar called cement. This cement has a true adhesive action in holding the blocks together.

Adhesives are commonly used in gluing wood, repairing broken china, and as bonding agents on tape. Important considerations to be made when selecting adhesives, include the kind of materials to be joined, their surfaces, the application method, and the service conditions expected of the bonded

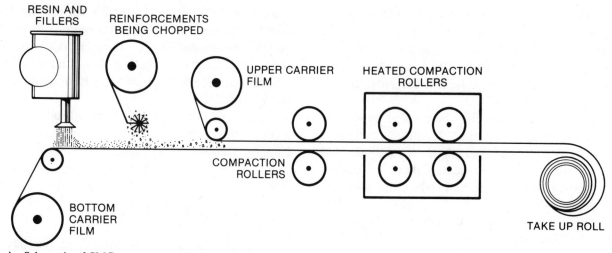

A. Schematic of SMC.

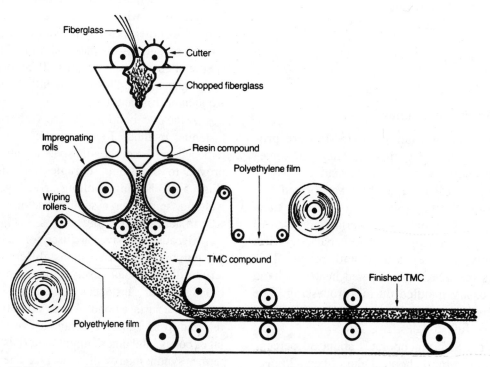

B. Schematic TMC (courtesy USS Chemicals Division).

Fig. 5-4. Production method for sheet molding compound.

Fig. 5-5. Adhesive properties are illustrated by the failure of the paper (rather than the hot-melt adhesive) in this aluminum-foil-to-paper laminate. (Henkel Corp.)

product. (Chapter 18 will explore bonding in greater detail.)

Types of Adhesion

There are two kinds of forces that contribute to *adhesion*: (a) mechanical forces, and (b) chemical forces.

Mechanical Adhesion. Mechanical adhesion has been defined as "adhesion between surfaces in which the adhesive holds the parts together by interlocking action." Adhesives are sold in various forms, however, they must be in a liquid or semiliquid state during the bonding operation. This ensures close contact with the *adherends* (surfaces being adhered). Physical or mechanical adhesion usually involves secondary (van der Waals') forces and simple mechanical interlocking on the surface. It does not result in any material flow between adherends. In organic materials, there may also be some primary forces at work. Metal-to-metal bonding is usually due largely to secondary and physical forces.

Chemical Adhesion. Chemical or specific adhesion has been defined as "adhesion between surfaces held together by valence forces of the same type as those that give rise to cohesion." The forces that hold the molecules of all materials together are referred to as *cohesive* forces. These forces include the strong primary valence bonds and the weaker secondary bonds.

In chemical adhesives, there is a strong valence attraction between the materials as the molecules flow together. During the welding of metals, for example, molten metal flows and there is a chemical cohesive action between the pieces.

It should be apparent that only by causing a softening or flow in the two materials can chemical bonding occur. If the surfaces are not caused to flow, only mechanical or physical forces hold the pieces together.

In joining thermoplastics, plastics or metals, either bonding method may be used. In the cohesive bonding of metals, heat must be applied to cause the surface molecules to flow and intermingle. In the cohesive bonding of plastics, solvents may be used on most of the thermoplastic polymers. These solvents soften the two surfaces being joined allowing strong valence bonding between molecules. Solvent cementing will be discussed in greater detail in Chapter 18.

A *solvent cement* is one that dissolves the surfaces of the plastics being joined. This forms strong intermolecular bonds as it evaporates.

A *monomeric cement* is a monomer of at least one of the plastics to be joined. It is catalyzed so that a bond is produced by polymerization in the joint. Either cohesive or adhesive bonding will occur, depending on the chemical composition of the materials being joined.

There are internal cohesive forces involved within all adhesives. These forces resist any attempt to pull them apart. Both cohesion and adhesion are involved when *contact cement*, a popular polymer adhesive, is used. This adhesive is applied to both surfaces and allowed to set. The two

surfaces are then pressed together, resulting in a strong *cohesive* bond between the two surfaces of the contact cement, while *adhesive* bonds form between the cement and the adherends.

If surfaces are properly prepared, nearly all materials may be bonded with an adhesive.

Types of Bonding Action

Physical or mechanical bonds:

  Wood to wood
  Paper to paper
  Leather to leather

Chemical or cohesive bonds:

  Metal to metal — when welded
  Ceramic to ceramic — when fired
  Plastics to plastics — when welded (mostly semi-crystalline) or solvent-cemented (mostly amorphous)

Adhesive Forms

Synthetic adhesives come in a number of physical forms: powders, films, dispersions, hot melts, pastes, liquids, and two-part components (Table 5-1). The adhesives on tape are generally

Table 5-1. Available Forms of Selected Plastics Adhesives

Plastic Adhesives	Available Forms
Thermosetting	
Casein	Po, F
Epoxy	Pa, D, F
Melamine formaldehyde	Po, F
Phenol formaldehyde	Po, F
Polyester	Po, F
Polyurethane	D, L, Po, F
Resorcinol formaldehyde	D, L, Po, F
Silicone	L, Po, Pa
Urea formaldehyde	D, Po, F
Thermoplastic	
Cellulose acetate	L, H, Po, F
Cellulose butyrate	L, Po, F
Cellulose, carboxymethyl	Po, L
Cellulose, ethyl	H, L
Cellulose, hydroxyethyl	Po, L
Cellulose, methyl	Po, L
Cellulose nitrate	Po, L
Polyamide	H, F
Polyethylene	H
Polymethyl methacrylate	L
Polystyrene	Po, H
Polyvinyl acetate	Pt, D, L
Polyvinyl alcohol	Po, D, L
Polyvinyl chloride	Pa, Po, L

Note: Po—Powder; F—Film; D—Dispersion; L—Liquid; Pa—Paste; H—Hot Melt; Pt—Permanently tacky

designed to remain tacky throughout their service life.

Adhesives are generally classified into four general types: 1) *Thermoplastic resins* are generally solvent-based or hot melts. Cyanoacrylate cures through a polymerization process. 2) *Thermosetting resins* cure with the aid of thermal energy or catalyst systems. A few are room-temperature-curing. 3) *Elastomeric* types are generally solvent-based and are used in the assembly of elastometers or where flexibility is required. 4) *Miscellaneous* adhesives are available as water-based systems. Most are polymer emulsions or rubber lattices.

Amino Resins. Casein and urea formaldehyde (amino resins) are used in the woodworking industries. Some of the urea resins are sold in liquid forms for use in the manufacture of plywood, particle boards, and hardboards.

Shell-molding is an important process used by foundry workers in casting metal parts. Phenolic and amino resins are used to bond the sand mold together.

Phenolic Resins. Phenol-formaldehyde (phenolic) resins are sold in liquid, powder, and film forms. The films, about 0.001 in. [0.025 mm] thick, are placed between the materials to be bonded. Moisture in the material or external steam causes the adhesive film to flow and liquify. The curing reaction takes place at temperatures in the 250 °F to 300 °F [120 °C to 150 °C] range. A reinforced film is often used. These films are usually thin, like tissue paper, and saturated with the adhesive. They are used in the same manner as unreinforced films but are easier to handle and apply.

Large quantities of phenolic resin adhesives are used in the manufacture of exterior plywood and tempered or smooth-finished hardboard.

Resorcinol-formaldehyde. Resorcinol-formaldehyde resins are another phenolic-based adhesive that usually comes in a liquid form. It is mixed with a powdered catalyst at the time of use. This resin has the advantages of curing at room temperature and being water- and heat-resistant. High grades of exterior and marine plywoods are bonded with these adhesives.

High-frequency heating greatly speeds the curing or polymerization time of many plastics used as adhesives. The high-frequency field excites the molecules of the adhesive, causing heat and rapid polymerization. Wood joints are often assembled by this method, using resorcinal-formaldehyde adhesives. (Fig. 5-6).

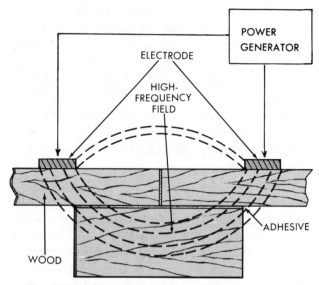

Fig. 5-6. High frequency (radio) waves may be used to heat adhesives.

Resin-bonded grinding wheels and sandpapers are made from abrasive grains and a plastics bonding agent. The grinding wheels are made from abrasive grains, powdered resin, and a liquid resin by a cold-molding process (discussed in Chapter 10). Figure 5-7 shows some typical sandpapers and grinding wheels bonded with phenolic or other resins.

Epoxy Resins. Epoxy resin adhesives are thermosetting plastics available in two-part paste components. Epoxy resin and either a powdered or a resinous catalyst are mixed to polymerize the resin. Heat is sometimes used to aid or speed this hardening process. Specially formulated one-part epoxies may be polymerized by the application of heat alone.

Epoxy adhesives have almost perfect adhesion to nearly all materials if the surfaces are properly

Fig. 5-7. Many grinding wheels and sandpapers are resin-bonded.

prepared. However, polyethylene, silicones, and fluorocarbons are among the most difficult to bond. The excellent adhesive properties of epoxies are used to mend broken china, bond copper to phenolic laminates in printed circuits, and to bond components in sandwich or skin-type structures (Fig. 5-8).

Acrylic Adhesives. Acrylic adhesives range from flexible to hard materials. The cyanoacrylate adhesive is a popular, rapid-setting adhesive that polymerizes when pressure is applied to the joint. This adhesive does not cure by solvent evaporation. N,N-dimethylformamide is a solvent.

Vinyl Adhesives. Vinyl adhesives include a variety of resins, elastomers, and plastics. Polyvinyl alcohol is a water-based adhesive used to bond paper, textiles, and leather. The interlayer of safety glass is polyvinyl butyral because it has excellent adhesion to glass. The electrical and insulation value of polyvinyl formal makes it ideal for wire enamels. Polyvinyl acetal excels as a bonding adhesive for metals.

The glue pots associated with the older animal glues have disappeared in favor of the more modern and easily used adhesives. One very popular modern adhesive is white glue, a polyvinyl acetate adhesive. This adhesive comes ready to use, in a fast-setting liquid form. This familiar adhesive is a dispersion of polyvinyl acetate in a solvent. Often the solvent is water, and as a result, it must be kept from freezing. Carpenters, artists, secretaries, and many other people make use of the adhesive properties of this material.

Cellulosic Adhesives. Cellulosic adhesives are popular and are available in solvent, hot-melt, and dry-powder forms. Duco Cement is a general purpose cellulose nitrate adhesive. It is waterproof, clear, and will adhere to wood, metal, glass, paper, and many plastics. Cellulose acetates and butyrates are familiar cements for plastics models.

Hot-melt Adhesives. Hot-melt materials are popular because they are easy to use, somewhat flexible, and obtain their highest adhesive qualities when cooled. Several thermoplastics are used as hot-melt adhesives, including polyethylene, polystyrene, and polyvinyl acetate.

Small sticks or rods of these plastics are heated in an electric gun. The hot plastics is forced from the gun nozzle onto the gluing surface. Figure 5-9 shows a leather strap being assembled using hot-melt plastics in an electrically heated applicator gun. The shoe industry is presently using this adhesive as an effective means of assembly of leather goods.

Small, rapidly assembled articles may be bonded with this method. Probably the most serious drawback to hot-melt adhesives is the difficulty in making large glue joints. The adhesive cools too quickly to ensure adhesion on large bonding surfaces.

(A) Various plastics adhesives.

(B) Two-part epoxy adhesive.

Fig. 5-8. Plastics adhesives are widely used. Industry uses epoxy, casein, urea, and resorcinol adhesives, while polyvinyl (white glue) adhesives are used in the home.

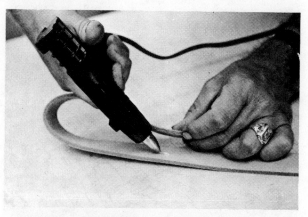

Fig. 5-9. This glue gun is used to apply hot-melt adhesives.

Adhesive-Faced Tapes. Adhesive-faced tapes are important members of the adhesive family. The adhesive coating is generally activated by one of the following methods: (a) solvents, such as water on gummed labels and stamps, (b) heat, such as iron-on fabric patches and heat-sealing package tapes and (c) pressure, such as masking tapes, labels, and other similar applications.

Other Plastics Adhesives. Caulking, sealing, patching, glazing, and putty compounds may be classified as plastics adhesives. For usefulness, they depend on their ability to bond effectively with the substrate to which they are applied. Their basic purpose is to keep moisture, air, or other agents out of cracks or small openings. These compounds include many elastomers as well as plastics (Fig. 5-10).

The term *caulking* is probably derived from the practice of forcing tar-soaked wicking into the seams of wooden boat hulls. Caulking implies that a compound is being forced into a crack.

The word *sealant*, a modern term, describes compounds that fill cracks. The words *putty, patching,* and *glazing compounds* usually mean sealant compounds, but describe materials with a special application. Putties and patching compounds are used to fill large cracks or openings. They are applied from bulk containers with a spatula or putty knife. These compounds contain a large amount of filler. They are less expensive and shrink less than other compounds. Sealing large cracks in wood, steel, and concrete are among their many typical uses. Glazing compounds are most often used to seal openings around windows. Because a great variety of construction materials are joined and because each material has a different expansion rate, sealants must remain flexible, non-drying, and adhesive (Fig. 5-11). Aluminum, for example, will expand about two and one-half times more than glass.

A modern and efficient method of applying sealants is by use of compressible tapes or extruder guns. Compressible tapes are rolls of sealant in ribbon form. They are used in the automotive industry for sealing metal joints and as adhesive sealants in window construction (Fig. 5-12).

Extruder guns are convenient applicators that use disposable or refillable containers of sealing compound. These devices push the sealant out of a nozzle during application (Figs. 5-13 and 5-14).

Some of the more common sealants include polysulfides, acrylics, polyurethanes, silicones, and both natural and synthetic rubber compounds.

The polymer polysulfide is a very effective sealant with a wide range of applications in the aircraft, electrical, and building industries (Fig. 5-15).

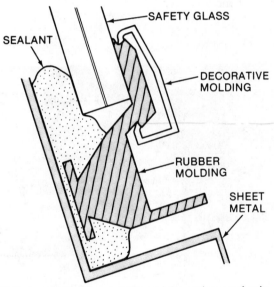

(A) Sealant with rubber molding and decorative metal strip.

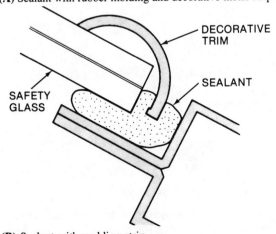

(B) Sealant with molding strip.

Fig. 5-11. Sealing methods used for automobile windshields.

Fig. 5-10. There are many caulks, sealants, and glazing materials available in cartridges for ease of application.

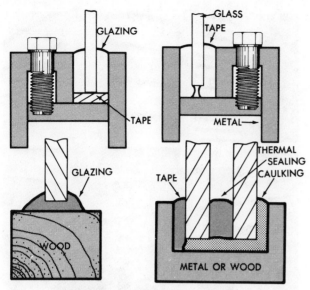

Fig. 5-12. Various sealing methods.

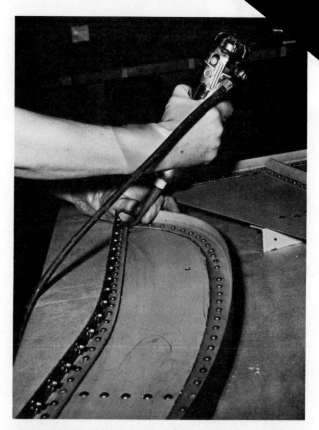

Fig. 5-14. Overlap angle joints, used in aerospace vehicles, are being sealed with LP® polysulfide base sealant. (Thiokol Corporation, Chemical Division)

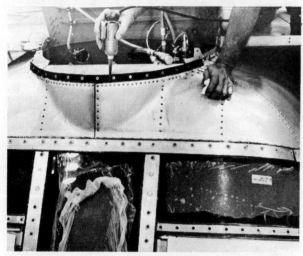

Fig. 5-13. A special applicator gun is used to apply LP® polysulfide base sealant for an aircraft motor housing. (Thiokol Corporation, Chemical Division)

Glazing compounds and putties may be made of acrylic compounds. They are easily applied and will not crack, sag, or break down.

Special formulations of epoxy and silicone compounds find uses as sealants around lavatories and bathtubs.

Silicone construction sealant is quite expensive, but it is easily used, cures in one hour, and remains permanently flexible.

Fig. 5-15. The aircraft and aerospace industries use polysulfide polymers in many sealing applications. (Thiokol Corporation, Chemical Division)

5-3 DIMENSIONAL OR PROFILE FORMS

Most of the dimensional or profile forms of plastics are made by the extrusion process. This familiar form is available in such structural shapes as rods, tubes, bars, and sheets. These shapes may be produced in continuous lengths that are then cut to standard sizes for handling and shipping. A few of the typical dimensional or profile shapes are shown in Figure 5-16.

Fig. 5-16. Some of the many extruded profile forms of plastics. (Fellows Corp.)

Fig. 5-17. This exterior siding is made of solid vinyl and has a wood-grain appearance. (Bird and Son, Inc.)

Not all dimensional or profile shapes are made by the extrusion process. The casting process may be used for tubes, rods, bars, or other shapes.

Since filaments and films have different product applications and are usually made with small cross-sectional areas, they will be covered separately. A plastics form close to films and filaments is the sheet. The term *sheet* refers to a plastics made in a long, flat, continuous form and greater than 0.01 in. [0.25 mm] in thickness. The obsolete term 10 mils is also used, indicating a thickness of 0.010 inches.

The word *sheeting* is used for flat pieces of plastics that are manufactured in a long continuous section. Usually, the term sheet is used for relatively small pieces of plastics sheeting cut to dimensional sizes.

Artists and hobbyists use the familiar sheet form because it is easily formed and shaped with hand tools. Sheets may be veneered or clad to metal and wood as durable materials for countertops, wall coverings, or exterior siding. Figure 5-17 shows vinyl siding for use on homes.

Sheets are thermoformed into various products, including picnic coolers, car bodies, toys, outdoor signs, luggage, and refrigerator liners. Sheets are commonly produced by forcing the melted plastics through an adjustable die opening. (See extrusion and calender molding in Chapter 11 and casting processes in Chapter 14.)

5-4 FILMS

The American Society for Testing and Materials (ASTM) has defined *film* as plastics sheeting 0.01 in. [0.25 mm] or less in thickness. Nearly every thermoplastic family is represented in the versatile form of film (Fig. 5-18). Common methods of making film include extrusion (both lip-die and blow film extrusion), calendering, and casting.

Uses of Films

Plastics films have three broad use categories: (a) packaging, (b) industrial uses, and (c) soft goods.

Packaging. The major use of film is in *packaging* applications. Meats, breads, produce, toys, frozen foods, housewares, and garments are only a small sample of the many packaging uses for plastics films. Boilable and sterilizable bags are growing in popularity for food packages.

Meats packaged in transparent polyvinyl chloride film can be seen by the consumer. The film also helps preserve the freshness and red color of the meat. Many films are semipermeable; that is, they allow certain materials to enter through the film walls. Oxygen must permeate the film to keep the appealing red color in meats.

Plastics bags have grown in popularity to package bread, cover garments, and contain trash.

Fig. 5-18. This polyurethane film, 0.008 mm [0.000 3 in] thick, demonstrates its tensile strength and puncture resistance. (BF Goodrich Chemical Division)

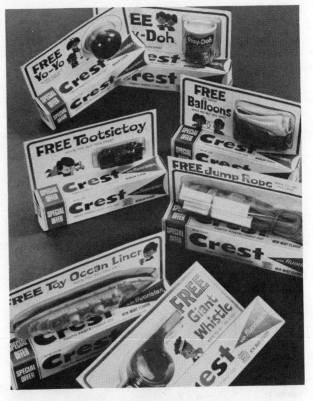

Fig. 5–19. Plastics (cellulosic) blister packaging helps these toy premiums attract attention to the merchandise. (Celanese Plastic Materials Co.)

Heat-sealing and vacuum-packaging film applications are often seen in supermarkets. Many products are packaged in plastics films that have been formed or shaped to help hold and display the product. The *blister packaging* of hardware and other products is a well-known example (Fig. 5-19).

Because polyvinyl alcohol film has a high water permeability, it is used to package soaps and other chemicals. These handy products release their contents when placed in water.

Industrial Uses. Plastics film is widely used for industrial applications. Film sheeting may be used as protective covering for growing plants and silage, and for other agricultural applications. Figure 5-20 shows a degradable plastics mulch being placed on the prepared soil. The film applicator is attached to a tractor. The front discs open up a shallow trough into which the edges of the film are pressed by the first pair of wheels. The spades cover the edges with soil and the rear wheels press the soil down. One farmer can apply about 32 000 ft^2 [3 000 m^2] of plastics film in 20 min.

Sheeting films provide moisture barriers in residential and commercial construction.

In industry, film is used for such varied purposes as magnetic recording and adhesive tapes. Because of the electrical resistance and dielectric properties of plastics, they are used as protective sheathing on electrical conductors. Plastics are used in photographic films and as laminates on paper and cardboard. Films are veneered or clad to metal and wood as durable building materials.

Soft Goods. Soft goods feel like fabric, drape like woven textiles, and are produced from a variety of films. These fabrics, or soft goods, may be either unsupported or supported films. The supported film may have a woven or bonded backing. Polyvinyl chloride, either supported or unsupported, is a very popular film used for raincoats, dresses, handbags, shoes, shower curtains, and wall coverings. These films can be embossed to look like woven fabric, leather, or other materials.

Foam-backed or supported films are used for automotive upholstery, balloonlike furniture, and other types of home furnishings.

The leather industry has probably felt the greatest impact of plastics soft goods. Plastics have replaced much of the real leather used in garments,

(A) Closeup view of film applicator.

(B) Rows of film mulch.

(C) Zucchini and eggplant being grown in rows of mulch.

Fig. 5–20. A degradable cellulosic plastics mulch is used in vegetable growing. (Princeton Chemical Research, Inc.)

handbags, luggage, and shoes. Naugahyde, a trade name for a fabric-supported polyvinyl film, is a soft but tough upholstery material.

Some film is laminated to a layer of foam. This makes a fabric that is used to produce garments with many uses for the sportsman, police, and other groups. There are two basic methods for making bonded textiles (Fig. 5-21).

5-5 FILAMENTS AND FIBERS

Everyone is an ultimate consumer of plastics filaments and fibers. Large volumes of these special plastics forms are used in the textile industry.

A *filament* is a single, long, slender strand of plastics. The fisherman is probably most familiar with monofilament fishing line. This single filament of plastics may be made in any desired length.

Yarns may be composed of either monofilament or multifilament strands of plastics.

The term *fiber* is used to describe all types of filaments; natural or plastics, monofilament or multifilament. Fibers are first spun or twisted into yarns. They are then woven into the finished fabrics, screening, or other products ready for consumer use.

Synthetic Fibers in Textiles

Many people are unaware of the major role that plastics, in the form of synthetic fibers, play in the textile industry today. Natural fibers such as wool, cotton, flax, and silk have been used throughout history. All commercial fibers used before the twentieth century were of natural origins. Thus, the synthetic fiber industry is relatively new. A popular use of plastics fiber is shown in Figure 5-22.

There are a number of reasons why synthetic textiles have grown in popularity.

- They may be produced year-round with uniform quality.
- They may be tailor-made to size and shape.
- They may be made to simulate leather and fur.
- They may be made to stretch or hold their shape (stretch fabrics and permanent press).
- They are mothproof, nonallergenic, mildew resistant, and free from odor when wet, and thus may be used for carpeting both indoors and out.

Denier. The fineness of a fiber is expressed by a unit called *denier*. One denier equals the mass in grams of 29 527 ft [9 000 m] of fiber. For example, 9 000 m of a 10-denier yarn weighs 0.35 oz. [10 g]. The name of the unit may have evolved from the name for a sixteenth-century French coin that was used as a standard for measuring the fineness of silk fibers.

Plastics fibers of the same fineness vary in denier because of differences in density. Although two filaments may have the same denier, one may

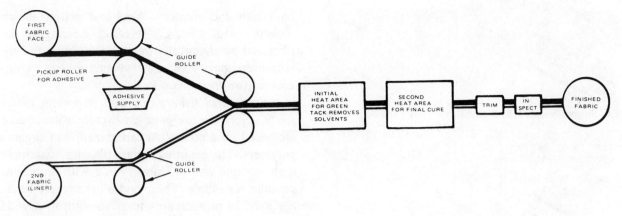

(A) Wet adhesive bonding.

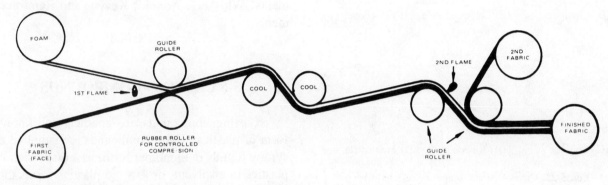

(B) Flame bonding of foam.

Fig. 5-21. Two methods of producing bonded textiles. (*American Fabrics* magazine)

Fig. 5-22. Lifelike wigs made of plastics fibers.

have a larger diameter because of a lower relative density.

To calculate the denier of the filaments in yarn, divide the yarn denier by the number of filaments:

$$\frac{80\text{-denier yarn}}{40 \text{ filaments}} = 2 \text{ denier for each filament.}$$

Remember: 9 000 m of 1-denier filament weighs 1 g.

9 000 m of 2-denier filament weighs 2 g.

Tex. The International Organization for Standards (ISO) has developed a universal system for designating linear density of textiles called the *tex*. The fabric industry has adopted the tex as a measure of linear density. In the tex system the yarn count is equal to the mass of yarn in g/km.

Non-Textile Fibers. Not all of the synthetic filaments are used by the textile industry. Some monofilaments are used for bristles in brooms, toothbrushes, and paint brushes. The relative shapes of these filaments vary, as shown in Figure 5-23.

If the fiber is to be pliable and soft, a fine filament is needed. If the fiber is to withstand crushing and be stiff, a thicker filament is used. Clothing fibers range from 2 to 10 denier per filament. Fibers for carpeting range from 15 to 30 denier in fineness per filament.

The cross-sectional shape of the fiber helps determine the texture of the finished product. Triangular and trilobal shapes impart to the synthetic fiber many of the properties of the natural fiber, silk. Many of the ribbon- and bean-shaped filaments resemble cotton fibers

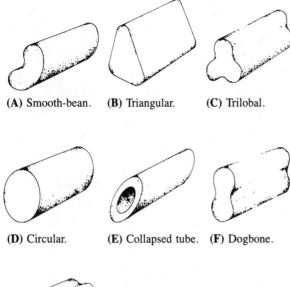

(A) Smooth-bean.   (B) Triangular.   (C) Trilobal.

(D) Circular.   (E) Collapsed tube.   (F) Dogbone.

(G) Lobular.   (H) Ribbon.   (I) Striated-flat.

Fig. 5-23. Cross-sectional shapes of selected fibers.

The future for plastics fibers will no doubt expand the many current uses of plastics. In addition to the manufacture of textiles, these fibers are used in automobile upholstery, marine nets, ropes, screening, carpets, draperies, and flock.

Some of the outstanding characteristics of plastics filaments may be shown by the performance of fibers used in carpeting. Carpets are used in homes, schools, hospitals, churches, hotels, and other areas of severe traffic wear. Carpet fibers are often exposed to wet soil, harsh sunlight, extreme atmospheric conditions, and insects yet they retain their color, resilience, and durability. They are nonallergenic, can be made in any color and many textures, and can be densely woven.

There are two complaints lodged against plastics fibers, flammability and low moisture absorbency. Fabrics made of wool and fiber glass are *fire retardant*. This means that they will burn slowly. Cotton and cellulosic plastics fibers are *flammable or inflammable*. Both terms mean the fabric will burn. Sparks or glowing embers will melt holes in most synthetic fabrics. Very sheer or napped fabrics are more likely to burn than tightly woven ones.

The fact that most plastics fibers absorb only small amounts of moisture has an important bearing on health and comfort. Wool and cotton absorb moisture with ease. By selective wearing of the fiber and by chemically treating the fibers, the absorptive capacity of most synthetics can be increased for greater comfort.

Structural Fibers. Several important fibers are produced for use in polymer composites. Some are based on a controlled carbonization of organic polymers. The resulting carbon fiber has extremely high strength and modulus, along with high temperature resistance. These and other organic fibers are used to provide structural strength to a wide variety of polymer composite products. See Filaments, Whiskers, Aramid, Kevlar, and Reinforcements.

5-6 CASTING COMPOUNDS

Casting compounds are another well-known form of plastics. These compounds are supplied as syrupy liquids in monomer form, as a dispersion of plastics in a solvent, or as solid plastics forms that are melted and cast like candle wax.

Casting of plastics is one of the simplest of all molding techniques. It needs no molding pressure. For this reason casting plastics are very popular as hobby and art material. Castings of unusual or complicated shapes are possible with this material. Glass, rubber, plaster, or metal molds may be used. Many of these molds are used only once.

Liquid Resins

Many plastics may be obtained in syrupy resin form. Polyester, epoxy, acrylic, and polystyrene casting resins are among the most important. Polymerization of these resins is carried out by adding catalysts, heat, or both.

Phenolic castings were part of the early development of the plastics industry. Leo Baekeland introduced numerous cast articles of Bakelite. Today, cast phenolic resins have replaced many ivory and horn items, and are used to produce most of the world's billiard balls.

Large amounts of fillers and reinforcements are generally used in casting polyesters. For example, cultured marble is a product that contains marble dust or calcium carbonate (limestone) as the filler and polyester plastics as the binding material. Some consider this product better than genuine marble. It is used to produce lamp bases, table tops, exterior veneers, statues, and other marble-like products (Fig. 5-24).

Literally thousands of items are made from clear polyester resins. Clear acrylic plastics are seen as cast rods, sheets, and tubing. A monomer of partially polymerized syrupy resin is used for this purpose.

Major uses of all casting resins and plastics have been in embedment, potting, and encapsulation. Polyester and acrylic resins are used to embed objects while polyester and epoxy compounds are sometimes used for potting and encapsulation. (See Casting Processes, Chapter 14 and Coating Processes, Chapter 17.)

Embedments are always placed in a transparent plastics (Fig. 5-25).

Embedding is a process whereby an object is completely encased in a polymer. The liquid resin completely surrounds the object. After polymerization, the casting is removed from the mold.

Objects are embedded for preservation, display, and study. In the biological sciences, animal and plant specimens often are embedded to help preserve them. This allows safe handling of the most fragile sample.

Potting is used to protect electrical and electronic components from harmful environments. The potting process completely encases the desired components in plastics, and the mold becomes part of the product (Fig. 5-26). Vacuum, pressure, or centrifugal force are frequently applied to ensure that all voids are filled with resin.

Fig. 5–24. This one-piece sink is a casting made with filled (marble dust) polyester. Acrylic and polyester resins are used as binders for many products.

Encapsulation is similar to potting. Encapsulation is a solventless covering on electrical components (Fig. 5-27). It does not fill all the voids. This envelope of plastics is usually achieved by dipping the object in the casting resin. After potting, many components are encapsulated.

Hot-Melt Plastics

Potting and encapsulation sometimes employ a second form of casting compounds. These are referred to as *hot-melt* plastics, and are already polymerized and in the solid plastics state. Many of the hot-melt adhesive plastics may be used. Such plastics as cellulose acetate, ethyl cellulose, polyolefins, polyamides, and others can be heated to a syrupy state, and then forced around the electrical components.

Plastisols and Organosols

The last group of casting compounds consists of dispersions of plastics in selected solvents, plasticizers, and other additives. Centrifugal, rotational, slush, and solvent castings are often made with these compounds. Organosols and plastisols are the plastics compounds most commonly associated with these processes.

Organosols are vinyl or polyamide dispersions in organic solvents and plasticizers (Fig. 5-28). *Plastisols* contain only finely ground plastics and plasticizers, and may be nearly 100 percent solids (Fig. 5-29). Organosols may be 50 to 90 percent solids. They contain both plasticizers and varying amounts of solvents. A plastisol may be converted to an organosol by the addition of selected solvents.

Both plastisols and organosols may contain thixotropic materials. When these agents are added, a paste form called a *plastigel* or *organogel* is formed. Plastigels and organogels are easily spread. They are probably associated most often with the plastics coatings on paper and fabrics.

Plastisols may be cast in hollow, open metal molds or in closed rotational molds and are converted to the final product by the addition of heat.

Organosols are cast and the solvents allowed to escape. The dry, unfused plastics is left on the substrate. Heat is then applied to fuse the plastics.

Many modified casting procedures such as dip casting are used by the plastics industry. *Dip casting*

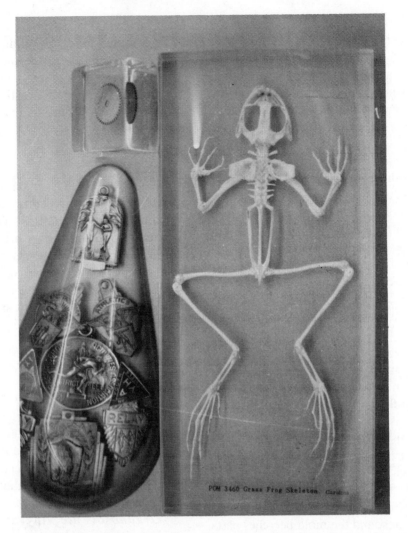

Fig. 5-25. Embedding in transparent polyester plastics is gaining wide use for objects such as classroom science samples.

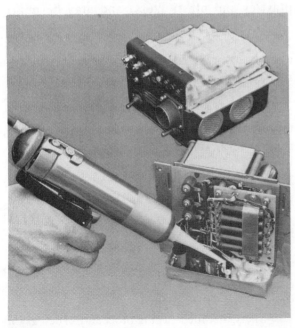

Fig. 5-26. Silicone elastomer is being used to pot an electronic unit. (Dow-Corning Corp.)

is a simple process in which a heated mold is dipped into liquid dispersions of plastics. The plastics melts and adheres to the hot metal surface as the mold is withdrawn. Additional curing in ovens is needed to ensure proper fusion of the plastics particles. After curing, the plastics is peeled from the mold.

5-7 COATINGS

Many plastics are used in the form of *coatings*. The main purpose of a coating is to form a film over the substrate to which it is applied. Coating and casting processes are similar and are sometimes confused. Casting compounds are applied and then *removed* from the molds when polymerized or cooled. Coatings are normally *thin protective layers* on a substrate and are *not removed*.

Every polymer and plastics has been considered for use in coatings. The paint, varnish, and

(A) Transformer being encapsulated with silicone elastomer.

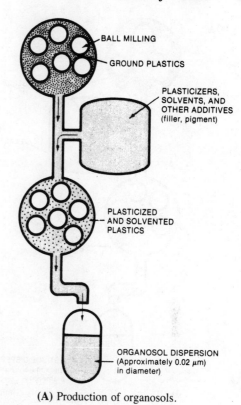

(A) Production of organosols.

(B) Electronic components are encapsulated (epoxy) into convenient modules.

Fig. 5-27. Examples of encapsulation. (Dow-Corning Corp.)

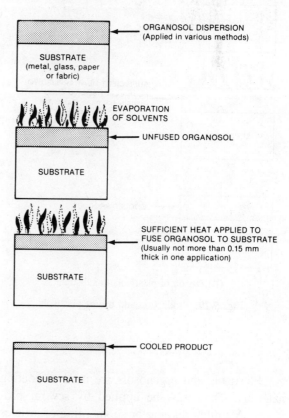

(B) Fusion of organosols on substrate.

Fig. 5-28. Production and use of organosols.

lacquer industries use large quantities of plastics compounds. Plastics-based paint, varnish, and lacquers are generally far superior to their natural counterparts. One-coat house paints, tough transparent varnishes that are both stain and water resistant, and quick-drying finishes are all benefits of the plastics age.

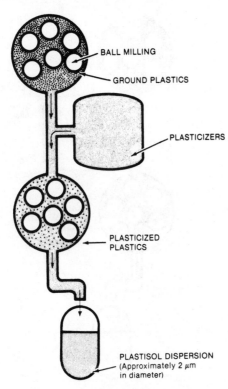

(A) Production of plastisols.

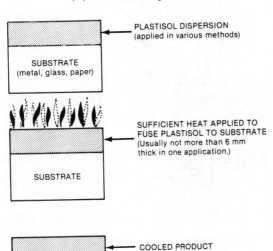

(B) Fusion of plastisols to substrate.

Fig. 5-29. Production and use of plastisols.

Plastisols and organosols are used as coating materials. They may be applied by several processes. A simple dipping or submerging of articles in plastics dispersions is easily done. It requires little equipment.

Many products are coated in this manner. They include coated fabric gloves, dish drainers, tool handles, bobby pins, and other special applications. (See Chapter 17, Coating Processes.)

5-8 CELLULAR OR FOAMED PLASTICS

Cellular, or foamed, rubber has existed for a number of years. Most thermoplastics also may be processed into a cellular form.

There are two basic types of foamed plastics that have cellular structure: (a) *Open-celled foams* are like sponges. The cells are interconnected and can hold liquids. (b) *Closed-celled foams* have cells completely closed by thin walls of plastics. Figure 5-30 shows both open and closed cellular plastics. In the closeup, the plastics on the top is closed-celled, while the one on the bottom is open-celled.

More than ten groups of foamed plastics are made commercially in the United States. Many of these foams may be obtained in rigid, semirigid, or flexible forms. Many may be made flame-retardant.

Foamed polystyrene, with the trade name *Styrofoam,* is probably the best-known of the foams. Nearly all of the foams are waterproof and have low densities. For these reasons, they are used for all types of thermal and acoustical insulation, as cores for sandwich structures, as electrical insulation, and for toys and containers (Fig. 5-31). (See Expansion Processes in Chapter 16.)

Fig. 5–30. Examples of closed-cell PS (top) and open-celled PV (bottom) cellular plastics.

5-9 COMPOSITES

The term *polymer composite* has evolved over a period of years. In general terms, it is a form of plastics.

A composite is composed of two or more materials and has properties that the component materials do not have by themselves. A precise definition of composite must be divided into two basic forms: 1) composite materials, and 2) composite structures. *Composite materials* are composed of a reinforcing structure, surrounded by a continuous polymer matrix. The structure must be capable of arbitrary variation to be considered a composite material. *Composite structures* exhibit a discontinuous matrix, in which the dissimilar materials are not capable of arbitrary variation. Most laminates fall into this category.

In either form, the reinforcements and polymer matrix contribute to a synergistic property change.

Polymer matrix composites may be divided into three major classifications: 1) fibrous, 2) laminar, and 3) particulate.

Fibrous

Fibrous composites are composed of reinforcing fibers in a polymer matrix. This type of composite is often referred to as reinforced plastics (RP) or fiber reinforced plastics (FRP). (See Fig. 5-32) *Reinforced plastics* fall in this classification.

There is some confusion about the use of the term *reinforced plastics*. The term is sometimes used to describe both laminated plastics and molded products in which the reinforcement is not laminated. The word *reinforced* should indicate that an additive or special ingredient (a reinforcement) is present in the molding compound. It does not refer to a specific molding or forming process. (See Fig. 5-33) Throughout this text, we will consider the term *reinforced* to mean a system or form of a composite. The word *reinforcing* will be used to describe the process of strengthening and/or making stronger. All fibrous and particulate composite classes will be placed into the processing category of reinforcing. (See Reinforcing Processes in Chapter 13 and Reinforcements in Chapter 6)

Laminar

Laminar composites are composed of layers of materials held together by the polymer matrix.

Fig. 5–31. Cellular (foamed) polystyrene has excellent thermal insulating qualities. (Sinclair-Koppers Co.)

Fig. 5-32. "Tuff pedals" are made of graphite and acetal homopolymer resin. (Du Pont Co.)

(A) First stage is layup of 32 mm [1¼ in.] thick single skin laminate.

Fig. 5-33. Three stages of construction for the glass-reinforced hull of the HMS *Wilton*. (Plastics Design Forum)

(B) Transverse and longitudinal stiffening frames are bonded in place in the hull.

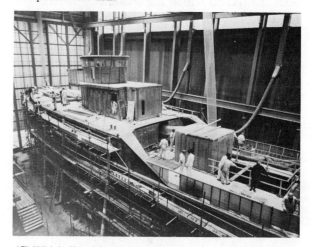

(C) With hull and framing in place, work on the decking and superstructure nears completion.

Laminar composites may be divided into two distinct classes: laminates and sandwiches. Figure 5-34 illustrates these two classes.

The verb *laminate* is used to describe the process of laminating, or bonding two or more layers of materials by either cohesive or adhesive action. (See Fig. 5-35.) These layers of material may function as reinforcing agents (much like the wire screen and steel rods used to strengthen concrete) or as decorative surfaces on furniture or table tops.

Laminate may also be used as an adjective referring to the laminae (layers), or as a noun referring to a laminated product.

The plastics industry also uses the terms to refer to other nonplastics composite laminates. Laminated wooden beams and plywood are familiar examples. Wood is the reinforcing agent and plastics is the matrix bonding agent.

Similar confusion surrounds the use of the terms *laminated* and *laminating*. In this text, the term *laminated* will describe a plastics form or product. *Laminating* will describe a process. All laminar composite classes of materials will be categorized as a laminating process. (See Laminating Processes in Chapter 12)

The plastics resin does not have to impregnate the material for the product to be considered a laminar composite. Sheets of metals, plastics,

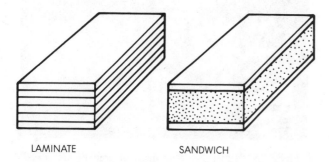

LAMINATE   SANDWICH

Fig. 5–34. Two distinct classes of laminar composites.

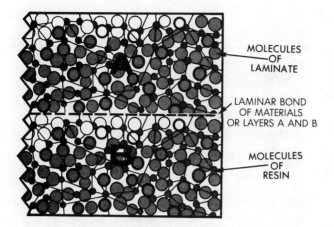

Fig. 5-35. Resin impregnated material.

Fig. 5–37. This prototype turboprop aircraft has an all-composites primary structure. It has the strength of titanium, but is somewhat lighter than aluminum. (Plastics Design Forum)

woods, or fabric are often used to produce special parts with improved properties. Figure 5-36 shows a composite radar dome (radome).

Laminar composites and other materials are used extensively in the production of aircraft structures, helicopter blades, and other aerospace applications. (Fig. 5-37) They are also used as structural materials for automobiles, furniture, bridges, and homes. Thin metallic layers of foil may be clad, or bonded, to layers of paper or fabric. These laminated products are strong, corrosion and electrical resistant printed circuit boards. (Fig. 5-38)

There are many uses of laminates in the textile field. Layers of cloth, plastics film, and foam are bonded together to produce special-purpose materials.

Many laminates are composed of layers of resin-impregnated materials. The pores (interstices)

Fig. 5–38. This circuit board is made of a phenolic high-pressure laminate.

Fig. 5-36. A radome of composite laminates protects the radar antenna of this capsule. (FMC Corporation)

of the material are filled with resin (Fig. 5-38). Once the resin is polymerized, there is a strong cohesive bond formed between layers. The resin has completely impregnated the reinforcing material.

For an impregnated material to be classed as a laminate, it must be composed of two or more resin-soaked layers. Particle board and fiberboard are not laminates. They are resin-impregnated wood or cellulose fibers, pressed into a board or sheet form and are classed as particulate composite.

Formica, Micarta, and Texolite are trade names for a familiar laminate form. These laminates are used as countertops, wall panels, or flooring. They are used where a decorative and durable surface is needed. Most of these laminates are composed of layers of resin-impregnated paper. The surface may be textured or patterned by several methods. It may contain metallic flakes or abrasives, or it may be embossed to simulate wood grain, fabric, or stone.

Sandwiches are defined by the ASTM as combinations of alternating dissimilar simple or composite materials, assembled and intimately fixed in relationship to each other to use the properties of each to specify structural advantages for the whole assembly.

Several possible core and facing materials can be used for sandwiches. As a rule, honeycomb, waffle, and cellular sandwiches are isotropic, with excellent thermal, acoustical, and strength-to-weight ratios. Corrugated core sandwiches are more anisotropic.

Sandwiches hold great promise for the future in housing. Figure 5-39 shows the roof of a prototype four-bedroom home that is mechanically fastened and needs no nails. The panels have rigid polyurethane-foam cores.

Although conventional in appearance, the home in Figure 5-40 was made by unconventional means. The 5 000 square feet (465 square metres) of panels for the walls and roof have an aluminum exterior and polyurethane foam cores. They were produced in a plant, trucked to the site, and assembled.

Many small prefabricated structures are now being made with a honeycomb sandwich or reinforced plastics (Fig. 5-41).

Sandwich construction is nothing new and is beginning to be accepted by the consumer. The mobile home, automobile, and aircraft industries use these sandwich structures. They are very strong for their mass. Figure 5-42 shows a paper honeycomb sandwich material strengthened by fiberglass-reinforced polyester. This sandwich material cuts fabrication and maintenance costs and reduces mass by over 20 percent. It is making inroads

(A) Installing panels.

(B) Sealing panels.

Fig. 5-39. Polyurethane-foamed roof panels require few nails.

into markets once dominated by aluminun, steel, wood, and other materials.

Particulate

Particulate composites consist of small particles dispersed in a polymer matrix. These particles are sometimes divided into *skeletal* and *flake structures*. A *skeletal structure* is composed of a continuous skeletal shape filled by one or more additional materials. *Flakes* describes the flat shape of the reinforcement. Flakes are generally oriented parallel to each other.

Particulates may have any shape, size, or configuration. The distinction between fillers and reinforcing agents will be discussed in Chapter 6.

Metallic flakes are placed in a polymer matrix to improve electrical properties, radiation shield-

(A) Installing wall panels.

(B) Construction of roof.

(C) Front view of completed home.

Fig. 5-40. A prefabricated home using panels of polyurethane foam and aluminum.

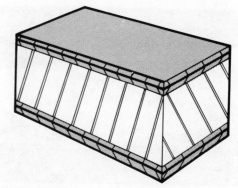

(A) Basic design.

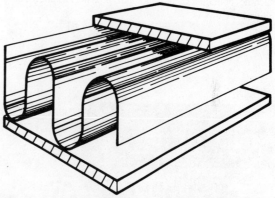

(B) Corrugated paper core.

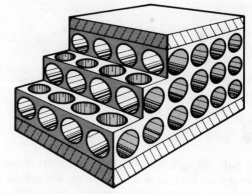

(C) Cellular plastics core.

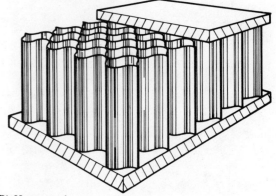

(D) Honeycomb core.

Fig. 5-41. Various types of core construction.

ing, and strength. Computers and other electronic equipment may be protected from the annoying or catastrophic effects of electromagnetic pulse (EMP) or electromagnetic interference (EMI) with the use of metallic particulate composite housings. (Fig. 5-43) Overlapping and touching metal flakes provide electrical conductivity through the composite.

Small particulates and a polymer matrix may be forced into the skeletal structure of cellular or foamed plastics. In one process, the plastics foam structure is washed away with solvents, leaving a light, strong skeletal structure.

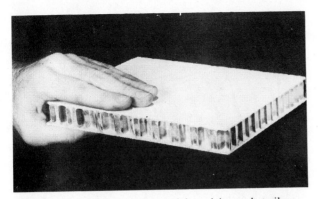

Fig. 5-42. A honeycomb material used in truck trailers, machine containers, and modular housing. (Panel-Comb Industries Corp.)

Fig. 5–43. Metallic particulate (ABS matrix) composite housings are used on many computers and peripheral equipment to minimize EMP and EMI effects. (NC Microproducts)

VOCABULARY

The following vocabulary words are found in this chapter. Look up the definition of any of these words you do not understand as they apply to plastics in the glossary, Appendix A.

Adhesion
Adhesive
ASTM
Cellular
Cement
Closed-cell
Cohesion
Composite
Denier
Electromagnetic interference (EMI)
Electromagnetic pulse (EMP)
Embedment
Encapsulation
Fibers
Filament
Foamed
Glue
Honeycomb
Hot melt
Impregnation
Laminar composite
Laminates
Laminating
Matrix
Molding compounds
Organosol
Open-celled
Particulate
Pellets
Sandwich
Substrate
Tex

5-1. Many of the available forms of plastics begin as ____?____ .

5-2. Pills, tablets, biscuits or premolds are popular names for ____?____ .

5-3. BMC stands for ____?____ .

5-4. Name a substance capable of holding materials together by surface attachment.

5-5. Name the plastics form that is composed of layers of bonded materials.

5-6. The terms reinforcing and laminating should be used to describe a ____?____ .

5-7. SMC stands for ____?____ .

5-8. The terms reinforced and laminated describes plastics ____?____ .

5-9. Open-celled and closed-celled are terms that describe ____?____ .

5-10. Name the process in which a heated mold is dipped into liquid dispersions.

5-11. Name the casting compounds that are vinyl dispersions in solvents.

5-12. Name the nine basic forms of plastics.

5-13. Which plastics form would be used to make large service station signs?

5-14. Which plastics form would be selected to make picture frames?

5-15. Formica, Micarta and Texolite are trade names for what plastics form?

5-16. In the process of ____?____ , there is an intermingling of molecules by chemical action.

5-17. Name four plastics adhesives used to bond wood:

5-18. Films are less than ____?____ mm in thickness.

5-19. Sheets are greater than ____?____ mm in thickness.

5-20. A process where an object is completely enclosed in transparent plastics is called ____?____ .

5-21. The process ____?____ is similar to ____?____ , except complete penetration of all voids in the object are filled before the resin polymerizes or cools.

5-22. Dispersions of plastics and plasticizers are called ____?____ .

5-23. What shortens molding times and helps remove moisture from resins and plastics?

5-24. Name two basic forces which attribute to adhesion:

5-25. Name the major application or use of film?

5-26. Name four products that often have plastics coating on them:

5-27. A term used to describe a cellular plastics in which there is a predominance of noninterconnecting cells is ____?____ .

5-28. The term ____?____ is used to indicate that a reinforcement additive is present in the molding compound.

5-29. Name four materials used with resins to produce composite laminate.

5-30. The mass per volume or relative density of a fiber is measured in a unit called ____?____ .

5-31. What plastics forms would be used in making the following products:
pocket comb
food packages
fingernail polish
aircraft structures

5-32. The term ____?____ is probably derived from the practice of forcing tar-soaked wicking into the seams of wooden boat hulls.

5-33. Most of the dimensional or profile forms of plastics are produced by the ____?____ process.

5-34. The most familiar trade name for foamed polystyrene is ____?____ .

5-35. Very small particles of plastics held in suspension is called a ____?____ .

Ingredients of Plastics

Resins and polymers may be radically altered by copolymerization, crosslinking, orientation, and processing.

In this chapter, you will become familiar with special additives that enhance the properties and characteristics of a polymer. Many improvements are engineered merely by the addition of special ingredients. As you probably know, various metals are often added to iron ore to produce special steel alloys. Similarly, there are many different ingredients that can be added to plastics to provide them with special or improved physical and mechanical characteristics.

Some of the reasons for including additives, reinforcements, and fillers are:

- to improve processability
- to reduce material costs
- to reduce shrinkage
- to permit higher curing temperatures by reducing or diluting reactive materials
- to improve surface finish
- to change the thermal properties such as expansion coefficient, flammability, and conductivity
- to improve electrical properties including conductivity or resistance
- to prevent degradation during fabrication and service
- to provide desirable color or tint
- to improve mechanical properties such as modulus, strength, hardness, abrasion resistance, and toughness
- to lower the coefficient of friction.

6-1 RESINS

Ingredients that may be added to plastics include various resins, a wide variety of fillers, and different kinds of reinforcements.

Resins

The nouns *plastics* and *resin* are often used interchangeably, but such use is incorrect. The resin in a plastics determines whether that plastics is thermosetting or thermoplastic. Resins do not become plastics until polymerization has occurred and the product in the finished state. Also the classification *plastics* does not include the large category of elastomers, such as synthetic rubbers. The term *polymers,* however, includes both plastics and elastomers (see Appendix E).

In a sense, alloys and blends can be considered ingredients of plastics. These are combinations of two or more polymers in which at least two of the polymers are present in concentrations greater than five percent. A plastics alloy is formed when two or more different polymers are physically mixed during the melt. Although there may be some bond attachment between chains, these should not be considered copolymers. The term *polyblend* sometimes refers to plastics which are modified by the addition of elastomers.

One of the major reasons for using additives in plastics is to improve selected properties for a specific product application. Alloys and polyblends may improve chemical resistance, weatherability, impact strength, and thermal properties. Some combinations may be synergistic while others result in reduced property benefits in some performance parameters. Table 6-1 lists selected alloys and blends.

6-2 ADDITIVES

As a rule, additives do not comprise more than ten percent (by weight) of the finished product. Considerable care must be used in selecting and using additives in resins and polymers. Manufacturers and suppliers will provide valuable information about the additives cleared by the FDA for use in food contact applications, about migration or deterioration of the additive over a period of time, and about the toxic or hazardous results of handling an additive or exposing workers to it. (See Chapter 3 on Health and Safety.)

Antioxidants, Antiozonants, and Ultraviolet Absorbers

Antioxidants are stabilizers that slow down or stop oxidation. Polypropylene and polyethylene are particularly susceptible to breakdown through oxidation. (See Stabilizers.)

Antiozonants are additives that help prevent breakdown by ozone gas in the atmosphere.

Polyolefins, polystyrene, polyvinyl chloride, ABS polyesters and polyurethanes are susceptible to ultraviolet solar breakdown.

Solar radiation of polymers may result in crazing, chalking, color changes, or loss of physical, electrical, and chemical properties. This weathering damage is caused when the polymer absorbs light energy. Ultraviolet light is the most destructive portion of the solar radiation striking plastics products. There may be enough energy involved to break chemical bonds between atoms. Polyester, PE, PS, and PMMA use UV stabilizers.

Table 6-1. Selected Alloys and Blends

ABS/PA	PA/elastomer	SAN/EPDM
AS/PC	PA/EMA	PPS/PTFE
ABS/PTFE	PA/EPDM	LDPE/EVA
ABS/PVC	PA/PE	EPDM/EEA
ABS/PSO	PC/PBT/elastomer	EPDM/HDPE/PP
ABS/SMA	PVC/elastomer	PET/PSO
ASA/PC	PVC/CPE	PET/elastomer
ASA/PMMA	PVC/EVA	PBT/PET
PVC/PMMA	SMA/PS	PC/SMA

The most widely used ultraviolet absorbers are 2-hydroxybenzophenones, 2-hydroxyphenyl-benzotriazoles and 2-cyanodiphenyl acrylates. Carbon black may block UV light in some polymers.

Antistatic Agents

Antistatic agents may be compounded into the plastics or applied to the product surface. These agents attract moisture from the air making the surface more conductive, which in turn dissipates static charges.

The most common antistatic agents are amines, quaternary ammonium compounds, organic phosphates, and polyethylene glycol esters. Concentrations of antistatic agents may exceed 2 percent, but application and FDA approval are the prime considerations in their use.

Colorants

Plastics possess a wide range of colors—a property that most other materials cannot exhibit. In fact, plastics have enjoyed a wide variety of uses because they are available in a multitude of colors.

There are four types of colorants used in plastics:

1. Dyes
2. Organic pigments
3. Inorganic pigments
4. Special-effect pigments

One of the basic differences between pigments and dyes is solubility. *Dyes* are more complex and color the material by forming chemical linkages with molecules. They have excellent clarity and optical properties, but poor thermal and light stability. Thousands of dyes used in plastics are synthesized from coal-tar chemicals.

Pigments are not soluble in common solvents or in the resin; therefore, they must be mixed and evenly dispersed within the resin. Migration, or bleeding, of these colorants during processing or during the life of the product may cause serious surface color defects. Color bleeding may stain skin or cause allergic reactions.

Organic pigments provide the most brilliant opaque colors available. Translucent and transparent colors achieved with organic pigments are not as brilliant as those produced with dyes. They are better than those possible with inorganic pigments, however.

Inorganic pigments are often quite simple chemicals, such as carbon (black), iron oxide (red), cobalt oxide (blue), cadmium sulfide (yellow), and lead sulfate (white). These metallic oxides and sulfides are easily dispersed in the resin. They do not produce colors as brilliant as those from organic pigments and dyes, but because of their inorganic structure, they resist light and heat more effectively. Most inorganic pigments are used in high concentrations to produce opaque colored plastics. Low concentrations of iron oxide pigment will produce a translucent color.

Special-effect pigments may be either organic or inorganic compounds. Colored glass is used in a fine powdered form, and is a heat- and light-stable pigment for plastics. Colored glass powder is effective in exterior uses because of its color stability and chemical resistance.

Flakes of aluminum, brass, copper, and even gold may be used to produce a striking metallic sheen. Iridescent plastics are used by the automotive industry in producing metallic finishes. When metallic powders are mixed with a colored plastics, a finish varying in highlights and reflective hues is produced. Pearl essence, either natural or synthetic, may be used to produce a brilliance where pearl luster is desired.

When energy is absorbed by a material, a portion of that energy may be released in the form of light. This light is radiated when the molecules and atoms have their electrons excited to such a state that they begin to lose energy in the form of photons, or light particles. If heat causes the electrons to release photons of light energy, the radiation is called *incandescence*.

When chemical, electrical, or light energy excites electrons, the radiation of light is called *luminescence*. Luminescent materials are often added to plastics for special effects. Luminescence is categorized into *fluorescence* and *phosphorescence* (Fig. 6-1).

Fluorescent materials emit light only when their electrons are being excited. These materials cease to emit light when the energy source exciting their electrons is removed. Fluorescent materials are made from sulfides of zinc, calcium, and magnesium. Fluorescent paint on instrument dials allows a pilot to read instruments with little visible light being emitted. Other uses for fluorescent materials include hunting jackets, protective helmets, gloves, life preservers, rain slickers, bicycle stripes, and road warning signs.

(A) Illuminated signs.

(B) Nonilluminated signs.

Fig. 6-1. Phosphorescent pigments glow in the dark after exposure to light.

Phosphorescent pigments possess an afterglow; that is, they continue to emit light for a limited time after the exciting force has been removed. The most common example of phosphorescence is the television picture tube which emits light when electrical energy excites the phosphorescent materials coating the inside of the face of the tube. Phosphorescent pigments used in plastics and paints are made from calcium sulfide or strontium sulfide.

Mesothorium and radium compounds are radioactive materials sometimes used for special luminescence. Note that there may be harmful effects from prolonged exposure to radioactive materials.

Because it is often necessary to mix colorants to achieve the desired shade of color hue, a chart of primary and secondary colors (Fig. 6-2) is of value. By mixing the primary colors yellow and blue, a secondary color, green, is produced. Yellow and green will yield a third color, yellow-green.

Coupling Agents

Coupling agents are sometimes called *promotors* (not to be confused with promoters). They are especially important in processing composites.

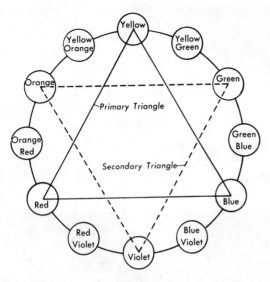

Fig. 6-2. Primary and secondary pigment colors on color chart.

Coupling agents are used as surface treatments to improve the interfacial bond between the matrix, reinforcements, fillers, or laminates. Without this treatment, many resins and polymers do not want to adhere to reinforcements or other substrates. Good adhesion is essential if the polymer matrix is to transfer stress from one fiber, particle, or laminar substrate to the next. Silane and titanate coupling agents are commonly used.

Curing Agents

Curing agents are a group of chemicals that cause cross linking. These chemicals cause the ends of the monomers to join, forming long polymer chains and crosslinkages.

Because resins may be partially polymerized systems (e.g. B-stage resins), other forms of energy may cause premature polymerization. *Inhibitors* (stabilizers) may be used to prolong storage and block polymerization.

Catalysts, sometimes called hardeners (more correctly initiators), are the chemicals which help cause the monomers to join and/or crosslink. Organic peroxides are used to polymerize and crosslink thermoplastics (PVC, PS, LDPE, EVA, and HDPE) as well as the familiar thermosetting polyesters.

The most widely used initiators are unstable peroxides or azo compounds. Benzoyl peroxide and methyl ethyl ketone peroxide are widely used organic initiators.

When catalysts are added, polymerization begins. Catalysts are little influenced by the inhibitors in the resin. As organic peroxides are added to poly-

ester resin, the polymerization reaction begins and yields exothermic heat. This formation of heat further speeds up the cross linking and polymerization.

Promoters (or *accelerators*) are additives that react in a manner opposite that of inhibitors, and they are often added to resins to aid in polymerization. Promoters react only when the catalyst is added. This reaction, which causes the polymerization, produces heat energy. A common promoter used with the catalyst methyl ethyl ketone peroxide is cobalt naphthanate. All promoters and peroxides should be handled with caution. *WARNING:* Peroxides may cause skin irritation and acid burns. If promoters and catalysts are added at the same time, a violent reaction may occur. Always thoroughly mix in the promoter, then add the desired amount of catalyst to the resin. Be sure that there is adequate ventilation and use personal protective wear.

Resins that have not been *prepromoted* generally have a longer shelf life. Remember that other forms of energy also cause polymerization. Heat, light or electrical energy may initiate this reaction.

Always store curing agents at their recommended storage temperature and in their original containers.

Flame Retardants

Flame retardants are often added to the basic resin. Most commercial flame-retarding chemicals are based on combinations of bromine, chlorine, antimony, boron, and phosphorus. Many of these retardants emit a fire-extinguishing gas (halogen) when heated. Others react by swelling or foaming, forming an insulation barrier against heat and flame (Fig. 6-3).

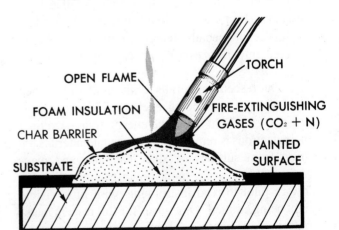

Fig. 6-3. This protective finish swells to form an insulating char barrier when heated. It also emits a fire-extinguishing gas to retard burning.

A char barrier helps to insulate and reduce the combustible material available. Some retardant additives create a cooling reaction by releasing water when heated.

According to the National Fire Prevention Association, more than ten billion dollars worth of damage is caused annually by fires. Generally, the fire record of plastics used for construction has been good. Continued expansion and growth of plastics in construction will depend on reducing flammability of materials and on well-defined fire test standards and building codes.

Foaming/Blowing Agents

The terms *foaming, blowing, frothed, cellular,* and *bubble* are sometimes used to cover a wide variety of compounds and processing techniques to make polymers with a cellular structure. (See Expansion Processes)

Chemical blowing agents such as azodicarbonamide are widely used to produce cellular HDPE, PP, ABS, PS, PVC, and EVA. This chemical decomposes at a specified temperature, released gases form the cellular structure.

Heat Stabilizers

Heat stabilizers are additives which retard the decomposition of a polymer caused by heat, light energy, oxidation, or mechanical shear. Metal salts of calcium/zinc are used in some PVC packaging applications. PVC, HDPE, PP, ABS, and PS may be stabilized with organophosphites.

Impact Modifiers

One or more monomers (usually elastomers) in varying amounts may be added to rigid plastics to improve (modify) impact properties, melt index, processibility, surface finish, and weather resistance. PVC is toughened by modification with ABS, CPE, EVA, or other elastomers. (See Alloys, Blends, Ethylene-ethyl acrylate, and Styrene-butadiene.)

Lubricants

Lubricants are needed ingredients for plastics. During the making of polymers, lubricants are added for three basic reasons. First, they help get rid of some of the friction between the resin and the manufacturing equipment. Second, lubricants aid in emulsifying other ingredients and provide internal lubrication for the resin. Third, some lubricants prevent the plastics from sticking to the mold surface during processing. After the products are taken from the mold, lubricants may exude from the plastics and prevent the products from adhering to each other, and may provide a nonsticking or slippery quality to the plastics surface.

Many different lubricants are used as ingredients in plastics. For example, waxes such as montan, carnauba, paraffin, and stearic acid, and metallic soaps such as the stearates of lead, cadmium, barium, calcium and zinc are used as lubricants (Table 6-2). Most of the lubricant is lost during the process of manufacturing the resin. Excess lubricant may slow polymerization or cause a

Table 6-2. Lubricants Chart

Plastics	Alcohol Esters	Amide Waxes	Complex Esters	Comb. Blends	Fatty Acids	Glycerol Esters	Metallic Stearates	Paraffin Waxes	Poly-ethylene Waxes
ABS		X				X	X		
Acetals	X			X					
Acrylics	X				X				
Alkyd							X		
Cellulosics	X	X				X	X		
Epoxy				X			X		
Ionomers		X							
Melamines				X			X		
Phenolics				X		X	X		
Polyamides	X				X				
Polyester				X		X	X		
Polyethylene		X							
Polypropylene		X				X	X		
Polystyrene		X		X	X		X		
Polyurethanes				X					
Polyvinyl chloride	X	X	X	X	X	X	X	X	X
Sulfones				X					

lubrication bloom seen as an irregular, cloudy patch on the plastics surface.

Some plastics exhibit nonstick and self-lubrication properties. Examples are fluorocarbons, polyamides, polyethyene, and the silicone plastics. They are sometimes used as lubricants in other polymers.

Remember, all additives must be carefully selected for toxic effects and desired service use.

Plasticizers

Plasticity refers to the ability of a material to flow or become fluid under force. A plasticizer is a chemical agent added to plastics to increase flexibility, reduce melt temperature, and lower viscosity. All of these properties aid in processing and molding.

Plasticizers act much like solvents by lowering the resin viscosity. They also act like lubricants by allowing slip to occur between molecules. Remember, van der Waals' bonds are only physical attractions and not chemical bonds, and plasticizers help neutralize most of these forces. Plasticizers, much like solvents, produce a more flexible polymer. They are not designed to evaporate from the polymer during normal service life, however.

Plasticizer leaching or loss is an important consideration. It is undesirable when in contact with food, pharmaceuticals or other products for consumption. Leaching and degassing may cause PVC hoses, upholstery, and other products to become stiff or brittle and to crack.

For best results, the plasticizer and polymer must have similar solubility parameters.

Camphor, the material Hyatt added to cellulose nitrate, may be considered a plasticizer. It made that plastics more moldable, more flexible, and less explosive.

Over five hundred different plasticizers are formulated to modify polymers. One of the most widely used plasticizers is dioctyl phthalate. Plasticizers are vital ingredients in plastics coatings, extrusions, moldings, adhesives, and films.

Some plastics are listed in Table 6-3.

Preservatives

Elastomers and heavily plasticized PVC are most susceptible to attack by microorganisms, insects, or rodents. As moisture stays or condenses on shower curtains, automobile tops, pool liners, water bed liners, cable coatings, tubing, etc., microbiological deterioration may occur. Antimicrobials, mildewicides, fungicides, and rodenticides may be used to provide the necessary protection in many polymers. The EPA and FDA carefully regulate the use and handling of all antimicrobials.

Processing Aids

There are a variety of additives used to improve processing behavior, increase production rates, or improve surface finish.

Antiblocking agents such as waxes exude to the surface and prevent two polymer surfaces from adhering. *Emulsifiers* are used to lower the surface tension between compounds. They act as detergents and wetting agents. Wetting agents used to lower viscosity are called *viscosity depressants*. They are used in plastisol compounds to assist in processing heavily filled materials or those that become too thick with age.

Solvents are added to resins for several reasons. Many natural resins are very viscous or hard;

Table 6-3. Compatibility of Selected Plasticizers and Resins

Plasticizer	Polyvinyl Acetate	Polyvinyl Chloride	Polyvinyl Butyral	Polystyrene	Cellulose Nitrate	Cellulose Acetate	Cellulose Acetate Butyrate	Ethyl Cellulose	Acrylic	Epoxy	Urethane	Polyamide
Butyl benzyl phthalate	C	C	C	C	C	P	C	C	C	C	C	C
Butyl cyclohexyl phthalate	C	C	C	C	C	P	C	C	C	C	C	C
Butyl decyl phthalate	I	C	C	C	C	I	C	C	C	P	C	P
Butyl octyl phthalate	I	C	P	C	C	I	C	C	C	P	C	C
Dioctyl phthalate	I	C	P	C	C	I	C	C	C	I	C	C
Diphenyl phthalate	C	C	C	C	C	I	P	C	C	P	P	P
Cresyl diphenyl phosphate	C	C	C	P	C	C	C	C	C	C	C	C
Methyl phthaly ethyl glycollate	C	P	C	C	C	C	C	C	C	C	C	I
N-Cyclohexyl-p-toluenesulfonamide	C	I	C	P	C	P	C	C	C	P	P	C
N-Ethyl-o,p-toluenesulfonamide	C	I	C	P	C	C	C	C	C	P	C	C
o,p-Toluenesulfonamide	C	I	C	P	C	C	C	C	C	P	C	C
Chlorinated biphenyls	C	P	C	C	C	I	C	C	P	C	C	C
Chlorinated paraffins	C	P	P	C	P	I	P	C	P	P	C	C
Didecyl adipate	I	C	I	C	C	I	C	C	I	I	P	C
Dioctyl adipate	I	C	C	C	C	I	C	C	I	I	P	C
Dioctyl azelate	I	C	P	C	C	I	C	C	P	I	P	C
Dioctyl sebacate	I	C	P	C	C	I	P	C	I	I	P	C

Notes:
C—Compatible
P—Partially compatible
I—Incompatible

therefore, they must be diluted or dissolved before processing. Resinous varnish and paints must be thinned with solvents for proper application.

Solvents may be considered a processing aid.

In solvent molding, the solvent holds the resin in solution while it is being applied to the mold. The solvent rapidly evaporates, leaving a layer of plastics film on the mold surface.

Solvents dissolve many thermoplastics; therefore, they are used for both identification and cementing purposes. Solvents are also useful for cleaning resins from tools and instruments.

Benzene, toluene, and other aromatic solvents will dissolve the natural oils of the skin. All chlorinated solvents are potentially toxic. Avoid breathing the fumes or allowing skin contact when using plastics additives. (See Chapter 3, Health and Safety.)

Stabilizers

Stabilizers include antioxidants, antiozonants, and ultraviolet absorbers.

Products must maintain their physical and chemical properties during normal service life, but there are a number of factors that may alter their properties. Degradation may be caused by heat or light energy, oxidation, or mechanical shear. If enough energy or force is applied, the chemical and physical bonds of the plastics may be destroyed. To help prevent this breakdown of resins and plastics, stabilizers are added during manufacture to prevent discoloration and decomposition during the storage (shelf) life and service life of the polymer.

Liquid or powder stabilizers are added to the resin to absorb energy, transfer energy to other molecules, or actually screen out harmful ultraviolet light rays. If chemical bonds are broken in the polymer, further reactions such as crosslinking and oxidation may occur.

Nearly all formulations of plastics must contain a small proportion of stabilizer to allow conventional processing. When a plastics is *plasticized,* or caused to become fluid by heat and pressure, thermal degradation may be stopped effectively with proper stabilizers.

There are many stabilizer types to consider when formulating the basic resin. Liquid or powder stabilizers must not be toxic if used in contact with food products. Food additive laws have imposed standards for plastics packaging of food products, and regulatory agencies including the U. S. Food and Drug Administration, the National Sanitation Foundation, and the Meat Inspection Division of the U. S. Department of Agriculture enforce the laws.

6-3 REINFORCEMENTS

Much of the strengthening of polymers to this point has been based on the branching, crosslinking, and other chain stiffening mechanisms controlled by the polymer chemist. Some important processing and designing techniques for strengthening polymer products will be discussed in later chapters.

Reinforcements are ingredients added to resins and polymers. These ingredients do not dissolve in the polymer matrix, and thus form composites. Among many reasons for adding reinforcements, one important reason is to produce dramatic improvements in the physical properties of the composite.

Reinforcements are often confused with fillers. Fillers, however, are only small particles and contribute only slightly to strength. Reinforcements, on the other hand, are ingredients that increase strength, impact resistance, and stiffness. One major reasaon for confusing the two is that some materials, asbsetos and glass, for example, may act as fillers, as reinforcements, or both.

Reinforcements may be divided into three major groups of materials: 1) fibrous, 2) particulate, and 3) lamina. These represent a diverse group of materials.

There are six general variables which influence the properties of the reinforced (fibrous, particulate, or lamina) composite materials and structures.

1. **Interface bond between matrix and reinforcements:** The matrix functions to transfer most of the stress to the (much stronger) reinforcements. In order to accomplish this task, there must be an excellent adhesion between the matrix and the reinforcement.
2. **Properties of the reinforcement:** It is assumed that the reinforcement is much stronger than the matrix. The actual properties of each reinforcement may vary by composition, shape, size, and number of defects. The production, handling, processing, surface enhancement, or hybridization can also determine properties for each type of reinforcement.
3. **Size and shape of the reinforcement:** Some shapes and sizes may help provide superior handling, loading, processing, packing orientation, or

adhesion in the matrix. Some fibers are so small that they are handled in bundles, while others are woven into cloth. Particulates are more likely to be randomly distributed than long fibers.

4. **Loading of the reinforcement:** Generally, mechanical strength of the composite depends on the amount of reinforcing agent it contains. A part containing 60 percent reinforcement and ten percent resin matrix is almost six times stronger than a part containing the opposite amounts of these two materials. Some glass filament wound composites may have up to 80 percent (by weight) loading by unidirectional orientation of the filament. Most reinforced thermoplastic composites contain less than 40 percent (by weight) reinforcements. (See Molding Techniques)

5. **Processing technique:** Some processing techniques allow the reinforcements to be more carefully aligned or oriented. These products are more anisotropic. During processing, reinforcements may be broken or damaged, resulting in lower mechanical properties. Depending on processing technique, particulate reinforcements and short fibers are more likely to have random rather than oriented placement in the matrix. This will result in an isotropic composite component.

6. **Alignment or distribution of the reinforcement:** Alignment or distribution of the reinforcement allows versatility in composites. The processor can align or orient the reinforcements to provide directional properties. In Figure 6-4, the parallel (anisotropic) alignment of continuous strands provided the highest strength; bidirectional (cloth) alignment provides a middle strength range, and random (mat) or isotropic alignment gives the lowest.

The basic structural element of laminar composites is the *lamina*. Lamina may be unidirectional fibers, woven cloth, or sheets of material. Because the individual layers act as a reinforcement, they may be included as an ingredient or additive. These laminar layers are more than just a processing technique. (See Laminating) The lamina selection, alignment, and composition constitute the performance properties of the laminar composite. (See Sandwich Composites) It should become clear that alignment of the reinforcements is the key to designing the composite with anisotropic or isotropic properties. As a rule, if all the reinforcements are placed parallel to each other (0° lay-up), the composite will be isotropic. (See Pultrusion) The exception is on mandrel or tank construction, where the composite part has anisotropic properties. The directional tensile strength properties achieved with different fiber reinforcement alignment is illustrated in Figure 6-5. The random chopped strand mat provides equal strength properties in all directions. The unidirectional fiber alignment has the highest strength parallel to (or in the direction of) the fiber. As this angle varies from 0° to 90°, the strength varies proportionally. Remember that the matrix must securely adhere to the reinforcement and prevent the reinforcement from buckling to transfer the applied stress.

There are six fibrous classes:

- glass
- carbonaceous
- polymer

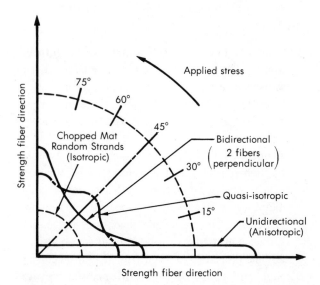

Fig. 6-4. Effect of alignment or distribution of the reinforcement.

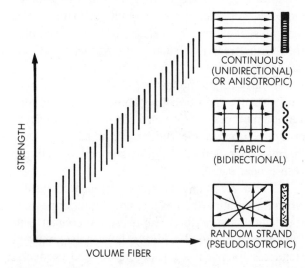

Fig. 6-5. Strength relation to reinforcement alignment and volume of fiber.

- inorganic
- metal
- hybrids

Glass Fibers

One of the most important reinforcing materials is fibrous glass (Tables 6-4 and 6-5). Thermosetting and thermoplastic resins make use of fibrous glass reinforcements.

Glass fibers are produced by several different methods. One common method of production involves pulling a strand of molten glass after it has been formed by a small orifice. The diameter of the strand is controlled by the pulling action.

Thje major constituent of glass is silica, but other ingredients allow the production of many types of fibrous glass. The most common type is E glass fiber, which has good electrical (E) properties and high strength. For chemical resistance, C glass is used. Both E and C glass have tensile strength exceeding 493 183 psi [3.4 GPa]. For low dielectric constant and density, D glass is used. Radomes are typical electronic applications.

For radiation protection, L glass contains lead oxide. For high strength uses, S glass is selected. It is about 20% stronger and stiffer than E-glass.

This glass has a tensile strength of more than 696 258 psi [4.8 GPa]. A high-modulus glass is M glass.

Figure 6-6 shows a reinforced plastics gear housing. It is lighter and stronger than the metal housing it replaces. A glass-filled diallyl phthalate drum pod and cap for a coffee vending machine is shown in Figure 6-7. Brewing can be performed in

Table 6-4. Properties of Thermoplastics: Unreinforced and Reinforced Materials

Property	Polyamide		Polystyrene*		Polycarbonate		Styrene-Acrylonitrile†		Polypropylene		Acetal		Linear Polyethylene	
	U	R	U	R	U	R	U	R	U	R	U	R	U	R
Tensile strength, MPa	82	206	59	97	62	138	76	124	35	46	69	86	23	76
Impact strength, notched, J/mm														
At 22.8 °C	0.048	0.202	0.016	0.133	0.106§	0.213§	0.024	0.160	0.069–0.112	0.128	3.20	0.160	—	0.240
At −40 °C	0.032	0.224	0.010	0.170	0.080§	0.213§	—	0.213	—	0.133	—	0.160	—	0.266
Tensile strength, MPa	2.75	—	2.75	8.34	2.2	11.71	3.58	10.34	1.37	3.10	2.75	5.58	0.82	6.20
Shear strength, MPa	66	97	—	62	63	83	—	86	33	34	65	62	—	38
Flexural strength, MPa	79	255	76	138	83	179	117	179	41 to 55	48	96	110	—	83
Compressive strength, MPa	34††	165	96	117	76	130	117	151	59	41	36	90	19 to 24	41
Deformation (27.58 MPa), %	2.5	0.4	1.6	0.6	0.3	0.1	—	0.3	—	6.0	—	1.0	—	0.4‡
Elongation, %	60.0	2.2	2.0	1.1	60–100	1.7	3.2	1.4	>200	3.6	9–15	1.5	60.0	3.5
Water absorption in 24 hr, %	1.5	0.6	0.03	0.07	0.3	0.09	0.2	0.15	0.01	0.05	0.20	1.1	0.01	0.04
Hardness, Rockwell	M79	E75 to 80	M70	E53	M70	E57	M83	E65	R101	M50	M94	M90	R64	R60
Relative density	1.14	1.52	1.05	1.28	1.2	1.52	1.07	1.36	0.90	1.05	1.43	1.7	0.96	1.30
Heat distortion temp. (at 1.82 MPa), °C	65.6	261	87.8	104.4	137.8	148.9	933	107	683	137.8	100	168.6	52.2	126.7
Coef. of thermal expansion, per °C × 10^{-6}	90	15	60	35	60	15	60	30	70	40	65	30	85	25
Dielectric strength (short time), V/mm	15,157	18,898	19,685	15,591	15,748	18,976	17,717	20,276	29,528	—	19,685	—	—	23,622
Volume resistivity, ohm-cm × 10^{15}	450	2.6	10.0	36.0	20.0	1.4	10^{16}	43.5	17.0	15.0	0.6	38.0	10^{15}	29.0
Dielectric constant at 60 Hz	4.1	4.5	2.6	3.1	3.1	3.8	3.0	3.6	2.3	—	—	—	2.3	2.9
Power factor at 60 Hz	0.0140	0.009	0.0030	0.0048	0.0009	0.0030	0.0085	0.005	—	—	—	—	—	0.001
Approximate cost, ¢/cm³	0.256	0.70	0.04	0.21	0.31	0.56	0.08	0.30	0.05	0.18	0.28	0.67	0.06	0.26

Notes: Columns marked "U" unreinforced, "R" reinforced. *Medium-flow, general-purpose grade. †Heat-resistant grade. §Impact values for polycarbonates are a function of thickness. ‡6.8 mPa load. ††At 1% deformation. Source: *Machine Design*, Plastics Reference Issue.

Table 6-5. Properties of Thermosetting Plastics: Glass-Fiber Reinforced Resins

Property	Base Resin				
	Polyester	Phenolic	Epoxy	Melamine	Polyurethane
Molding quality	Excellent	Good	Excellent	Good	Good
Compression molding					
Temperature, °C	76.7–160	137.8–176.7	148.9–165.6	137.8–171.1	148.9–204.4
Pressure, MPa	1.72–13.78	13.78–27.58	2.06–34.47	13.78–55.15	0.689–34.47
Mold shrinkage, mm/mm	0.0–0.05	0.002–0.025	0.025–0.05	0.025–0.100	0.228–0.762
Relative density	1.35–2.3	1.75–1.95	1.8–2.0	1.8–2.0	1.11–1.25
Tensile strength, MPa	173–206	35–69	97–206	35–69	31–55
Elongation, %	0.5–5.0	0.02	4	—	10–650
Modulus of elasticity, Pa	0.55–1.38	2.28	2.09	1.65	—
Compressive strength, MPa	103–206	117–179	206–262	138–241	138
Flexural strength, MPa	69–276	69–414	138–179	103–159	48–62
Impact, Izod, J/mm	0.1–0.5	0.5–2.5	0.4–0.75	0.2–0.3	No break
Hardness, Rockwell	M70–M120	M95–M100	M100–M108	—	M28–R60
Thermal expansion, per °C	$5–13(\times 10^{-4})$	4×10^{-4}	$2.8–7.6(\times 10^{-4})$	3.8×10^{-4}	$25–51(\times 10^{-4})$
Volume resistivity					
(at 50% RH, 23 °C), ohm-cm	1×10^{14}	7×10^{12}	3.8×10^{15}	2×10^{11}	$2 \times 10^{11}–10^{14}$
Dielectric strength, V/mm	13 780–19 685	5 512–14 567	14 173	6 693–11 811	12 992–35 433
Dielectric constant					
At 60 Hz	3.8–6.0	7.1	5.5	9.7–11.1	5.4–7.6
At 1 kHz	4.0–6.0	6.9	—	—	5.6–7.6
Dissipation factor					
At 60 Hz	0.01–0.04	0.05	0.087	0.14–0.23	0.015–0.048
At 1 kHz	0.01–0.05	0.02	—	—	0.043–0.060
Water absorption, %	0.01–1.0	0.1–1.2	0.05–0.095	0.9–21	0.7–0.9
Sunlight (change)	Slight	Darkens	Slight	Slight	None to slight
Chemical resistance	Fair*	Fair*	Excellent	Very good†	Fair
Machining qualities	Good	—	Good	Good	Good

*Attacked by strong acids or alkalies. †Attacked by strong acids. Source: *Machine Design*, Plastics Reference Issue.

a matter of seconds, since the resin-molded drum can withstand extreme heat. The drum also resists staining, which can affect the coffee flavor.

Rovings are long strands of fibrous glass that may be easily cut and applied to resins. A roving is made up of many strands of glass resembling a loosely twisted or stranded rope. (Fig. 6-8) and *chopped strands* (Fig. 6-9) are among the least costly forms of glass reinforcement. Chopped strands range in length from $\frac{1}{8}$ to 2 in. [3 to 50 mm]. Figure 6-10 shows the production of chopped strands from rovings.

Milled fibers are less than $\frac{1}{16}$ in. [1.5 mm] in length, and are produced by hammer milling glass strands (Fig. 6-11). Milled fibers are added to a resin as a premix to increase viscosity and product strength (Fig. 6-12).

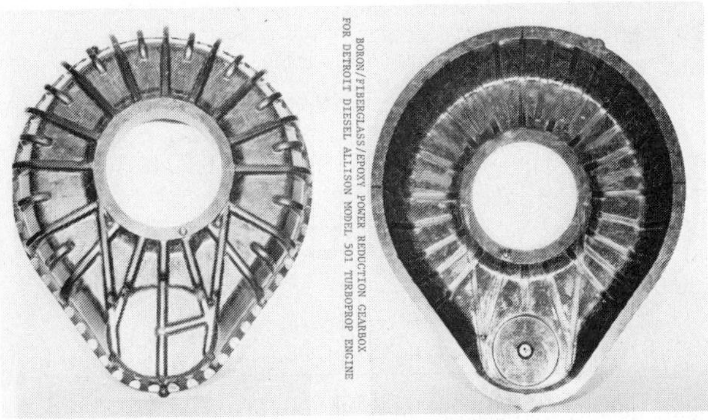

Fig. 6-6. This boron-epoxy gear housing has replaced a metal housing. (Allison Division, Detroit Diesel)

Fig. 6-7. A drum pod and cap for a coffee vending machine, made from glass-filled diallyl phthalate. (FMC Corporation)

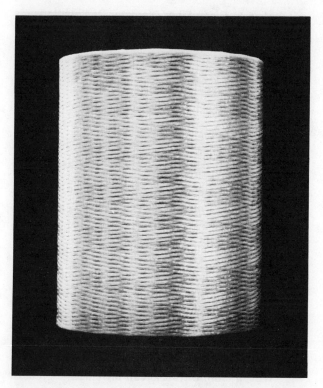

Fig. 6-8. Fibrous glass roving. (Reichhold)

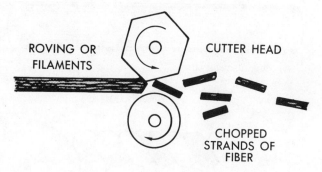

Fig. 6-10. Production of chopped glass strands.

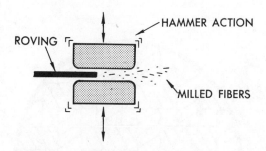

Fig. 6-11. Production of milled glass fibers.

Fig. 6-12. Milled glass fibers may be used as reinforcements or as fillers. (Reichhold)

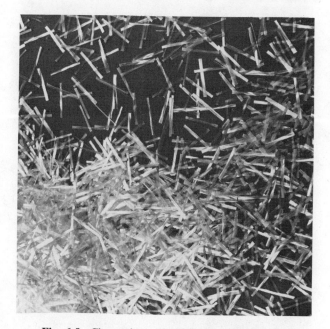

Fig. 6-9. Chopped strands of fibrous glass. (PPG)

would be:

E = Electrical glass

C = Continuous filament

G = filament diameter of 9 μm (See Table 6-6)

150 = 1/100 of the total approximate bare yardage in a pound or 1500 yards

2/2 = single strands were twisted and two of the twisted strands were plied together (S or Z may be used to designate the type of twist)

Yarns resemble rovings but are twisted like a rope (Fig. 6-13). Reinforcing yarns are used in the fabrication of large liquid tank containers.

Fiber glass yarn product nomenclature is based on a letter-number system. (See Fig. 6-14) For example, a yarn designated as ECG 2/2 2.8

(A) Monofilament yarn.

(B) Multifilament yarn.

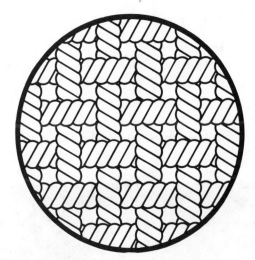

(C) Woven yarn fabric.

Fig. 6-13. Yarns.

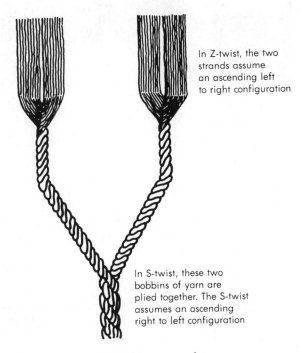

In Z-twist, the two strands assume an ascending left to right configuration

In S-twist, these two bobbins of yarn are plied together. The S-twist assumes an ascending right to left configuration

Fig. 6-14. Yarn nomenclature.

Table 6-6. Glass Fiber Diameter Designations

Filament Designation	Filament Diameter in μm	(inches)
C	4.50	0.000 175
D	5.00	0.000 225
DE	6.00	0.000 250
E	7.00	0.000 275
G	9.10	0.000 375
H	11.12	0.000 425
K	13.14	0.000 525

2.85 = number of turns per inch in the twist of final yarn with S twist

There were two basic strands in the yarn and two twisted strands. Thus,

$$\frac{15\,000}{2 \times 2} = 3\,750 \text{ yards per pound of yarn}$$

Reinforcing *mats* are nondirectional pieces of chopped strands. They are held together by a resinous binder or by mechanical stitching called *needling* (Fig. 6-15).

Woven *cloth* or fabric can provide the greatest physical strength of all the fibrous forms but is

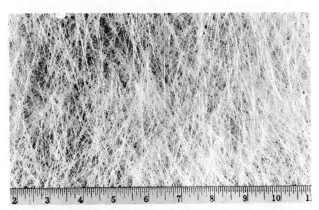

(A) Resin-bonded.

(B) Stitched (needled).

Fig. 6-15. Fibrous glass mats. (Owens-Corning Fiberglas Corp.)

about 50 percent more costly than other forms. Standard rovings may be woven into fabric form (known as woven roving) and used for thick reinforcements.

There are several types of woven glass fabrics. Glass fiber yarns are woven into several basic patterns as shown in Fig. 6-16. Figure 6-17 shows three different forms of fibrous glass reinforcements.

The *warp* yarns run in the direction of the fabric (lengthwise) and the filler or *pick* (weft) yarns generally run crosswise or at right angles to the warp. Graphite, boron, or hybrid fibers may be used in combinations with glass to design special performance properties.

Tapes and three-dimensional (stocking shaped) forms are also produced from fibers.

A hollow glass fiber nearly 40 percent lighter than solid fibers is produced for special uses.

Carbonaceous Fibers

Carbon fibers are usually made by oxidizing, carbonizing and graphitizing an organic fiber.

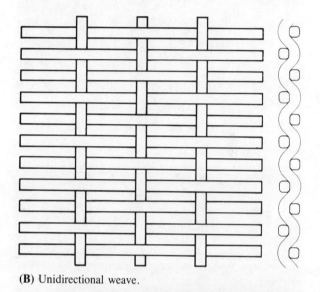

(B) Unidirectional weave.

(A) Plain (square) weave (cloth).

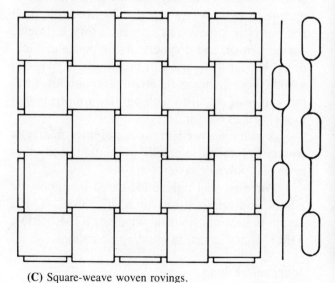

(C) Square-weave woven rovings.

Fig. 6-16. Weave patterns and roving. (Owens-Corning Fiberglas Corp.)

(D) Multifilament wound or twisted roving used in making heavy woven fibrous glass products. (Owens-Corning Fiberglas Corp.)

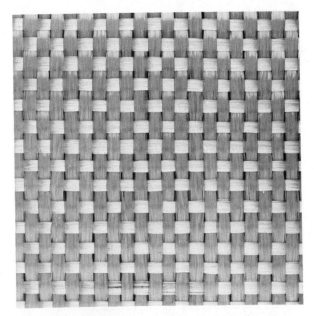

(A) Woven roving.

(B) Fine strand mat.

Fig. 6-17. Some of the many forms of fibrous glass reinforcements. (PPG)

(C) Combination woven roving and mat.

Rayon and polyacrylonitrile (PAN) are currently used. Pitch fibers can also be produced directly from oil and coal. Ordinary pitches are isotropic in character and must be oriented to be useful as reinforcing agents. The terms *carbon* and *graphite* are used interchangeably, but there is a distinction. Carbon (PAN) fibers are about 95% carbon, while graphite fibers are graphitized at much higher temperatures and result in a carbon elemental analysis of 99%.

Once the organic materials have been driven off (pyrolized and stretched into filaments), the result is a high-strength, high-modulus, low-density fiber.

Polymer Fibers

For years cotton and silk filaments were used as reinforcements in belting, tires, gears, and other products. Synthetic polymers of polyester, PA, PAN, PVA, CA, and others are currently used. Kevlar aramid is an aromatic polyamide polymer fiber with nearly twice the stiffness and about half the density of glass. Kevlar is a trademark of DuPont. Aramid is the generic name for a series of Kevlar fibers. Unlike carbon fibers, Kevlar fibers do not conduct electricity, nor are they electrically opaque to radio waves. Kevlar 29 fibers are used for ballistic protection, ropes, army helmets, coated fabrics, and a variety of composite applications. Kevlar 49 is used in boat hulls, flywheels, v-belts, hoses, composite armor, and aircraft structures. It is of equal strength but has a much higher modulus than Kevlar 29.

A common high-strength polymer matrix is epoxy. Polyesters, phenolic polyimide, and other resin and polymer systems are used.

Polyester and polyamide-based thermoplastic fibers find applications in BMC, SMC, TMC, layup, pultrusion, filament winding, RTM, RRIM, TERTM, and injection molding operations.

Inorganic Fibers

Inorganic fibers are a class of short crystalline fibers. They are sometimes called crystal *whisker* fibers. Crystal whisker fibers are made of aluminum oxide, beryllium oxide, magnesium oxide, potassium titanate, silicon carbide, titanium boride, and other materials. (Fig. 6-18) Potassium titanate whiskers are used in large quantities to strengthen composites in thermoplastic matrices. Inorganic continuous boron fibers are stronger than carbon and may be used in a polymer and aluminum matrix. Boron in an epoxy matrix is used to make many composite parts for military and civilian aircraft.

These fibers are very costly to make with present technologies; however, they display tensile strengths greater than 5 802 146 psi [40 GPa]. Research into use of these reinforcements in dental plastics fillings, turbine compressor blades, and special deep water equipment has shown encouraging results. Figure 6-19 shows a helicopter tail rotor shaft made of boron-reinforced epoxy.

Carbon and graphite fibers may exceed glass in strength. They are finding many uses as self-lubricating materials, heat-resistant reentry bodies, blades for turbines and helicopters, and valve-packing compounds. Figure 6-20 shows carbon and glass fibers used together to reinforce an injection molded nylon racquetball racquet.

Fig. 6-19. Rotor driveshaft for helicopter, with end fittings and bearing supplements. (Bell Helicopter Co.)

Fig. 6-18. Submicron ceramic whiskers grown in a fibrous ball. There is a higher concentration of fibers near the center of the ball. The fibers range from as small as two billionths of a meter up to 50 billionths. The minute diameter and length of these fibers are advantageous in injection molding, permitting greater processing speed with minimal fiber damage. (J. M. Huber Corp.)

Fig. 6-20. A combination of glass fiber and carbon fiber is used to reinforce this injection-molded nylon racquetball racquet.

Ceramic fibers have high tensile strengths and low thermal expansion. Some fibers may reach a tensile strength of 2 030 750 psi [14 GPa]. Present applications for ceramic fibers included dental fillings, special electronics, and spacecraft research. See tensile strength of whisker fibers in Figure 6-21.

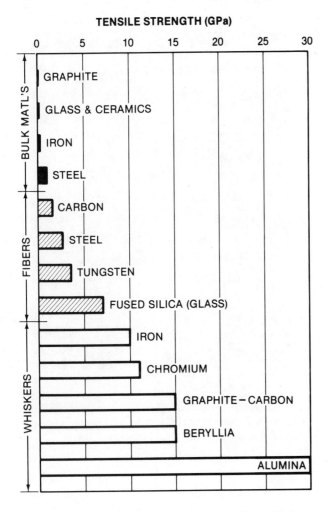

TENSILE STRENGTH (GPa)

Fig. 6-21. Tensile strengths of various materials in bulk, fiber, and whisker forms.

Metal Fibers

Steel, aluminum, and other metals are drawn into continuous filaments. They do not compare with the strength, density, and other properties exhibited by other fibers. Metal fibers are used for added strength, heat transfer, and electrical conductivity.

Hybrid Fibers

Hybrid fibers refer to a special available form of fibers. Two or more fibers may be combined (hybridization) to tailor the reinforcement to the needs of the designer. Hybrid fibers provide diverse properties and many possible material combinations. They can maximize performance, minimize cost, or improve any deficiency of the other fiber component (a synergistic effect). Glass and carbon fibers are used together to increase impact strength and toughness, prevent galvanic action, and reduce cost of a 100 percent carbon composite. When these fibers are placed in a matrix, the composite—not the fibers—is called a *hybrid*. A composite of metal foils or metal composite plies stacked in a specified orientation and sequence is called a *super hybrid*. (See Laminar)

Properties of the most commonly used fiber reinforcing agents are shown in Table 6-7.

Table 6-7. Properties of the Most Commonly Used Fiber Reinforcing Agents (Metallic and Nonmetallic)

Fiber	Relative Density	Tensile Strength Ultimate (MPa)	Tensile Modulus of Elasticity Modulus (GPa)
Aluminum	2.70	620	73.0
Aluminum oxide	3.97	689	323.0
Aluminum silica	3.90	4130	100.0
Aramid (Kevlar 49)	1.4	276	131.0
Asbestos	2.50	1380	172.0
Beryllium	1.84	1310	303.0
Beryllium carbide	2.44	1030	310.0
Beryllium oxide	3.03	517	352.0
Boron-Tungsten boride	2.30	3450	441.0
Carbon	1.76	2760	200.0
Glass, E-glass	2.54	3450	72.0
S-glass	2.49	4820	85.0
Graphite	1.50	2760	345.0
Molybdenum	10.20	1380	358.0
Polyamide	1.14	827	2.8
Polyester	1.40	689	4.1
Quartz (fused silica)	2.20	900	70.0
Steel	7.87	4130	200.0
Tantalum	16.60	620	193.0
Titanium	4.72	1930	114.0
Tungsten	19.30	4270	400.0

6-4 FILLERS

The term *filler* is not entirely appropriate for this diverse class of ingredients. Filler was originally selected to describe any additive used to "fill" space in the polymer and lower cost. The word *extender* is equally inappropriate since some fillers are more expensive than the polymer matrix. The terms *dilutant* and *enhancer* are sometimes used to describe the addition of fillers. Ambiguity of terms and overlap of function add to the confusion. The term *filler* has come to mean any minute particle from various sources, functions, composition, and morphology. Fillers can be saucer-, sphere-, or needle-shaped, or irregular in shape. (Fig. 6-22)

According to the ASTM, a filler is a relatively inert material added to a plastic to modify its strength, permanence, working properties or other qualities, or to lower costs.

Fillers may be either organic or inorganic ingredients of plastics or resins. They may increase bulk or viscosity, replace more costly ingredients, reduce mold shrinkage, and improve the physical properties of the composite item. The size and shape of the filler greatly influence the composite. Flakes or fibers have aspect ratios which make them resist movement or realignment; thus, they improve strength. Spheres have no aspect ratio and produce composites with isotropic properties. Metallic flakes are used in particulate composites to form an electrical barrier or layer in the polymer matrix. The main types of fillers and their functions are shown in Table 6-8.

Fillers may improve processability, product appearance, and other factors. One of the most widely used fillers is wood flour, which is obtained by grinding waste wood stock into a fine granular state. This powdered filler is often added to phenolic res-

Table 6-8. Principal Types of Fillers

Filler	Bulk	Processability	Thermal Resistance	Electrical Resistance	Stiffness	Chemical Resistance	Hardness	Reinforcement	Electrical Conductivity	Thermal Conductivity	Lubricity	Moisture Resistance	Impact Strength	Tensile Strength	Dimensional Stability
Organic															
Wood flour	X	X												X	X
Shell flour	X	X										X		X	X
Alpha cellulose (wood pulp)	X				X	X							X		
Sisal fibers	X				X	X	X	X	X			X	X	X	X
Macerated paper	X				X								X		
Macerated fabric	X					X							X		
Lignin	X	X													
Keratin (feathers, hair)	X					X							X		
Chopped rayon			X	X	X		X	X	X			X	X	X	X
Chopped Nylon			X	X	X	X	X	X	X		X		X	X	X
Chopped Orlon			X	X	X	X	X	X	X			X	X	X	X
Powdered coal	X			X		X	X						X		
Inorganic															
Asbestos	X		X	X	X	X	X					X			X
Mica	X		X	X	X	X	X				X	X			X
Quartz			X	X	X	X						X	X		
Glass flakes		X	X	X	X	X	X	X			X	X	X		
Chopped glass fibers			X	X	X	X	X	X			X	X	X	X	
Milled glass fibers	X	X	X	X	X	X	X	X			X	X	X	X	X
Diatomaceous earth	X	X	X	X	X		X						X		X
Clay	X	X	X	X	X		X						X		X
Calcium silicate			X	X		X		X			X	X			X
Calcium carbonate			X	X		X		X					X		
Alumina trihydrate			X		X	X		X				X			
Aluminum powder				X			X	X	X	X			X		
Bronze powder				X			X	X	X	X			X		
Talc	X	X	X	X	X	X	X				X		X	·	X

This table does not indicate the degree of improvement of function. The prime function will also vary between thermosetting and thermoplastic resins. This table is to be used only as a guide to the selection of fillers.

ins to reduce brittleness and resin cost, and improve product finish.

In most molding operations, the volume of fillers does not exceed 40 percent, but as little as 10 percent resin may be used for molding large desk tops, trays, and particle boards. In the foundry industry, as little as 3 percent resin is used to adhere sand together in the shell molding process.

One product, called *cultured marble,* uses inorganic marble dust and polyester resin to produce items that have the appearance of genuine marble. This product has the advantage of being stain resistant, and it is easily produced in many colors, shapes, or sizes.

Many organic fillers cannot withstand high temperatures, that is, they have low heat resistance.

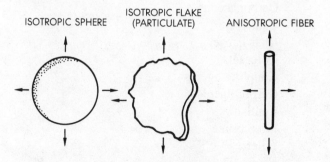

Fig. 6-22. Spheres are isotropic but have no aspect ratio. Isotropic particulates have uniform mechanical properties in the plane of the flake. Fibers have lower aspect ratios but are anisotropic.

To improve heat resistance, a silica filler is used such as sand, quartz, tripoli, diatomaceous earth, and asbestos.

Diatomaceous earth consists of the fossilized remains of microscopic organisms (diatoms). This filler provides improved compressive strength in rigid polyurethane foam.

A filler as fine as cigarette smoke (0.007–0.050 μm) is fumed silica. This submicroscopic silica is added to resins to achieve *thixotropy*. Thixotropy is a state of a material that is gel-like at rest but fluid when agitated. Other thixotropic fillers may be made from very fine powders of polyvinyl chloride, china clay, alumina, calcium carbonate and other silicates. Cab-O-Sil and Sylodex are trade names for two commercial thixotropic fillers. Thixotropic fillers may be added to either thermosetting or thermoplastic resins. They are used to thicken the resin, improve strength, suspend other additives, improve the flow properties of powders, and decrease costs. (See Figure 6-23)

Thixotropic fillers increase viscosity (internal resistance to flow), and thus are desirable in paints, adhesives, and other compounds applied to vertical surfaces. These fillers may be added to polyester resin in fabrication of inclined or vertical surfaces. Thixotropic fillers can also act as emulsifiers to prevent separation of two or more liquids. Both oil and water-based additives may be added and held in an emulsified state.

Fillers such as steel, brass, graphite, and aluminum are added to resins to produce electrically conductive moldings, or to give added strength. Plastics with these fillers may be electroplated. Plastics containing powdered lead are used as neutron and gamma-ray shields.

Wax, graphite, brass, or glass are sometimes added to provide self-lubricating qualities to plastics gears, bearings, and slides.

Glass, especially, is used in plastics for several reasons. It is easily added, relatively inexpensive, improves physical properties, and may be colored. Colored glass has optical advantages (especially color stability) over other chemical colorants. Tiny, hollow glass spheres called *microballoons* (also called microspheres) are used as a filler in producing low-density composites.

Carbon black is a common filler for polyethylene. It improves mechanical strength by acting as a reinforcement, and it also provides some protection from the sun's ultraviolet rays.

VOCABULARY

The following vocabulary words are found in this chapter. Look up the definition of any of these words you do not understand as they apply to plastics, in the glossary, Appendix A.

Antiblocking
Antioxidant
Antiozonants
Antistatic
Carbonaceous
Catalysts
Char barrier
Colorants
Coupling agents
Curing agents
Emulsifiers
Exothermic
Filler
Flame retardants
Fluorescence

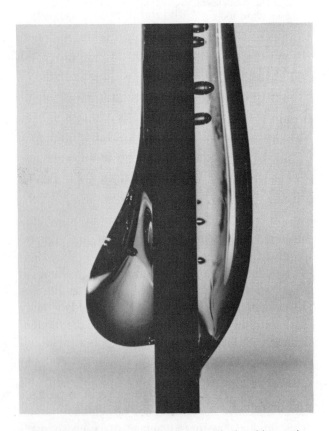

Fig. 6-23. The resin on the right has had a thixotropic agent added. The one on the left will sag and run without this ingredient. (Cabot Corp.)

Foaming/blowing agents
Glass, type C
Glass, type E
Impact modifiers
Inhibitor
Initiator
Lamina
Lubrication bloom
Luminescence
Microballoons (microspheres)
Phosphorescence
Photon

Plasticizer
Primary colors
Promoters
Reinforcements
Roving
Secondary colors
Stabilizer
Strand
Thixotropy
Warp
Whiskers
Yarn

6-1. What term is used for softening a plastics during processing causing it to flow and be reshaped.

6-2. To improve or extend the properties of plastics, ___?___ are used.

6-3. Name four types of colorants:

6-4. The major difference between dyes and pigments is that dyes ___?___ in the plastics material.

6-5. A major disadvantage of organic dyes is their poor ___?___ and ___?___ stability.

6-6. Polyvinyl chloride commonly has plasticizers added to provide ___?___ .

6-7. What name is given to chemical resistant glass fibers.

6-8. Single crystals used as reinforcements are called ___?___ .

6-9. The ___?___ determines whether a plastics is thermoplastics or thermosetting.

6-10. Thixotropic fillers increase ___?___ .

6-11. One of the most widely used plasticizers is ___?___ .

6-12. Name the ___?___ light that is destructive to plastics.

6-13. Polymerization is initiated by the use of ___?___ .

6-14. Chemical ___?___ are sometimes used to produce cellular plastics.

6-15. Name three fillers that may be added to plastics to provide self-lubricating qualities?

6-16. Name four reasons glass fillers are added to resins.

6-17. Name an inexpensive filler used in polyethylene that provides some protection from the sun's rays.

6-18. Identify the special stabilizers that retard or inhibit degradation through oxidation.

6-19. Radical initiators such as ___?___ are curing agents used to cure unsaturated polyester.

6-20. Name three functions fillers provide in plastics.

6-21. A disadvantage of ___?___ fillers is that they can not withstand high processing temperatures.

6-22. Long fibrous ingredients that increase strength, impact resistance and stiffness are called ___?___ .

6-23. Name four reasons why glass fiber is often selected as a reinforcement.

6-24. Long rope-like strands of fibrous glass are called ___?___ .

6-25. To lower viscosity and aid in processing ___?___ and ___?___ are added to resins.

6-26. Zinc stearate is a ___?___ additive that acts as a lubricant in molding.

6-27. To increase flexibility, reduce melt temperature and lower viscosity ___?___ are added to resins.

6-28. Small particles that contribute only slightly to strength are called ___?___ .

6-29. To help prevent discoloration and decomposition of resins and plastics ___?___ are added.

6-30. Crazing, chalking, color changes and loss of properties may be caused by ___?___ .

6-31. Name the four types of colorants used in plastics.

6-32. When chemical, electrical or light energy excites electrons, the radiation of light from the plastics is called ___?___ .

6-33. Lubrication bloom is caused by excess ___?___ .

6-34. Because the color is distributed throughout the product, ___?___ plastics are superior to painted plastics.

6-35. To obtain a low density composite, hollow spheres or ___?___ may be used.

6-36. Which additive will produce the stronger composite, carbon black or graphite fibers?

6-37. Which is stronger, a composite made by curing SMC or a filament wound structure?

6-38. If you were asked to design a chemical tank for exterior use of polyethylene, name two ingredients you may want to add to the resin.

6-39. Pre-promoted polyester resins should not be stored for a period of time, in a ___?___ container or in a ___?___ place.

6-40. Static charges may be dissipated by the addition of ___?___ agents.

Thermoplastics and Their Properties

It is important that you become familiar with thermoplastic plastics. In this chapter, look for the outstanding properties and applications of each plastics since plastics properties affect product design, processing, economics, and service.

This chapter treats individual groups of thermoplastics in alphabetical order:

- acetals (polyacetals)
- acrylics
- acrylic-styrene-acrylonitrile
- acrylonitrile-chlorinated polyethylene styrene
- cellulosics
- chlorinated polyethers
- coumarone-indene plastics
- ethylene acid
- ethylene-ethyl acrylate
- ethylene-methyl acrylate
- ethylene-vinyl acetate
- fluorplastics
- ionomers
- nitrile barrier plastics
- phenoxy
- polyallomers
- polyamides
- polyarylsulfone
- polyetheretherketone
- polyetherimide
- polycarbonates
- polyphenylene ether
- polymethylpentene
- polyolefins
- polyphenylene oxides
- polystyrene
- polysulfones
- polyvinyls

- styrene-maleic anhydride
- styrene-butadiene plastics
- thermoplastic polyesters
- thermoplastic polyimides.

7-1 POLYACETAL PLASTICS (POM)

A highly reactive gas, formaldehyde (CH_2O), may be polymerized in a number of ways. Formaldehyde, or *methanal,* is the simplest of the aldehyde group of chemicals. The ending for the aldehydes is *al,* derived from the first syllable of aldehyde.

Simple polymers based on formaldehyde have been known since 1859. The first polyformaldehyde was put on the market by Du Pont in 1960. The polyformaldehyde (polyacetal) polymer is basically a linear, highly crystalline, long molecular structure. The term *acetal* refers to the oxygen atom that joins the repeating units of the polymer structure. Polyoxymethylene (POM) is the correct chemical term for this polymer. Acetal is a generic term.

A number of initiators or catalysts are used to polymerize the basic polyacetal resin including acids, bases, metallic compounds, cobalt, and nickel.

The polyacetal structure is:

$$H-O-(CH_2-O-CH_2-O)nH : R$$
$$OR$$

$$n\,\overset{\displaystyle H}{\underset{\displaystyle H}{C}} {=\!=} O \rightarrow CH_2\,CH_2\,CH_2\,CH_2\,CH_2\,CH_2\,R$$

$$\underset{\displaystyle O\quad O\quad O\quad O\quad O\quad O}{\big\backslash\!/\ \big\backslash\!/\ \big\backslash\!/\ \big\backslash\!/\ \big\backslash\!/\ \big\backslash\!/}$$

R = Ether or Ester

The best known polyformaldehyde plastics is the oxymethylene linear structure with attached terminal groups. There are, however, many miscellaneous aldehyde-derived polymers.

Thermal and chemical resistance is increased when esters or ethers are attached as terminal groups. Both esters and ethers are relatively inert toward most chemical reagents. They are chemically compatible in organic chemical reactions.

$$\underset{\textit{Water}}{\text{H-O-H}} \qquad \underset{\textit{Ether}}{\text{R-O-R}} \qquad \underset{\substack{\textit{Carboxylic} \\ \textit{Acid}}}{\overset{\overset{\textstyle O}{\|}}{\text{R-C-OH}}} \qquad \underset{\textit{Ester}}{\overset{\overset{\textstyle O}{\|}}{\text{R-C-OR}}}$$

The formulas above show some of the structural relationships between water, ethers, carboxylic acids, and esters.

Close packing and short bond lengths are typical of polyacetal plastics, providing a hard, rigid, dimensionally stable material. Polyacetals have high resistance to organic chemicals and a wide temperature range (Fig. 7-1).

Polyacetals are easy to fabricate, offer properties not found in metals, and are competitive in cost and performance with many nonferrous metals. They are similar to polyamides in many respects. Acetals provide superior fatigue endurance, creep resistance, stiffness, and water resistance. They are among the strongest and stiffest thermoplastics and may be filled for even greater strength, dimensional stability, abrasion resistance and improved electrical properties (Fig. 7-2).

At room temperatures, polyacetals are resistant to most chemicals, stains, and organic solvents. These include tea, beet juice, oils, and household detergents. Hot coffee, however, will usually cause staining. Resistance to strong acids, strong alkalies, and oxidizing agents is poor. Copolymerization and filling improves the chemical resistance of the material.

Moisture and thermal resistance are characteristic of acetal polymers. For this reason, they are used for plumbing fixtures, pump impellers, conveyor belts, aerosol stem valves, and shower heads.

Acetals must be protected from prolonged exposure to ultraviolet light. Such exposure causes surface chalking, reduced molecular mass, and slow degradation. Painting, plating, and/or filling with carbon black or ultraviolet-absorbing chemicals will protect acetal products for outdoor use.

Acetals are available in pellet or powder form for processing in conventional injection molding, blow molding, and extrusion machines. Because of the highly crystalline structure of polyacetals, it is not possible to make optically transparent film. The operator must have adequate ventilation when processing polyacetal materials because upon degrading at high temperature, acetals release a toxic and potentially lethal gas.

Table 7-1 gives some of the most important properties of acetal plastics while a list of six advantages and disadvantages follows.

Fig. 7-1. This automobile door handle made of polyacetal plastics will remain strong and keep its glossy finish even though it will be exposed to all kinds of weather and the ultra-violet rays of the sun. (Du Pont Co.)

Fig. 7-2. These high-stress videocassette parts are molded from acetal plastics. (Du Pont Co.)

Table 7-1. Properties of Acetals

Property	Acetal (Homopolymer)	Acetal (20% Glass-Filled)
Molding qualities	Excellent	Good to Excellent
Relative density	1.42	1.56
Tensile strength, MPa	68.9	58.6–75.8
(psi)	(10 000)	(8 500–11 000)
Compressive strength (10% defl.), MPa	124	124
(psi)	(18 000)	(18 000)
Impact strength, Izod, J/mm	0.07 (Inj) 0.115 (Ext)	0.04
(ft lb/in)	1.4 (Inj) 2.3 (Ext)	(0.8)
Hardness, Rockwell	M94, R120	M75–M90
Thermal expansion, $(10^{-4}/°C)$	20.6	9–20.6
Resistance to heat, °C	90	85–105
(°F)	(195)	(185–220)
Dielectric strength, V/mm	14 960	22 835
Dielectric constant (at 60 Hz)	3.7	3.9
Dissipation factor (at 60 Hz)	3.7	3.9
Arc resistance, s	129	136
Water absorption (24 h), %	0.25	0.25–0.29
Burning rate, mm/min	Slow to 28	20–25.4
(in/min)	(Slow to 1.1)	(0.8–1.0)
Effect of sunlight	Chalks slightly	Chalks slightly
Effect of acids	Resists some	Resists some
Effect of alkalies	Resists some	Resists some
Effect of solvents	Excellent resistance	Excellent resistance
Machining qualities	Excellent	Good to fair
Optical qualities	Translucent to opaque	Opaque

Advantages of Polyacetals

1. High tensile strength with rigidity and toughness
2. Excellent dimensional stability
3. Glossy molded surfaces
4. Low static and coefficient of friction
5. Retention of electrical and mechanical properties to 248 °F [120 °C]
6. Low gas and vapor permeability

Disadvantages of Polyacetals

1. Poor resistance to acids and bases
2. Subject to UV degradation
3. Flammable
4. Unsuitable for contact with food
5. Difficult to bond
6. Toxic, releases fumes upon degradation

7-2 ACRYLICS

In 1901, Otto Rohm reported much of the research that later led to the commercial exploitation of acrylics. Dr. Rohm, pursuing his research in Germany, took an active part in the first commercial development of polyacrylates in 1927. By 1931, there was a Rohm and Haas Company plant operating in the United States. Most of these early materials were used as coatings or for aircraft components. For example, acrylics were used for windshields and bubble turrets on aircraft during World War II. From these early compounds, an extensive group of monomers has become available, and commercial applications of these polymers have grown steadily.

The term *acrylic* includes acrylic and methacrylic esters, acids, and other derivatives. The principal acid and ester monomers are shown in Table 7-2. In order to avoid possible confusion, the basic acrylic formula is shown, with possible side groups R_1 and R_2, in Figure 7-3.

There are many monomer possibilities and preparation methods. The most important is the commercial preparation of methyl methacrylate from acetone cyanohydrin. These homomonomers and comonomers may be polymerized by various commercial methods, including bulk, solution, emulsion, suspension, and granulation polymerization. In all cases, an organic peroxide catalyst is used to start polymerization. Many of the molding powders are made by emulsion methods. Bulk polymerization is used for casting sheets and profile shapes.

The versatility of acrylic monomers in processing, copolymerization, and ultimate, or final state, properties has contributed to their wide use.

Table 7-2. Principal Acid and Ester Monomers

Acrylic acid	Methyl acrylate	Ethyl acrylate	*n*-Butyl acrylate	Isobutyl acrylate	2-Ethylhexyl acrylate
$CH_2{=}CHCOOH$	$CH_2{=}CHCOOCH_3$	$CH_2{=}CHCOOC_2H_5$	$CH_2{=}CHCOOC_4H_9$	$CH_2{=}CHCOOCH_2CH(CH_3)_2$	$CH_2{=}CHCOOCH_2CH(C_2H_5)C_4H_9$

Methacrylic acid	Methyl methacrylate	Ethyl methacrylate	*n*-Butyl methacrylate	Isobutyl methacrylate	Lauryl methacrylate
$CH_2{=}CCOOH$	$CH_2{=}CCOOCH_3$	$CH_2{=}CCOOC_2H_5$	$CH_2{=}CCOOC_4H_9$	$CH_2{=}CCOOCH_2CH(CH_3)_2$	$CH_2{=}CCOO(CH_2)_nCH_3$
$\quad\quad CH_3$	$\quad\quad CH_3$	$\quad\quad CH_3$	$\quad\quad CH_3$	$\quad\quad CH_3$	$\quad\quad CH_3$

Stearyl methacrylate	2-Hydroxyethyl methacrylate	Hydroxypropyl methacrylate	2-Dimethylaminoethyl methacrylate	2-*t*-Butylaminoethyl methacrylate
$CH_2{=}CCOO(CH_2)_6CH_3$	$CH_2{=}CCOOCH_2CH_2OH$	$CH_2{=}CCOO(C_3H_6)OH_4$	$CH_2{=}CCOOCH_2CH_2N(CH_3)_2$	$CH_2{=}CCOOCH_2CH_2NHC(CH_3)_3$
$\quad\quad CH_3$	$\quad\quad CH_3$	$\quad\quad CH_3$	$\quad\quad CH_3$	$\quad\quad CH_3$

$$CH_2{=}C{\Big\langle}\begin{array}{l} R_1 \\ COOR_2 \end{array}$$

(A) Basic acrylic formula.

$$CH_2{=}C{\Big\langle}\begin{array}{l} H \\ COOH \end{array}$$

(B) Hydrogen replaces R_1 and R_2 to produce acrylic acid.

$$CH_2{=}C{\Big\langle}\begin{array}{l} CH_3 \\ COOH \end{array}$$

(C) Methyl group replaces R_1 to produce methacrylic acid.

Fig. 7-3. Acrylic formula, with two possible radical replacements.

Table 7-3 gives some of the basic properties of acrylics.

Polymethyl methacrylate is an atactic, amorphous, transparent thermoplastic. Because of its high transparency (about 92 percent), it is used for many optical applications (Fig. 7-4A). It is a good electrical insulator for low frequencies, and has very good resistance to weathering (Fig. 7-4B). Outdoor advertising signs are a familiar use of acrylics.

Polymethyl methacrylate is a standard material for automobile taillight lenses and covers (Fig. 7-4C). This material is used for aircraft windshields and cockpit covers, and for bubble bodies on helicopters.

(C) Taillight lenses.

(A) Contact lenses.

(B) Panels on building.

(D) Wall paints.

Fig. 7-4. Acrylics have many useful properties.

Polymethyl methacrylate may be produced by any usual thermoplastic process. It may be fabricated by solvent cementing. Cast and extruded sheets and profile shapes are popular forms. Sheet forms are widely used for room dividers and skylight domes, and as a substitute for glass in windows. There has been wide use of these plastics in the paint industry in the form of emulsions (Fig. 7-4D). Emulsion acrylics are popular as a clear, hard, and glossy "wax" coating for floors. Acrylic-based adhesives are available with a wide range of uses and properties. These adhesives are transparent and available in solvent-based (air-drying), hot-melt, or pressure-sensitive forms. Figure 7-5 shows an acrylic sealant being applied directly to an oily aluminum window frame under water.

Glass reinforced sheets are used to produce sanitary ware, vanities, tubs, and counters. Protective liquid coatings, known as gel coats, may be used with reinforcements as cover stock. Heavily filled and reinforced resins formulated to crosslink into a thermosetting matrix are used to produce marble-like bathroom fixtures and furniture.

Well-known trade names for polymethyl methacrylate are Plexiglas, Lucite, and Acrylite. From the following lists you can see there are more advantages (11) of acrylics than disadvantages (5).

Advantages of Acrylics

1. Wide range of colors
2. Outstanding optical clarity
3. Slow-burning, releasing little or no smoke
4. Excellent weatherability and ultraviolet resistance
5. Ease of fabrication
6. Excellent electrical properties
7. Unaffected by food and human tissues
8. Rigidity with good impact strength
9. High gloss and good *feel*
10. Excellent dimensional stability and low mold shrinkage
11. Stretch forming improves biaxial toughness

Disadvantages of Acrylics

1. Poor solvent resistance
2. Possibility of stress cracking
3. Combustibility
4. Limited continuous service temperature of 93 °C [200 °F]
5. Inflexibility

Polyacrylates

Polyacrylates are transparent, resistant to chemicals and weather, and have a low softening point. Applications include films, adhesives, and surface coatings for paper and textiles. They are usually copolymer compositions. Polyethyl acrylate may be crosslinked to form thermosetting elastomers. Polyacrylate monomers are used as plasticizers for other vinyl polymers.

Acrylic esters may be obtained from the reaction of ethylene cyanohydrin with sulphuric acid and an alcohol.

$$HO \cdot CH_2 \cdot CH_2 \cdot CN \xrightarrow[H_2SO_4]{ROH} CH_2 : CH \cdot CO \cdot O \cdot R$$

Acrylic and polyvinyl chloride (PMMA/PVC) may be alloyed to produce a tough, durable sheet, easily thermoformed into signs, aircraft trays, and public service seating. (Fig. 7-6)

Polyacrylonitrile and Polymethacrylonitrile

The elastomers and fibers produced from polyacrylonitrile and polymethacrylonitrile materials were merely laboratory curiosities before World War II. Since that time there has been a rapid expansion of the use of acrylonitrile often as the main ingredient in acrylic fibers. These polymers are copolymerized and stretched to orient the molecular chain. Orlon and Dynel, classified as *modacrylic fibers*, contain less than 85 percent acrylonitrile. Modacrylic fibers contain at least 35 percent acrylonitrile units. *Acrylic fibers* such as Acrilan, Creslan, and Zefran contain more than 85 percent acrylonitrile.

Unmodified polyacrylonitrile is only slightly thermoplastic and is difficult to mold because its hydrogen bonds resist flow. Copolymers of styrene,

Fig. 7-5. An acrylic sealant being applied under water. (Cabot Corp.)

Fig. 7-6. This continuous cast, thermoformed acrylic sheet offers superior impact strength and sunlight resistance. (United States Steel)

ethyl acrylates, methacrylates, and other monomers are extruded into the amorphous fiber form. At this point, however, the fiber is too weak to be of value. It is then stretched to produce a greater degree of crystallization. Tensile strength is greatly increased as a result of this molecular orientation. (See Nitrile barrier plastics.)

Monomers of acrylonitrile and methacrylonitrile are shown below.

$$CH_2{=}CHCN$$
Acrylonitrile

$$CH_2{=}C{-}CN$$
$$|$$
$$CH_3$$
Methacrylonitrile

Acrylic-Styrene-Acrylonitrile (ASA)

The terpolymer of acrylic-styrene-acrylonitrile may vary in the percentages of each component to enhance or tailor a specific property. The excellent surface gloss makes it attractive as a cover layer for coextrusion over ABS, PC, or PVC. Exterior applications include signs, downspouts, siding, recreational equipment, camper tops, ATV bodies, and gutters.

This material is also blended with PC (ASA/PC) and alloyed with PVC (ASA/PVC) and PMMA (ASA/PMMA) to enhance specific properties. ASA/PMMA alloys have outstanding weatherability, gloss, and toughness. They are used to produce spas and hot tubs. (See Alloy and Blend)

Acrylonitrile-Butadiene-Styrene (ABS)

ABS polymers are opaque thermoplastic resins formed by the polymerization of acrylonitrile-butadiene-styrene monomers. Because they possess such a diverse combination of properties, many experts classify them as a family of plastics. They are, however, terpolymers ("ter" meaning three) of three monomers and *not* a distinct family. Their development resulted from research efforts on synthetic rubber during and after World War II. The proportions of the three ingredients may vary, which accounts for the great number of possible properties.

The three ingredients are shown below. Acrylonitrile is also known as vinyl cyanide and acrylic nitrile.

$$CH_2{=}CHCN \qquad CH_2{=}CH{-}CH{=}CH_2$$
Acrylonitrile   *Butadiene*

$$\text{⌬}{-}CH{=}CH_2$$
Styrene

Below is a representative structure for acrylonitrile-butadiene-styrene.

Acrylonitrile-Butadiene-Styrene

Graft polymerization techniques are commonly used to make various grades of this material.

The resins are hygroscopic (moisture-absorbing). Predrying before molding is advisable. ABS materials can be produced on all thermoplastic processing machines.

ABS materials are characterized by resistance to chemicals, heat, and impact. They are used for appliance housings, light luggage, camera bodies, pipe, power tool housings, automotive trim, battery cases, tool boxes, packing crates, radio cases, cabinets, and various furniture components. They may be electroplates and used in automotive, appliance, and housewares applications (Fig. 7-7).

Table 7-3 lists some of the properties of ABS materials and a reference list of nine advantages and five disadvantages follows.

Advantages of ABS

1. Ease of fabrication and coloring
2. High impact resistance with toughness and rigidity
3. Good electrical properties
4. Excellent adhesion by metal coatings
5. Fairly good weather resistance and high gloss
6. Ease of forming by conventional thermoplastic methods
7. Good chemical resistance
8 Light weight
9. Very low moisture absorption

Disadvantages of ABS

1. Poor solvent resistance
2. Subject to attack by organic materials of low molecular mass
3. Low dielectric strength
4. Only low elongations available
5. Low continuous service temperature

Fig. 7-7. The bodies of these electric-powered vehicles used by the U.S. Postal Services are made of ABS thermoplastic. (Borg-Warner Chemicals, Inc.)

Table 7-3. Properties of Acrylics

Property	Methyl Methacrylate (Molding)	Acrylic–PVC Copolymer (Molding)	ABS (High Impact)
Molding qualities	Excellent		Good to excellent
Relative density	1.17–1.20	1.30	1.01–1.04
Tensile strength, MPa	48–76	38	30–53
(psi)	(7 000–11 000)	(5 500)	(4 500–7 500)
Compressive strength, MPa	83–125	43	30–55
(psi)	(12 000–18 000)	(6 200)	(4 500–8 000)
Impact strength, Izod J/mm	0.015–0.025	0.75	0.25–0.4*
(ft lb/in)	(0.3–0.5)	(15)	(5.0–8.0)†
Hardness, Rockwell	M85–M105	R104	R75–R105
Thermal expansion, $10^{-4}/°C$	12–23	12–29	24–33
Resistance to heat, °C	60–94	60–98	60–98
(°F)	(140–200)	(140–210)	(140–210)
Dielectric strength, V/mm	15 800–20 000	15 800	13 800–18 000
Dielectric constant (at 60 Hz)	3.3–3.9	4	2.4–5.0
Dissipation factor (at 60 Hz)	0.04–0.06	0.04	0.003–0.008
Arc resistance, s	No track	25	50–85
Water absorption (24 h), %	0.1–0.4	0.13	0.20–0.45
Burning rate, mm/min	Slow 0.5–30	Nonburning	Slow to self-extinguishing
(in/min)	(Slow 0.6–1.2)		
Effect of sunlight	Nil	Nil	Yellows
Effect of acids	Attacked by strong oxidizing acids	Slight	Attacked by strong oxidizing acids
Effect of alkalies	Attacked	None	None
Effect of solvents	Soluble in ketones, esters, and aromatic and chlorinated hydro-carbons	Attacked by ketones esters, and aromatic and chlorinated hydro-carbons	Soluble in ketones and esters
Machining qualities	Good to excellent	Excellent	Excellent
Optical qualities	Transparent to opaque	Opaque	Translucent

Notes: *At 23 °C, 3 × 12 mm bar
†At 73 °F, 1/8 × 1/2 in bar

Acrylonitrile-Chlorinated polyethylene-Styrene (ACS)

Because of the chlorine content, this terpolymer surpasses ABS in flame-retardant properties, weatherability, and service temperatures. Applications include office machine housing, appliance cases, and electrical connectors.

ABS/PA blends have excellent chemical and temperature resistance for underhood automobile components. ABS/PC alloys fill the price and performance gap between polycarbonate and ABS. Typical applications include typewriter housing, headlight rings, institutional food trays, and appliance housings. ABS/PVC alloys are used for air conditioner fans, grills, luggage shells, and computer housings because of their outstanding impact strength, toughness, and cost. ABS/EVA alloys have good impact and stress crack resistance. The elastomer content in ABS/EPDM improves low-temperature impact and modulus.

7-3 CELLULOSICS

Cellulose ($C_6H_{10}O_5$) is the material that composes the framework, or cell walls, of all plants. It is our oldest, most familiar, most useful industrial raw material because cellulose is abundant everywhere in one form or another. Plants are also a very inexpensive raw material. From them, we produce shelter, clothing, and food. Cereal straws and grass are composed of nearly 40 percent cellulose, wood is about 50 percent cellulose, and cotton may be nearly 98 percent cellulose. Wood and cotton are major industrial sources of this material. Long-chained molecules of repeating glucose units are referred to as "chemically modified natural plastics."

The chemical structure of cellulose is shown in Figure 7-8. Each cellulose molecule contains three hydroxyl (OH) groups at which different groups may attach to form various cellulosic plastics. Cellulose can undergo reaction at the ether linkage between the units.

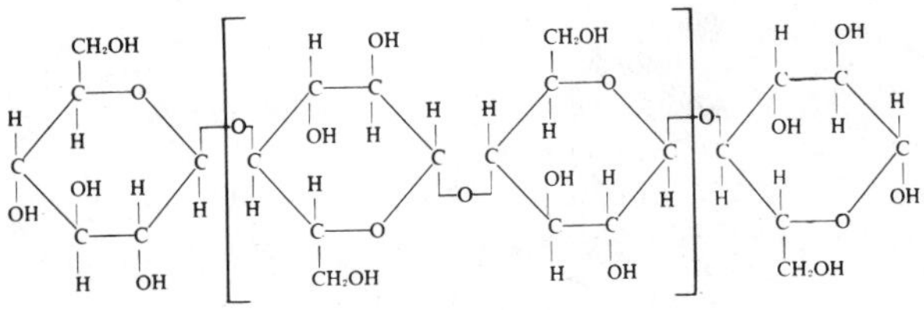

The term *cellulosics* refers to plastics derived from cellulose, a family that consists of many separate and distinct types of plastics. The relationship of cellulose to many plastics and applications is shown in Figure 7-9.

There are three large groups of cellulosic plastics. *Regenerated cellulose* is first chemically changed to a soluble material then reconverted by chemical means into its original substance. *Cellulose esters* are formed when various acids react with the hydroxyl (OH) groups of the cellulose. *Cellulose ethers* are compounds derived by the alkylation of cellulose.

Regenerated Cellulose

Regenerated cellulose products are cellophane, viscose rayon, and cuprammonium rayon (no longer of commercial importance).

In its natural form, cellulose is insoluble and incapable of flow by melting. Even in the powdered form, it retains a fibrous structure.

There is evidence that in 1857 cellulose was found to be dissolvable in an ammoniacal solution of copper oxide. By 1897, Germany was commercially producing a fibrous yarn by spinning this solution into an acid or alkaline coagulating bath. Any remaining copper ions were removed by additional acid baths. This expensive process was called the *cuprammonium* process (for copper and ammonia), and the fiber was known as *cuprammonium rayon*. Newer synthetic fibers with equally desirable properties are less costly to produce, therefore cuprammonium rayon has lost its popularity.

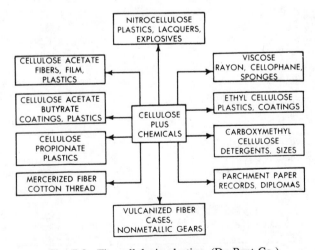

Fig. 7-9. The cellulosic plastics. (Du Pont Co.)

Fig. 7-8. Chemical structure of cellulose.

In 1892, C. F. Cross and E. J. Bevan of England produced a different cellulose fiber. They treated alkali cellulose (cellulose treated with caustic soda) with carbon disulfide to form a xanthate. Cellulose xanthate is soluble in water to give a viscous solution (hence the name) called *viscose*. Viscose was then extruded through spineret openings into a coagulating solution of sulfuric acid and sodium sulfate. The regenerated fiber is called *viscose rayon*. Rayon has become an accepted generic name for fibers composed of regenerated cellulose. It is still found as a clothing fabric and has some use as tire cord.

Production patents were granted to J. F. Brandenberger of France in 1908 for an extruded, regenerated cellulose film called *cellophane*. Like viscose rayon, the xanthate solution is regenerated by coagulation in an acid bath. After the cellophane is dried, it is usually given a water resistant coating of ethyl acetate or cellulose nitrate. Cellophane films, coated and uncoated, are used for packaging food products and pharmaceuticals.

Cellulose Esters

Among the esters of cellulose are cellulose nitrate, cellulose acetate, cellulose butyrate, and cellulose acetate propionate. In this group of plastics, acids are used to react with the hydroxyl (OH) groups to form esters.

Professor Bracconot of France first carried out the nitration of cellulose in 1832. The discovery that mixing nitric and sulfuric acids with cotton will produce nitrocellulose (or cellulose nitrate) was made by C. F. Schonbein of England. This material was a useful military explosive but it found little commercial value as a plastics. Figure 7-10 shows the nitration of cellulose.

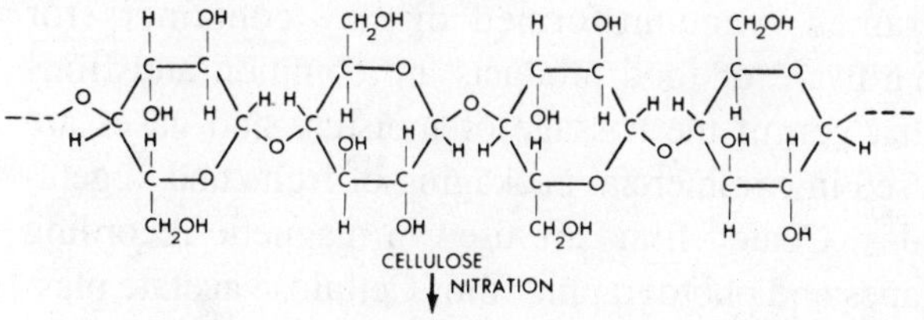

Fig. 7-10. Nitration of cellulose used to produce nitrocellulose.

At the London International Exhibition in 1862, Alexander Parks of England was awarded a bronze medal for his new plastics material called *Parkesine*. It was composed of cellulose nitrate with a plasticizer of castor oil.

In the United States, John Wesley Hyatt created the same material while seeking a substitute for ivory billiard balls. His experiments followed the earlier work of the American chemist Maynard. Maynard had dissolved cellulose nitrate in ethyl alcohol and ether to form a product used as a bandage for wounds. He gave this solution the name *Collodion*. When the Collodion solution was spread over a wound, the solvent evaporated to leave a thin, protective film. One story of Hyatt's discovery of the material that became known as *Celluloid* tells of his accidently spilling camphor on some pyroxylin (cellulose nitrate) sheets and noticing the improved properties. Another version tells that he had been treating a wound covered by Collodion with a camphor solution and discovered a change in the cellulose nitrate product.

In 1870, Hyatt and his brother patented the process of treating cellulose nitrate with camphor and by 1872, Celluloid became a commercial success. Products made entirely of nitrocellulose were highly explosive, became brittle, and suffered greatly from shrinkage. The use of camphor as a plasticizer eliminated many of these disadvantages. Celluloid is made from pyroxylin (nitrated cellulose), camphor, and alcohol. It is highly combustible but not explosive.

Cellulose Nitrate (CN)

Cellulose nitrate was once used in photographic film, bicycle parts, toys, knife handles, and table-tennis balls (Fig. 7-11). Today, it is seldom used because of the difficulty in processing and its high flammability. Cellulose nitrate cannot be injection or compression molded. It is usually extruded or cast into large blocks from which sheets are sliced. Films are made by continuous casting of a cellulose solution on a smooth surface. As the solvents evaporate, the film is removed from the surface and placed on drying rollers. Sheets and films may be vacuum-processed. Table-tennis balls and a few novelty items still may be made of cellulose nitrate plastics. Cellulose nitrate esters are found in lacquers for metal and wood finishes and are common ingredients for aerosol paints and fingernail polishes.

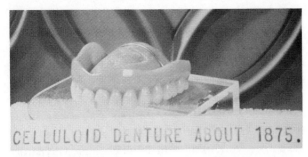

(A) Dental application.

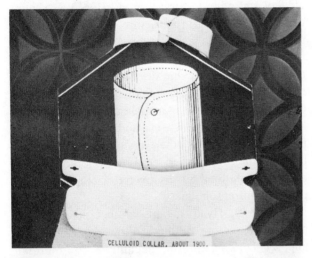

(B) Use in men's clothing.

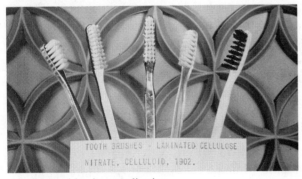

(C) Personal hygiene application.

(D) Film base, which led to growth of photography as a popular hobby.

Fig. 7-11. Some early uses of Celluloid. (Celanese Plastic Materials Co.)

acetic acid and acetic anhydride, using sulfuric acid as a catalyst. The acetate or acetyl group or radical (CH_3CO) is the source of chemical reaction with the hydroxyl (OH) groups. The structure of cellulose triacetate is as follows:

Cellulose Triactate

Cellulose acetates exhibit poor heat, electrical, weathering, and chemical resistance. They are fairly inexpensive and may be transparent or colored. Their main uses are as films and sheets in the packaging and display industries. They are fabricated by nearly all the thermoplastic processes and molded into brush handles, combs, and spectacle frames. Vacuum-formed display containers for hardware or food products are common and films that permit the passage of moisture and gases are used in commercial packaging of fruits and vegetables. Coated films are used in magnetic recording tapes and photographic film. Cellulose acetate plastics are made into fibers for use in textiles. They are also employed as lacquers in the coating industry. Table 7-4 gives some of the properties of cellulose acetate.

Cellulose Acetate (CA)

Cellulose acetate is the most useful of the cellulosic plastics. During World War I, the British secured the aid of Henry and Camille Dreyfus of Switzerland to start large-scale production of cellulose acetate. Cellulose acetate provided a fire-retardant lacquer for the fabric-covered airplanes then in use. By 1929, commercial grades of molding powder, fibers, sheets, and tubes were being made in the United States.

Basic methods for making this material resemble those used in making cellulose nitrate. Acetylation of cellulose is carried out in a mixture of

Cellulose Acetate Butyrate (CAB)

Cellulose acetate butyrate was developed in the mid-1930s by the Hercules Powder Company and

Table 7-4. Properties of Cellulosics

Property	Ethyl Cellulose (Molding)	Cellulose Acetate (Molding)	Cellulose Propionate (Molding)
Molding qualities	Excellent	Excellent	Excellent
Relative density	1.09–1.17	1.22–1.34	1.17–1.24
Tensile strength, MPa	13.8–41.4	13–62	14–53.8
(psi)	(2 000–8 000)	(1 900–9 000)	(2 000–22 000)
Compressive strength, MPa	69–241	14–248	165.5–152
(psi)	(10 000–35 000)	(2 000–36 000)	(2 400–22 000)
Impact strength, Izod, J/mm	0.1–0.43	0.02–0.26	0.025–0.58
(psi)	(2.0–8.5)	(0.4–5.2)	(0.5–11.5)
Hardness, Rockwell	R50–R115	R34–R125	R10–R122
Thermal expansion, $10^{-4}/°C$	25–50	20–46	28–43
Resistance to heat, °C	46–85	60–105	68–105
(°F)	(115–185)	(140–220)	(155–220)
Dielectric strength, V/mm	13 800–19 685	9 840–23 620	11 810–17 715
Dielectric constant (at 60 Hz)	3.0–4.2	3.5–7.5	3.7–4.3
Dissipation factor (at 60 Hz)	0.005–0.020	0.01–0.06	0.01–0.04
Arc resistance, s	60–80	50–310	175–190
Water absorption (24 h), %	0.8–1.8	1.7–6.5	1.2–2.8
Burning rate, mm/min	Slow	Slow to self-extinguishing	Slow 25–33
(in/min)			(1.0–1.3)
Effect of sunlight	Slight	Slight	Slight
Effect of acids	Decomposes	Decomposes	Decomposes
Effect of alkalies	Slight	Decomposes	Decomposes
Effect of solvents	Soluble in ketones, esters, chlorinated hydrocarbons, aromatic hydrocarbons	Soluble in ketones, esters, chlorinated hydrocarbons, aromatic hydrocarbons	Soluble in ketones, esters, chlorinated hydrocarbons, aromatic hydrocarbons
Machining qualities	Good	Excellent	Excellent
Optical qualities	Transparent to opaque	Transparent to opaque	Transparent to opaque

Eastman Chemical. This material is produced by reacting cellulose with a mixture of sulfuric and acetic acids. *Esterification* is completed when the cellulose is reacted with butyric acid and acetic anhydride. The reaction is much like the making of cellulose acetate, except that butyric acid is also used. The product that results has acetyl groups (CH_3CO) and butyl groups ($CH_3CH_2CH_2CH$) in the repeating cellulose unit. This product has improved dimensional stability, weathers well, and is more chemical and moisture resistant than cellulose acetate.

Cellulose acetate butyrate is used for tabulator keys of office machines, automobile parts, tool handles, display signs, strippable coatings, steering wheels, tubes, pipes, and packaging components. Probably the most familiar application of this material is in screwdriver handles. (Fig. 7-12)

Cellulose Acetate Propionate

Cellulose acetate propionate (also called simply cellulose propionate) was developed by the Celanese Plastics Company in 1931. It found little use until materials shortages developed during World War II. It is made like other acetates with the addition of propionic acid (CH_3CH_2COOH) in place of acetic anhydride. Its general properties are similar to those of cellulose acetate butyrate. However, it exhibits superior heat resistance and lower moisture absorption. The main uses of cellulose acetate propionate include pens, automotive parts, brush handles, steering wheels, toys, novelties, and film packaging displays. Below is a list of nine advantages and four disadvantages of cellulosic esters.

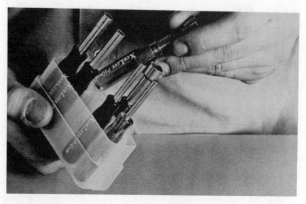

Fig. 7-12. Cellulose butyrate and cellulose acetate are used for the handles of these tools. (Eastman Chemical Products, Inc.)

Advantages of Cellulosic Esters

1. Forms glossy moldings by thermoplastic methods
2. Exceptional clarity (butyrates and propionates)
3. Toughness, even at low temperatures
4. Excellent colorability
5. Nonpetrochemical base
6. Wide range of processing characteristics
7. Resists stress cracking
8. Outstanding weatherability (butyrates)
9. Slow burning (except cellulose nitrate)

Disadvantages of Cellulosic Esters

1. Poor solvent resistance
2. Poor resistance to alkaline materials
3. Relatively low compressive strength
4. Flammability

Cellulose Ethers

Among the ethers of cellulose are ethyl cellulose, methyl cellulose, hydroxymethyl cellulose, carboxymethyl cellulose, and benzyl cellulose. Their manufacture generally involves the preparation of alkali cellulose and other reactants as shown in Figure 7-13.

Ethyl cellulose (EC) is the most important of the cellulose ethers and is the only one used as a plastics material. Basic research and patents were established by Dreyfus in 1912. By 1934, the Hercules Powder Company offered commercial grades in the United States.

Alkali cellulose (soda cellulose) is treated with ethyl chloride to form ethyl cellulose. The radical substitution (*etherification*) of this ethoxy may cause the final product to vary over a wide range of properties. In etherification the hydrogen atoms of the hydroxyl groups are replaced by ethyl (C_2H_5) groups (Fig. 7-14). This cellulose plastic is tough, flexible, and moisture resistant. Table 7-5 gives some of the properties of ethyl cellulose.

Ethyl cellulose is used in football helmets, flashlight cases, furniture trim, cosmetic packages, tool handles, and blister packages. It has been used

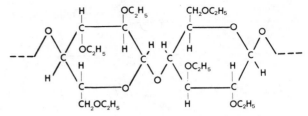

Fig. 7-14. Ethyl cellulose (fully ethylated).

for protective coatings on bowling pins and in the formulations of paint, varnish, and lacquers. Ethyl cellulose is a common ingredient in hair sprays. It is often used as a hot melt for strippable coatings. These coatings protect metal parts against corrosion and marring during shipment and storage.

Methyl cellulose is prepared like ethyl cellulose, using methyl chloride or methyl sulphate instead of ethyl chloride. In etherification, the hydrogen of the OH group is replaced by methyl (CH_3) groups:

$$R(ONa)_{3n} + CH_3Cl \rightarrow R(OCH)_{3n}$$
Methyl cellulose

Methyl cellulose finds a wide number of uses. It is water soluble and edible. It is used as a thickening emulsifier in cosmetics and adhesives, and is a well-known wallpaper adhesive and fabric-sizing material. It is useful for thickening and emulsifying water-based paints, salad dressings, ice cream, cake mixes, pie filling, crackers, and other products. In pharmaceuticals, it is used to coat pills and as contact lens solutions (Fig. 7-15).

Hydroxyethyl cellulose is produced by reacting the alkali cellulose with ethylene oxide. It has many of the same applications as methyl cellulose. In the schematic equation below, R represents the cellulose skeleton.

$$R(ONa)_{3n} + CH_2CH_2 \rightarrow R(OCH_2CH_2OH)_{3n}$$
Hydroxyethyl cellulose

At one time, *benzyl cellulose* was used for moldings and extrusions. In the United States, however, it is unable to compete with other polymers.

Carboxymethyl cellulose (sometimes called sodium carboxymethyl cellulose) is made from alkali cellulose and sodium chloracetate. Like methyl cellulose, it is water soluble and used as a sizing, gum, or emulsifying agent. Carboxymethyl

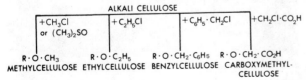

Fig. 7-13. Cellulosic plastics from alkali cellulose and other reactants. The *R* in these formulas represents the cellulose skeleton.

Fig. 7-15. Methyl cellulose is used as a coating for pharmaceuticals.

Table 7-5. Properties of Chlorinated Polyether

Property	Chlorinated Polyether
Molding qualities	Excellent
Relative density	1.4
Tensile strength, MPa	41
(psi)	(6 000)
Impact strength, Izod, J/mm	0.002
(ft lb/in)	(0.4)
Hardness, Rockwell	R100
Thermal expansion, $10^{-4}/°C$	20
Resistance to heat, °C	143
(°F)	(290)
Dielectric strength, V/mm	16 000
Dielectric constant (at 60 Hz)	3.1
Dissipation factor (at 60 Hz)	0.01
Water absorption (24 h), %	0.01
Burning rate, mm/min	Self-extinguishing
Effect of sunlight	Slight
Effect of acids	Attacked by oxidizing acids
Effect of alkalies	None
Effect of solvents	Resistant
Machining qualities	Excellent
Optical qualities	Translucent to opaque

cellulose may be found in foods, pharmaceuticals, and coatings. It is a first rate water-soluble suspending agent for lotions, jelly bases, ointments, toothpastes, paints, and soaps. It is used to coat pills, paper, and textiles.

$$R(ONa)_{3n} + ClCH_2 \cdot COONa \rightarrow R(OCH_2COONa)_{3n} + NaCl$$
Sodium Carboxymethyl Cellulose

7-4 CHLORINATED POLYETHERS

In 1959, the Hercules Powder Company introduced *chlorinated polyethers* with the trade name Penton. In 1972, however, Hercules discontinued production and sale. This thermoplastic material was produced by chlorinating pentaerythritol. The resulting dichloromethyl oxycyclobutane was polymerized into a crystalline, linear product. This material was over 45 percent chlorine by mass (Fig. 7-16).

Chlorinated polyethers may be processed on thermoplastic equipment. These materials are high-performance, high-priced plastics (Table 7-5). They have been coated on metal substrates by fluidized-bed, flame-spraying, or solvent processes. Molded parts possess high strength, heat resistance, excel-

lent electrical and chemical resistance, and low water absorption. Although the high price restricted their wider use, they did find use as coatings for valves, pumps, and meters. Molded parts included chemical meter components, pipe lining, valves, laboratory equipment, and electrical insulation. At degrading temperatures, lethal chlorine gas is released, however.

To date, there are no plans for another producer to make this plastics with its distinct chemical nature.

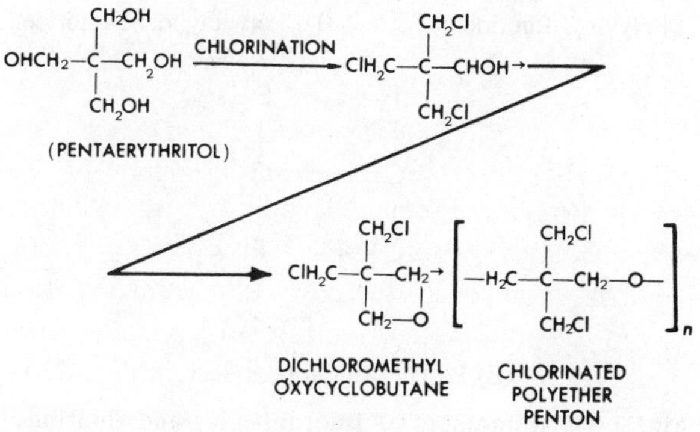

7-5 COUMARONE-INDENE PLASTICS

Coumarone and indene are obtained from the fractionation of coal tar, but are seldom separated. These inexpensive products resemble styrene in chemical structure. Coumarone and indene may be polymerized by the ionic catalytic action of sulfuric acid. A wide range of products, from sticky resins to brittle plastics, may be obtained by varying the coumarone-indene ratio or by copolymerization with other polymers.

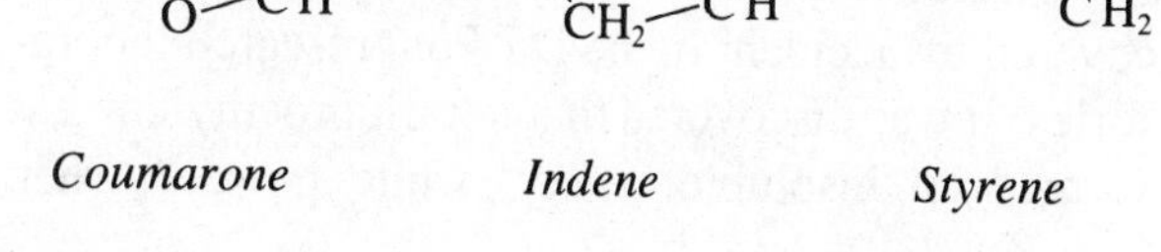

Fig. 7-16. Production of Penthol, a chlorinated polyether.

Coumarone   *Indene*   *Styrene*

Although available since long before World War II, they are not available as molding compounds and have found only limited uses. They are used as binders, modifiers, or extenders for other polymers and compounds. The largest quantities of coumarone and indene are used in the making of paints or varnishes and as binders in flooring tiles and mats.

The traits of these compounds vary greatly. Coumarone-indene copolymers are good electrical insulators. They are soluble in hydrocarbons, ketones, and esters. They range from light to dark in color and are inexpensive to make. They are true thermoplastic materials with a softening point of 100° to 120 °F [<35 °C to >50°C]. Applications include printing inks, coatings for paper, adhesives, encapsulation compounds, some battery boxes, brake linings, caulking compounds, chewing gum, concrete-curing compounds, and emulsion binders.

7-6 FLUOROPLASTICS

Elements in the seventh column of the Periodic Table are closely related. These elements (fluorine, chlorine, bromine, iodine, and astatine) are called *halogens* from the Greek word that means "salt-producing." Chlorine is found in common table salt. All of these elements are electronegative (they can attract and hold valence electrons) having only seven electrons in their outermost shell. Fluorine and chlorine are gases not found in a pure or free state. Fluorine is the most reactive element known. Large quantities of fluorine are needed for processes connected with nuclear energy technology, such as isolation of uranium metal.

Compounds containing fluorine are commonly called fluorocarbons although in the strictest definition, *fluorocarbon* should be used to refer to compounds containing only fluorine and carbon.

The French chemist Moissan isolated pure fluorine in 1886, but it remained a laboratory curiosity until 1930. In 1931, the trade name Freon was announced. This fluorocarbon is a compound of carbon, chlorine, and fluorine, for example, CCl_2F_2. Freons are used extensively as refrigerants. Fluorocarbons are also used in the formation of polymeric materials.

In 1938, the first polyfluorocarbon was discovered by accident in the Du Pont research laboratories. It was discovered that tetrafluoroethylene gas formed an insoluble, waxy, white powder when

stored in steel cylinders. As a result of this chance discovery, a number of fluorocarbon polymers have been developed.

The term *fluoroplastic* is used to describe alkene-like structures that have some or all of the hydrogen atoms replaced by fluorine.

Polyethylene   *Polytetrafluoroethylene*

It is the presence of the fluorine atoms that provides the unique properties characteristic of the fluoroplastic family. These properties are directly related to the high carbon-to-fluorine bonding energy and to the high electronegativity of the fluorine atoms. Thermal stability and solvent, electrical, and chemical resistance are weakened if fluorine (F) atoms are replaced with hydrogen (H) or chlorine (Cl) atoms. The C—H and C—Cl bonds are weaker than C—F bonds and are more vulnerable to chemical attack and thermal decomposition.

The major fluoroplastics are shown in Figure 7-17. There are only two types of fluorocarbon

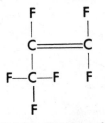

(A) Polychlorotrifluoroethylene.   **(B)** Polytetrafluoroethylene.

(C) Polyvinyl fluoride.   **(D)** Polyvinylidene fluoride.

(E) Polyhexafluoropropylene.

Fig. 7-17. Monomers of fluoroplastics and fluorine-containing polymers.

plastics: polytetrafluoroethylene (PTFE) and poly-tetrafluoropropylene (FEP or fluorinated ethylene propylene). The others must be considered copolymers or fluorine-containing polymers.

Polytetrafluoroethylene (PTFE)

Polytetrafluoroethylene $(CF_2 = CF_2)_n$ accounts for nearly 90 percent (by volume) of the fluorinated plastics. The monomer tetrafluoroethylene is obtained by pyrolysis of chlorodifluoromethane. Tetrafluoroethylene is polymerized in the presence of water and a peroxide catalyst under high pressure. PTFE is a highly crystalline, waxy thermoplastic material with a service temperature of −450 °F to +550 °F [−268 °C to +288 °C]. The high bonding strength and compact interlocking of fluorine atoms about the carbon backbone prevent PTFE processing by the usual thermoplastic methods. At present, it cannot be plasticized to aid processing. Most of the material is made into preforms and sintered.

Sintering is a special fabricating technique used for metals and plastics. The powdered material is pressed into a mold at a temperature just below its melting or degradation point until the particles are fused (sintered) together. The mass as a whole does not melt in this process. Sintered moldings may be machined. Special formulations may be extruded in the form of rods, tubes, and fibers by using organic dispersions of the polymer. These are later vaporized as the product is sintered. Coagulated suspensoids may be used in much the same way. Presintered grades of this material may be extruded through extremely long compacting and sintering zones of special dies. Many films, tapes, and coatings are cast, dipped, or sprayed from PTFE dispersions by a drying and sintering process. Films and tapes may also be cut or sliced from sheet stock.

Teflon is a familiar trade name for homopolymers and copolymers of polytetrafluoroethylene. Its antistick property (low coefficient of friction) makes it a very useful coating. Teflon is applied to many metallic substrates, including cookware (Fig. 7-18). There is no known solvent for these materials. They may be chemically etched and adhesive-bonded with contact or epoxy adhesives. Films may be heat-sealed together, but not to other materials.

Fluorocarbons have greater mass than hydrocarbons, because fluorine has an atomic mass of 18.998 4 and hydrogen only 1.007 97. Fluoroplastics are thus heavier than other plastics. Relative densities range from 2.0 to 2.3.

(A) Cookware and utensils coated with Teflon-II.

(B) Tools coated with Teflon-S.

Fig. 7-18. Some applications of Teflon. (Chemplast, Inc.)

PTFE requires special fabricating techniques. Its chemical inertness, unique weathering resistance, excellent electrical insulation characteristics, high heat resistance, low coefficient of friction, and nonadhesive properties have led to numerous uses. Parts coated with PTFE have such a low coefficient of friction they release and slide easily, and require no lubrication. Saw blades, cookware, utensils, snow shovels, bakery equipment, and bearings are common applications. Aerosol spray dispersions of micron-sized particles of polytetrafluoroethylene are used as a lubricant and antistick agent for metal, glass, or plastics substrates. Many profile shapes are used for chemical, mechanical, and electrical applications (Fig. 7-19). Shrinkable tubing is used to cover rollers, springs, glass, and electrical parts. Skived or extruded tapes and films may be used for seals, packing, and gasket materials. Bridges, pipes, tunnels, and buildings may rest on slip joints, expansion plates, or bushing or bearing pads of polytetrafluoroethylene (Fig. 7-20).

The excellent electrical insulating and low dissipation factors of polytetrafluoroethylene make

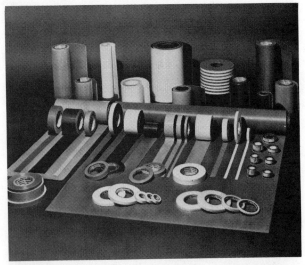

(A) A variety of tapes for many uses.

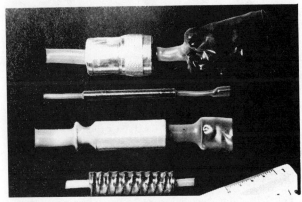

(B) Shrinkable protective tubing.

(C) Various rods and tubes.

Fig. 7-19. Various PTFE applications. (Chemplast, Inc.)

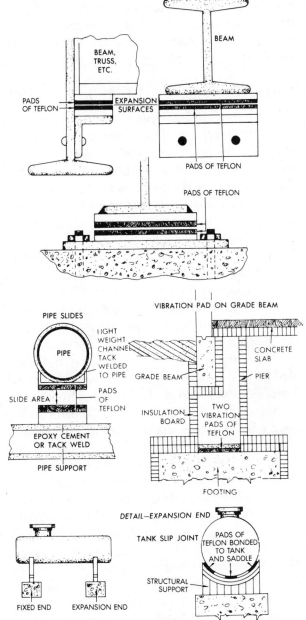

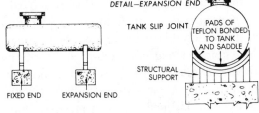

Fig. 7-20. Teflon pads have many construction uses. (Du Pont Co.)

it useful for wire and cable insulation, coaxial wire spacers, laminates for printed circuits, and many other electrical applications. Six advantages and eight disadvantages of PTFE are shown below.

Advantages of Polytetrafluoroethylene (PTFE)

1. Nonflammable
2. Outstanding chemical and solvent resistance
3. Excellent weatherability
4. Low friction coefficient (antistick property)
5. Wide thermal service range
6. Very good electrical properties

Disadvantages of Polytetrafluoroethylene

1. Not processable by common thermoplastic methods
2. Toxic in thermal degradation
3. Subject to creep
4. Permeable
5. Requires high processing temperatures
6. Low strength
7. High density
8. Comparatively high cost

Polyfluoroethylenepropylene (PFEP) or (FEP)

In 1965, DuPont announced another Teflon fluoroplastic wholly composed of fluorine and carbon atoms. Polyfluoroethylenepropylene (PFEP or FEP) is made by copolymerizing tetrafluoroethylene with hexafluoropropylene (Fig. 7-21).

The partial disruption of the polymer chain by the propylene-like groups $CF_3CF{=}CF_2$ reduces the melting point and viscosity of FEP resins. Polyfluoroethylenepropylene may be processed with normal thermoplastic methods. This reduces production costs of items previously molded of PTFE. Because of pendant CF_3 groups this copolymer is less crystalline, more processable, and transparent in films up to 0.25 mm [0.01 in.] thick.

Commercial PFEP plastics possess properties similar to those of PTFE. They are chemically inert, have very good electrical insulation properties, and a somewhat greater impact strength. Service temperatures may exceed 400 °F [205 °C]. Polyfluoroethylenepropylene plastics are used extensively by the military and aircraft and aerospace industries for electrical insulation and high reliability at cryogenic temperatures. They are used for lining chutes, pipes, and tubes, and coating objects where a low coefficient of friction or non-adhesive characteristics are required. PFEP is molded into parts including gaskets, gears, impellers, printed circuits, pipes, fittings, valves, expansion plates, bearings, and other profile shapes (Fig. 7-22).

As early as 1933, both Germany and the United States were making a fluoroplastic material in connection with atomic bomb research. It was used in handling uranium fluoride, a compound of uranium with fluorine.

The following list shows six advantages and disadvantages of polyfluoroethylenepropylene.

Fig. 7-22. Shrinkable PFEP antistick covers applied to rollers. (Chemplast, Inc.)

Advantages of Polyfluoroethylenepropylene (PFEP)

1. Processable by normal thermoplastic methods
2. Resistant to chemicals (including oxidizing agents)
3. Excellent solvent resistance
4. Antistick characteristics
5. Nonflammable
6. Low coefficient of friction, dielectric constant, mold shrinkage, and water absorption.

Disadvantages of Polyfluoroethylenepropylene

1. Comparatively high cost
2. High density
3. Subject to creep
4. Low compressive and tensile strength
5. Low stiffness
6. Toxic upon thermal decomposition

Polychlorotrifluoroethylene (PCTFE) or (CTFE)

Polychlorotrifluoroethylene is produced in various formulations. Chlorine atoms are substituted for fluorine in the carbon chain.

$$-CF_2-CF- \atop \qquad\quad | \atop \qquad\quad Cl$$

Polychlorotrifluoroethylene (PCTFE)

Monomers are obtained by fluorinating hexachloroethane, then dehalogenating (controlled removal of the halogen, chlorine) with zinc in alcohol:

$$CCl_3CCl_3 \xrightarrow[\text{HF}]{\text{Anhydrous}} CCl_2FCClF_2 \xrightarrow[\substack{\text{boiling} \\ \text{ethyl} \\ \text{alcohol}}]{\text{Zinc}} CClF$$

Hexachloroethane

$$= CF_2 + Cl_2$$

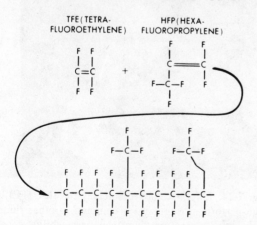

Fig. 7-21. Manufacture of polyfluoroethylenepropylene.

The polymerization is similar to PTFE, in that it is accomplished in an aqueous emulsion and suspension. During bulk polymerization, peroxide or Ziegler-type catalysts are used.

$$n\mathrm{CF_2 = CFCl} \xrightarrow{\text{polymerize}} \mathrm{(-CF_2CFCl-)}_n$$
Chlorotrifluoroethylene   Polychlorotrifluoroethylene

The addition of the chlorine atoms to the carbon chain allows processing by normal thermoplastic equipment. The chlorine presence also allows selected chemicals to attack and break the partially crystalline polymer chain. PCTFE may be produced in an optically clear form depending on the degree of crystallinity. Copolymerization with vinylidene fluoride or other fluoroplastics provides varying degrees of chemical inertness, thermal stability, and other unique properties.

Polychlorotrifluoroethylene is harder, more flexible, and possesses a higher tensile strength than PTFE. It is more expensive than PTFE and has a service temperature from $-400\ °F$ to $+400\ °F$ $[-240\ °C$ to $+205\ °C]$. Introduction of the chlorine atom lowers their electrical insulating properties and raises the coefficient of friction. Although it is more expensive than polytetrafluoroethylene, it finds similar uses. These include insulation for wire, cable, printed circuit boards, and electronic components. The property of chemical resistance is best used in producing transparent windows for chemicals, seals, gaskets, O-rings, and pipe lining, as well as pharmaceutical and lubricant packaging (Fig. 7-23). Dispersions and films may be used in coating reactors, storage tanks, valve bodies, fittings, and pipes. Films may be sealed by thermal or ultrasonic techniques. Epoxy adhesives may be used on chemically etched surfaces.

In Figure 7-24, the tough plastics ethyl-ene-chlorotrifluoroethylene (E-CTFE) is shown. The lists below show eight advantages and five disadvantages of polychlorotrifluoroethylene.

Advantages of Polychlorotrifluoroethylene (PCTFE or CTFE)

1. Excellent solvent resistance
2. Optically clear
3. Zero moisture absorption
4. Self-extinguishing
5. Low permeability
6. Creep resistance better than PTFE or PFEP

(A) Insulation for wires and cables.

(B) Ball-valve seats.

Fig. 7-23. Two uses of PCTFE. (Chemplast, Inc.)

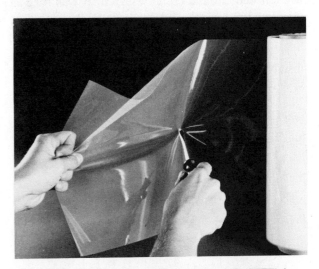

Fig. 7-24. Ethylenechlorotrifluoroethylene (E-CTFE) has high-performance properties common to other fluoropolymers. It surpasses them in impermeability, tensile strength, and resistance to abrasion. (Chemplast, Inc.)

7. Very low coefficient of friction
8. Good low temperature capability

Disadvantages of Polychlorotrifluoroethylene

1. Lower electrical properties than PTFE
2. More difficult to mold than PFEP
3. Crystallizes upon slow cooling
4. Less solvent resistance than PTFE and PFEP
5. Higher coefficient of friction than PTFE and PFEP

Polyvinyl Fluoride (PVF)

The earliest preparation of vinyl fluoride gas (in 1900) was considered impossible to polymerize. In 1958, however, Du Pont announced the polymerization of vinyl fluoride (PVF). In 1933, monomer resins were prepared in Germany by the reaction of hydrogen fluoride with acetylene using selected catalysts:

$$HF + CH \equiv CH \xrightarrow[\text{catalysts}]{} CH_2 = CHF$$

Although the monomer has been known to chemists for some time, it is hard to manufacture or polymerize. Polymerization is carried out using peroxide catalysts in various aqueous solutions at high pressures.

Polyvinyl fluoride may be processed by normal thermoplastic methods. These plastics are strong, tough, flexible, and transparent. They have outstanding weather resistance. PVF has good electrical and chemical resistance, with a service temperature near 300 °F [150 °C].

Uses include protective coatings and surfaces for exterior use, finishes for plywood, sealing tapes, packaging of corrosive chemicals, and many electrical insulating applications. Coatings may be applied to automotive parts, lawn mower housings, house shutters, gutters, and metal siding. Four advantages and disadvantages are shown below. (Also see polyvinyls)

Advantages of Polyvinyl Fluoride

1. Processable by thermoplastic methods
2. Low permeability
3. Flame retardance
4. Good solvent resistance

Disadvantages of Polyvinyl Fluoride

1. Lower thermal capability than highly fluorinated polymers
2. Toxic (in thermal decomposition)
3. High dipole bond
4. Subject to attack by strong acids

Polyvinylidene Fluoride (PVDF)

Closely resembling polyvinyl fluoride is polyvinylidene fluoride (PVF_2). It became available in 1961 and was produced by the Pennwalt Chemical Corporation under the trade name Kynar. Polyvinylidene fluoride is polymerized thermally by the dehydrohalogenation (removal of hydrogen and chlorine atoms) of chlorodifluoroethane under pressure:

$$CH_3CClF_2 \xrightarrow{500\text{-}1700\ °C} CH_2 = CF_2 + HCl$$

Like polyvinyl fluoride, PVDF does not have the chemical resistance of PTFE or PCTFE. The alternating CH_2 and CF_2 groups in its backbone contribute to its tough, flexible characteristics. The presence of the hydrogen atoms reduces chemical resistance and allows solvent cementing and degradation. PVDF materials are processed by thermoplastic methods, and they are ultrasonically and thermally sealed. PVDF finds wide use in film and coating forms because of its toughness, optical properties, and resistance to abrasion, chemicals, and ultraviolet radiation. Service temperatures range from −80 °F to +300 °F [−62 °C to +150 °C]. A familiar use is the coating seen on aluminum siding and roofs. Molded parts may include such items as valves, impellers, chemical tubing, ducting, and electronic components (Fig. 7-25). Six advantages and three disadvantages of polyvinylidene fluoride are listed below. (Also see polyvinyls)

Advantages of Polyvinylidene Fluoride

1. Processable by thermoplastic methods
2. Low creep
3. Excellent weatherability
4. Nonflammability
5. Abrasion resistance better than PTFE
6. Good solvent resistance

Disadvantages of Polyvinylidene Fluoride

1. Lower thermal capability and chemical resistance than PTFE or PCTFE
2. Toxic in thermal decomposition
3. High dipole

Perfluoroalkoxy (PFA)

In 1972, Du Pont offered perfluoroalkoxy (PFA) under the trade name of Tefzel. It is produced from polymerization of perfluoroalkoxyethylene.

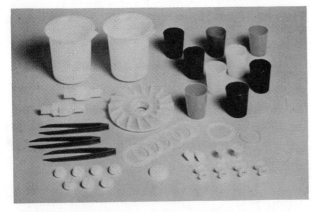

(A) Lab equipment, spools, impellers, containers, and gaskets.

(B) Film, sheet, rods, moldings, and coatings.

Fig. 7-25. Polyvinylidene fluoride has many uses. (KREHA)

PFA may be processed by conventional thermoplastics methods. This plastics has properties similar to PTFE and PFEP and is available in pellet, film, sheet, rod, and powder forms.

Uses include dielectric and electrical insulators, coatings, and liners for valves, pipes, and pumps. The following list includes seven advantages and six disadvantages of PFA.

Advantages of Perfluoroalkoxy (PFA)

1. Higher temperature capability than PFEP
2. Excellent resistance to chemicals (including oxidizing agents)
3. Excellent solvent resistance
4. Antistick characteristics
5. Nonflammability
6. Low coefficient of friction
7. Processable by thermoplastic methods

Disadvantages of Perfluoroalkoxy

1. Comparatively high cost
2. High density

3. Subject to creep
4. Low compressive and tensile strength
5. Low stiffness
6. Toxic in thermal decomposition

Other Fluoroplastics

There are numerous other fluorine-containing polymers and copolymers. Chlorotrifluoroethylene/vinylidene fluoride is used in fabricating O-rings and gaskets:

$$\left[\begin{array}{cccc} F & H & F & F \\ | & | & | & | \\ -C- & C- & C- & C- \\ | & | & | & | \\ F & H & Cl & F \end{array} \right]_n$$

Hexafluoropropylene/vinylidene fluoride is an outstanding oil- and grease-resistant elastomer for O-rings, seals and gaskets.

The chemical structure of two fluoroacrylate elastometers is shown in Figure 7-26. Polytrifluoronitrosomethane, fluorine-containing silicones and polyesters, and other fluorine-containing polymers are also being produced.

The linear chained copolymer ethylene-chlorotrifluoroethylene (ECTFE) has high performance properties common to other fluoroplastics. It surpasses them in permeability, tensile strength, wear resistance, and creep. Uses include release agents, tank linings, and dielectrics.

The copolymer ethylene-tetrafluoroethylene (ETFE) has properties and applications similar to those of ECTFE.

At degradation temperatures, toxic fluorine gas is released. (See Chapter 3, Health and Safety.) The properties of the basic fluoroplastics are listed in Table 7-6.

7-7 IONOMERS

In 1964, Du Pont introduced a new material, known as *ionomers*, that had characteristics of both thermoplastics and thermosets. Ionic bonding

$$(CH_2-CH)_n \qquad (CH_2-CH)_n$$
$$| \qquad\qquad |$$
$$C=O \qquad\qquad C=O$$
$$| \qquad\qquad |$$
$$O \qquad\qquad O$$
$$| \qquad\qquad |$$
$$CH_2 \qquad\qquad CH_2$$
$$| \qquad\qquad |$$
$$C_3F_7 \qquad CF_2-CF_2-O-CF_3$$

Fig. 7-26. Two fluoroacrylate elastomers.

Table 7-6. Properties of Fluoroplastics

Property	Polytetra-fluoroethylene (PTFE)	Polyfluoro-ethylenepro-pylene (PFEP)	Polychloro-trifluoroethylene (PCTFE)	Polyvinyl Fluoride (PVF)	Polyvinylidene Fluoride (PVF$_2$)
Molding qualities	Excellent	Excellent	Excellent	Excellent	Excellent
Relative density	2.14–2.2	2.12–2.17	2.1–2.2		1.75–1.78
Tensile strength, MPa	14–35	19–21	30–40	58–124	40–50
(psi)	(2 000–5 000)	(2 700–3 100)	(4 500–6 000)		(5 500–7 400)
Compressive strength, MPa	12–14		30–50	60	60
(psi)	(1 700–2 000)		(4 500–7 400)	(8 680)	(8 680)
Impact strength, Izod, J/mm	0.40	No break	0.125–0.135	0.18–0.2	0.18–0.2
(ft lb/in)	(8)		(2.5–2.7)	(3.6–4.0)	(3.6–4.0)
Hardness	Shore D50–D65	Rockwell R25	Rockwell R75–R95	Shore D80	Shore D80
Thermal expansion, 10^{-4}/°C	25	20–27	11–18	70	22
Resistance to heat, °C	287	205	175–199	149	149
(°F)	(550)	(400)	(350–390)	(300)	(300)
Dielectric strength, V/mm	19 000	19 500–23 500	19 500–23 500	10 000	10 000
Dielectric constant (at 60 Hz)	2.1	2.1	2.24–2.8	8.4	8.4
Dissipation factor (at 60 Hz)	0.000 2	<0.000 3	0.001 2	0.049	0.049
Arc resistance, s	300	165+	360	50–70	50–70
Water absorption (24 h), %	0.00	0.01	0.00	0.04	0.04
Burning rate, mm/min	None	None	None	Self-extin-guishing	Self-extin-guishing
Effect of sunlight	None	None	None	Slight bleach	Slight
Effect of acids	None	None	None	Attacked by fuming sulfuric	Attacked by fuming sulfuric
Effect of alkalies	None	None	None	None	None
Effect of solvents	None	None	None	Resists most	Resists most
Machining qualities	Excellent	Excellent	Excellent	Excellent	Excellent
Optical qualities	Opaque	Transparent	Translucent to opaque	Transparent	Transparent to translucent

is seldom found in plastics but is a characteristic of ionomers. Ionomers possess chains similar to polyethylene with ion cross-links of sodium, potassium, or similar ions (Fig. 7-27). In this material both organic and inorganic compounds are linked together, and because the cross-linking is basically ionic, the weaker bonds are easily broken on heating. This material may therefore be processed as a thermoplastics. At atmospheric temperatures, the plastics has properties usually associated with linked polymers.

The basic ionomer chain is made by polymerization of ethylene and methacrylic acid. Other inexpensive polymer chains may be developed with similar cross-links.

Since they combine ionic and covalent forces in their molecular structure, ionomers can exist in a number of physical states and have a number of physical properties. They may be processed and reprocessed by any thermoplastic technique. They are more expensive than polyethylene, but possess a higher moisture vapor permeability than polyethylene. Ionomers are available in transparent forms.

Uses of ionomers include safety glasses, shields, bumper guards, toys, containers, packaging films, electrical insulation, and coatings for paper, bowling pins, or other substrates (Fig. 7-28). In the shoe industry they are used for inner liners and as soles and heels. Ionomers are coextruded with polyester films to produce a heat-sealable layer while improving package durability.

Ionomers are used for a number of composite market applications. Laminated or coextruded films are used as tear-open pouches for pharma-

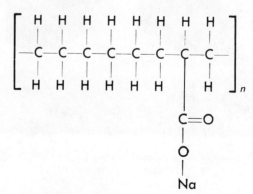

Fig. 7-27. An example of ionomer structure.

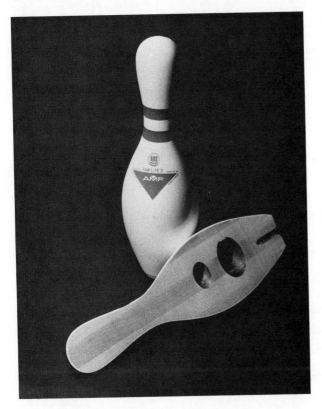

(A) Bowling pins coated with ionomer last longer than pins covered with other protective materials.

(B) Foam injection-molded bumper guards of ionomer are stronger and lighter than solid constructions.

Fig. 7-28. Ionomer applications. (Du Pont Co.)

ceutical and food packaging. Metal foil laminates and heat-sealable skin and blister packaging continue to grow. Foamed applications include bumper guards, shoe components, wrestling mats, and ski lift seat pads. Ionomer coatings on golf balls and bowling pins extend the service lives of these products.

Table 7-7 gives some properties of ionomers while the following list shows seven advantages and four disadvantages of ionomers.

Advantages of Ionomers

1. Outstanding abrasion resistance
2. Excellent shock resistance, even at low temperatures
3. Good electrical traits
4. High melt strength
5. Abrasion resistance
6. Do not dissolve in common solvents
7. Excellent transparent colors

Disadvantages of Ionomers

1. Some swelling from detergent-alcohol mixtures
2. Must be stabilized for exterior use
3. Service temperature of 161 °F [72 °C]
4. Must compete with the less expensive polyolefins

7-8 NITRILE BARRIER PLASTICS

Formulations of copolymers with a nitrile (C≡N) functionality of over 50% are called nitrile polymers.

Table 7-7. Properties of Ionomers

Property	Ionomer
Molding qualities	Excellent
Relative density	0.93–0.96
Tensile strength, MPa	24–35
(psi)	(3 500–5 000)
Izod, impact, J/mm	0.3–0.75
(ft lb/in.)	(6–15)
Hardness, Shore	D50–D65
Thermal expansion, 10^{-4}/°C	30
Resistance to heat, °C	70–105
(°F)	(160–220)
Dielectric strength, V/mm	35 000–40 000
Dielectric constant (at 60 Hz)	2.4–2.5
Dissipation factor (at 60 Hz)	0.001–0.003
Arc resistance, s	90
Water absorption (24 h), %	0.1–1.4
Burning rate, mm/min	Very slow
Effect of sunlight	Requires stabilizers
Effect of acids	Attacked by oxidizing acids
Effect of alkalies	Very resistant
Effect of solvents	Very resistant
Machining qualities	Fair to good
Optical qualities	Transparent

Table 7-8. Estimated Compositions of Nitrile Barrier Plastics

Composition	Borg-Warner Cycopac 930	Monsanto Lo Pac	Sohio Barex	DuPont NR–16	ICI LPT*
Acrylonitrile, %	65–75	65–75	65–75	65–75	65–75
Butadiene, %	5–10		5–8		
Methyl acrylate, %			20–25		
Methyl methacrylate, %				3–5	
Styrene, %	20–30	25–35		25–35	25–35

*LPT–Low permeability thermoplastics

These plastics offer very low permeability and form a barrier against gases and odors. This property is the result of the high nitrile content. Their transparency and processibility make them useful as containers.

Each maker of nitrile barrier plastics varies the formulation.

Most combinations are based on acrylonitrile (AN or methacrylonitrile). Some formulations may reach 75% AN. All formulations are amorphous, with a slight yellow tint. Approximate monomer compositions of nitrile barrier are shown in Table 7-8. The Borg-Warner product, *Cyclopac 930,* contains more than 64 percent acrylonitrile, 6 percent butadiene, and 21 percent styrene.

Although heat-sensitive, nitrile barrier plastics have been used with all thermoplastics processing techniques (Table 7-9).

These plastics have found uses as beverage containers, food packages, and containers for many nonfood fluids.

Table 7-9. Properties of Nitrile Barrier Plastics

Property	Nitrile Barrier (Standard)
Molding qualities	Good
Relative density	1.15
Tensile strength, MPa	62
(psi)	(9 000)
Impact strength, Izod, J/mm	0.075–0.2
(ft lb/in)	(1.5–4)
Hardness, Rockwell	M72–78
Thermal expansion, $10^{-4}/°C$	16.89
Resistance to heat, °C	70–100
(°F)	(158–212)
Dielectric strength, V/mm	8 660
Dielectric constant (at 60 Hz)	4.55
Dissipation factor (at 100 Hz)	0.07
Water absorption (24 h), %	0.28
Burning rate, mm/min	—
Effect of sunlight	Slight yellowing
Effect of acids	None to attacked
Effect of alkalies	None to attacked
Effect of solvents	Dissolves in acetonitrile
Machining qualities	Good
Optical qualities	Transparent

The AN residual monomers of food containers cannot exceed 0.10 p.p.m

7-9 PHENOXY

A family of resins based on bisphenol A and epichlorohydrin was introduced in 1962 by Union Carbide. These resins were called *phenoxy* but could be classed as polyhydroxyethers. The plastics structure resembles polycarbonates, and the material has similar properties.

Phenoxy resins are made and sold as thermoplastic epoxide resins. They may be processed on normal thermoplastic machinery, with product service temperatures in excess of 170 °F [75 °C].

Phenoxy resins, because of the reactive hydroxyl groups, may be cross-linked. Cross-linking agents may include diisocyanates, anhydrides, triazines, and melamines.

Homopolymers have good creep resistance, high elongation, low moisture absoprtion, low gas transmission, and high rigidity, tensile strength, and ductility. They find applications as clear or colored protective coatings, molded electronic parts, pipe for gas and crude oil, sports equipment, appliance housings, cosmetic cases, adhesives, and containers for food or drugs.

Table 7-10 gives some of the properties of phenoxy.

7-10 POLYALLOMERS

In 1962, a plastics distinctly different from simple copolymers of polyethylene and polypropylene was produced by Eastman. The process,

Table 7-10. Properties of Phenoxy

Property	Phenoxy
Molding qualities	Good
Relative density	1.18–1.3
Tensile strength, MPa	62–65
(psi)	(9 000–9 500)
Impact strength, Izod, J/mm	0.125
(ft lb/in)	(2.5)
Resistance to heat, °C	80
(°F)	(175)
Dielectric constant (at 60 Hz)	4.1
Dissipation factor (at 60 Hz)	0.001
Water absorption (24 h), %	0.13
Burning rate	Self-extinguishing
Effect of acids	Resistant
Effect of alkalies	Resistant
Effect of solvents	Soluble in ketones
Machining qualities	Good
Optical qualities	Translucent to opaque

called *allomerism*, is conducted by alternately polymerizing ethylene and propylene monomers. Allomerism is a variation in chemical composition without a change in crystalline form. The plastics exhibits the crystallinity usually associated with the homopolymers of ethylene and propylene. The term *polyallomer* is used to distinguish this alternately segmented plastics from homopolymers and co-polymers of ethylene and propylene.

Although highly crystalline, polyallomers may have a relative density as low as 0.896. They are available in various formulations. Properties include high stiffness, impact strength, and abrasion resistance. Their flexibility has been used in the production of hinged boxes, looseleaf binders, and various other folders (Fig. 7-29). Polyallomers may be used as food containers or films over a wide range of service temperatures. They find limited use where other polyolefins are marginal.

Processing may be done on all normal thermoplastic equipment. Like polyethylene, polyal-

Fig. 7-29. The covers of these notebooks are made of poly-allomer plastics.

lomers are not cohesively bonded but may be welded.

Table 7-11 lists some of the properties of polyallomers.

7-11 POLYAMIDES (PA)

From research that began in 1928, Wallace Hume Carothers and his colleagues concluded that linear polyesters were not suitable for commercial fiber production. Carothers did succeed in producing polyesters of high molecular mass and oriented them by elongation under tension. These fibers were still inadequate and could not be successfully spun. The amino acids present in silk, a natural fiber, prompted Carothers to study synthetic polyamides. Of many formulations of amino acids, diamines, and dibasic acids, several showed promise as possible fibers. By 1938, the first commercially developed polyamide was introduced by Du Pont. It was 6,6 polyamide and was given the trade name *Nylon*. This condensation plastics was called Nylon 6,6 (also written as 66 or 6/6) because both the acid and the amine contain six carbon atoms.

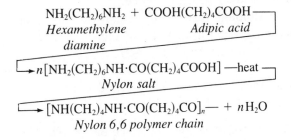

$$NH_2(CH_2)_6NH_2 + COOH(CH_2)_4COOH$$
$$\textit{Hexamethylene} \qquad \textit{Adipic acid}$$
$$\textit{diamine}$$

$$n[NH_2(CH_2)_6NH\cdot CO(CH_2)_4COOH] - heat$$
$$\textit{Nylon salt}$$

$$[NH(CH_2)_4NH\cdot CO(CH_2)_4CO]_n - + nH_2O$$
$$\textit{Nylon 6,6 polymer chain}$$

Table 7-11. Properties of Polyallomers

Property	Polyallomer (Homopolymer)
Molding qualities	Excellent
Relative density	0.896–0.899
Tensile strength, MPa	20–27
(psi)	(3 000–3 850)
Impact strength, Izod, J/mm	8.5–12.5
(ft lb/in)	(170–250)
Hardness, Rockwell	R50–R85
Thermal expansion, 10^{-4}/°C	21–25
Resistance to heat, °C	50–95
(°F)	(124–200)
Dielectric strength, V/mm	32 000–36 000
Dielectric constant (at 60 Hz)	2.3–2.8
Dissipation factor (at 60 Hz)	0.000 5
Water absorption (24 h), %	0.01
Burning rate	Slow
Effect of sunlight	Slight—should be protected
Effect of acids	Very resistant
Effect of alkalies	Very resistant
Effect of solvents	Very resistant
Machining qualities	Good
Optical qualities	Transparent

Nylon has come to mean any polyamide that can be processed into filaments, fibers, films, and molded parts.

The repeating—CONH—(amide) link is present in a series of linear, thermoplastic Nylons.

- Nylon 6—Polycaprolactam:

$$[NH(CH_2)_5CO]_x$$

- Nylon 6,6—Polyhexamethyleneadipamide:

$$[NH(CH_2)_6NHCO(CH_2)_4CO]_x$$

- Nylon 6,10—Polyhexamethylenesebacamide:

$$[NH(CH_2)_6NHCO(CH_2)_8CO]_x$$

- Nylon 11—Poly(11-aminoundecanoic acid)

$$[NH(CH_2)_{10}CO]_x$$

- Nylon 12—Poly(12-aminododecanoic acid)

$$[NH(CH_2)_{11}CO]_x$$

There are many other types of nylon currently available, including Nylon 8, 9, 46, and copolymers from more sophisticated diamines and acids. In the United States, Nylon 6,6 and Nylon 6 are by far the most used. Properties of Nylons may also be changed by introducing additives. Amino-containing polyamide resins will react with a number of materials, and cross-linked, thermosetting reactions are possible.

Although developed primarily as a fiber, polyamides find use as molding compounds, extrusions, coatings, adhesives, and casting materials. Acetals and fluorocarbons share some of the same properties and uses as polyamide. Polyamide resins are costly. They are selected when other resins will not meet service requirements. Acetal resins are superior in fatigue endurance, creep resistance, water resistance. Nylons have met increased competition from these resins.

Molding compounds were offered in 1941 and have grown in applications. They are among the toughest of plastics materials. Nylons are self-lubricating, impervious to most chemicals, and highly impermeable to oxygen. They are not attacked by fungi or bacteria. Polyamides may be used as food containers.

The largest applications of homopolymer molding compounds (Nylons 6; 6,6; 6,10; 11; and 12) include gears, cams, bearings, valve seats, combs, furniture casters, and door catches. They are used where wear resistance, quiet operation, and low coefficients of friction are needed.

Because of their crystalline structure, polyamide products have a milky-opaque appearance (Fig. 7-30). Transparent films may be obtained from Nylon 6 and 6,6 if they are cooled very rapidly. Polyamides are clear amorphous materials when melted. On cooling, they crystallize and become cloudy. This crystallinity contributes to stiffness, strength, and heat resistance. Polyamides are harder to process than other thermoplastic materials. All thermoplastic processing equipment may be used, but fairly high processing temperatures are needed. The melting point of a polyamide is abrupt or sharp; that is, they do not soften or melt over a broad range of temperatures (Fig. 7-31). When sufficient energy has overcome the crystalline and molecular attractions, they suddenly become liquid and may be processed. Because all

Fig. 7-30. Polyamide products have a milky-opaque appearance as seen in this valve and fitting. (Du Pont Co.)

Fig. 7-31. An automobile radiator can be made of polyamide because it will not melt over a wide temperature range. (BASF)

Nylons absorb water, they are dried before molding. This ensures desirable physical properties in the molded product.

Extruded and blown films are used to package oils, grease, cheese, bacon, and other products where low gas permeability is essential. The high service temperatures of Nylon film are used for boil- and bake-in-the-bag food products. Although polyamides are hygroscopic (absorb water), they find many applications as electrical insulators.

Nylon 11 may be used as a protective coating on metal substrates. Polyamides are used in a powder form by spraying or fluidized bed processes. Typical uses include rollers, shafts, panel slides, runners, pump impellers, and bearings. Water dispersions and organic solvents of polyamide resin permit certain adhesive and coating applications on paper, wood, and fabrics.

Polyamide-based adhesives may be of the hot-melt or solution type. Hot-melt adhesives are simply heated above the melting point and applied. Amino-polyamide resins may be reacted with epoxy or phenolic resins to produce a thermosetting adhesive. These adhesives find use in bonding wood, paper laminates, and aluminum, and in adhering copper to printed circuit boards. They are used as flexible adhesives for bread wrappers, dried soup packets, cigarette packages, and bookbindings. Polyamide-epoxy combinations are used as two-part systems in casting applications, including potting and encapsulation of electrical components. When combined with pigments and other modifying agents, polyamides may be used as printing inks. The uses of polyamides in textiles and carpets are well known and need little discussion.

Clothing, light tents, shower curtains, and umbrellas are among products made from Nylon. The monofilaments, multifilaments, and staple fibers are made by melt spinning. This is followed by cold drawing to increase tensile strength and elasticity. Monofilaments are used in fishing lines, surgical sutures, tire cords, rope, sports equipment, brushes, artificial human hair, and synthetic animal fur.

Polyamides are easily machined (Fig. 7-32), but drilled or reamed holes are likely to be slightly undersized, due to the resiliency of the material.

Cementing polyamide is difficult because of its solvent resistance; however, phenols and formic acid are specific solvents that are used in cementing polyamides. Epoxy resins are also employed for this purpose.

Fig. 7-32. This polyamide part was machined in 18.5 seconds on an automatic lathe. The same part in light metal would take more than twice as long to machine; a free-cutting steel part, 13 times as long. (BASF)

Some of the basic traits of polyamides are shown in Table 7-12. Seven advantages and five disadvantages are listed below.

Advantages of Polyamide (Nylon)

1. Tough, strong and impact resistant
2. Low coefficient of friction
3. Abrasion resistance
4. High temperature resistance
5. Processable by thermoplastic methods
6. Good solvent resistance
7. Resistant to bases

Disadvantages of Polyamide

1. High moisture absorption with related dimensional instability
2. Subject to attack by strong acids and oxidizing agents
3. Requires ultraviolet stabilization
4. High shrinkage in molded sections
5. Electrical and mechanical properties influenced by moisture content

7-12 POLYCARBONATES (PC)

An important material in the production of plastics is phenol. It is used in producing phenolic, polyamide, epoxy, polyphenylene oxide, and polycarbonate resins.

Table 7-12. Properties of Polyamides

Property	Nylon 6,6 (Unfilled)	Nylon 6,10 (Unfilled)	Nylon 6,10 (Glass-Filled)
Molding qualities	Excellent	Excellent	Excellent
Relative density	1.13–1.15	1.09	1.17–1.52
Tensile strength, MPa	62–82	58–60	89–240
(psi)	(9 000–12 000)	(8 500–8 600)	(13 000–35 000)
Compressive strength, MPa	46–86	46–90	90–165
(psi)	(6 700–12 500)	(6 700–13 000)	(13 000–24 000)
Impact strength, Izod, J/mm	0.05–0.1	0.06	0.06–0.3
(ft lb/in)	(1.0–2.0)	(1.2)	(1.2–6)
Hardness, Rockwell	R108–R120	R111	M94, E75
Thermal expansion, $10^{-4}/°C$	20	23	3–8
Resistance to heat, °C	80–150	80–120	150–205
(°F)	(180–300)	(180–250)	(300–400)
Dielectric strength, V/mm	15 000–185 000	13 500–19 000	16 000–20 000
Dielectric constant (at 60 Hz)	4.0–4.6	3.9	4.0–4.6
Dissipation factor (at 60 Hz)	0.014–0.040	0.04	0.001–0.025
Arc resistance, s	130–140	100–140	92–148
Water absorption, (24 h), %	1.5	0.4	0.2–2
Burning rate, mm/min	Self-extinguishing	Self-extinguishing	Self-extinguishing
Effect of sunlight	Discolors slightly	Discolors slightly	Discolors slightly
Effect of acids	Attacked	Attacked	Attacked
Effect of alkalies	Resistant	None	None
Effect of solvents	Dissolved by phenol & formic acid	Dissolved by phenols	Dissolved by phenols
Machining qualities	Excellent	Fair	Fair
Optical qualities	Translucent to opaque	Translucent to opaque	Translucent to opaque

Phenol is a compound that has one hydroxyl group attached to an aromatic ring. It is sometimes called *monohydroxy benzene,* C_6H_5OH.

Bisphenol A (two phenol and acetone), a vital ingredient in the production of polycarbonates, may be prepared by combining acetone with phenol (Fig. 7-33). Bisphenol A is sometimes referred to as diphenylol propane or bis-dimethylmethane.

Polycarbonates are linear, amorphous polyesters because they contain esters of carbonic acid and an aromatic bisphenol.

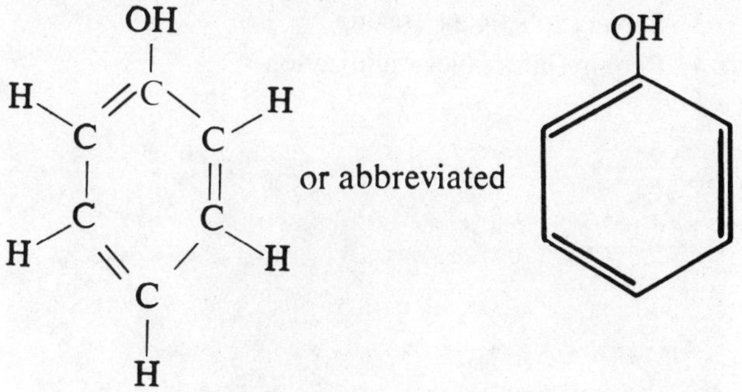

or abbreviated

Phenol or monohydroxy benzene

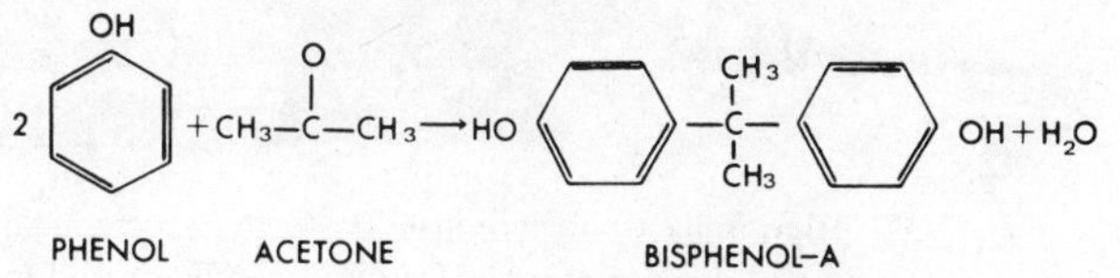

Another important material used in the production of polycarbonate is phosgene. Phosgene, a poisonous gas, was used in World War I.

As early as 1898, A. Einhorn prepared a polycarbonate material from the reaction of resorcinol and phosgene. Both W. H. Carothers and F. J. Natta performed research on a number of polycarbonates using ester reactions.

Research continued after World War II in Germany by Farbenfabriken Bayer and in the United States by General Electric. By 1957, both had arrived at the production of polycarbonates produced from bisphenol A. Volume production in the United States did not begin until 1959.

There are two general methods of preparing polycarbonates. The most common method is to react purified bisphenol A with phosgene under alkaline conditions (Fig. 7-34). An alternate method involves the reaction of purified bisphenol A with diphenyl carbonate (meta-carbonate) in the presence of catalysts, under a vacuum (Fig. 7-35).

Purity of the bisphenol A is vital if the plastics is to possess high-clarity and long linear chains with no cross-linking substances.

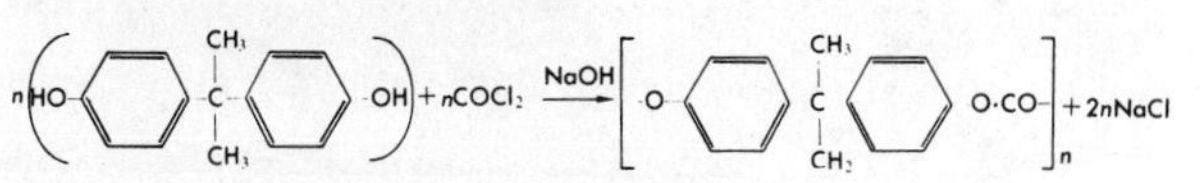

Fig. 7-33. Preparation of bisphenol-A.

Fig. 7-34. The first method used to prepare polycarbonates.

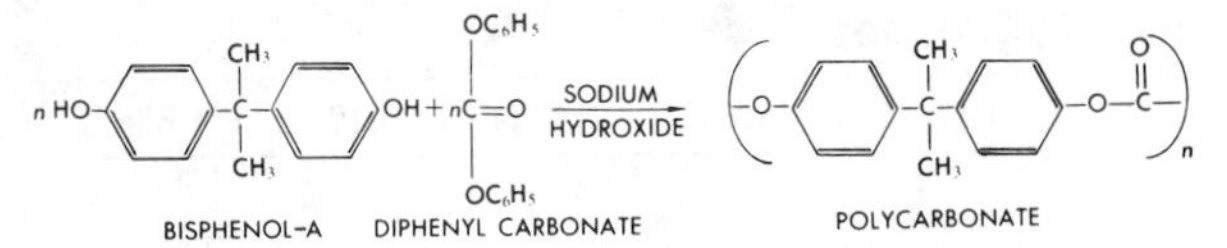

Fig. 7-35. The second method used to prepare polycarbonates.

The phosgenation process is preferred, since it may be carried out at low temperatures using simple technology and equipment. The process requires the recovery of solvents and inorganic salts. In either method, however, the product is comparatively high in cost.

Polycarbonates may be processed by all usual themoplastic methods. Their resistance to heat and their high melt temperatures does require higher processing temperatures. The molding temperature is very critical and must be accurately controlled to produce usable products. Polycarbonates are sensitive to hydrolysis at high processing temperatures. Compounds should be dried, or vented barrel equipment used, since water will cause bubbles and other blemishes on parts. The unique properties of polycarbonate are due to the carbonate groups and the presence of benzene rings in the long, repeating molecular chain. Polycarbonate properties include: high impact strength, transparency, excellent creep resistance, wide temperature limits, high dimensional stability, good electrical characteristics, and self extinguishing behavior. Tough transparent grades are used in lenses, films, windshields, light fixtures, containers, appliance components and tool housings (Fig. 7-36). Temperature resistance is made use of in hot-dish handles, coffee pots, popcorn popper lids, hair dryers, and appliance housings. These plastics have excellent properties from −275 °F to +270 °F [−170 °C to +132 °C]. Polycarbonates supply the impact and flexural strength needed in pump impellers, safety helmets, beverage dispensers, small appliances, trays, signs, aircraft parts, cameras, and various packaging and film uses. Coextruded packages are used for ovenable frozen food trays and microwave pouches. Polycarbonate parts also have very good dimensional stability. Glass-filled grades have improved impact, moisture, and chemical resistance (Fig. 7-37).

Most aromatic solvents, esters, and ketones will attack polycarbonates. Chlorinated hydrocarbons are used as solvent cements for cohesive bonds.

There are several hundred variations on the polycarbonate structure. The structure may be changed by substituting various radicals as side groups or by separating the benzene rings by more than one carbon atom. Some possible structural combinations are shown in Figure 7-38.

Some of the properties of polycarbonates are listed in Table 7-13. A list of five advantages and four disadvantages follows.

Advantages of Polycarbonates

1. High impact strength
2. Excellent creep resistance
3. Available in transparent grades
4. Continuous application temperature over 120 °C (248 °F)
5. Very good dimensional stability

Disadvantages of Polycarbonates

1. High processing temperatures
2. Poor resistance to alkalies
3. Subject to solvent crazing
4. Require ultraviolet stabilization

Fig. 7-36. Polycarbonate used in products around the home.

Fig. 7-37. High-impact polycarbonate has better mechanical properties and equivalent electrical properties when used to replace glass insulators. (H. K. Porter Co., Inc.)

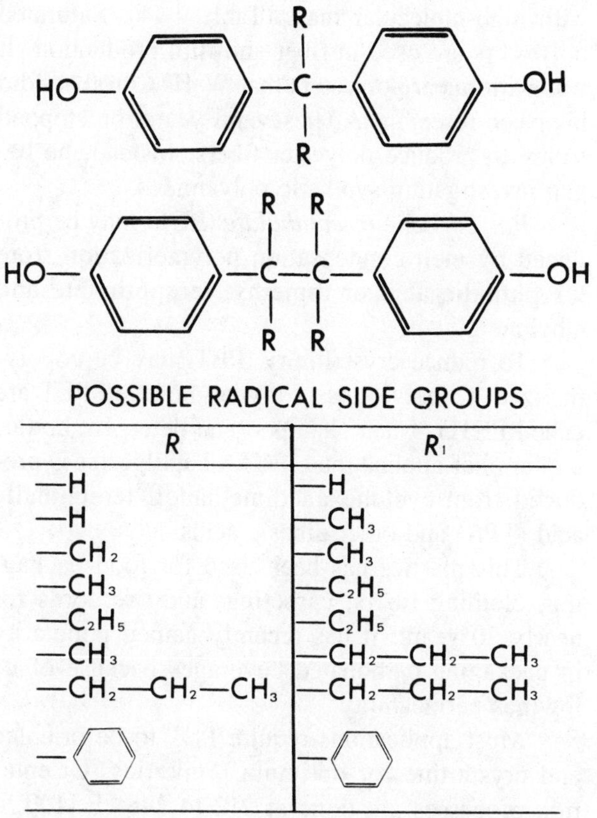

POSSIBLE RADICAL SIDE GROUPS

R	R_1
—H	—H
—H	—CH_3
—CH_2	—CH_3
—CH_3	—C_2H_5
—C_2H_5	—C_2H_5
—CH_3	—CH_2—CH_2—CH_3
—CH_2—CH_2—CH_3	—CH_2—CH_2—CH_3
— ⬡	— ⬡

Fig. 7-38. Possible combinations of polycarbonates.

7-13 POLYETHERETHERKETONE (PEEK)

The wholly aromatic structure of PEEK contributes to the high temperature resistance of this crystalline thermoplastic. The basic repeating unit is shown in Figure 7-39. PEEK can be melt-processed in conventional thermoplastic equipment. Applications include coatings on wire and high temperature composites for aerospace and aircraft components.

7-14 POLYETHERIMIDE (PEI)

Polyetherimide (PEI) is an amorphous thermoplastic based on repeating ether and imide units. The general chemical structure of these polymers is shown in Figure 7-40. Reinforced and filled grades improve strength, temperature resistance, and creep. All grades are processed on conventional equipment. Typical applications include jet engine components, ovenable cookware, flexible circuitry, composite structures for aircraft, and food packaging.

Table 7-13. Properties of Polycarbonates

Property	Polycarbonate (Unfilled)	Polycarbonate (10–40% Glass-Filled)
Molding qualities	Good to excellent	Very good
Relative density	1.2	1.24–1.52
Tensile strength, MPa	55–65	83–172
(psi)	(8 000–9 500)	(12 000–25 000)
Compressive strength, MPa	71–75	90–145
(psi)	(10 300–10 800)	(13 000–21 000)
Impact strength, Izod (ft lb/in.)	0.6–0.9	0.06–0.325
J/mm	(12–18)	(1.2–6.5)
Measurement of bar	12.7 × 3.175 mm	6.35 × 12.7 mm
Hardness, Rockwell	M73–78, R115, R125	M88–M95
Thermal expansion, $10^{-4}/°C$	16.8	4.3–10
Resistance to heat, °C	120	135
(°F)	(250)	(275)
Dielectric strength, V/mm	15 500	18 000
Dielectric constant (at 60 Hz)	2.97–3.17	3.0–3.53
Dissipation factor (at 60 Hz)	0.000 9	0.000 9–0.001 3
Arc resistance, s	10–120	5–120
Water absorption, (24 h), %	0.15–0.18	0.07–0.20
Burning rate, mm/min	Self-extinguishing	Slow 20–30
(in/min)		(0.8–1.2)
Effect of sunlight.	Slight	Slight
Effect of acids	Attacked slowly	Attacked by oxidizing acids
Effect of alkalies	Attacked	Attacked
Effect of solvents	Soluble in aromatic and chlorinated hydrocarbons	Soluble in aromatic and chlorinated hydrocarbons
Machining qualities	Excellent	Fair
Optical qualities	Transparent to opaque	Translucent to opaque

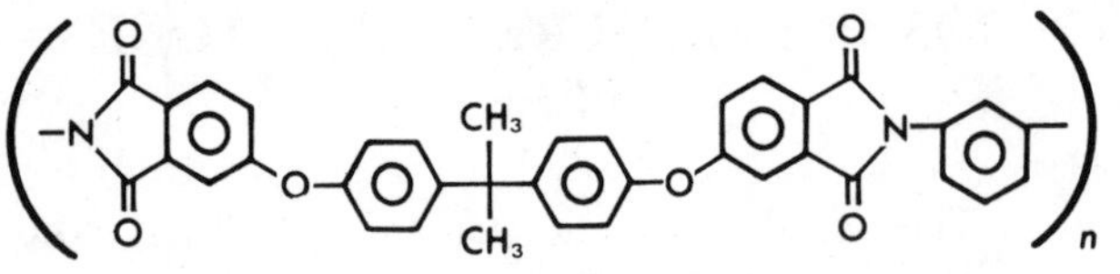

Fig. 7–39. The general chemical structure of polyetherimide.

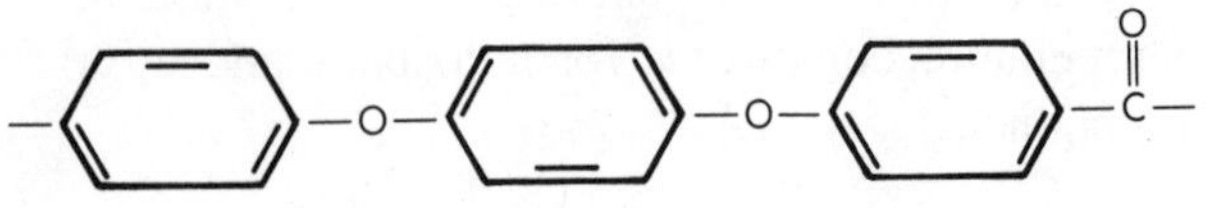

Fig. 7–40. Chemical structure of PEEK.

7-15 THERMOPLASTIC POLYESTERS

The thermoplastic polyester group of plastics includes saturated polyesters and aromatic polyesters. Familiar trade names of saturated polyesters are Dacron and Mylar.

Saturated Polyesters

Saturated polyesters are based on the reaction of terephthalic acid ($C_6H_4(COOH)_2$) and ethylene glycol ($(CH_2)_2(OH)_2$). They are linear polymers with high-molecular mass (Table 7-14). Saturated polyesters are used in fiber and film production. It was with linear polyesters that W. H. Carothers did his basic research. After several years, he stopped trying to produce polyester fibers. Instead, he began investigating synthetic polyamides.

Polyethylene terephthalate (PET) may be produced by melt condensation polymerization from terephthalic acid or dimethyl terephthalate and ethylene glycol.

To reduce crystallinity, PET may be copolymerized. Copolyesters of glycol-modified PET are called PETG. Clear shampoo and detergent bottles are familiar applications. PCTA copolyester is produced from cyclohexanedimethanol, terephthalic acid (TPA) and other dibasic acids.

This plastics has been used for food packaging, clothing fibers, carpeting, and tire cords for nearly 20 years. It has recently gained popularity in packaging carbonated beverages because of its low gas permeability.

Most applications require PET to be oriented and crystalline for optimum properties. Orientation processes are done at 212 to 248 °F [100 to 120 °C], or slightly above the glass-transition temperature (T_g).

Table 7-14. Properties of Thermoplastic Polyesters: Saturated Polyester

Property	Polyethylene Terephthalate (PETP or PET)	Polybutylene Terephthalate (PBTP) (Unfilled)	Linear Aromatic (Decomposes at 550)	Linear Aromatic (Injection grade)
Molding qualities	Good	Good	Sinters	Good
Relative density	1.34–1.39	1.31–1.38	1.45	1.39
Tensile strength, MPa	59–72	56	17	20
(psi)	(8 550–10 500)	(8 100)	(2 500)	(2 900)
Compressive strength, MPa	76–128	59–100	76–105	68
(psi)	(11 025–18 130)	(8 550–14 500)	(11 025–15 230)	(9 860)
Impact Strength, Izod (ft lb/in)	0.01–0.04	0.04–0.05		0.08
J/mm	(0.2–0.8)	(0.8–1)		(1.6)
Hardness	Rockwell M94–M101	Rockwell M68–M98	Shore D88	—
Thermal expansion, 10^{-4}/°C	15.2–24	15.5	7.1	7.36
Resistance to heat, °C	80–120	50–90		280
(°F)	(176–248)	(122–194)		(536)
Dielectric strength, V/mm	13 780–15 750	16 500		13 750
Dielectric constant (at 60 Hz)	3.65	3.29	3.22	
Dissipation factor (at 60 Hz)	0.005 5		0.004 6	
Arc resistance, s	40–120	75–192		100
Water absorption (24 h), %	0.02	0.08		0.01
Burning rate, mm/min	Slow burning	10		
Effect of sunlight	Discolors slightly	Discolors	None	
Effect of acids	Attacked by oxidizing acids	Attacked	Slight	Slight
Effect of alkalies	Attacked	Attacked	Attacked	Attacked
Effect of solvents	Attacked by halogen hydrocarbons	Resistant	Resistant	Resistant
Machining qualities	Excellent	Fair	Fair	Good
Optical qualities	Transparent to opaque	Opaque	Opaque	Translucent

PET is used for synthetic fibers, photographic film, videotape, dual ovenable containers, computer and magnetic tapes, and numerous beverage bottles, including distilled spirits. Reinforced and filled grades are used in gears, cowl vent grills, electrical switches, and sporting goods.

Polybutylene terephthalate (PBT) or polytetramethylene terephthalate (PTMT) were introduced in 1962. Ethylene glycol, an automobile antifreeze, is also one of the main materials in production of polyester fibers. The original development began in England by I.C.I. By 1953, Du Pont had purchased rights to develop Dacron fibers. Extrusion and injection grades were on the market by 1969.

Saturated (unreactive) polyesters do not undergo any cross-linking. These linear polyesters are thermoplastic. Clothing and draperies are common uses of these fibers. Industrial uses may include reinforcements for belting or tires.

Polyester films are used for recording tape, dielectric insulators, photographic film, and boil-in-the-bag food products (Fig. 7-41).

Because of their thermoplastic nature, compounds based on saturated PET and PBT may be injection or extrusion molded.

(A) These vegetables are packaged in films designed for "boil-in-the-bag" use.

(B) Polyester "bake-in-the-bag" film helps retain moisture and flavor, and eliminates need for cleanup.

Fig. 7-41. Some uses of polyester films.

Well-known trade names include Teryl, Dacron, and Kodel fibers, and Mylar film. Other uses include gears, distributor caps, rotors, appliance housings, pulleys, switch parts, furniture, fender extensions and packaging. Following is a list of two advantages and three disadvantages of saturated polyesters.

Advantages of Saturated Polyesters

1. Tough and rigid
2. Processable by thermoplastic methods

Disadvantages of Saturated Polyesters

1. Subject to attack by acids and bases
2. Low thermal resistance
3. Poor solvent resistance

Aromatic Polyesters

Oxybenzoyl polyesters were introduced by Carborundum in 1971 and 1974 under the trade names Ekonol and Ekcel. Both materials are linear chains of *p*-oxybenzoyl units. Because Ekonol does not melt below its decomposition temeprature, it must be sintered, compression molded, or plasma-sprayed. Ekcel can be processed with injection and extrusion equipment. High-temperature stability, stiffness, and thermal conductivity are important properties.

Some formulations can be melt-processed, but might require processing temperatures between 572 °–842 °F (300 °–400 °C). Members of this class of materials are sometimes called nematic, anisotropic, liquid crystal polymer (LCP), or self-reinforcing polymers. These terms try to describe the formation of tightly packed fibrous chains during the melt phase. It is the fibrous chain that gives the polymer its self-reinforcing qualities. Parts must be designed to accommodate the anisotropic characteristics of LCP. Applications include chemical pumps, dual ovenable cookware, engine parts, and aerospace components.

Typical uses include bearings, seals, valve seats, rotors, high performance aerospace and automotive parts, electrical insulation components, and coatings for pans.

In 1978, *polyarylate* was introduced. This light-amber colored plastics is made from iso- and terephthalic acid and bisphenol A.

Polyarylates are aromatic polyester thermoplastic materials. The term *aryl* refers to a phenyl group derived from an aromatic compound.

A number of alloy and filled grades are available.

Polyarylate must be dried before injection or extrusion molding. It has excellent ultraviolet, thermal, and heat-deflection resistance. Applications include glazing, appliance housings, electrical connectors and lighting fixtures, exterior glazing, halogen lamp lenses, and selected microwave cookware.

Typical properties are shown in Table 7-15.

7-16 THERMOPLASTIC POLYIMIDES

Polyimides were developed by Du Pont in 1962. They are obtained from condensation polymerization of an aromatic dianhydride and an aromatic diamine (Fig. 7-42). Aromatic polyimides are linear and thermoplastic, and are hard to process. They can be molded by allowing enough time for flow to occur once glass transition temperature is exceeded. Many polyimides do not melt, but must be fabricated by machining or other forming methods. (See Thermosetting Polyimides in Chapter 8.)

Addition polymerization provides plastics with slightly lower heat resistance than does condensation polymerization.

Polyimides compete with various fluorocarbons for applications requiring low friction, good strength, toughness, high dielectric strength, and heat resistance. They possess good resistance to radiation, but are surpassed in chemical resistance by fluoroplastics. Polyimide is attacked by strong alkaline solutions, hydrazine, nitrogen dioxide, and secondary amine compounds.

Polyimides, though costly and hard to process, are used in the making of aerospace, electronics, nuclear power, and office and industrial equipment. Other parts made include valve seats, gaskets, piston rings, thrust washers, and bushings. Films are made by a casting process (usually from prepolymer form). They are used for laminates, dielectrics, and coatings.

In Figure 7-43, polyimide in a hot liquid, electrostatic spray form is being applied to electric skillets and houseware items. After curing and baking at 550 °F [290 °C] polyimide forms a hard and flexible glossy finish similar to porcelain.

Prolonged contact with this resin and its reducers may cause serious cracking or workers' skin. The solvents are no more toxic than other aromatics.

Table 7-16 gives some properties of polyimides. The following list gives six advantages and disadvantages of polyimides.

Table 7-15. Properties of Polyarylate

Property	Polyarylate
Molding qualities	Good
Relative density	1.21
Tensile strength, MPa	48–75
(psi)	(6 962–10 879)
Impact strength, Izod (6 mm),	
J/mm	0.24
(ft lb/in)	(4)
Hardness, Rockwell	R105
Deflection temperature (at	
1.82 MPa or 264 psi), °C	280
(°F)	(536)
Water absorption (24h), %	0.01
Optical, refractive index	1.64

(A) Colored polyimide finishes are used on cookware.

(B) Electric skillets are coated on an automatic spray line.

Fig. 7-43. Examples of polyimide coating. (DeBeers Labs, Inc.)

Fig. 7-42. Basic polyimide structure.

Table 7-16. Properties of Polyimides

Property	Polyimide (Unfilled)
Molding qualities	Good
Relative density	1.43
Tensile strength, MPa	70
(psi)	(10 000)
Compressive strength, MPa	>165
(psi)	(>24 000)
Impact Strength, Izod, J/mm	0.045
(ft lb/in)	(0.9)
Hardness, Rockwell	E45–E58
Resistance to heat, °C	300
(°F)	(570)
Dielectric strength, V/mm	22 000
Dielectric constant (at 60 Hz)	3.4
Arc resistance, s	230
Water absorption (24 h), %	0.32
Burning rate	Nonburning
Effect of acids	Resistant
Effect of alkalies	Attacked
Effect of solvents	Resistant
Machining qualities	Excellent
Optical qualities	Opaque

Advantages of Polyimide

1. Short-exposure temperature capability of 600 °F to 700 °F [315 to 371 °C]
2. Excellent barrier
3. Very good electrical properties
4. Excellent solvent and wear resistance
5. Good adhesion capability
6. Especially suitable for composite fabrication

Disadvantages of Polyimide

1. Difficulty of fabrication
2. Hygroscopic (moisture-absorbing)
3. Subject to attack by alkalies
4. Comparatively high cost
5. Dark color
6. Most types have volatiles or contain solvents that must be vented during cure

Polyamide-imide (PAI)

An amorphous member of the polyimide family is polyamide-imide. It was marketed in 1972 under the trade name Torlon by Amoco Chemicals. This material contains aromatic rings and a nitrogen linkage, as shown in Figure 7-44. Polyamide-imide has striking properties (Table 7-17). This material can withstand continuous temperatures of 500 °F [260 °C]. Because of its low coefficient of friction, excellent service temperature, and dimensional stability, polyamide-imide

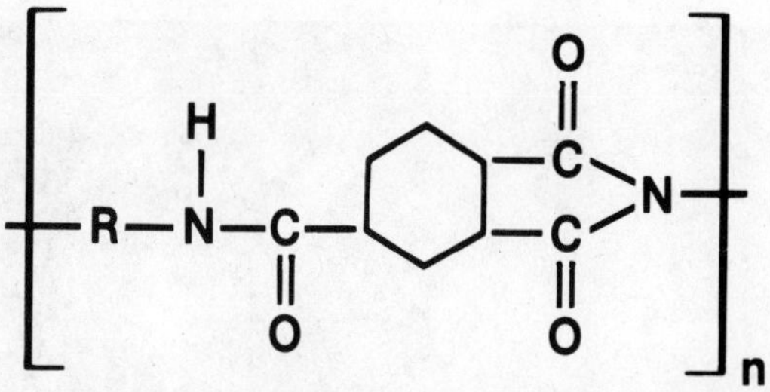

Fig. 7–44. General structural formula of polyamide-imide.

may be melt-processed into aerospace equipment, gears, valves, films, laminates, finishes, adhesives, and jet engine components. (Fig. 7-45.)

7-17 POLYMETHYLPENTENE

This plastics is reported to be an isotactically arranged aliphatic polyolefin of 4-methylpentene-1. Polymethylpentene was developed in the laboratory as early as 1955. It did not gain commercial value until Imperial Chemical Industries, Ltd., announced it under the tradename of TPX in 1965.

Ziegler-type catalysts are used to polymerize 4-methylpentene-1 at atmospheric pressures (Fig. 7-46). After polymerization, catalyst residues are removed by washing with methyl alcohol. The material is then compounded with stabilizers, pigments, fillers, or other additives into a granular form.

Table 7-17. Properties of Polyamide-imide

Property	Poly(amide-imide) (Unfilled)
Molding qualities	Excellent
Relative density	1.41
Tensile strength, MPa	185
(psi)	(26 830)
Compressive strength, MPa	275
(psi)	(39 900)
Impact strength, Izod J/mm	0.125
(ft lb/in)	(2.5)
Hardness, Rockwell	E78
Thermal expansion, 10^{-4}/°C	9.144
Resistance to heat, °C	260
(°F)	(500)
Dielectric strength, V/mm	>400
Dielectric constant (at 60 Hz)	3.5
Arc resistance, s	125
Water absorption (24 h), %	0.28
Burning rate	Nonburning
Effect of acids	Very resistant
Effect of alkalies	Very resistant
Effect of solvents	Very resistant
Machining qualities	Excellent

Fig. 7–45. The piston skirt, number two ring, connecting rods, wrist pins, intake valves, valve spring retainers, tappets, timing gears, and other parts are made of (Torlon) polyamide-imide for a Polimotor Lola 2 liter engine that muscles up 318 horsepower at 9500 rpm, but weighs only 168 pounds. (Amoco Chemicals)

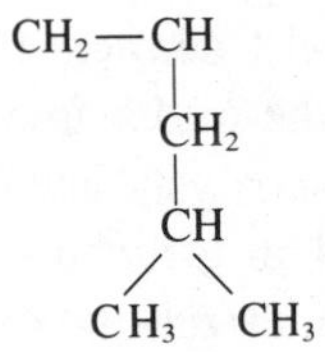

CH$_2$—CH
        |
       CH$_2$
        |
       CH
      /   \
   CH$_3$ CH$_3$

Fig. 7-46. Poly (4-methylpentene-1)

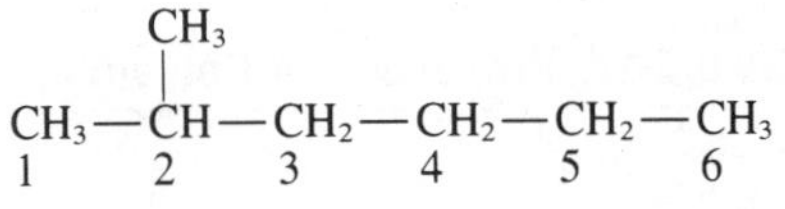

CH$_3$
            |
CH$_3$—CH—CH$_2$—CH$_2$—CH$_2$—CH$_3$
  1   2   3   4   5   6

(A) 2-methylhexane.

 1   2   3
CH$_3$—CH$_2$—CH—CH$_3$
                |
              4 CH$_2$
                |
              5 CH$_3$

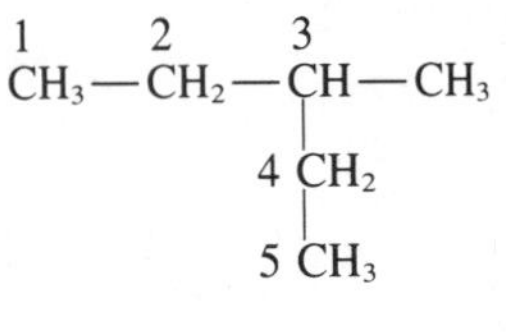

(B) 3-methylpentane.

Fig. 7-47. Continuous-chain formulas with carbon atoms numbered.

Formulas for this type of plastics are shown in Figure 7-47. To avoid confusion, the carbon atoms of the continuous chain must be numbered. This has been done in the formulas shown.

Copolymerization with other olefin units (including hexene-1, octene-1, decene-1, and octadecene-1) can offer enhanced optical and mechanical properties.

Commerical poly (4-methylpentene-1) has a relatively high service temperature, which may exceed 320 °F [160 °C]. Although the plastics is nearly 50 percent crystalline, it has a light-transmission value of 90 percent. Spherulite growth may be retarded by rapid cooling of the molded mass. The open packing of the crystalline structure gives polymethylpentene a low relative density of 0.83. This is close to the theoretical minimum for thermoplastics.

Polymethylpentene may be processed on normal thermoplastic equipment, at processing temperatures that may exceed 470 °F [245 °C].

In spite of its high cost, this plastics has found uses in chemical plants, autoclavable medical equipment, lighting diffusers, encapsulation of electronic components, lenses, and metalized reflectors (Fig. 7-48). A well-known use is the packaging of bake-in-the-bag and boil-in-the-bag foods. These packages are used in the home and in catering services for airlines or manufacturing plants. Packaged foods may be boiled in water or cooked in normal or microwave ovens. Transparency is useful in showing materials in dispensing equipment.

Other side-branched polyolefins are possible. Three possible polymers are shown in Figure 7-49. The branched side chains increase stiffness and lead to higher melting points. Polyvinyl cyclohex-

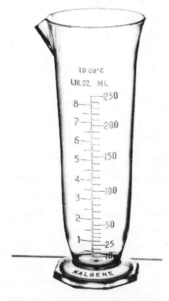

Fig. 7-48. The clarity, chemical resistance, and toughness of polymethylpentene make it suitable for laboratory items. (ICI, Ltd.)

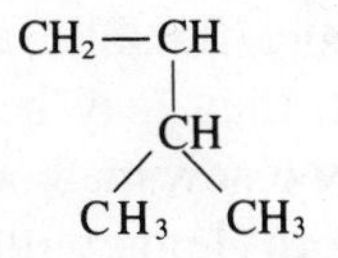

$$CH_2 — CH$$
$$|$$
$$CH$$
$$CH_3 \quad CH_3$$

(A) Poly (3-methylbutene-1).

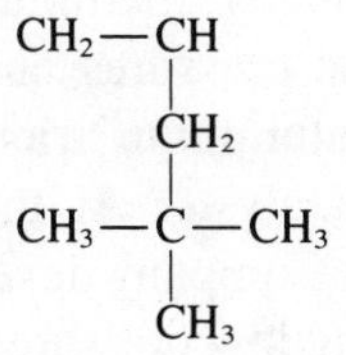

$$CH_2 — CH$$
$$|$$
$$CH_2$$
$$|$$
$$CH_3 — C — CH_3$$
$$|$$
$$CH_3$$

(B) Poly (4,4-dimethylpentene-1).

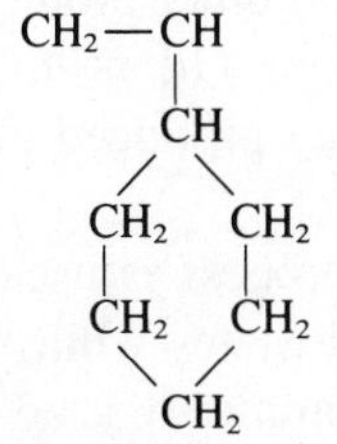

$$CH_2 — CH$$
$$|$$
$$CH$$
$$CH_2 \quad CH_2$$
$$CH_2 \quad CH_2$$
$$CH_2$$

(C) Poly (vinylcyclohexane).

Fig. 7-49. Side-branched polyolefin polymers.

Table 7-18. Properties of Polymethylpentene

Property	Polymethylpentene (Unfilled)
Molding qualities	Excellent
Relative density	0.83
Tensile strength, MPa	25–28
(psi)	(3 500–4 000)
Impact strength, Izod, J/mm	0.02–0.08
(ft lb/in)	(0.4–1.6)
Hardness, Rockwell	L67–74
Thermal expansion, $10^{-4}/°C$	29.7
Resistance to heat, °C	120–160
(°F)	(250–320)
Dielectric strength, V/mm	28 000
Dielectric constant (at 60 Hz)	2.12
Dissipation factor (at 60 Hz)	0.000 7
Water absorption (24 h), %	0.01
Burning rate, mm/min	25
(in/min)	(1.0)
Effect of sunlight	Crazes
Effect of acids	Attacked by oxidizing agents
Effect of alkalies	Resistant
Effect of solvents	Attacked by chlorinated aromatics
Machining qualities	Good
Optical qualities	Transparent to opaque

ane melts at about 640 °F [338 °C]. Table 7-18 gives some of the properties of polymethylpentene. A list of five advantages and two disadvantages of polymethylpentene follows.

Advantages of Polymethylpentene

1. Minimum density (lower than polyethylene)
2. High light-transmission value (90%)
3. Excellent dielectric, volume resistivity, and power factor
4. Higher melting point than polyethylene
5. Good chemical resistance

Disadvantages of Polymethylpentene

1. Must be stabilized against most radiation sources
2. More costly than polyethylene

7-18 POLYOLEFINS: POLYETHYLENE (PE)

Ethylene gas is a member of an important group of unsaturated, aliphatic hydrocarbons called *olefins*, or *alkenes*. An *ethenic* refers to ethylene materials. The word olefin means oil forming.

It was originally given to ethylene because oil was formed when ethylene was treated with chlorine. The term olefin now applies to all hydrocarbons with linear carbon-to-carbon double bonds. Olefins are highly reactive because of this carbon-to-carbon double bond. Some of the major olefin monomers are shown in Table 7-19.

Table 7-19. The Principal Olefin Monomers

Chemical Formula	Olefin Name
$$\begin{array}{cc} H & H \\ | & | \\ C & = C \\ | & | \\ H & H \end{array}$$	Ethylene
$$\begin{array}{cc} H & H \\ | & | \\ C & = C \\ | & | \\ CH_3 & H \end{array}$$	Propylene
$$\begin{array}{cc} H & H \\ | & | \\ C & = C \\ | & | \\ C_2H_5 & H \end{array}$$	Butene-1
$$\begin{array}{cc} H & H \\ | & | \\ C & = C \\ | & | \\ H_2C & H \\ | & \\ H—C—CH_3 & \\ | & \\ CH_3 & \end{array}$$	4-Methylpentene

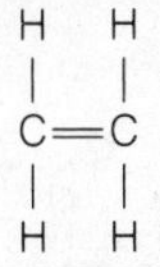

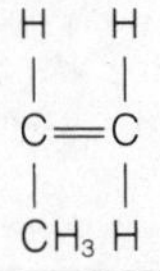

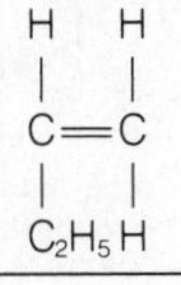

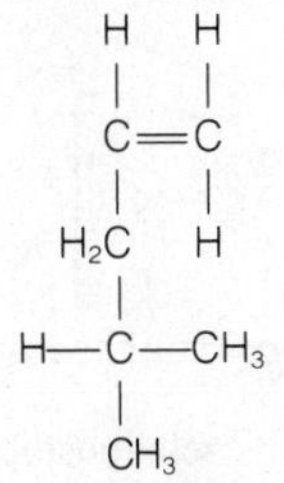

In the United States, ethylene gas is readily produced by cracking higher hydrocarbons of natural gas or petroleum. The importance and relationship of ethylene to other polymers is shown in Figure 7-50.

Between 1879 and 1900, several chemists experimented with linear polyethylene polymers. In 1900, E. Bamberger and F. Tschirner used the expensive material diazomethane to produce a linear polyethylene which they called "polymethylene."

$$2n\left(\begin{array}{c} CH_2 \\ N\!\!=\!\!N \end{array}\right) \rightarrow -(\!-CH_2\!-CH_2\!-)_n\!- \; + \; 2n \cdot N_2$$
Diazomethane   *Polyethylene*

W. H. Carothers and co-workers reported producing polyethylene of low molecular mass in 1930. The commercial feasibility of polyethylene resulted from research by Dr. E. W. Fawcett and Dr. R. O. Gibson of the Imperial Chemical Industries (ICI) in England. Their discovery, in 1933, was a result of investigating the reaction of benzaldehyde and ethylene (obtained from coal) under high pressure and temperature. In September, 1939, ICI began commercial production of polyethylene, and the demands of World War II used all the polyethylene produced as insulation of high-frequency radar cables. By 1943, the United States was producing polyethylene by the high-pressure methods developed by ICI. These early, low-density materials were highly branched, with a disorderly arrangement of molecular chains. Low-density mate-

rials are softer, more flexible, and melt at lower temperatures, thus, they may be more easily processed. By 1954, two new methods were developed for making polyethylene with higher relative densities of 0.91 to 0.97.

One process, developed by Karl Ziegler and associates in Germany, permitted polymerization of ethylene at low pressures and temperatures in the presence of aluminum triethyl and titanium tetrachloride as catalysts. At the same time, the Phillips Petroleum Company developed a polymerization process using low pressures with a chromium trioxide promoted, silica-alumina catalyst. The conversion of ethylene to polyethylene may also be done with a catalyst of molybdenum oxide on an alumina support and other promoters, a process developed by Standard Oil of Indiana. Only small quantities have been produced in the United States with this process.

The Ziegler process is used more extensively outside the United States while the Phillips Petroleum process is commonly used by U. S. firms.

Polyethylene can be produced with branched or linear chains (Fig. 7-51) by either the high-pressure (ICI) or low-pressure (Ziegler, Phillips, Standard Oil) methods. The differentiation of polymer type based on pressures used for polymerization is not employed today. The American Society for Testing and Materials (ASTM) has divided polyethylenes into four groups:

- Type 1   (Branched) 0.910–0.925 (low density)
- Type 2   0.926–0.940 (medium density)
- Type 3   0.941–0.959 (high density)
- Type 4   (Linear) 0.969 and above (high density to ultrahigh density homopolymers)

From Figure 7-52, it can be seen that the physical properties of low-density (branched) and high-density (linear) polyethylenes are different. Low-density polyethylene has a crystallinity of 60 to 70 percent. Higher-density polymers may vary in crystallinity from 75 to 90 percent. (Fig. 7-53).

With increased density the properties of stiffness, softening point, tensile strength, crystallinity, and creep resistance are increased. Increased density reduces impact strength, elongation, flexibility, and transparency.

Polyethylene properties may be controlled and identified by molecular mass and its distribution. Molecular mass and its distribution may have the effects shown in Table 7-20.

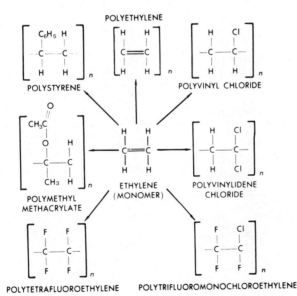

Fig. 7-50. The ethylene monomer and its relationship to other monomer resins.

MONOMER

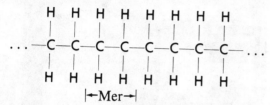

(A) Monomers of ethylene.

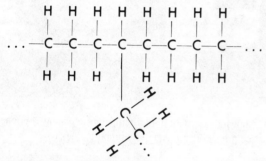

←Mer→

(B) Polymer containing many C_2H_4 mers.

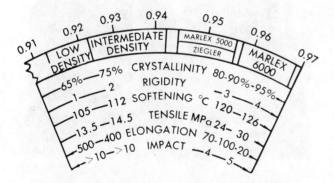

(C) Polymer with branching.

Fig. 7-51. Addition polymerization of ethylene. The original double bond of the ethylene monomer is broken, forming two bonds to connect adjacent mers.

(A) Electron micrograph shows intercrystalline links bridging radial arms of polyethylene spherulite.

(B) Small platelet crystals of polyethylene grown on intercrystalline links may be seen in this electron micrograph.

Figure 7-54 is a schematic representation comparing a polymer with a narrow molecular-mass distribution to a polymer with broad molecular-mass distribution. Molecular-mass distribution is the ratio of large, medium, and small molecular chains in the resin. If the resin is composed of chains that are all of close to the average length,

Fig. 7-52. Density range of polyethylene. (Phillips Petroleum Co.)

	0.92	0.93	0.94	0.95	0.96
0.91	LOW DENSITY	INTERMEDIATE DENSITY		MARLEX 5000 ZIEGLER	MARLEX 6000	0.97
	65%—	75% CRYSTALLINITY	80-90%-95%-		
	1	2 RIGIDITY	—3	—4
	105—	112 SOFTENING °C 120	—126
	13.5—	14.5 TENSILE MPa 24—	30
	500—	400 ELONGATION 70-100-20
	>10—	>10 IMPACT —4	—5

(C) Lamellar crystals of polyethylene were formed by depositing polymer from solution of links.

Fig. 7-53. Closeup views of polyethylene. (Bell Telephone Laboratories)

Table 7-20. Property Changes Caused by Molecular Mass and Distribution

Property	As Average Molecular Mass Increases (Melt Index Decreases)	As Molecular Mass Distribution Broadens
Melt viscosity	Increases	
Tensile strength at rupture	Increases	No significant change
Elongation at rupture	Increases	No significant change
Resistance to creep	Increases	Increases
Impact strength	Increases	Increases
Resistance to low temperature brittleness	Increases	Increases
Environmental stress cracking resistance	Increases	Increases
Softening temperature		Increases

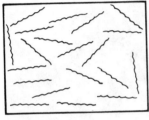 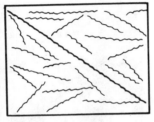

(A) Narrow.   *(B) Broad.*

Fig. 7-54. Molecular mass (weight) distribution.

the molecular mass distribution is called *narrow*. Molecular chains of average length can flow past each other more easily than large ones.

A melt-index device (Fig. 7-55) is used to measure the melt flow at a specified temperature and pressure. This melt index depends on molecular mass and its distribution. As melt index goes down, the melt viscosity, tensile strength, elongation, and impact strength increase. For many processing methods, it is desirable to have the hot resin flow very easily indicating use of a resin with a high melt index. Polyethylenes with high molecular mass have a low melt index. By varying density, molecular mass, and molecular mass distribution, polyethylene may be produced with a wide variety of properties.

Polyethylene may be cross-linked to convert a thermoplastic material to a thermoset. Such conversion, after forming, opens up many new possibilities. This cross-linking may be accomplished by chemical agents (usually peroxides) or by irradiation. There is growing commercial use of irradiation to cause branching of polyethylene products. (See Chapter 17, *Radiation Processing*.)

(A) Melt Indexer with digital temperature display, optional elapsed time indicator and mass support.

(B) Sieglaff-McKelvey rheometer.

Fig. 7-55. These machines are used to test the flow of plastics materials.

Radiation cross-linkage is rapid and leaves no objectionable residues. Irradiated parts may be exposed to temperatures in excess of 250 °F [500 °F], as shown in Figure 7-56.

Excessive radiation may reverse the cross-linking effect by breaking the main links in the molecular chain. Stabilizers or pigments of carbon

Fig. 7-56. Controlled radiation treatment can improve the heat-resistance of polyethylene. The center container was so treated and held its shape at 350 °F [175 °C].

black must be used to absorb or block the damaging effects of ultraviolet radiation on polyethylene.

Because of low price, processing ease, and a broad range of properties, polyethylene has become the most used plastics. As one of the lightest thermoplastics, it may be chosen where costs are based on cubic mass. Very good electrical and chemical resistance have led to wide use of polyethylene in wire coatings and dielectrics. Polyethylene is also used as containers, tanks, pipes, and coatings where chemical agents are present. At room temperature, there is no solvent for polyethylene; however, it is easily welded.

Polyethylene may be easily processed by all thermoplastic methods. Probably the largest use is in the making of containers and film consumed by the packaging industry. Blow-molded containers are seen in every supermarket. These containers replace heavier ones of glass and metal (Fig. 7-57). Tough plastics bags for packaging foods and films for packaging fresh fruit, frozen

foods, and bakery products are but a few of the many uses for polyethylene.

Lower-density polyethylene films are made with good clarity by quick-chilling of the melt as it emerges from the die. As the hot amorphous melt is quickly chilled, wide-spread crystallization does not have time to occur. Low-density films are used to package new garments, such as shirts and sweaters, as well as sheets and blankets. At the laundry and dry cleaner, very thin films are used for packaging. Higher-density films are used where greater heat resistance is required, as for boil-in-the-bag food packs. Silage covers, reservoir linings, seedbed covers, moisture barriers, and covering for harvested crops are only a few uses in construction, agriculture, or horticulture.

Although polyethylene is a good moisture barrier, it has a high gas permeability. It should not be used under vacuum or for transporting gaseous materials. Although permeable to oxygen and carbon dioxide, films used for meats and some produce may require small vent holes to allow ventilation. Oxygen keeps the meat looking red and prevents moisture from condensing on packaged produce. Heat sealing and shrink wrapping are easily done with these films. Electronic or radio-frequency heat sealing is difficult because of the low electrical dissipation (power) factor of polyethylene.

Polyethylene is used to coat paper, cardboard, and fabrics to improve their wet strength as well as other properties. Coated materials may then be heat sealed. The packaging of milk is a well-known use. Powdered forms are used for dip coating, flame spraying, and fluidized-bed coating where a layer impervious to chemicals and moisture is required.

Injection molded toys, small appliance housings, garbage cans, freezer containers, and artificial flowers benefit from polyethylene's toughness, chemical inertness, and low service temperatures (Fig. 7-58).

Extruded polyethylene pipe and ducting is used in chemical plants and for some domestic cold water service. Corrugated drain pipe is replacing clay or concrete pipe and tiles because it is less costly, faster to install, and lighter. Monofilaments find uses for ropes and fishing nets and are woven for lawn chairs. Polyethylene is widely used as electrical wire and cable covering. Irradiated or chemically cross-linked films are used as dielectrics in winding electrical coils. They also have limited packaging applications.

Polyethylene may be foamed by several methods. A foaming agent that breaks down and releases

Fig. 7-57. Polyethylene beverage containers. (Uniloy Blow-molding Machinery Division, Hoover International)

Fig. 7-58. Blow-molded guitar case made of polyethylene. (Chemplex Co.)

(A) Ice cream freezer.   **(B)** Drinking glass.

(C) Decorative wagon wheel.

Fig. 7-59. Foamed polyethylene products. (Phillips Petroleum Co.)

a gas during the molding operation is preferred for commercial uses. A physical foaming method consists of introducing a gas, such as nitrogen, into the molten resin under pressure. While in the mold and under atmospheric pressure, the gas-filled polyethylene expands. Azodicarbonamide may be used for chemical foaming of either low- or high-density resins. Foams may be selected as gasket material and dielectric materials in coaxial cables. Cross-linked foams are suitable for cushioning, packaging, or flotation, while structural foam is used for furniture components and internal panels in automobiles. Low-density foams find use in wrestling mats, athletic padding, and flotation equipment. Figure 7-59 shows some uses of foamed polyethylene.

Table 7-21 gives properties of low-, medium-, and high-density polyethylene. A list of six advantages and five disadvantages of polyethylene follows.

Advantages of Polyethylene

1. Low cost (except UHMWPE)
2. Excellent dielectric properties
3. Moisture resistance
4. Very good chemical resistance
5. Available in food grades
6. Processable by all thermoplastic methods (except HMWHPE and UHMWPE)

Disadvantages of Polyethylene

1. High thermal expansion
2. Poor weathering resistance
3. Subject to stress cracking (except UHMWPE)
4. Difficulty in bonding
5. Flammable

Adding fillers, reinforcements, or other monomers may also change properties. Some of the common comonomers are shown in Table 7-22.

A number of new polymerization techniques has expanded the potential application of polyethylene.

Very Low Density Polyethylene (VLDPE)

This linear, nonpolar polyethylene is produced by copolymerization of ethylene and other alpha olefins. Densities range from 0.890 to 0.915. VLDPE is easily processed into disposable gloves, shrink packages, vacuum cleaner hoses, tubing, squeeze tubes, bottles, shrink wrap, diaper film liners and other health care products.

Linear Low Density Polyethylene (LLDPE)

Production of linear low-density polyethylene is controlled through catalyst selection and regulation of reactor conditions. Densities range from 0.916 to 0.930. These plastics contain little if any

Table 7-21. Properties of Polyethylene

Property	Low-Density Polyethylene	Medium-Density Polyethylene	High Density Polyethylene
Molding qualities	Excellent	Excellent	Excellent
Relative density	0.910–0.925	0.926–0.940	0.941–0.965
Tensile strength, MPa	4–16	8.24	20–38
(psi)	(600–2 300)	(1 200–3 500)	(3 100–5 500)
Compressive strength, MPa			19–25
(psi)			(2 700–3 600)
Impact strength, Izod, J/mm	No break	0.025–0.8	0.025–1.0
(ft lb/in)		(.05–16)	(0.5–20)
Hardness, Shore	D41–D46	D50–D60	D60–D70
	R10	R15	
Thermal expansion, $10^{-4}/°C$	25–50	35–40	28–33
Resistance to heat, °C	80–100	105–120	120
(°F)	(180–212)	(220–250)	(250)
Dielectric strength, V/mm	18 000–39 000	18 000–39 000	18 000–20 000
Dielectric constant (at 60 Hz)	2.25–2.35	2.25–2.35	2.30–2.35
Dissipation factor (at 60 Hz)	0.000 5	0.000 5	0.000 5
Arc resistance, s	135–160	200–235	
Water absorption (24 h), %	0.015	0.01	0.01
Burning rate, mm/min	Slow 26	Slow 25–26	Slow 25–26
(in./min)	(1.04)	(1–1.04)	(1–1.04)
Effect of sunlight	Crazes — must be stabilized	Crazes — must be stabilized	Crazes — must be stabilized
Effect of acids	Oxidizing acids	Oxidizing acids	Oxidizing acids
Effect of alkalies	Resistant	Resistant	Resistant
Effect of solvents	Resistant (below 60 °C)	Resistant (below 60 °C)	Resistant (below 60 °C)
Machining qualities	Good	Good	Excellent
Optical qualities	Transparent to opaque	Transparent to opaque	Transparent to opaque

Table 7-22. Common Comonomers Used with Olefins

Formula	Name
H H | | C = C | | C_4H_9 H	1-Hexene
H H | | C = C | | O H || O=C—CH_3	Vinyl acetate
H H | | C = C | H | O=C—O—CH_3	Methyl acrylate
H H | | C = C | H | O=C—OH	Acrylic acid

chain branching. As a result, these plastics exhibit good flex life, low warpage, and improved stress-crack resistance. Films for ice, trash, garment, and produce bags are tough and puncture and tear resistant.

High Molecular Weight-High Density Polyethylene (HMW-HDPE)

High molecular weight-high density polyethylenes are linear polymers with molecular mass ranging from 200 000 to 500 000. Propylene, butene and hexene are common monomers. High molecular mass results in toughness, chemical resistance, impact strength and high abrasion resistance. High melt viscosities require special attention to equipment and mold designs. Densities are 0.941 or greater. Trash liners, grocery bags, industrial pipe, gas tanks, and shipping containers are familiar applications.

Ultrahigh Molecular Weight Polyethylene (UHMWPE)

Ultrahigh molecular weight polyethylenes have molecular weights from 3 to 6 million. This accounts for their high wear resistance, chemical inertness, and low coefficient of friction. These materials do not melt of flow like other polyethylenes. Processing is similar to methods used with polytetrafluoroethylene. (PTFE)

Sintering produces products with microporosity. Ram extrusion and compression molding are major forming methods.

Applications include chemical pump parts, seals, surgical implants, pen tips and butcher-block cutting surfaces.

Ethylene acid

A wide variety of properties similar to those of LDPE may be produced by varying the pendent carboxyl groups on the polyethylene chain. These carboxyl groups reduce polymer crystallinity, thus improving clarity, lowering the temperature required to heat seal, and improving adhesion to other substrates.

Molds should be designed to accommodate the adhesive qualities. Processing equipment should be corrosion resistant.

The FDA allows up to 25% acrylic acid and 20% methacrylic acid for ethylene copolymers in contact with food.

Most applications of ethylene acid and copolymers are for packaging of foods, coated papers, and composite foil pouches and cans.

Ethylene-ethyl acrylate (EEA)

By varying the ethyl acrylate pendent groups in the ethylene chain, properties may vary from rubbery to tough polyethylene-like polymers. The ethyl group on the PE chain lowers crystallinity.

Applications include hot melt adhesives, shrink wrap, produce bags, bag-in-box products, and wire coating.

Ethylene-methyl acrylate (EMA)

This copolymer is produced by the addition of methyl acrylate (40% by weight) monomer with ethylene gas.

EMA is a tough, thermally stable olefin with good elastomeric characteristics. Typical applications include disposable medical gloves, tough, heat-sealable layers, and coating for composite packaging.

EMA polymers meet the FDA and USDA requirements for use in food packaging. (See Elastomers, Appendix E)

Ethylene-vinyl acetate (EVA)

A wide variety of properties is available from this family of thermoplastic polymers. The vinyl acetate is copolymerized in varying amounts from 5–50% by weight onto the ethylene chain. If the vinyl acetate side groups exceed 50%, they are considered vinyl acetate-ethylene (VAE).

Typical EVA applications include hot melts, flexible toys, beverage and medical tubing, shrink wrap, produce bags, and numerous coatings.

7-19 POLYOLEFINS: POLYPROPYLENE

Until 1954, most attempts to produce plastics from polyolefins had little commercial success, and only the polyethylene family was commercially important. It was in 1955 that Italian scientist F. J. Natta announced the discovery of sterospecific polypropylene. The word *sterospecific* indicates that the molecules are arranged in a definite order in space. This is in contrast to branched or random arrangements. Natta called this regular, arranged material *isotactic polypropylene*. Experimenting with Ziegler-type catalysts, he replaced the titanium tetrachloride in $Al(C_2H_5) + TiCl_4$ with the sterospecific catalyst titanium trichloride. This led to the commercial production of polypropylene.

It is not surprising that polypropylene and polyethylene have many of the same properties. They are similar in origin and manufacture. Polypropylene has become a strong competitor of polyethylene.

Polypropylene gas, $CH_3—CH=CH_2$, is cheaper than ethylene. It is obtained from high-temperature cracking of petroleum hydrocarbons and propane. The basic structural unit of polypropylene is:

$$\left(\begin{array}{cc} CH_3 & H \\ | & | \\ -C & -C- \\ | & | \\ H & H \end{array}\right)_n$$

Figure 7-60 shows the stereostatic arrangements of polypropylene. In Figure 7-60A, the molecular chains show a high degree of order with all the CH_3 groups along one side. Atactic polymers are rubbery, transparent materials of limited commercial value. Atactic and syndiotactic plastics grades are more impact resistant than isotactic grades. Both syndiotactic and atactic structures may be present in small quantities in isotactic plastics. Commercially available polypropylene is about 90 to 95 percent isotactic.

The general physical properties of polypropylene are similar to those of high-density polyethylene. However, polyethylene and polypropylene differ in four important respects:

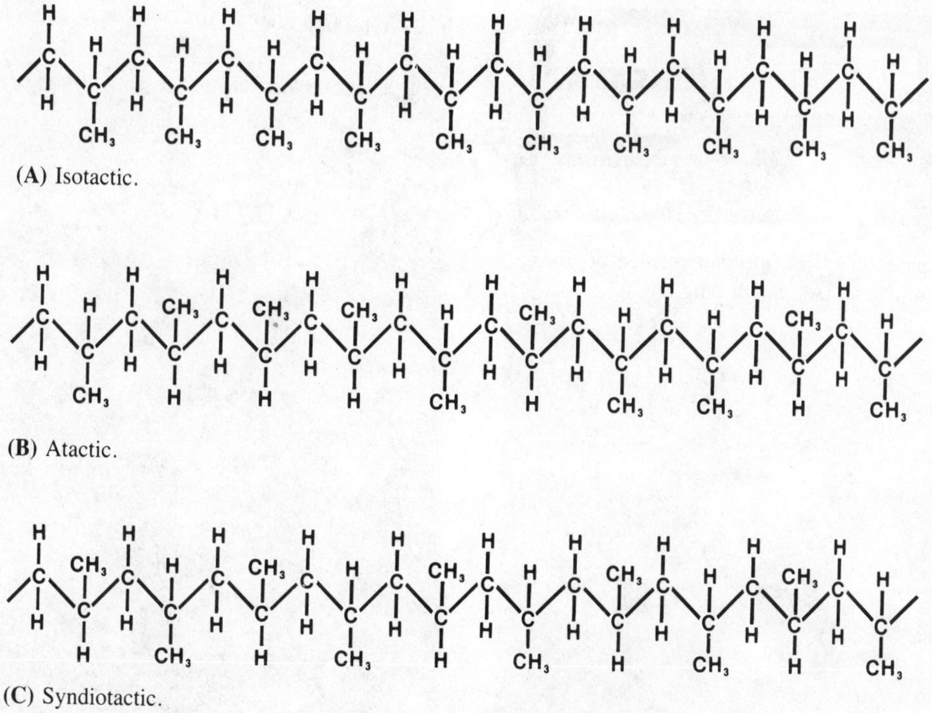

(A) Isotactic.

(B) Atactic.

(C) Syndiotactic.

Fig. 7-60. Stereotactic arrangements of polypropylene.

1. Polypropylene has a relative density of 0.90; polyethylene has relative densities of 0.941 to 0.965
2. The service temperature of polypropylene is higher
3. Polypropylene is harder, more rigid, and has a higher brittle point
4. Polypropylene is more resistant to environmental stress cracking (Fig. 7-61)

Fig. 7-61. Chair seats and backs molded of polypropylene. (Exxon Chemical Co.)

The electrical and chemical properties of the two materials are very similar. Polypropylene is more susceptible to oxidation and degrades at elevated temperatures.

Polypropylene may also be made with a variety of properties by adding fillers, reinforcements, or blends of special monomers (Fig. 7-62). It is easily processed in all conventional thermoplastic equipment. It cannot be cemented by cohesive means, but is readily welded.

Polypropylene is competitive with polyethylene for many uses. It has the advantage of a higher service temperature (Fig. 7-63). Typical uses include sterilizable hospital items, dishes, appliance parts, dishwasher components, containers, items incorporating integral hinges, automotive ducts, and trim. Extruded and cold-drawn monofilaments find use as rotproof ropes that will float on water. Some fibers are finding increasing uses for textiles and for outdoor or automotive carpeting. It may be used as tough packaging film or as electrical insulation on wire and cable. Slit film fiber, a process known as filibration, is widely used in producing ropes and fibers from polypropylene. It is coextrusion-blow-molded into numerous food containers. See Figure 7-64. Also see Barrier plastics, ethylene-vinyl, and polyvinylidene chloride or polyvinyl alcohol.

(A) Instrument panel that was injection molded of glass reinforced polypropylene. (AC Spark Plug Div.)

Fig. 7-64. Multilayered containers provide an effective oxygen barrier, allowing plastics to be used for foods such as tomato-based products, fruit juices, dressings, sauces, pickles, jams, and jellies. The container is made of a layer of ethylene-vinyl alcohol copolymer (EVOH) protected by inner and outer layers of polypropylene (PP). (Continental Can Co.)

(B) Pump housing, impeller, magnetic housing, and volute are made of glass fiber reinforced polypropylene. (Fiberfill Div., Dart Industries)

Fig. 7-62. Some uses of polypropylene.

(A) Coffee pot.   (B) Sterilizable hospitalware.

(C) Other gas- and steam-sterilizable hospital items.   (D) Washer-dryer combination.

Fig. 7-63. The high service temperature of polypropylene permits wide application.

Because of first rate abrasion resistance, high service temperature, and potentially lower cost, foamed polypropylene is finding growing use. Cellular polypropylene is foamed in much the same manner as polyethylene.

Table 7-23 gives some of the properties of polypropylene. Eleven advantages and six disadvantages of polypropylene are listed below.

Advantages of Polypropylene

1. Processable by all thermoplastic methods
2. Low coefficient of friction
3. Excellent electrical insulation
4. Good fatigue resistance
5. Excellent moisture resistance
6. First rate abrasion resistance
7. Good grade availability
8. Service temperature to 126 °C [260 °F]

Table 7-23. Properties of Polypropylene

Property	Polypropylene Homopolymer (Unmodified)	Polypropylene (Glass-Reinforced)
Molding qualities	Excellent	Excellent
Relative density	0.902–0.906	1.05–1.24
Tensile strength, MPa	31–38	42–62
(psi)	(4 500–5 500)	(6 000–9 000)
Compressive strength, MPa	38–55	38–48
(psi)	(5 500–8 000)	(5 500–7 000)
Impact strength, Izod, J/mm	0.025–0.1	0.05–0.25
(ft lb/in)	(0.5–2)	(1–5)
Hardness, Rockwell	R85–R110	R90
Thermal expansion, 10^{-4}/°C	14.7–25.9	7.4–13.2
Resistance to heat, °C	110–150	150–160
(°F)	(225–300)	(300–320)
Dielectric strength, V/mm	20 000–26 000	20 000–25 500
Dielectric constant (at 60 Hz)	2.2–2.6	2.37
Dissipation factor (at 60 Hz)	0.000 5	0.002 2
Arc resistance, s	138–185	74
Water absorption (24 h), %	0.01	0.01–0.05
Burning rate	Slow	Slow-nonburning
Effect of sunlight	Crazes—must be stabilized	Crazes—must be stabilized
Effect of acids	Oxidizing acids	Slowly attacked by oxidizing acids
Effect of alkalies	Resistant	Resistant
Effect of solvents	Resistant	Resistant
	(below 80 °C)	(below 80 °C)
Machining qualities	Good	Fair
Optical qualities	Transparent to opaque	Opaque

9. Very good chemical resistance
10. Excellent flexural strength
11. Good impact strength

Disadvantages of Polypropylene

1. Broken down by ultraviolet radiation
2. Poor weatherability
3. Flammable, (flame-retarded grades are available)
4. Subject to attack by chlorinated solvents and aromatics
5. Difficult to bond
6. Oxidative breakdown accelerated by several metals

7-20 POLYOLEFINS: POLYBUTYLENE (PB)

A polyolefin, called polybutylene (PB), was introduced by Witco Chemical Corporation in 1974. It has ethyl side groups in the linear backbone. This linear isostatic material may exist in a variety of crystalline forms. Upon cooling, the material is less than 30 percent crystalline. During aging and before complete crystalline transformation, many postforming techniques may be used. Crystallinity then varies from 50 to 55 percent after cooling. Polybutylene may be formed by conventional thermoplastic techniques.

Major uses include high performance films, tank liners, and pipes. It is used as hot melt adhesives and coextruded as moisture barrier and heat sealable packages. Table 7-24 gives some properties of polybutylene.

7-21 POLYPHENYLENE OXIDES

This family of materials should probably be called *polyphenylene*. Several plastics have been developed by separating the benzene ring backbone of polyphenylene with other molecules, making these plastics more flexible and able to be molded by usual thermoplastic methods. Polyphenylene with no benzene-ring separation is very brittle, insoluble, and infusible.

Polyphenylene

Poly(phenylene oxide)

Table 7-24. Properties of Polybutylene

Property	Polybutylene (Molding grades)
Molding qualities	Good
Relative density	0.908–0.917
Tensile strength MPa	26–30
(psi)	(3 770–4 350)
Impact strength, Izod, J/mm	No break
Hardness, Shore	D55–D65
Thermal expansion, 10^{-4}/°C	—
Resistance to heat, °C	<110
(°F)	(<230)
Dielectric constant (at 60 Hz)	2.55
Dissipation factor (at 60 Hz)	0.000 5
Water absorption (24 h), %	<0.01–0.026
Burning rate, mm/min	45.7
(in/min)	(1.8)
Effect of sunlight	Crazes
Effect of acids	Attacked by oxidizing acids
Effect of alkalies	Very resistant
Effect of solvents	Resistant
Machining qualities	Good
Optical qualities	Translucent

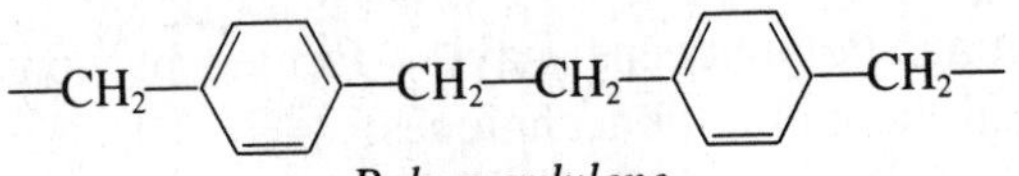

Poly-p-xylylene

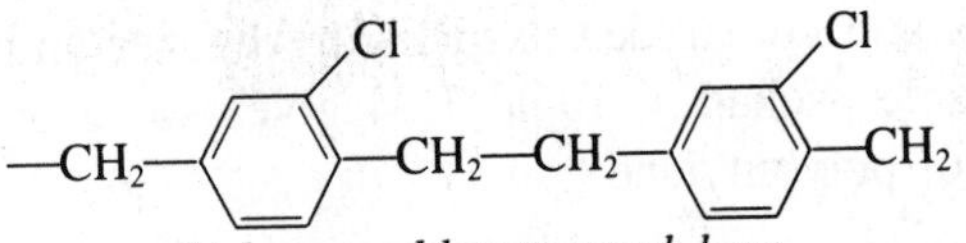

Polymonochloroparaxylylene

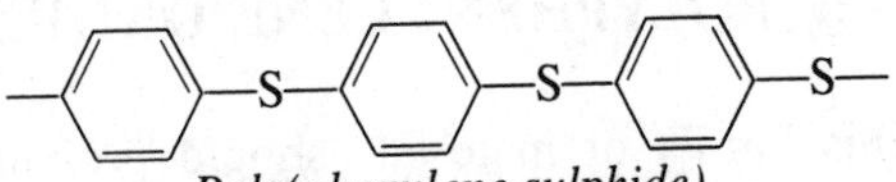

Poly(phenylene sulphide)

Three advantages and disadvantages of polyphenylenes are listed below.

Advantages of Polyphenylene

1. Excellent solvent resistance
2. Good radiation resistance
3. High thermal and oxidative stability

Disadvantages of Polyphenylene

1. Difficult to process
2. Comparatively expensive
3. Limited availability

Polyphenylene Oxide (PPO)

Union Carbide brought out a heat-resisting plastics called polyphenylene oxide in 1964. It may be prepared by the catalytic oxidation of 2,6-dimethyl phenol (Fig. 7-65).

Similar materials have been prepared making use of ethyl, isopropyl, or other alkyl groups. In 1965, the General Electric Co. introduced poly-2, 6-dimethyl-1, 4-phenylene ether as a polyphenylene oxide material. Then in 1966, General Electric announced another, similar thermoplastic with the tradename Noryl. This material is a physical blend of polyphenylene oxide and high impact polystyrene with a large diversity in formulations and properties.

Because Noryl costs less and has properties similar to polyphenylene oxides, many uses are the same (Fig. 7-66). This modified phenylene oxide material (Noryl) may be processed by normal thermoplastic equipment with processing temperatures from 375 °F to 575 °F [190 °C to 300 °C]. Modified phenylene oxide parts may be welded, heat-sealed, or solvent-cemented with chloroform and ethylene dichloride. Filled, reinforced, and flame-retardant grades are used as alternatives to die-cast metals, PC, PA, and polyesters. Video display terminals, pump impellers, radomes, small appliance housings, and instrument panels are typical applications. The following list shows five advantages and one disadvantage of polyphenylene oxide.

Advantages of Polyphenylene Oxide

1. Good fatigue strength and impact strength
2. Can be metal-plated
3. Thermally and oxidatively stable
4. Resistant to radiation
5. Processable by thermoplastics methods

Polyphenylene Ether (PPE)

Polyphenylene polymers and alloys belong to a group of aromatic polyethers. To be useful, these polyethers are alloyed with PS to lower melt vis-

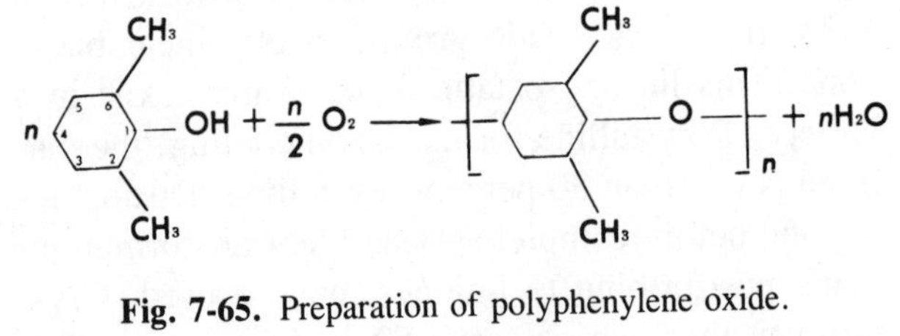

Fig. 7-65. Preparation of polyphenylene oxide.

(A) Housing for desktop computer.

(B) Injection-molded housing of electric pinking shears.

Fig. 7-66. Some applications of polyphenylene oxide. (General Electric Co.)

cosity and allow conventional processing. Without this PS separation, the polymer is very brittle, insoluble, and infusible. These copolymers are used for small appliance housing and electrical components.

Disadvantage of Polyphenylene Oxide

1. Comparatively high cost

Parylenes

In 1965, Union Carbide introduced poly-*p*-xylylene under the trade name *Parylene* (Fig. 7-67). Its primary market is in coating and film applications.

$$\left(\begin{array}{c} \overset{\displaystyle H}{\underset{\displaystyle H}{|}} \\ -C- \\ | \end{array} \bigcirc \begin{array}{c} \overset{\displaystyle H}{\underset{\displaystyle H}{|}} \\ -C- \\ | \end{array} \right)_n$$

Fig. 7-67. Structure of poly-para-xylylene.

Parylene C (polymonochloroparaxylylene) offers improved permeability to moisture and gases. Parylenes are not formed as other thermoplastics are. They are polymerized as coatings on the surface of the product. The process is similar to vacuum metalizing. Table 7-25 lists some properties of parylenes.

Polyphenylene Sulfide (PPS)

In 1968, the Phillips Petroleum Company announced a material known as polyphenylene sulfide, with the tradename Ryton. The material is available as thermoplastic or thermosetting compounds. Cross-linking is achieved by thermal or chemical means.

This rigid, crystalline polymer with benzene rings and sulfur links exhibits outstanding high temperature stability and chemical and abrasion resistance. Computer components, range components, hair dryers, submersible pump enclosures, and small appliance housings are typical uses. It is used as an adhesive, laminating resin, and as coatings for electrical parts.

Table 7-26 gives some of the traits of three polyphenylene oxides and below are listed six advantages plus four disadvantages of polyphenylene sulfide.

Advantages of Polyphenylene Sulfide

1. Capable of extended usage at 450 °F [232 °C]
2. Good solvent and chemical resistance
3. Good radiation resistance
4. Excellent dimensional stability
5. Nonflammable
6. Low water absorption

Disadvantages of Polyphenylene Sulfide

1. Hard to process (high melt temperature)
2. Comparatively high cost
3. Fillers needed for good impact strength
4. Subject to attack by chlorinated hydrocarbons

Polyaryl Ethers

Polyaryl ethers, polyaryl sulfone, and phenylene oxide have good physical and mechanical properties, good heat deflection temperatures, high impact strength, and good chemical resistance.

There are three different chemical groups that link the phenylene structure — isopropylidene, ether, and sulfone. The ether linkage and the car-

Table 7-25. Properties of Parylene

Property	Polyparaxylyene	Polymonochloro-paraxylyene
Molding qualities	Special process	Special process
Relative density	1.11	1.289
Tensile strength, MPa	44.8	68.9
(psi)	(6 500)	(9 995)
Thermal expansion, $10^{-4}/°C$	17.52	8.89
Resistance to heat °C	94	116
(°F)	(201)	(240)
Water absorption (24 h), %	0.06	0.01
Effect of solvents	Insoluble in most	Insoluble in most
Optical qualities	Transparent	Transparent

Table 7-26. Properties of Polyphenylene Oxide

Property	Polyphenylene Oxide (Unfilled)	Noryl SE-1 SE-100	Polyphenylene Sulfides
Molding qualities	Excellent	Excellent	Excellent
Relative density	1.06–1.10	1.06–1.10	1.34
Tensile strength, MPa	54–66	54–66	75
(psi)	(7 800–9 600)	(7 800–9 600)	(10 800)
Compressive strength, MPa	110–113	110–113	
(psi)	(16 000–16 400)	(16 000–16 400)	
Impact strength, Izod, J/mm	0.25*	0.25*	0.015 at 24 °C
			0.5 at 150 °C[†]
(ft lb/in)	(5.0)*	(5.0)*	(0.3 at 75 °F)
			(1.0 at 300 °F)
Hardness, Rockwell	R115–R119	R115–R119	R124
Thermal expansion, $10^{-4}/°C$	13.2	8.4–9.4	14
Resistance to heat °C	80–105	100–130	205–260
(°F)	(175–220)	(212–265)	(400–500)
Dielectric strength, V/mm	15 500–21 500	15 500–21 500	23 500
Dielectric constant (at 60 Hz)	2.64	2.64–2.65	3.11
Dissipation factor (at 60 Hz)	0.000 4	0.000 6–0.000 7	
Arc resistance, s	75		
Water absorption (24 h), %	0.066		0.02
Burning rate	Self-ex., nondrip	Self-ex., nondrip	Nonburning
Effect of sunlight	Colors may fade	Colors may fade	
Effect of acids	None		Attacked by oxidizing acids
Effect of alkalies	None		None
Effect of solvents	Soluble in some aromatics	Soluble in some aromatics	Resistant
Machining qualities	Excellent	Excellent	Excellent
Optical qualities	Opaque	Opaque	Opaque

*bar size 12.7 × 3.1 mm (½ × ⅛ in)
[†]bar size 12.7 × 7.25 mm (½ × ¼ in)

bon of the isopropylidene group impart toughness and flexibility to the plastics. See polyphenylene oxide, and polyphenylene ether.

In 1972, Uniroyal introduced a polyaryl ether plastics under the trade name of Arylon T.

Polyaryl ethers are prepared from aromatic compounds containing no sulfur links. The resulting polymer is more easily processed and service temperatures may be in excess of 170 °F [75 °C]. Uses include business machine parts, helmets, snowmobile parts, pipes, valves, and appliance components.

Table 7-27 gives properties of polyaryl ether.

7-22 POLYSTYRENE (PS)

Styrene is one of the oldest known vinyl compounds, however, industrial exploitation of this material did not begin until the late 1920's. This

Table 7-27. Properties of Polyaryl Ether

Property	Polyaryl Ether (Unfilled)
Molding qualities	Excellent
Relative density	1.14
Tensile strength, MPa	52
(psi)	(7 500)
Compressive strength, MPa	110
(psi)	(16 000)
Impact strength, Izod, J/mm	0.4
(ft lb/in)	(8)
Bar size, mm	12.7 × 7.25
in	½ × ¼
Hardness, Rockwell	R117
Thermal expansion, $10^{-4}/°C$	16.5
Resistance to heat, °C	120–130
(°F)	(250–270)
Dielectric strength, V/mm	16 930
Dielectric constant (at 60 Hz)	3.14
Dissipation factor (at 60 Hz)	0.006
Arc resistance, s	180
Water absorption (24 h), %	0.25
Burning rate	Slow
Effect of sunlight	Slight, yellows
Effect of acids	Resistant
Effect of alkalies	None
Effect of solvents	Soluble in ketones, esters, chlorinated aromatics
Machining qualities	Excellent
Optical qualities	Translucent to opaque

simple aromatic compound had been isolated as early as 1839 by the German chemist Edward Simon. Early monomer solutions were obtained from such natural resins as storax and dragon's blood (a resin from the fruit of the Malayan rattan palm). In 1851, French chemist M. Berthelot reported the production of styrene monomers by passing benzene and ethylene through a red-hot tube. This dehydrogenation of ethyl benzene is the basis of today's commercial methods.

By 1925, polystyrene (PS) was commercially available in Germany and the United States. For Germany, polystyrene became one of the most vital plastics during World War II. Germany had already embarked upon large-scale synthetic rubber production. Styrene was an essential ingredient for the production of styrene-butadiene rubber. When the natural sources of rubber were cut off in 1941, the United Stated began a crash program for the production of rubber from butadiene and styrene. This synthetic rubber became known as Government Rubber-Styrene (GR-S). There is still a large demand for styrene-butadiene synthetic rubber.

Styrene is chemically known as vinyl benzene, with the formula:

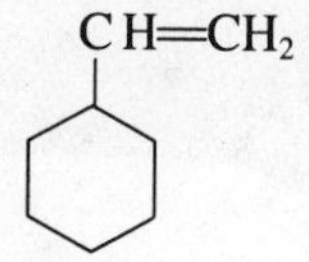

Styrene

In pure form, this aromatic vinyl compound will slowly polymerize by addition at room temperature. The monomer is obtained commercially from ethyl benzene (Fig. 7-68).

Styrene may be polymerized by bulk, solvent, emulsion, or suspension polymerization. Organic peroxides are used to speed the process.

Polystyrene is an atactic, amorphous thermoplastic with the formula shown in Figure 7-69. It is inexpensive, hard, rigid, transparent, easily molded and possesses good electrical and moisture resistance (Fig. 7-70). Physical properties vary, depending on the molecular mass distribution, processing, and additives.

Polystyrene may be processed by all normal thermoplastic processes and may be solvent-cemented. Some common uses include wall tile, electrical parts, blister packages, lenses, bottle caps, small jars, vacuum-formed refrigerator liners, containers of all kinds, and transparent display boxes. Thin films and sheet stock have a metallic ring when struck or dropped. These forms are used in packaging foods and other items, such as some cigarette packets. Children see the use of polystyrene in model kits and toys. Adults may be aware of this material in inexpensive dishes, utensils, and glasses. Filaments are extruded and deliberately stretched or drawn to orient the molecular chains. This orientation adds tensile strength in the direc-

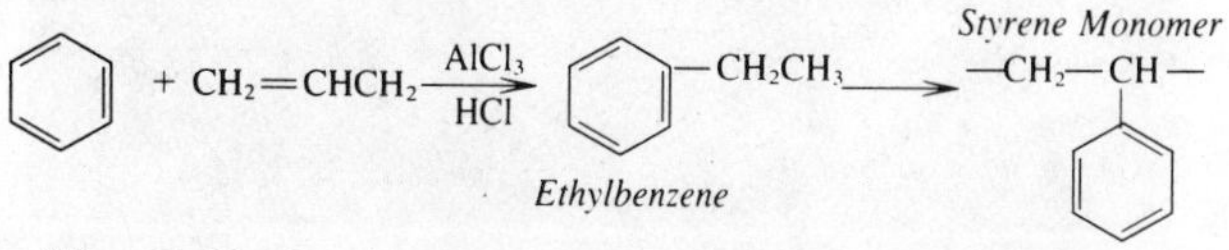

Fig. 7-68. Production of the vinyl benzene (styrene) monomer.

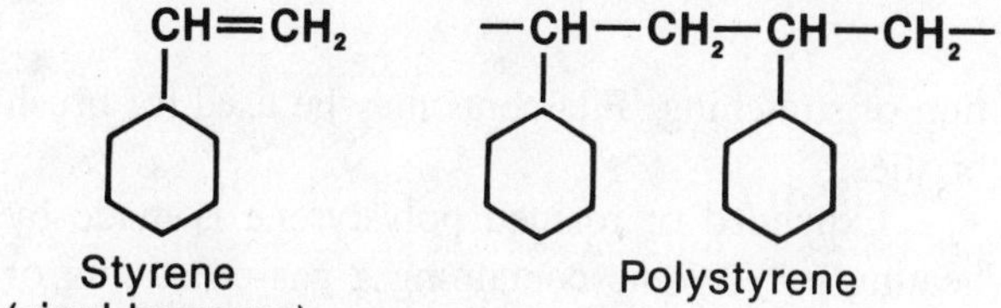

Fig. 7-69. Polymerization of styrene.

(A) Drum table with marble top.

(B) Automobile heater housing. (BASF)

Fig. 7-70. Some polystyrene objects.

tion of stretching. Filaments may be used for brush bristles.

Expanded or foamed polystyrene is made by heating polystyrene containing a gas-producing or *blowing* agent. The foaming is accomplished by blending a volatile liquid such as methylene chloride, propylene, butylene, or fluorocarbons into the hot melt. As the mixture emerges from the extruder,

the blowing agents release gaseous products resulting in a low density cellular material.

Expanded polystyrene (EPS) is produced from polystyrene beads containing an entrapped blowing agent. These agents may be pentane, neopentane, or petroluem ether. Upon either pre-expanding or final molding, the blowing agent volatizes causing the individual beads to expand and fuse together. Steam or other heat sources are used to cause the expansion. Both expanded and foamed forms have a closed cellular structure so that they can be used as flotation devices. Because of its low thermal conductivity, this material has found widespread use as thermal insulation (Fig. 7-71) used in refrigerators, cold storage rooms, freezer display cases, and building walls. It has the added advantage of being moistureproof. There are many packaging uses because of its thermal insulation value and shock absorption characteristics. Packing in cellular polystyrene can save on shipping and breaking costs (Fig. 7-72).

Foamed and expanded polystyrene sheets may be thermoformed. They are made into such familiar packaging items as egg cartons and meat or produce trays. Molded drinking cups, glasses, and *ice* chests are commonly used items.

Polystyrenes cannot withstand prolonged heat above 150 °F [65 °C] without distorting, thus they are not good exterior materials. Special grades and additives can correct this problem. Polystyrenes reinforced with glass fiber are used in automotive assemblies, business machines, and appliance housings.

The properties of polystyrene can be considerably varied by copolymerization and other

Fig. 7-71. Expanded polystyrene beads. (Sinclair-Koppers Co.)

Fig. 7-72. Cellular polystyrene has many packaging uses. (Sinclair-Koppers Co.)

modifications. Styrene-butadiene rubber has been mentioned above. Polystyrene is used in sporting goods, toys, wire and cable sheathing, shoe soles, and tires. Two of the most useful copolymers (terpolymers) are styrene-acrylonitrile and styrene-acrylonitrile-butadiene (ABS).

Table 7-28 gives some of the properties of polystyrene while below are listed nine advantages and six disadvantages of polystyrene.

Advantages of Polystyrene

1. Optical clarity
2. Light mass
3. High gloss
4. Excellent electrical properties
5. Good grades available
6. Processable by all thermoplastic methods
7. Low cost
8. Good dimensional stability
9. Good rigidity

Disadvantages of Polystyrene

1. Flammable (retarded grades available)
2. Poor weatherability
3. Poor solvent resistance
4. Brittleness of homopolymers
5. Subject to stress and environmental cracking
6. Poor thermal stability

Styrene-Acrylonitrile (SAN)

Acrylonitrile (CH_2=CHCHN) is copolymerized with styrene (C_6H_6), giving products a higher resistance than polystyrene to various solvents, fats, and other compounds (Fig. 7-73). These products are suitable for components requiring impact strength and chemical resistance and are used in vacuum cleaners and kitchen equipment.

Table 7-28. Properties of Polystyrene

Property	Polystyrene (Unfilled)	Impact- and Heat-Resistant Polystyrene	Polystyrene (20–30% Glass-Filled)
Molding qualities	Excellent	Excellent	Excellent
Relative density	1.04–1.09	1.04–1.10	1.20–1.33
Tensile strength, MPa	35–83	10–48	62–104
(psi)	(5 000–12 000)	(1 500–7 000)	(9 000–15 000)
Compressive strength, MPa	80–110	28–62	93–124
(psi)	(11 500–16 000)	(4 000–9 000)	(13 500–18 000)
Impact strength, Izod, J/mm	0.0125–0.02	0.025–0.55	0.02–0.22
(ft lb/in)	(0.25–0.40)	(0.5–11)	(0.4–4.5)
Hardness, Rockwell	M65–M80	M20–M80, R50–R100	M70–M95
Thermal expansion, $10^{-4}/°C$	15.2–20	8.5–53	4.5–11
Resistance to heat °C	65–78	60–80	82–95
(°F)	(150–170)	(140–175)	(180–200)
Dielectric strength, V/mm	19 500–27 500	11 500–23 500	13 500–16 500
Dielectric constant (at 60 Hz)	2.45–2.65	2.45–4.75	
Dissipation factor (at 60 Hz)	0.000 1–0.000 3	0.000 4–0.002	0.004–0.014
Arc resistance, s	60–80	10–20	25–40
Water absorption (24 h), %	0.03–0.10	0.05–0.6	0.05–0.10
Burning rate	Slow	Slow	Slow—nonburning
Effect of sunlight	Yellows slightly	Yellows slightly	Yellows slightly
Effect of acids	Oxidizing acids	Oxidizing acids	Oxidizing acids
Effect of alkalies	None	None	Resistant
Effect of solvents	Soluble in aromatic and chlorinated hydrocarbons	Soluble in aromatic and chlorinated hydrocarbons	Soluble in aromatic and chlorinated hydrocarbons
Machining qualities	Good	Good	Good
Optical qualities	Transparent	Translucent to opaque	Translucent to opaque

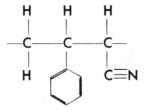

Fig. 7-73. SAN mer.

These products are suitable for components requiring impact strength and chemical resistance and are used in vacuum cleaners and kitchen equipment.

Styrene-acrylonitrile (SAN) copolymers may have about 20 to 30 percent acrylonitrile content. By varying the proportions of each monomer, a wide range of properties and processability may be obtained. A slight yellow cast is typical of SAN, due to the copolymerization of acrylonitrile with the styrene member.

This copolymer is easily molded and processed. SAN-type materials inherently absorb more moisture than polystyrene and may result in molding defects such as silver streaking. Predrying is advised.

Methyl ethyl ketone, trichloroethylene, and methylene chloride are among the effective solvents for SAN.

This tough, heat-resistant plastics is used for telephone parts, containers, decorative panels, blender bowls, syringes, refrigerator compartments, food packages, and lenses (Fig. 7-74).

Table 7-29 gives some of the properties of SAN. Three advantages and disadvantages of styrene-acrylonitrile (SAN) are listed below.

Advantages of Styrene-Acrylonitrile (SAN)

1. Processable by thermoplastic methods
2. Rigid and transparent
3. Improved solvent resistance over polystyrene

Disadvantages of Styrene-Acrylonitrile

1. Higher water absorption than polystyrene
2. Low thermal capability
3. Low impact strength

(Olefin-modified) Styrene-Acrylonitrile (OSA)

A tough, heat and weather resistant polymer is produced by tailoring the molecular weight and monomer ratios of saturated olefinic elastomer with styrene and acrylonitrile. It is used almost exclusively as a coextrudant over other substrates. Topper covers, boat hulls, and decorative wood and metal construction panels are typical applications.

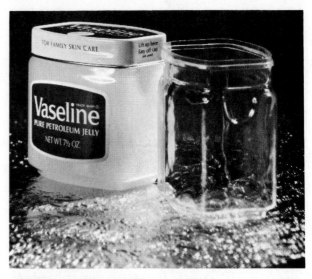

(A) Jar for petroleum jelly.

(B) Housing for air precleaner.

Fig. 7-74. Two uses of transparent SAN. (Monsanto Co.)

Styrene-Butadiene Plastics (SBP)

This amorphous copolymer consists of two blocks of styrene repeating units separated by a block of butadiene. This is in contrast to styrene butadiene (SBR), which is thermosetting.

These polymers are ideally suited for packaging applications including cups, deli containers, meat trays, jars, bottles, skin packages, and overwrap. Reinforced and filled grades are used in tool handles, office equipment housings, medical devices, and toys.

Table 7-29. Properties of SAN

Property	SAN (Unfilled)
Molding qualities	Good
Relative density	1.075–1.1
Tensile strength, MPa	62–83
(psi)	(9 000–12 000)
Compressive strength, MPa	97–117
(psi)	(14 000–17 000)
Impact strength, Izod, J/mm	0.01–0.02
(ft lb/in)	(0.35–0.50)
Hardness, Rockwell	M80–M90
Thermal expansion, 10^{-4}/°C	9–9.7
Resistance to heat, °C	60–96
(°F)	(140–205)
Dielectric strength, V/mm	15 750–19 685
Dielectric constant (at 60 Hz)	2.6–3.4
Dissipation factor (at 60 Hz)	0.006–0.008
Arc resistance, s	100–150
Water absorption (24 h), %	0.20–0.30
Burning rate	Slow to Self extinguishing
Effect of sunlight	Yellows
Effect of acids	Attacked by oxidizing agents
Effect of alkalies	None
Effect of solvents	Soluble in ketones and esters
Machining qualities	Good
Optical qualities	Transparent

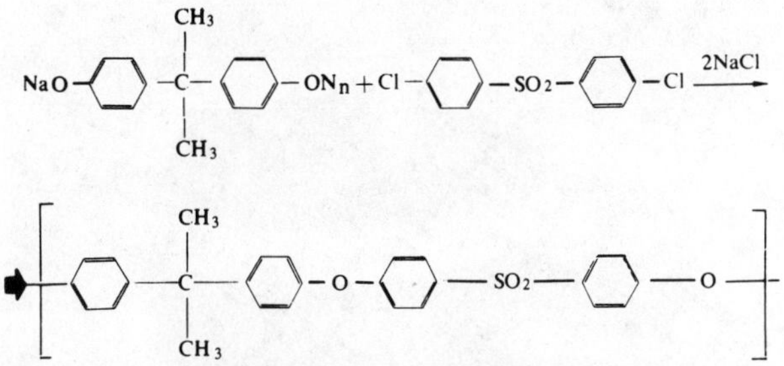

Fig. 7-75. Basic repeating structure of polysulfone.

Styrene-maleic anhydride (SMA)

This thermoplastic is distinguished by higher heat resistance than the parent styrenic and ABS families. SMA is obtained by the copolymerization of maleic anhydride and styrene. Butadiene is sometimes terpolymerized to produce impact-modified versions. Applications include vacuum cleaner housings, mirror housings, thermoformed headliners, fan blades, heater ducts, and food service trays.

7-23 POLYSULFONES

In 1965, Union Carbide introduced a linear, heat-resistant thermoplastic called polysulfone. The basic repeating structure consists of benzene rings joined by a sulfone group (SO_2), an isopropylidene group (CH_3CH_3C), and also an ether linkage (O).

One basic polysulfone is made by mixing bisphenol A with chlorobenzene and dimethyl sulfoxide in a caustic soda solution. The resulting condensation polymerization is shown in Figure 7-75. The light amber color of the plastics is a result of the addition of methyl chloride, which ends polymerization. The outstanding thermal and oxidation resistance is the result of the benzene-to-sulfone linkages. Polysulfone can be processed by all normal methods. It must be dried before use and may require processing temperatures in excess of 700 °F [370 °C]. Service temperatures range from −150° to 345 °F [−100° to +175 °C].

Polysulfone can be machined, heat-sealed, or solvent-cemented with dimethyl formamide or dimethyl acetamide.

Polysulfones are competitive with many thermosets. They may be processed in rapid-cycle thermoplastics equipment. The polysulfones have excellent mechanical, electrical, and thermal properties. They are used for hot-water pipes, alkaline battery cases, distributor caps, face shields for astronauts, electrical circuit breakers, appliance housings, dishwasher impellers, autoclavable hospital equipment, interior components for aerospace craft, shower heads, lenses, and numerous electrical insulating components (Fig. 7-76). When used outdoors, polysulfones should be painted or electroplated to prevent degradation. Five advantages and four disadvantages of polysulfones are found in the following list.

Advantages of Polysulfones

1. Good thermal stability
2. Excellent high-temperature creep resistance
3. Transparent
4. Tough and rigid
5. Processable by thermoplastic methods

Disadvantages of Polysulfones

1. Subject to attack by many solvents
2. Poor weatherability
3. Subject to stress cracking
4. High processing temperature

Polyarylsulfone

Polyarylsulfone is a high-temperature, amorphous thermoplastic introduced in 1983. It offers

(A) Immersible cornpopper has molded polysulfone cover.

(B) Microwave oven cookware made of polysulfone.

(C) Polysulfone thermos pitcher liner.

Fig. 7-76. Applications of polysulfone. (Union Carbide)

properties similar to other aromatic sulfones. Uses include circuit boards, high temperature bobbins, sight glasses, lamp housings, electrical connectors and housings, and panels of composite materials for numerous transporation components.

Polysulfones have been prepared with a variety of bisphenols with methylene, sulfide, or oxygen linkages. In polyaryl sulfone, the bisphenol groups are linked by ether and sulfone groups.

There are no isopropylene (aliphatic) groups present. The term *aryl* refers to a phenyl group derived from an aromatic compound. In naming these compounds, if more than one hydrogen is substituted in the aryl group, a numbering system is normally used. Three possible disubstituted benzenes are shown in Fig. 7-77. The basic properties of polysulfone are given in Table 7-30.

Polyethersulfone (PES)

This plastics, with outstanding oxidation and thermal resistance, was introduced in 1973. Polyethersulfone has performed well under creep and stress forces, at temperatures above 390 °F [200 °C]. It is characterized by an absence of aliphatic groups, and is an amorphous structure. Polyethersulfone is very resistant to both acids and alkalies, but it is attacked by ketones, esters, and some halogenated and aromatic hydrocarbons. The basic monomer unit is shown in Figure 7-78.

The distinguishing properties of PES are its high temperature performance, good mechanical strength, and low flammability.

Polyethersulfone has found uses in aerospace components, sterilizable medical components, and oven windows. Compounded grades extend their useful temperatures and improve mechanical properties. They have been used as adhesives and may be plated. Table 7-31 gives properties of this plastics.

Polyphenylsulfone (PPSO)

The sulfone that best resists stress cracking is polyphenylsulfone. Introduced in 1976, polyphenylsulfone is an amorphous structure with very high impact strength that will withstand a con-

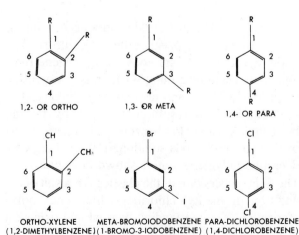

ORTHO-XYLENE   META-BROMOIODOBENZENE   PARA-DICHLOROBENZENE
(1,2-DIMETHYLBENZENE) (1-BROMO-3-IODOBENZENE) (1,4-DICHLOROBENZENE)

Fig. 7-77. Three possible disubstituted benzenes.

Table 7-30. Properties of Polysulfones

Property	Polysulfone (Unfilled)	Polyary Sulfone (Unfilled)
Molding qualities	Excellent	Excellent
Relative density	1.24	1.36
Tensile strength, MPa	70	90
(psi)	(10 200)	(13 000)
Compressive strength, MPa	96	123
(psi)	(13 900)	(17 900)
Impact strength, Izod, J/mm	0.06; bar 7.25 mm	0.25
(ft lb/in)	(1.3); (bar ¼ in)	(5)
Hardness, Rockwell	M69, R120	M110
Thermal expansion, $10^{-4}/°C$	13.2–14.2	11.9
Resistance to heat, °C	150–175	260
(°F)	(300–345)	(500)
Dielectric strength, V/mm	16 730	13 800
Dielectric constant (at 60 Hz)	3.14	3.94
Dissipation factor (at 60 Hz)	0.000 8	0.003
Arc resistance, s	75–122	67
Water absorption (24 h), %	0.22	1.8
Burning rate	Self-extinguishing	Self-extinguishing
Effect of sunlight	Strength loss, yellows slightly	Slight
Effect of acids	None	None
Effect of alkalies	None	None
Effect of solvents	Partly soluble in aromatic hydrocarbons	Soluble in highly polar solvents
Machining qualities	Excellent	Excellent
Optical qualities	Transparent to opaque	Opaque

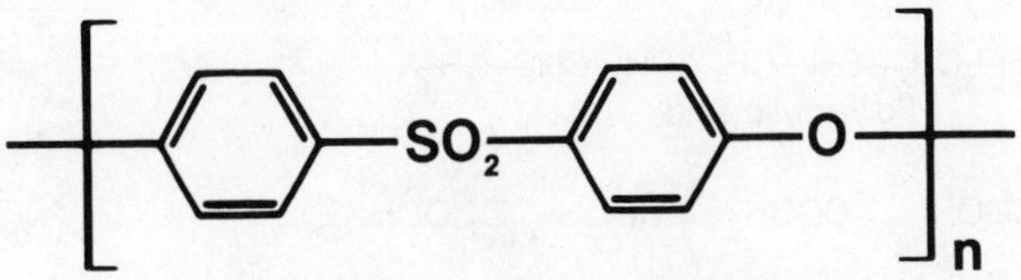

Fig. 7-78. Basic repeating unit of polyethersulfone.

tinuous service temperature of 375 °F [190 °C]. Uses include semiconductor carriers, valves, circuit boards and aerospace components.

7-24 POLYVINYLS

There is a large and varied group of addition polymers that chemists refer to as vinyls. These have the formulas:

$$CH_2{=}CH{-}R \quad \text{or} \quad CH_2{=}C\overset{\displaystyle R}{\underset{\displaystyle R}{|}}$$

Radicals (R) may be attached to this repeating vinyl group as side groups to form several polymers related to each other. Addition polymers are shown in Table 7-32 with the radical side groups attached.

Table 7-31. Properties of Polyethersulfone

Property	Polyethersulfone (Unfilled)
Molding qualities	Excellent
Relative density	1.37
Tensile strength, MPa	84
(psi)	(12 180)
Impact strength, Izod, J/mm	0.08
(ft lb/in)	(1.6)
Hardness, Rockwell	M88
Thermal expansion, $10^{-4}/°C$	13–97
Resistance to heat, °C	150
(°F)	(300)
Dielectric strength, V/mm	15 750
Dielectric constant (at 60 Hz)	3.5
Dissipation factor (at 60 Hz)	0.001
Arc resistance, s	65–75
Water absorption (24 h), %	0.43
Effect of sunlight	Yellows
Effect of acids	None
Effect of alkalies	None
Effect of solvents	Attacked by aromatic hydrocarbons
Machining qualities	Excellent
Optical qualities	Transparent

Through common usage, the *vinyl plastics* are those polymers with the vinyl name. Many authorities limit their discussion to include only polyvinyl chloride and polyvinyl acetate. Polyvinyl homopolymers or copolymers may include polyvinyl

Table 7-32. Monofunctional Monomers and Their Polymers

Monomer	Polymer
$CH_2{=}CH_2$ — Ethylene	$\rightarrow -CH_2-CH_2-CH_2-CH_2-CH_2-CH_2-CH_2-CH_2-$ — Polyethylene
$CH_2{=}CH$ with $O-COCH_3$ — Vinyl acetate	$\rightarrow -CH_2-CH-CH_2-CH-CH_2-CH-CH_2-CH-\ldots$ each CH bearing $O-COCH_3$ — Polyvinyl acetate
$CH_2{=}CH$ with Cl — Vinyl chloride	$\rightarrow -CH_2-CH-CH_2-CH-CH_2-CH-CH_2-CH-\ldots$ each CH bearing Cl — Polyvinyl chloride
$CH_2{=}CH$ with C_6H_5 — Styrene (Vinyl benzene)	$\rightarrow -CH_2-CH-CH_2-CH-CH_2-CH-CH_2-CH-\ldots$ each CH bearing C_6H_5 — Polystyrene
$CH_2{=}C$ with Cl above and Cl below — Vinylidene chloride	$\rightarrow -CH_2-CH-CH_2-CH-CH_2-CH-CH_2-CH-\ldots$ each CH bearing Cl above and Cl below — Polyvinylidene chloride
$CH_2{=}CH$ with $COOH$ — Acrylic acid	$\rightarrow -CH_2-CH-CH_2-CH-CH_2-CH-CH_2-CH-\ldots$ each CH bearing $COOH$ — Polyacrylic acid
$CH_2{=}C$ with $COOH$ above and CH_3 below — Methacrylic acid	$\rightarrow -CH_2-CH-CH_2-CH-CH_2-CH-CH_2-CH-\ldots$ each CH bearing $COOH$ above and CH_3 below — Polymethacrylic acid
$CH_2{=}C$ with CH_3 above and CH_3 below — Isobutylene	$\rightarrow -CH_2-CH-CH_2-CH-CH_2-CH-CH_2-CH-\ldots$ each CH bearing CH_3 above and CH_3 below — Polyisobutylene

chloride, polyvinyl acetate, polyvinyl alcohol, polyvinyl butyral, polyvinyl acetal, and polyvinylidene chloride. Fluorinated vinyls are discussed with other fluorine-containing polymers.

The history of polyvinyls may be traced to as early as 1835. French chemist V. Regnault reported that a white residue could be synthesized from ethylene dichloride in an alcohol solution. This tough white residue was again reported in 1872 by E. Baumann. It occurred while reacting acetylene and hydrogen bromide in sunlight. In both cases, sunlight was the polymerizing catalyst that pro-

duced the white residue. In 1912, Russian chemist I. Ostromislensky reported the same sunlight polymerization of vinyl chloride and vinyl bromide. Commercial patents were granted in several countries for the manufacture of vinyl chloride by 1930.

In 1933, W. L. Semon of the B. F. Goodrich Company added a plasticizer, tritolyl phosphate, to polyvinyl chloride compounds. The resulting polymer mass could be easily molded and processed without substantial decomposition.

Germany, Great Britain, and the United States produced plasticized polyvinyl chloride (PVC) commercially during World War II. It was largely used in place of rubber.

Today, polyvinyl chloride is the leading plastics produced in Europe while in the United States it ranks second after polyethylene. The polyvinyl chloride molecule (C_2H_3Cl) is like that of polyethylene, and Figure 7-79 shows this similarity.

Polyvinyl Chloride

The basic raw ingredient of polyvinyl chloride, depending on availability, is acetylene or ethylene gas. Ethylene is the chief source in the United States. During its manufacture, polymerization may be started by peroxides, azo compounds, persulfates, ultraviolet light, or radioactive sources. For addition polymerization, the double bonds of the monomers must be broken by the use of heat, light, pressure, or a catalyst system.

The uses of polyvinyl chloride plastics may be broadened by the addition of plasticizers, fillers, reinforcements, lubricants, and stabilizers. They may be formulated into flexible, rigid, elastomeric, or foamed compounds.

Polyvinyl chloride is most widely used in flexible film and sheet forms. These films and sheets are competitive with other films for collapsible containers, drum liners, sacks, and packages. Washable wallpapers and certain clothing such as handbags, rainwear, coats, and dresses are other uses. Sheets are made into chemical tanks and ductwork of all types. They are easily fabricated by welding, heat-sealing, or solvent-cementing with mixtures of ketones or aromatic hydrocarbons.

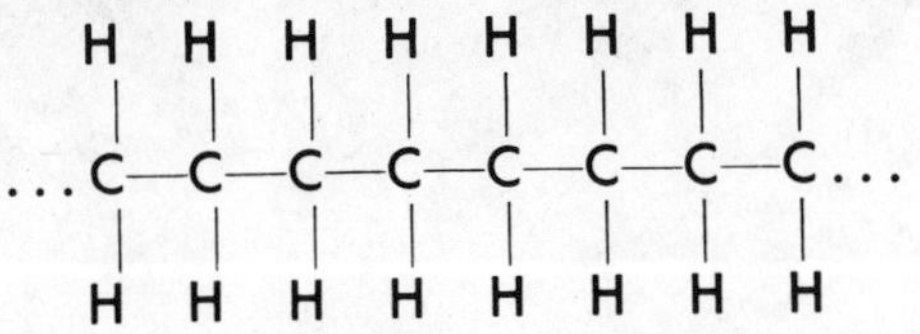

Extruded profile shapes of both rigid and flexible polyvinyl chloride find use as architectural moldings, seals, gaskets, gutters, exterior siding, garden hose, and moldings for movable partitions (Fig. 7-80). Injection-molded soles for footwear are popular in several countries.

Organosols and plastisols are liquid or paste dispersions or emulsions of polyvinyl chloride. They are used for coating various substrates, including metal, wood, plastics, and fabrics. They may be applied by dipping, spraying, spreading, or slush and rotational casting. Laminates of polyvinyl film, foam, and fabric are used for upholstery materials. Dip coatings are found on tool handles, sink drainers, and other substrates as a protective layer. Slush and rotational casting of polyvinyls are used to produce hollow articles such as balls, dolls, and large containers. Heavily filled polyvinyls and their copolymers are used in the production of floor coverings and tiles. Foams have found limited application in the textile and carpeting industries. Large quantities of PVC are used as blow-molded containers and as extruded coverings for electrical wire (Fig. 7-81).

Generally, materials in the vinyl family are flame, water, chemical, electrical, and abrasion resistant. They have good weatherability and may be transparent. To aid processing and provide various properties, polyvinyls are commonly plasticized. Both plasticized and unplasticized PVC compounds are available. Unplasticized grades are used in chemical plants and building industries. Plasticized grades are more flexible and soft. With increases in plasticizer, there is more bleeding or migration of the plasticizer chemical into adjacent materials. This is of prime importance in packaging of food products and medical supplies.

All thermoplastic processing techniques are employed with vinyls.

Polyvinyl chloride (PVC) is the most used and commonly thought-of vinyl. Other homopolymer and copolymer polyvinyls are finding increasing use, however. The following list shows six advantages and five disadvantages of polyvinyl chloride.

(A) Polyethylene.

MONOMER

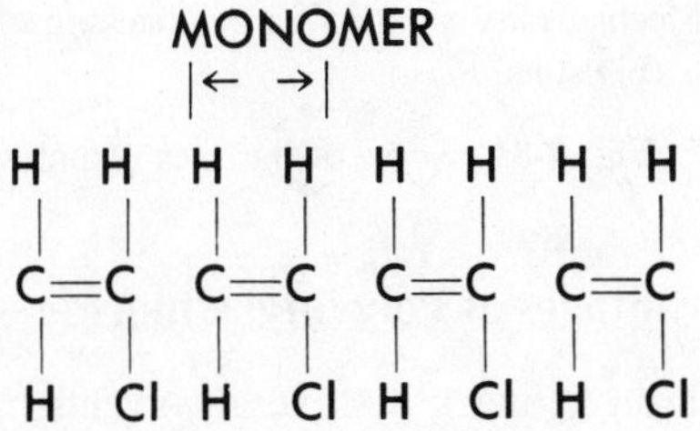

(B) Vinyl chloride.

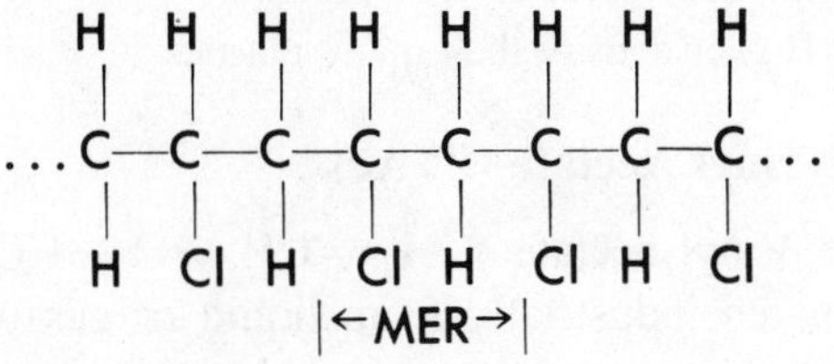

(C) Polyvinyl chloride.

Fig. 7-79. Similarity of polyethylene and polyvinyl chloride.

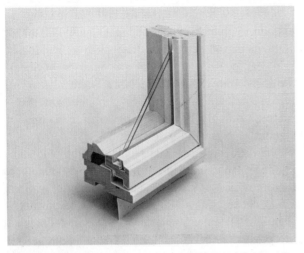

(A) Polyvinyl coating on wood for weatherproof finish and durability. (Perma-shield)

(B) Solid vinyl gutter and downspout system. (Bird & Son)

Fig. 7-80. Polyvinyl coatings have many uses.

Advantages of Polyvinyl Chloride (PVC)

1. Processable by thermoplastic methods
2. Wide range of flexibility (by varying levels of plasticizers)
3. Nonflammable
4. Dimensional stability
5. Comparatively low cost
6. Good resistance to weathering

(A) Pipe and plumbing fittings.

(B) Foamed flotation devices.

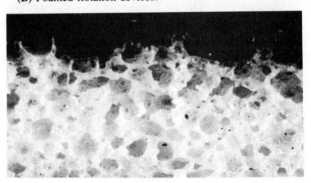

(C) Mechanically foamed carpet backing. Magnified 10 times. (Firestone Plastics Co.)

Fig. 7-81. Some of the uses of polyvinyl.

Disadvantages of Polyvinyl Chloride

1. Subject to attack by several solvents
2. Limited thermal capability
3. Thermal decomposition evolves HCl
4. Stained by sulfur compounds
5. Higher density than many plastics

Polyvinyl Acetate (PVAc)

Vinyl acetate ($CH_2=CH-O-COCH_3$) is prepared industrially from liquid or gaseous reactions of acetic acid and acetylene. Homopolymers find only limited uses because of excessive cold flow and low softening point. They are used in paints, adhesives, and various textile finishing operations. Polyvinyl acetates are usually in an emulsion form. *White glues* are familiar polyvinyl acetate emulsions. Moisture absorption characteristics are high with chosen alcohols and ketones as solvents. Remoistenable adhesives and hot-melt formulations are other well-known uses. Polyvinyl acetates are used as binder emulsions in some paint formulations. Their resistance to degradation by sunlight makes them useful for interior or exterior coatings. Other uses may include emulsion binders in paper, cardboard, portland cements, textiles, and chewing-gum bases.

Some of the best-known commercial products are copolymers of polyvinyl chloride and polyvinyl acetate used for floor coverings and for modern phonograph records (Fig. 7-82). These vinyl records have several advantages over polystyrene and the older shellac discs. Two advantages and three disadvantages of polyvinyl acetate are shown on the following page.

Advantages of Polyvinyl Acetate

1. Excellent for forming into films
2. Heat-sealable

Disadvantages of Polyvinyl Acetate

1. Low thermal stability
2. Poor solvent resistance
3. Poor chemical resistance

Polyvinyl Formal

Polyvinyl formal is generally produced from polyvinyl acetate, formaldehyde, and other additives that change the alcohol side groups on the chain to *formal* side groups. Polyvinyl formal finds its greatest uses as coatings of metal containers and as electrical wire enamels.

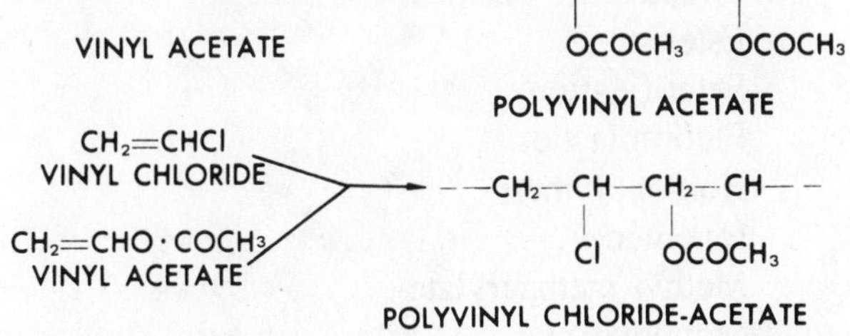

Fig. 7-82. Production of polyvinyl acetate and polyvinyl chloride-acetate.

Polyvinyl Alcohol (PVAl)

Polyvinyl alcohol is a useful derivative produced from the alcoholysis of polyvinyl acetate (Fig. 7-83). Methyl alcohol (methanol) is used in the process. Polyvinyl alcohol (PVA) is both alcohol- and water-soluble. The properties vary depending on the concentration of polyvinyl acetate that remains in the alcohol solution. Polyvinyl alcohol may be used as a binder and adhesive for paper, ceramics, cosmetics, and textiles. It finds use as water-soluble packages for soap, bleaches, and disinfectants. It is a useful mold-releasing agent used in the manufacture of reinforced plastics products. There has been only limited use of polyvinyl alcohol for moldings and fibers.

Polyvinyl Acetate

One more useful derivative of polyvinyl acetate is polyvinyl acetal. It is produced from the treatment of polyvinyl alcohol (from polyvinyl acetate) with an acetaldehyde (Fig. 7-84). Polyvinyl acetal materials find limited use as adhesives, surface coatings, films, moldings, or textile modifiers.

Polyvinyl Butyral (PVB)

Polyvinyl butyral is produced from polyvinyl alcohol (Fig. 7-85). This plastics is used as interlayer film in laminated safety glass.

Polyvinylidene Chloride (PVDC)

In 1839, a substance similar to vinyl chloride was discovered, but it contained one more chlorine atom (Fig. 7-86). This material has become commercially important as vinylidene chloride ($H_2C=CCl_2$).

Polyvinylidene chloride is costly and hard to process; thus, it is normally found as a copolymer

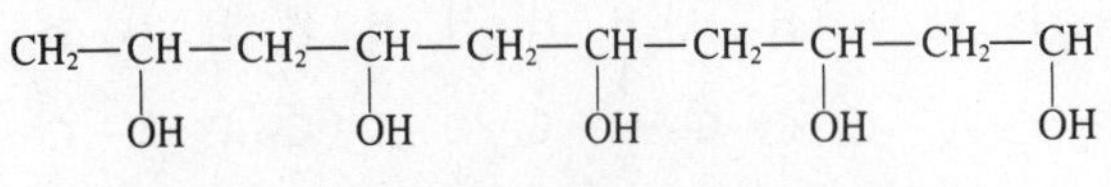

Fig. 7-83. Polyvinyl alcohol (OH side groups).

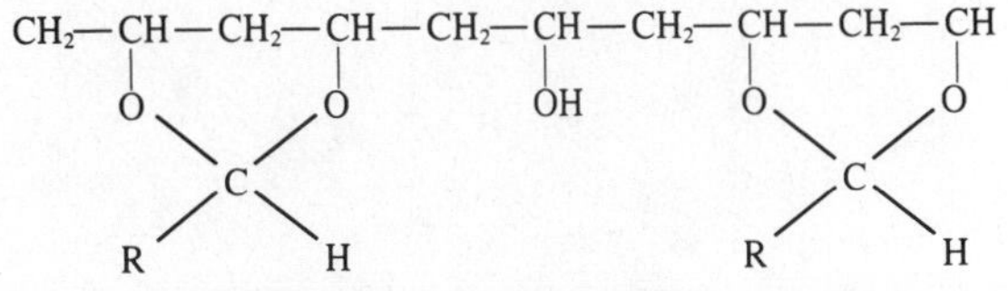

Fig. 7-84. A polyvinyl acetal.

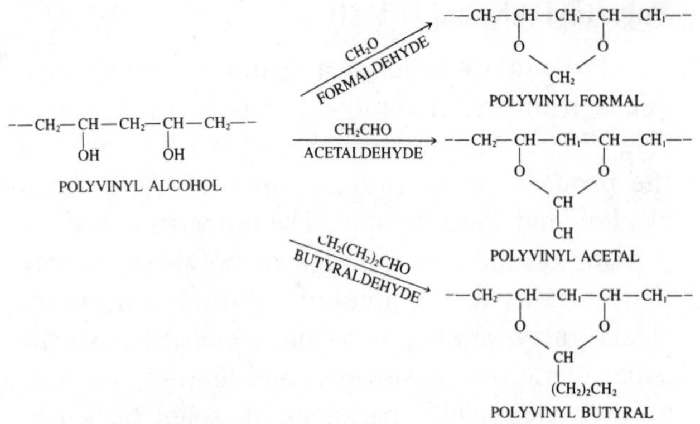

POLYVINYL ALCOHOL

CH₂O / FORMALDEHYDE → POLYVINYL FORMAL

CH₃CHO / ACETALDEHYDE → POLYVINYL ACETAL

CH₃(CH₂)₂CHO / BUTYRALDEHYDE → POLYVINYL BUTYRAL

Fig. 7-85. Production of polyvinyl butyral.

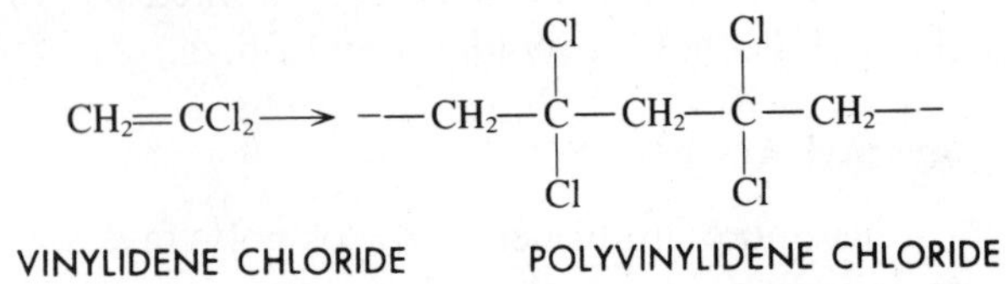

$$CH_2{=}CCl_2 \longrightarrow$$

VINYLIDENE CHLORIDE   POLYVINYLIDENE CHLORIDE

Fig. 7-86. Polymerization of vinylidene chloride.

with vinyl chloride, acrylonitrile, or acrylate esters. A well-known food wrapping film, Saran, is a copolymer of vinylidene chloride and acrylonitrile. It exhibits clarity and toughness, and permits little gas or moisture transmission.

The chief use of polyvinylidene chloride copolymers (Fig. 7-87) is in coating and film packaging; however, they have found some success as fibers for carpeting, automobile seat upholstery, draperies, and awning textiles. Their chemical inertness allows them to be used as pipe, pipe fittings, pipe linings, and filters.

There are many other polyvinyl polymers that warrant further research and study. Polyvinyl carbazole is used for dielectrics, polyvinyl pyrrolidone is used as a blood plasma substitute, and polyvinyl ethers are used as adhesives. Polyvinyl ureas, poly-

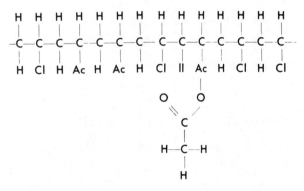

Fig. 7-87. Copolymerization of vinyl chloride and vinyl acetate.

vinyl isocyanates, and polyvinyl chloroacetate have been explored for commercial use.

All chlorinated or chlorine-containing polymers may emit toxic chlorine gas upon high temperature breakdown. Adequate venting should be provided to protect the operator during processing.

Table 7-33 gives some properties of polyvinyl plastics, while a list of four advantages and two disadvantages of polyvinylidene chloride is given below.

Advantages of Polyvinylidene Chloride

1. Low water permeability
2. Approved for food wrap by FDA
3. Processable by thermoplastic methods
4. Nonflammable

Disadvantages of Polyvinylidene Chloride

1. Lower strength than PVC
2. Subject to creep

VOCABULARY

The following vocabulary words are found in this chapter. Look up the definition of any of these words you do not understand as they apply to plastics, in the glossary, Appendix A.

Acetals (poly)
Acrylic
Acrylonitrile
Acrylonitrile-butadiene-styrene
Allyl
Amids
Amines
Azo group
Cellulosics
Chlorinated polyether
Coumarone
Cracking
Diamines
Dibasic acid
Engineering plastics
Ester
Esterification
Fluoroplastics
Homopolymers
Ionomer
Methyl methacrylate
Monohydric
Phenoxy
Polyolefin

Table 7-33. Properties of Polyvinyls

Property	Rigid Vinyl Chloride (PVC)	PVC-Acetate (Copolymer)	Vinylidene Chloride Compound
Molding qualities	Good	Good	Excellent
Relative density	1.30–1.45	1.16–1.18	1.65–1.72
Tensile strength, MPa	34–62	17–28	21–34
(psi)	(5 000–9 000)	(2 500–4 000)	(3 000–5 000)
Compressive strength MPA	55–90		14–19
(psi)	(8 000–13 000)		(2 000–2 700)
Impact strength, Izod, J/mm	0.02–1.0		0.06–0.05
(ft lb/in)	(0.4–20.0)		(0.3–1.0)
Hardness, Rockwell	M110–M120	R35–R40	M50–M65
Thermal expansion 10^{-4}/°C	12.7–47		48.3
Resistance to heat °C	65–80	55–60	70–90
(°F)	(150–175)	(130–140)	(160–200)
Dielectric strength, V/mm	15 750–19 700	12 000–15 750	15 750–23 500
Dielectric constant (at 60 Hz)	3.2–3.6	3.5–4.5	4.5–6.0
Dissipation factor (at 60 Hz)	0.007–0.020		0.030–0.045
Arc resistance, s	60–80		
Water absorption (24 h), %	0.07–0.4	3.0+	0.1
Burning rate	Self-extinguishing	Self-extinguishing	Self-extinguishing
Effect of sunlight	Needs stabilizer	Needs stabilizer	Slight
Effect of acids	None to slight	None to slight	Resistant
Effect of alkalies	None	None	Resistant
Effect of solvents	Soluble in ketones and esters	Soluble in ketones and esters	None to slight
Machining qualities	Excellent		Good
Optical qualities	Transparent	Transparent	Transparent

Polymethyl methacrylate	Polyamide
Polymethylpentene	Polycarbonate
Polyphenylene oxide	Polyethylene
Polypropylene	Polyimide
Polystyrene	Polysulfone
Polyacrylate	Polyvinyls
Polyallomer	Radical

7-1. Name ten (10) families of thermoplastic plastics.

7-2. Name four thermoplastics that have low coefficients of friction.

7-3. A widely used white adhesive is made of ___?___ .

7-4. Name the resin that might be selected as best for molding coffee pots, and to serve as sterilizable hospital equipment.

7-5. What three thermoplastics materials are used the most?

7-6. Name the liquid or paste dispersions of polyvinyl chloride.

7-7. What word is used to describe a material consisting of only one type of monomer?

7-8. Which plastics would you select for a transparent machine guard or police shield?

7-9. Which plastics would be used for window glazing?

7-10. Lucite is one trade name for ___?___ thermoplastic material.

7-11. By replacing one hydrogen atom in the ethylene monomer with chlorine, the resulting monomer is called ___?___ .

7-12. The best mechanical properties of polyethylene are provided by the ___?___ possible molecular mass (weight).

7-13. To aid or improve processability, add flexibility, and reduce stiffness of HMW polyvinyl chloride plastics, ___?___ may be added.

7-14. An advantage of crystallinity in a polymer is ___?___ service temperatures, and better resistance to chemical attack.

7-15. Polystyrene is crystal clear, brittle and has poor solvent resistance. Is polystyrene amorphous or crystalline in structure?

7-16. Polyamides will ___?___ moisture from the atmosphere; thus lowering stiffness, strength and hardness.

7-17. Name three polymers that may be included in the polyolefin family.

7-18. Name three very popular copolymers.

7-19. Name four thermoplastic plastics that you might select for high service temperatures.

7-20. Wood, cereal straws, grasses and other plants may be a source of ___?___ and ___?___ used to produce plastics.

7-21. Name three broad applications for nonplastics cellulose derivatives.

7-22. Because of its superior impact strength ___?___ is preferred over polystyrene for extruded sheet, pipe and some molded parts.

7-23. Transparency and weather resistance make ___?___ sheets an excellent choice for signs.

7-24. Bubbles and haze in molded polycarbonate parts may be caused by ___?___ .

7-25. The numbers 6, 6 and 6, 10 refer to the number of ___?___ atoms connecting the acid and amino groups in ___?___ .

7-26. Special extrusion, coating and sintering process must be used to make products of ___?___ .

7-27. The fluorinated polymer, ___?___ is useful over the widest temperature range.

7-28. The relative density of ___?___ show they are the heaviest of all homopolymers.

7-29. The two largest markets for the vinyl copolymer is in producing extruded ___?___ and ___?___ .

7-30. An indication of molecular mass, crystalinity and processibility of polyethelyne may be measured by viscosity or ___?___ .

7-31. The additive, ___?___ or antioxidents allows us to use polyethylene outdoors.

7-32. In ___?___ molecular arrangement, all the methyl units are located on one side of the carbon chain in polypropylene.

7-33. If you wanted a denser, more crystalline polypropylene with a higher melting point, would you select an isotactic, atactic or syndiotactic molecular arrangement?

7-34. List four classes or groups of polyethylene plastics relative to molecular mass (weight).

7-35. Many thermoplastics are ___?___ structural plastics, but this term usually denotes polyvinyl chloride.

7-36. Polyacrylonitrile is not molded because hydrogen bonds prevent polymer ___?___ at molding temperatures.

7-37. Which would have the greater crystal formations: low density or high density polyethylene?

7-38. Which would be more flexible: a random copolymer of ethylene and propylene, or high density polyethylene?

7-39. Would linear or cross-linked polyethylene have more thermal stability?

7-40. Does high density or low density polyethylene have a greater volume?

7-41. Would you select high density polyethylene or polypropylene to make a coffee pot?

7-42. You would not select polystyrene for most toy applications because it is too ___?___ .

7-43. The impact strength of polystyrene is improved with the addition of the alloy ___?___ .

7-44. Which plastics would you select for the following outstanding properties?
optical clarity
temperature resistance
impact resistance
low coefficient of friction
weather resistance
lightness

7-45. Styrene is chemically known as ___?___ .

7-46. Compounds containing two amino groups are called ___?___ .

7-47. Which plastics would you select for each of the following product applications?
rain gutters
implants for human body
auto seat covers
window glazing
textiles for slacks
bread packages
pocket comb
disposable ice cream dish

7-48. The characteristic ending for aldehydes is ___?___ .

7-49. ABS polymers are terpolymers meaning ___?___ monomers.

7-50. Ionic bonding is seldom found in plastics; however, ___?___ have ion cross-links.

7-51. The process, called ___?___ , is conducted by alternately polymerizing ethylene and propylene monomers.

Thermosetting Plastics

Your study of plastics resins is not complete until you become familiar with thermosetting plastics. Look for the outstanding properties and applications of each plastics described in this chapter because plastics properties affect product design, processing, economics, and service.

You should remember that thermosetting materials undergo a chemical reaction and become "set." In general, most are high molecular weight materials resulting in a hard, brittle plastics.

This chapter treats individual groups of thermosetting plastics. These are *allylics, alkyds, amino plastics (urea-formaledhyde* and *melamine-formaldehyde), casein, epoxy, furan, phenolics, unsaturated polyesters, polyurethane* and *silicones*.

8-1 ALKYDS

In the past, there was some confusion about ester-based alkyd resins. The term *alkyd* was once used strictly for unsaturated polyesters modified with fatty acids or vegetable oils. These resins find uses in paints and other coatings. Today, *alkyd molding compounds* refer to unsaturated polyesters modified by a nonvolatile monomer (such as diallyl phthalate) and various fillers. The compounds are formed into granular, rope, nodular, putty, and log shapes to allow continuous automatic molding (Fig. 8-1).

To obtain a resin suitable for molding compounds, the cross-linking mechanism must be modified so that rapid cure will take place in the mold. The use of initiators in the resin compound also speeds up the polymerization of the double bonds. These resins should not be confused with saturated polyester molding compounds, which are linear and thermoplastic.

R. H. Kienle coined the word *alkyd* from the "al" in *al*cohol and the "cid" from a*cid*. It is usually pronounced aĺ-kid. Alkyd resins may be produced by reacting phthalic acid, ethylene glycol, and the fatty acids of various oils such as linseed oil, soybean oil, or tung oil (Fig. 8-2). Kienle combined fatty acids with unsaturated esters in 1927, while searching for a better electrical insulating resin for General Electric. The need for materials in World War II greatly speeded up the interest in alkyd finishing resins.

Alkyd resins may have a structure like the one shown in Figure 8-3. When this resin is applied to a substrate, heat or oxidizing agents are used to begin cross-linking.

One expert estimates that nearly half of the surface coatings used in the United States are alkyd polyester coatings. The increasing use of silicone and acrylic latex coatings may greatly reduce that figure.

Alkyd coatings are of value because of their relatively low cost, durability, heat resistance, and adaptability. They may be changed to meet special coating needs. To increase durability and abrasion resistance, rosin may be used to modify alkyd resins. Phenolics and epoxy resins may improve hardness and resistance to chemicals and water. Styrene monomers added to the resin may extend flexibility of the finished coating and serve as cross-linking agents.

Alkyd resins are used in *oil-based* house paint, baking enamel, farm implement paint, emulsion paint, porch and deck enamel, spar varnish, and chlorinated rubber paint. Moldified alkyd resins

Fig. 8-1. Alkyd bulk molding compound in several different shapes. (Allied Chemical Corp.)

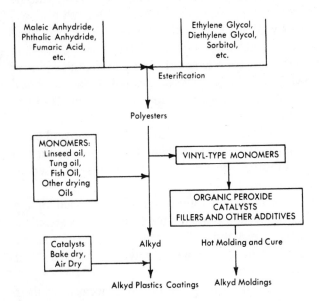

Fig. 8-2. Production of alkyd coatings and moldings.

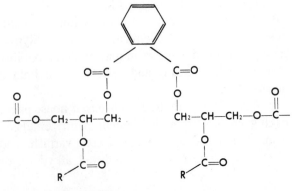

Fig. 8-3. Long alkyd resin chain with unsaturated groups.

may produce special coatings, such as the *wrinkle* and *hammer* finishes often used on equipment and machinery.

Alkyd baking finishes are referred to as *heat-convertible*. They polymerize or harden when heated. Resins that harden when exposed to air are called *air-convertible* resins.

Alkyd resins are used as plasticizers for various plastics, vehicles for printing inks, binders for abrasives and oils, and special adhesives for wood, rubber, glass, leather, and textiles.

New processing technologies and uses for thermosetting alkyd molding compounds are also of current commercial importance.

Alkyd molding compounds are processed in compression, transfer, and reciprocating-screw equipment. By using initiators such a benzoyl peroxide or tertiary butyl hydroperoxide, the unsaturated resins can be made to cross-link at molding temperatures (greater than 50 °C or 120 °F). Molding cycles may be less than twenty seconds.

Typical uses of alkyd molding compounds include appliance housings, utensil handles, billiard balls, circuit breakers, switches, motor cases, and capacitor and commutator parts (Fig. 8-4). They have been used in electrical applications where less costly phenolic or amino resins are not suitable.

Alkyd compounds in putty form are used to encapsulate electronic and electrical components. Different resin formulations, fillers, and processing techniques provide a wide range of physical characteristics.

Table 8-1 gives some of the properties of alkyd plastics.

8-2 ALLYLICS

Allylic resins usually involve the esterification of allyl alcohol and a dibasic acid (Fig. 8-5). The pungent odor of allyl alcohol (CH_2=$CHCH_2OH$) has been known to science since 1856. The name *allyl* was coined from the Latin word *allium*, meaning garlic. These resins did not become commercially useful until 1955 although one allyl resin was used in 1941 as a low-pressure laminating resin.

Allylics and polyesters have common applications, properties, and historical background of development. For this reason, allylics are sometimes erroneously included in the study of polyesters. Allyl monomers are used as cross-linking agents in polyesters, which adds to the confusion. Allylics are

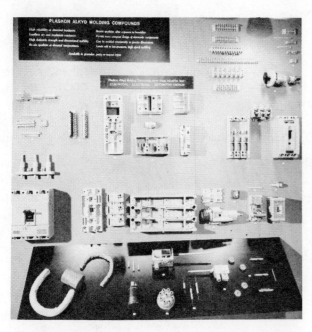

(A) Electrical, electronic, and automotive parts. (Allied Chemical Corp.)

Fig. 8-4. Various alkyd products.

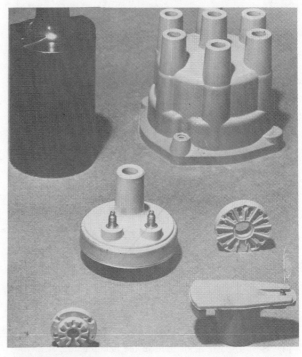

(B) Other automotive parts.

Table 8-1. Properties of Alkyds

Property	Alkyd, (Glass-Filled)	Alkyd Molding Compound (Filled)
Molding qualities	Excellent	Excellent
Relative density	2.12–2.15	1.65–2.30
Tensile strenfth, MPa	28–64	21–62
(psi)	(4 000–9 500)	(3 000–9 000)
Compressive strength, MPa	103–221	83–262
(psi)	(15 000–32 000)	(12 000–38 000)
Impact strength, Izod, J/mm	0.03–0.5	0.015–0.025
(ft lb/in)	(0.60–10)	(0.30–0.50)
Hardness, Barcol	60–70	55–80
Rockwell	E98	E95
Thermal expansion, 10^{-4}/°C	3.8–6.35	5.08–12.7
Resistance to heat, °C	230	150–230
(°F)	(450)	(300–450)
Dielectric strength, V/mm	9 845–20 870	13 800–17 720
Dielectric constant (at 60 Hz)	5.7	5.1–7.5
Dissipation factor (at 60 Hz)	0.010	0.009–0.06
Arc resistance, s	150–210	75–240
Water absorption (24 h), %	0.05–0.25	0.05–0.50
Burning rate	Slow to nonburning	Slow to nonburning
Effect of sunlight	None	None
Effect of acids	Fair	None
Effect of alkalies	Fair	Attacked
Effect of solvents	Fair to good	Fair to good
Machining qualities	Poor to fair	Poor to fair
Optical qualities	Opaque	Opaque

a distinct family of plastics based on monohydric alcohols while the chemical basis for polyesters is polyhydric alcohols.

Allylics are unique in that they may form pre-polymers (partly polymerized resins). They may be homopolymerized or copolymerized. Monoallyl esters may be produced as either thermosetting or thermoplastic resins. The saturated monoallyl esters (thermoplastic) are sometimes used as co-polymerizing agents in alkyd and vinyl resins.

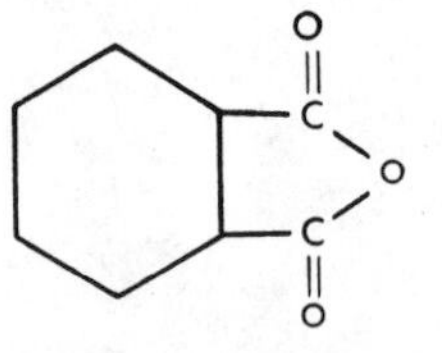

(A) Phthalic anhydride.

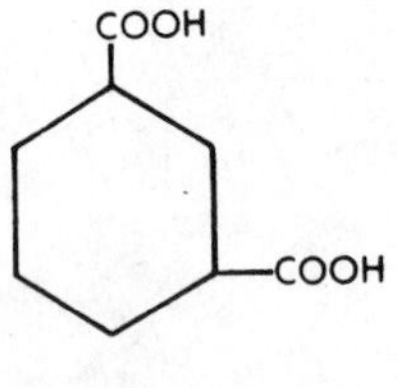

(B) Isophthalic acid.

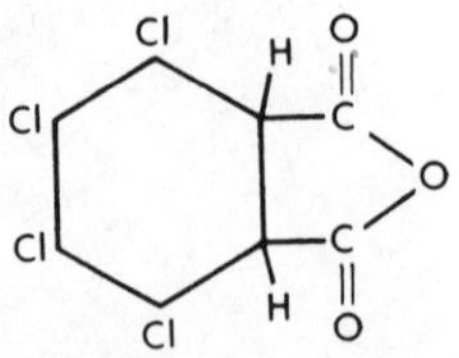

(C) Tetrachlorophthalic acid.

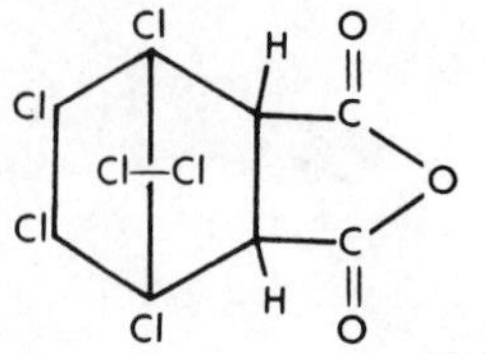

(D) Chlorendic anhydride.

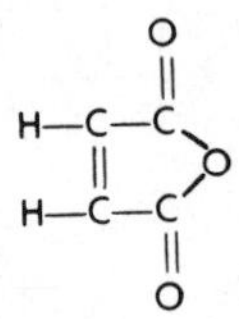

(E) Maleic anhydride.

Fig. 8-5. Dibasic acids used in the manufacture of allylic monomers.

Unsaturated monoallyl esters have been used to produce simple polymers, including allyl acrylate, allyl chloroacrylate, allyl methacrylate, allyl crotonate, allyl cinnamate, allyl cinnamalacetate, allyl furoate, and allyl furfurylacrylate.

Simple polymers have been produced from diallyl esters including diallyl maleate (DAM), diallyl oxalate, diallyl succinate, diallyl sebacate, diallyl pththalate (DAP), diallylcarbonate, diethylene glycol bis-allyl carbonate, diallyl isophthalate (DAIP), and others (Fig. 8-6). The most widely used commercial compounds are DAP (Fig. 8-6A) and DAIP (Fig. 8-6B).

Diallyl resins are usually supplied as monomers or prepolymers. Both forms are converted to the fully polymerized thermosetting plastics by the addition of selected peroxide catalysts. Benzoyl peroxide or *tert*-butyl perbenzoate are two often-used catalysts for polymerization of allylic resin compounds. These resins and compounds may be catalyzed and stored for more than a year, if kept at low ambient temperatures. When the material is subjected to normal temperatures of molds, presses, or ovens, complete cure is reached.

The diethylene glycol bis-allyl carbonate resin monomer (Fig. 8-6E) used for laminating and optical castings is illustrated below. This resin could be polymerized with benzoyl peroxide at a temperature of 180 °F [82 °C].

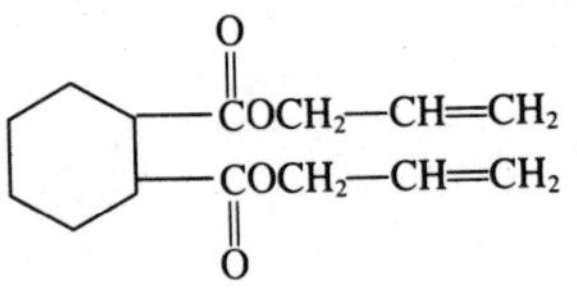

(A) Diallyl phthalate (ortho). 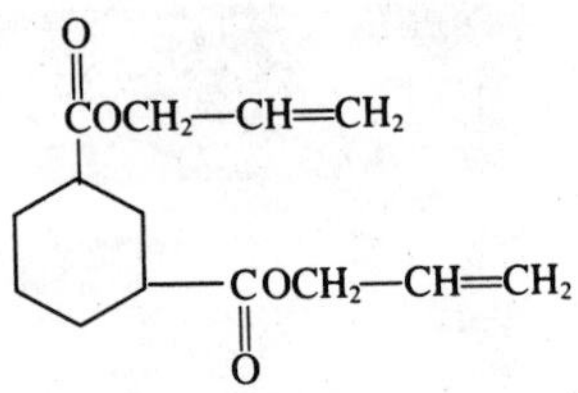 **(B)** Diallyl isophthalate (meta).

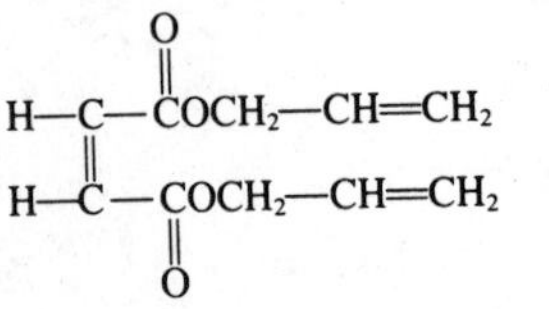

(C) Diallyl maleate.

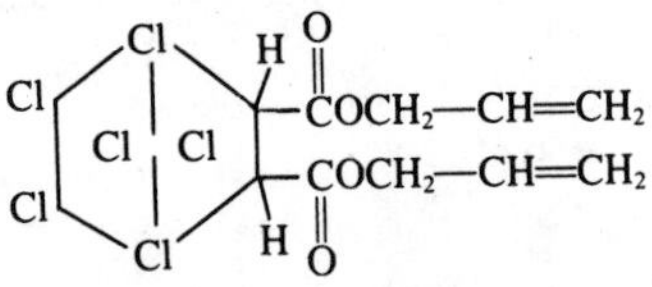

(D) Diallyl chlorendate.

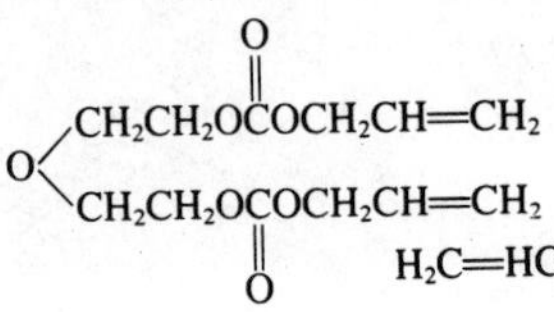

(E) Diethylene glycol bis-(allyl carbonate).

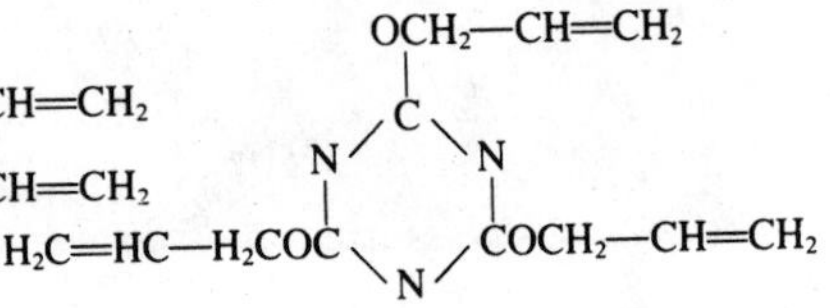

(F) Triallyl cyanurate.

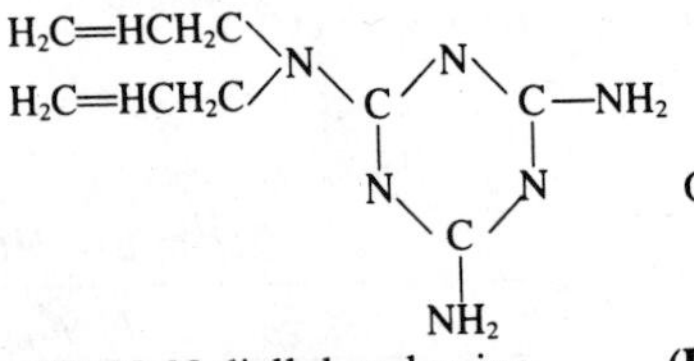

(G) N, N-diallyl melamine.

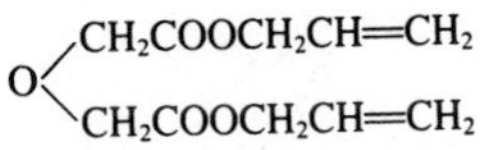

(H) Diallyl diglycollate.

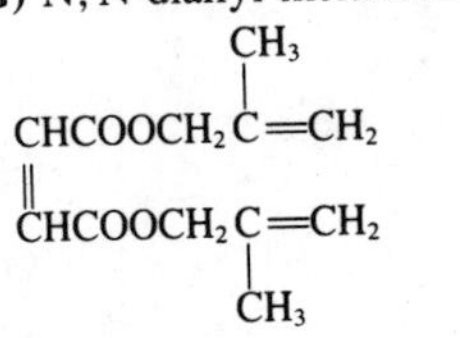

(I) Dimethallyl maleate.

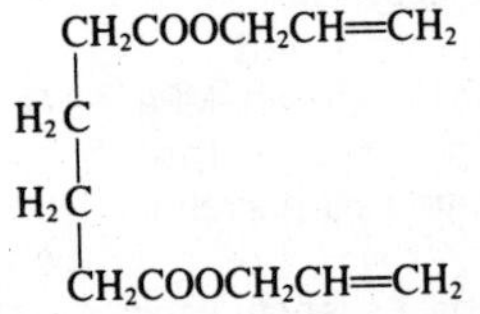

(J) Diallyl adipate.

Fig. 8-6. Structural formulas of some commercial allylic monomers.

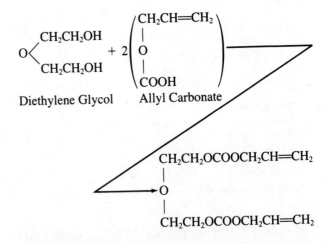

Diethylene Glycol   Allyl Carbonate

Diallyl phthalate may be used as a coating and laminating material but it is probably most noted for its principal use as a molding compound. Diallyl phthalate will polymerize and cross-link because of two available double bonds. (See Fig. 8-6A).

Good mechanical, chemical, thermal, and electrical properties are only a few of the attractive attributes of diallyl phthalate (Fig. 8-7). It also offers long shelf life and ease of handling. Radiation and ablation resistance make these materials useful in space environments.

Nearly all allylic molding compounds include catalysts, fillers, and reinforcements and are usually blended into puttylike premixes. High cost is a factor that limits use of allylics to applications where· their properties are vital. Compounds may be compression- or transfer-molded. Some meet the requirements for special high-speed injection machines, low-pressure encapsulation, and extrusion.

Some allylic resin formulations are used in producing laminates. Monomer resins are used to preimpregnate wood, paper, fabrics, or other materials for lamination. Some improve various properties in paper and garments. Crease-, water-, and fade-resistance may be enhanced by the addition of diallyl phthalate monomers to selected fabrics. For furniture and paneling, preimpregnated decorative laminates or overlays are bonded to cores of less

costly materials. Melamine and allylic resin-bound laminates have many of the same properties.

Allylic monomers and prepolymers also find use in prepregs (wet layup laminates). These prepregs consist of fillers, catalysts, and reinforcements. They are combined just prior to cure. Some are preshaped to ease molding. Allylic prepregs may be prepared in advance and stored until needed. Premix and prepreg molded parts have good flexural and impact strength, and their surface finishes are excellent.

Allylic monomers are used as crosslinking agents for polyesters, alkyds, polyurethane foams, and other unsaturated polymers. They are useful because the basic allylic monomer (homopolymer) does not polymerize at room temperature. This allows materials to be stored for very long periods. At temperatures of 150 °C [300 °F] and above, diallyl phthalate monomers cause polyesters to cross-link. These compounds may be molded at faster rates than those cross-linked with styrene.

Acrylics (methyl methacrylate) cross-linked with diallyl phthalate monomers have good surface hardness and elasticity.

Allylics have been used in vacuum impregnation to seal the pores of metal castings, ceramics, and other compositions. Other uses include the impregnation of reinforcing tapes that are used to wrap motor armatures. Allylics are used to coat electrical parts and to encapsulate electronic devices.

Table 8-2 gives some of the basic properties of allylic (diallyl phthalate) plastics. Four advantages and three disadvantages are listed below.

Advantages of Allylic Esters (Allyls)

1. Excellent moisture resistance
2. Availability of low-burning and self-extinguishing grades
3. Service temperatures as high as 204–232 °C [400–450 °F]
4. Good chemical resistance

Disadvantages of Allylic Esters (Allyls)

1. High cost (compared to alkyds)
2. Excessive shrinkage during curing
3. Not usable with phenols and oxidizing acids

8-3 AMINO PLASTICS

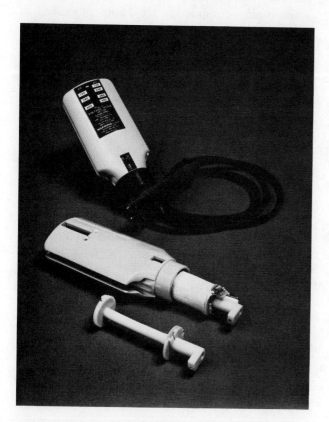

Fig. 8-7. This hand-held voltage tester has a solenoid coil bobbin made of a strong diallyl phthalate molding compound.

Various polymers have been produced by interaction of amines or amides with aldehydes. The

Table 8-2.   Properties of Allylic Plastics

Property	Diallyl Phthalate Compounds		Diallyl Isophthalate
	Glass-Filled	Mineral-Filled	
Molding qualities	Excellent	Excellent	Excellent
Relative density	1.61–1.78	1.65–1.68	1.264
Tensile strength, MPa	41.4–75.8	34.5–60	30
(psi)	(6 000–11 000)	(5 000–8 700)	(4 300)
Compressive strength, MPa	172–241	138–221	
(psi)	(25 000–35 000)	(20 000–32 000)	
Impact strength, Izod, J/mm	0.02–0.75	0.015–0.225	0.01–0.015
(ft lb/in)	(0.4–15)	(0.3–0.45)	(0.2–0.3)
Hardness, Rockwell	E80–E87	E61	M238
Thermal expansion, $10^{-4}/°C$	2.5–9	2.5–10.9	
Resistance to heat, °C	150–205	150–205	150–205
(°F)	(300–400)	(300–400)	(300–400)
Dielectric strength, V/mm	15 550–17 717	15 550–16 535	16 615
Dielectric constant (at 60 Hz)	4.3–4.6	5.2	3.4
Dissipation factor (at 60 Hz)	0.01–0.05	0.03–0.06	0.008
Arc resistance, s	125–180	140–190	123–128
Water absorption (24 h), %	0.12–0.35	0.2–0.5	0.1
Burning rate	Self-extinguishing to nonburning	Self-extinguishing to nonburning	Self-extinguishing to nonburning
Effect of sunlight	None	None	None
Effect of acids	Slight	Slight	Slight
Effect of alkalies	Slight	Slight	Slight
Effect of solvents	None	None	None
Machining qualities	Fair	Fair	Good
Optical qualities	Opaque	Opaque	Transparent

two most significant and commercially useful amino plastics are produced by the condensation of urea-formaldehyde and melamine-formaldehyde.

Urea-formaldehyde polymers may have been produced as early as 1884. In Germany, Gold-schmidt and his colleagues pointed their efforts toward making moldable amino plastics.

In the United States, urea-formaldehyde resins were commercially produced in 1920. Thio-urea-formaldehyde and melamine-formaldehyde resins were being produced in the period of 1934–1939.

Urea-Formaldehyde (UF)

Reaction between the white crystalline solid, urea (NH_2CONH_2), and aqueous solutions of form-aldehyde (formalin) produces urea resins, which may be modified by the addition of other reagents. For complete polymerization and cross-linkage of this thermosetting resin, heat or catalysts and heat are usually needed during the molding operation (Fig. 8-8).

Urea-formaldehyde polymerization is shown in Figure 8-9. The excess water molecule is a result of condensation polymerization.

Many products once produced with urea-formaldehyde plastics are now being made of thermoplastic materials. Faster production cycles

and higher output are possible with thermoplastics. Some amino compounds are being processed in specially designed injection molding equipment. This approach increases output and makes these products able to compete with most thermoplastics.

Resins with a urea-formaldehyde base are made into molding compounds that contain several ingredients including resin, filler, pigment, catalyst, stabilizer, plasticizer, and lubricant. Such compounds speed up curing and production rates.

Molding compounds are produced with latent acid catalysts added to the resin base. Such catalysts react at molding temperatures. Many urea-formaldehyde molding compounds have prepromoters or catalysts added causing them to have a limited storage life; therefore, they should be kept in a cool place. Stabilizers may be added to help control latent catalyst reaction. Lubricants are added to

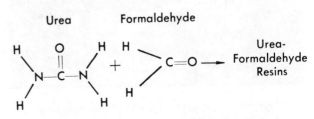

Fig. 8-8. Polymerization of urea and formaldehyde.

(A) Urea-formaldehyde resin.

(B) Urea-formaldehyde plastics.

Fig. 8-9. Formation of urea-formaldehyde resins.

improve molding quality. Plasticizers improve flow properties and help reduce cure shrinkage. Although the base resin is water-clear, color pigments (transparent, translucent, or opaque) may be used. If color is not needed, fillers such as alpha-cellulose (bleached cellulose) fiber, macerated fabric, or wood flour may be added to improve molding and physical characteristics and lower the cost.

Urea-formaldehyde plastics have many industrial uses because they have outstanding molding qualities and are low in cost. They are used in electrical and electronic applications where good arc and tracking resistance is needed. They provide good dielectric properties and are unaffected by common organic solvents, greases, oils, weak acids, alkalies, or other hostile chemical environments. Urea compounds do not attract dust by static electricity charges. They will not burn or soften when exposed to an open flame and they have very good dimensional stability when filled.

Urea-formaldehyde molding compounds are used to make bottle caps and electrical or thermal insulating materials.

Urea-formaldehyde products do not impart taste or odors to foods and beverages. They are preferred materials for appliance knobs, dials, handles, pushbuttons, toaster bases, and end plates. Wall plates, switch toggles, receptacles, fixtures,

circuit breakers, and switch housings are only a few of many electrical insulating uses (Fig. 8-10).

One of the largest current uses of urea-formaldehyde resins is in adhesives for furniture, plywood, and chipboard. Chipboard is made by combining 10 percent resin binder with wood chips which is then pressed into flat sheets. The product has no grain so it is free to expand in all directions and does not warp. Water resistance is poor, however. Plywood bonded with this resin is suitable only for interior use.

Urea-formaldehyde resins may be foamed and cured into a plastics state. Such foams are inexpensive, and various densities may be produced. They are easily produced by vigorously whipping a mixture of resin, catalyst, and a foaming detergent. Another foaming method involves introducing a chemical agent that generates a gas (usually carbon dioxide) while the resin is curing. These foams have found uses as thermal insulating materials in buildings and in refrigerators, and as low-density cores

(A) Terminal blocks for electrical connections.

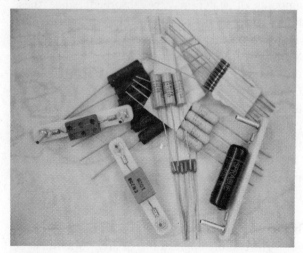

(B) Insulating blocks and packing for electronic components.

Fig. 8-10. Applications of urea-formaldehyde compounds.

(A) Flower-arranging block.

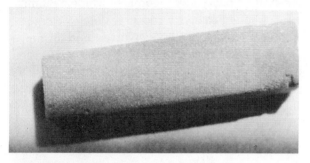

(B) Thermal insulation material.

Fig. 8-11. Uses of open-celled or foamed urea-formaldehyde.

for structural sandwich construction (Fig. 8-11). Undercured or partially polymerized foams have caused allergy and flu-like symptoms in some people. Degassing of molded furniture parts and wall insulation are a major source of this problem.

Urea-formaldehyde foams may be made flame-resistant, but at the expense of foam density. Large amounts of water may be absorbed, since these foams are open-celled (spongelike) structures. This ability to absorb water is made use of by florists. Stems of cut flowers are inserted into water-soaked urea foam used as a base for flower arrangements. Ground foam has been used as artificial snow on television and in theatrical productions. The open-celled structure may also be filled with kerosene and used as a lighting agent for fireplaces.

Urea resins find extensive use in the textile, paper, and coating industries. The existence of drip-dry fabrics is due largely to these resins.

Urea resins or powders are sometimes used as binders for foundry cores and shell molds.

The coating applications of urea-formaldehyde resins are limited. They can be applied only to substrates that can withstand curing temperatures of 100° to 350 °F [40° to 175 °C]. These resins are com-

bined with a compatible polyester (alkyd) resin to produce enamels. From 5 to 50 percent urea resin is added to the alkyd-based resin. These surface coatings are outstanding in hardness, toughness, gloss, color stability, and outdoor durability.

These surface coatings may be seen on refrigerators, washing machines, stoves, signs, venetian blinds, metal cabinets, and many machines. At one time, they were widely applied to automobile bodies, however the curing or baking time required has made this uneconomical for mass production. The baking time depends on the temperature and the proportion of amino resin to alkyd resin.

Urea resins modified with furfuryl alcohol have been used successfully in the manufacture of coated abrasive papers. (See Furan, below.)

Urea-formaldehyde plastics are molded with ease in compression and transfer molding machines. They may also be processed by reciprocating-screw injection molding machines. Depending on the grade of resin and the filler used, allowance should be made for shrinkage after removal from the mold. Improved dimensional stability can be had by post-conditioning the product in an oven.

Table 8-3 gives some properties of alpha-cellulose filled urea-formaldehyde plastics. Five advantages and four disadvantages of urea-formaldehyde are listed below.

Table 8-3.   Properties of Urea-Formaldehyde

Property	Urea-Formaldehyde (Alpha-Cellulose Filled)
Molding qualities	Excellent
Relative density	1.47–1.52
Tensile strength, MPa	38–90
(psi)	(5 500–13 000)
Compressive strength, MPa	172–310
(psi)	(25 000–45 000)
Impact strength, Izod, J/mm	0.012 5–0.02
(ft lb/in)	(0.25–0.40)
Hardness, Rockwell	M110–M120
Thermal expansion, 10^{-4}/°C	5.6–9.1
Resistance to heat, °C	80
(°F)	(170)
Dielectric strength, V/mm	11 810–15 750
Dielectric constant (at 60 Hz)	7.0–9.5
Dissipation factor (at 60 Hz)	0.035–0.043
Arc resistance, s	80–150
Water absorption (24 h), %	0.4–0.8
Burning rate	Self-extinguishing
Effect of sunlight	Grays
Effect of acids	None to decomposes
Effect of alkalies	Slight to decomposes
Effect of solvents	None to slight
Machining qualities	Fair
Optical qualities	Transparent to opaque

Advantages of Urea-Formaldehyde

1. Good hardness and scratch resistance
2. Comparatively low cost
3. Wide color range
4. Self-extinguishing
5. Good solvent resistance

Disadvantages of Urea-Formaldehyde

1. Must be filled for successful molding
2. Long-term oxidation resistance is poor
3. Attacked by strong acids and bases
4. Uncured, partially polymerized plastics and foams degas

Melamine-Formaldehyde (MF)

Until about 1939, melamine-formaldehyde was an expensive laboratory curiosity. Melamine ($C_3H_6N_6$) is a white crystalline solid. Combining it with formaldehyde results in the formation of a compound referred to as methylol derivative (Fig. 8-12). With additional formaldehyde, the formulation will react to produce tri-, tetra-, penta-, and hexamethylol-melamine. The formation of trimethylol melamine is shown in Figure 8-13.

Commercial melamine resins may be obtained without acid catalysts, but both thermal energy and catalysts are used to speed polymerization and cure. Polymerization of urea and melamine resins is a condensation reaction, producing water, which will evaporate or escape from the molding cavity.

Resins of formaldehyde-plus benzoguanamine ($C_3H_4N_5C_6H_5$) or thiourea ($CS(NH_2)_2$) are of minor commercial importance. The reaction of formaldehyde and compounds such as aniline, dicyandiamide, ethylene urea, and sulfonamide provide more complex resins for varied applications.

In pure form, amino *resins* are colorless and soluble in warm solutions of water and methanol.

Amino resins, urea plastics, and melamine plastics are frequently grouped together as one entity. Their chemical structures, properties, and applications have much in common. But melamine-formaldehyde products are better than the urea-formaldehyde plastics in several respects.

Melamine products are harder and more water-resistant. They may be combined with a greater variety of fillers that allow the making of products with better heat, scratch, stain, water and chemical resistance. Melamine products are also much more expensive than urea products.

Probably the largest single use of melamine-formaldehyde is the manufacture of tableware (Fig. 8-14). For this application, molding powders are normally filled with alpha-cellulose. Asbestos or other fillers are sometimes used for handles and utensil housings.

Melamine resin is widely used for surface coatings and decorative laminates. Paper-based laminates are sold under the trade names of Formica and Micarta. Photographic prints on fabrics or paper impregnated with melamine resin are placed on a base or core material and cured in a large press. Kraft paper impregnated with phenolic resin is commonly used for the base material since it is durable and is compatible with melamine resin. This base also costs less than multiple layers of melamine-impregnated paper. There is a broad spectrum of uses for these laminates, including surfacing wood, metal, plaster, and hardboard. A familiar use of these laminates is surfacing for kitchen counters and table tops.

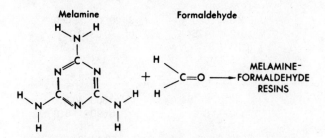

Fig. 8-12. Formation of melamine-formaldehyde resins.

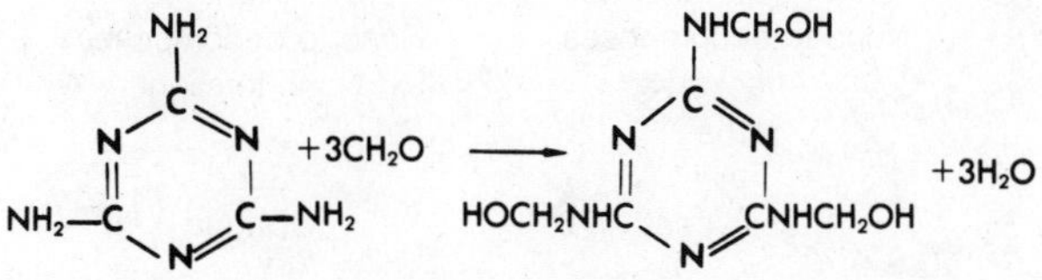

Fig. 8-14. Melamine plastics canister set.

Fig. 8-13. Formation of trimethylol melamine.

A 3 percent melamine resin solution may be added during the preparation of paper pulp to improve the wet strength of paper. Paper with this resin binder has a wet strength nearly as great as its dry strength. Crush and folding resistance is also greatly increased without added brittleness.

Finishes resistant to water, chemicals, alkali, grease, and heat are formulated of amino resins. Heat or heat and catalysts are needed to cure the amino resins. These finishes are seen on stoves, washing machines, and other appliances.

Urea or melamine resins may be made compatible with other plastics to produce finishes of outstanding merit. Polyester (alkyd) resins or phenolic resins may be combined to produce finishes with the best features of both resins. These finishes are sometimes used where a hard, tough, mar-resistant finish is required.

Melamine resins are often used to make waterproof exterior plywood and marine plywood, and for other adhesive applications requiring a light-colored, nonstaining adhesive. Catalysts, heat, or high-frequency energy are used to cure melamine adhesives in plywood and panel assemblies.

Melamine-formaldehyde resins are employed commercially for textile finishes. The well-known drip-dry fabrics and fabrics with permanent glazing, rotproofing, and shrinkage control owe their existence largely to such resins. Melamine and silicone resins are used to produce waterproof fabrics.

Melamine-formaldehyde compounds are easily molded in normal compression and transfer molding machines. Special reciprocating-screw injection machines are also used.

Table 8-4 gives some of the properties of melamine-formaldehyde plastics while listed below are five advantages and three disadvantages of melamine-formaldehyde.

Advantages of Melamine-Formaldehyde

1. Good hardness and scratch resistance
2. Comparatively low cost
3. Wide color range
4. Self-extinguishing
5. Good solvent resistance

Disadvantages of Melamine-Formaldehyde

1. Must be filled for successful molding
2. Poor long-term oxidation resistance
3. Subject to attack by strong acids and bases

Table 8-4. Properties of Melamine-Formaldehyde

Property	No Filler	Alpha-Cellulose Filled	Glass-Fiber Filled
Molding qualities	Good	Excellent	Good
Relative density	1.48	1.47–1.52	1.8–2.0
Tensile strength, MPa		48–90	34–69
(psi)		(7 000–13 000)	(5 000–10 000)
Compressive strength, MPa	276–310	276–310	138–241
(psi)	(40 000–45 000)	(40 000–45 000)	(20 000–35 000)
Impact strength, Izod, J/mm		0.012–0.0175	0.03–0.9
(ft lb/in)		(0.24–0.35)	(0.6–18.0)
Hardness, Rockwell		M115–M125	M120
Thermal expansion, 10^{-4}/°C		10	3.8–4.3
Resistance to heat, °C	99	99	150–205
(°F)	(210)	(210)	(300–400)
Dielectric strength, V/mm		10 630–11 810	6 690–11 810
Dielectric constant (at 60 Hz)		6.2–7.6	9.7–11.1
Dissipation factor (at 60 Hz)		0.030–0.083	0.14–0.23
Arc resistance, s	100–145	110–140	180
Water absorption (24 h), %	0.3–0.5	0.1–0.6	0.09–0.21
Burning rate	Self-extinguishing	Nonburning	Self-extinguishing
Effect of sunlight	Color fades	Slight color change	Slight
Effect of acids	None to decomposes	None to decomposes	None to decomposes
Effect of alkalies		Attacked	None to slight
Effect of solvents	None	None	None
Machining qualities		Fair	Good
Optical qualities	Opalescent	Translucent	Opaque

8-3 CASEIN

Casein plastics are sometimes classed as natural polymers and are referred to as *protein plastics* by many.

Casein is a protein found in a number of sources including animal hair, feathers, bones, and industrial wastes. There has been little interest in these sources, and only skimmed milk is now of commercial interest in the making of casein.

The history of protein-derived plastics can probably be said to date from the work of W. Krische, a German printer, and Adolf Spitteler of Bavaria around 1895. At that time, there was a demand in Germany for what may be described as a *white blackboard*. These boards were thought to possess better optical properties than those with a black surface. In 1897, Spitteler and Krische, in attempting to develop such a product, produced a casein plastics that could be hardened with formaldehyde. Until then casein plastics had proved unsatisfactory because they were soluble in water. Galaith (milkstone), Erinoid, and Ameroid are trade names of those early protein plastics.

Casein is not coagulated by heat but must be precipitated from milk by the action of rennin enzymes or acids. This powerful coagulant causes the milk to separate into solids (curds) and a liquid (whey). After the whey is removed, the curd containing the protein is washed, dried, and made into a powder. When kneaded with water, the doughlike material may be shaped or molded. A simple drying operation then causes much shrinkage. Casein is thermoplastic while being molded. Molded products may be made water resistant by soaking in a formalin solution that creates links holding the casein molecules together. The long linear chain of casein molecules is known as the *polypeptide chain*. There are a number of other peptide groups and possible side reactions. No single formula could represent the interaction between casein and formaldehyde. A very simplified reaction is shown in Figure 8-15.

It is doubtful that casein will gain popularity because it is costly to make and the raw material has value as a food. Casein plastics are seriously affected by humid conditions and cannot be used as electrical insulators. The lengthy hardening process and poor resistance to decomposition by heat make them unsuitable for modern processing rates. There are only limited commercial uses of casein plastics in the United States since they offer no property advantages over synthetic polymers, and production costs are high.

Casein plastics are used to a limited extent for buttons, buckles, knitting needles, umbrella handles, and other novelty items. They may be reinforced and filled, or obtained in transparent colors. Casein has held some of its appeal because it may be colored to imitate onyx, ivory, and imitation horn. Casein is most widely used in the stabilization of rubber latex emulsion, preparation of medical compounds and food products, preparation of paints and adhesives, and the sizing of paper and textiles. Casein finds other uses in insecticides, soaps, pottery, inks, and as modifiers in other plastics.

Films and fibers may be produced from these plastics. The woollike fibers are warm, soft, and have properties that compare favorably with natural wool. The films are generally of little use except as handy forms for the coating of paper and other materials. Casein glue is a well-known wood adhesive.

Table 8-5 gives some of the properties of casein plastics. Two advantages and disadvantages of casein follow.

Table 8-5. Properties of Casein

Property	Casein Formaldehyde (Unfilled)
Molding qualities	Excellent
Relative density	1.33–1.35
Tensile strength, MPa	48–79
(psi)	(7 000–10 000)
Compressive strength, MPa	186–344
(psi)	(27 000–50 000)
Impact strength, Izod, J/mm	0.045–0.06
(ft lb/in)	(0.9–1.20)
Hardness, Rockwell	M26–M30
Resistance to heat, °C	135–175
(°F)	(275–350)
Dielectric strength, V/mm	15 500–27 500
Dielectric constant (at 60 Hz)	6.1–6.8
Dissipation factor (at 60 Hz)	0.052
Arc resistance, s	Poor
Water absorption (24 h), %	7–14
Burning rate	Slow
Effect of sunlight	Yellows
Effect of acids	Decomposes
Effect of alkalies	Decomposes
Effect of solvents	Slight
Machining qualities	Good
Optical qualities	Translucent to opaque

$$\underset{|}{CO} \qquad \underset{|}{CO} \qquad \qquad \underset{|}{CO} \qquad \underset{|}{CO}$$
$$NH + CH_2O + NH \longrightarrow N-CH_2-N$$

Advantages of Casein

1. Produced from nonpetrochemical sources
2. Excellent molding qualities and colorability

Disadvantages of Casein

1. Poor resistance to acids and alkalies, yellows in sunlight
2. High water absorption

8-4 EPOXY (EP)

Hundreds of patents have been granted for the commercial uses of epoxide resins. One of the first descriptions of polyepoxides is a German patent by I. G. Farbenindustrie in 1939.

In 1943, the Ciba Company developed an epoxide resin of commercial significance in the United States. By 1948, a variety of commercial coating and adhesive applications were discovered.

Epoxy resins are thermosetting plastics. There are several thermoplastic epoxy resins used for coatings and adhesives. Many different epoxy resin structures available today are derived from bisphenol acetate and epichlorohydrin.

Bisphenol A (bisphenol acetate) is made by the condensation of acetone with phenol (Fig. 8-16). Epichlorohydrin-based epoxies are widely used because of availability and lower cost. The epichlorohydrin structure is obtained by chlorination of propylene:

$$CH_2 \overset{O}{\overset{\diagup\diagdown}{-}} CH - CH_2Cl$$

It will become evident that the epoxy group, for which the plastics family is named, has a triangular structure:

$$CH_2 \overset{O}{\overset{\diagup\diagdown}{-}} CH \ldots R$$

Epoxy structures are usually terminated by this epoxide structure but many other molecular structures may terminate the long molecular chain. A linear epoxy polymer may be formed when bisphenol A and epichlorohydrin are reacted (Fig. 8-17). In some literature, these polymers are called polyethers.

A typical structural formula of epoxy resin based on bisphenol A may be represented as shown in Figure 8-18. Other intermediate epoxy-based resins are possible, but are too numerous to mention.

Epoxy resins are cured, as a rule, by adding catalysts or reactive hardeners. Members of the aliphatic and aromatic amine family are commonly used hardening agents. Various acid anhydrides are also used to polymerize the epoxide chain.

Epoxy resins will polymerize and cross-link as thermal energy is added. Catalysts and heat are often used to reach a desired degree of polymerization.

Single-component epoxy resins may contain latent catalysts. These react when enough heat is applied. There is a practical shelf-life expectancy for all epoxy resins.

Reinforced epoxy resins are very strong. They have good dimensional stability and service temperatures as high as 600 °F [315 °C]. Preimpregnated reinforcing materials are used to produce products by hand-layup, vacuum-bag, or filament-

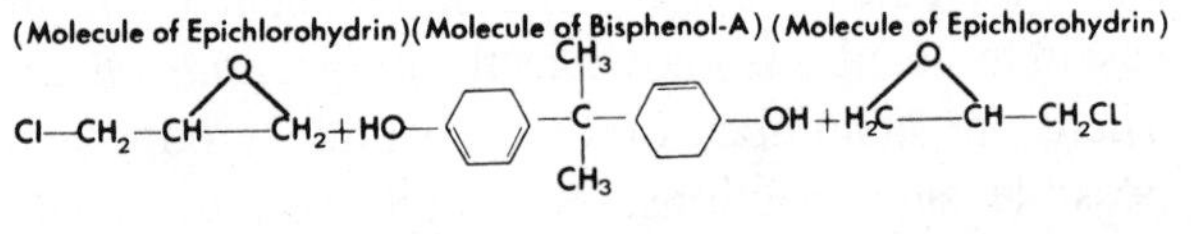

Fig. 8-17. Formation of linear epoxy polymer.

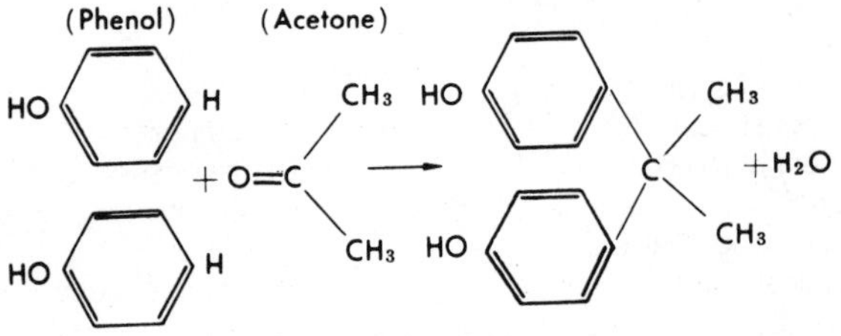

Fig. 8-16. Production of bisphenol-A.

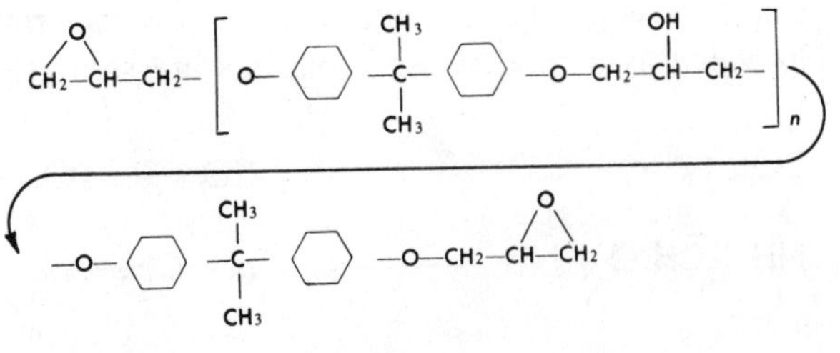

Fig. 8-18. Bisphenol-A-based epoxy resin.

Fig. 8–19. Continuous fibrous glass in a matrix of epoxy is used in this composite filament wound rocket motor case. (Structural Composite Industries)

(A) Epoxy adhesives, pressure-injected into hairline cracks in concrete, restore the original load-bearing strength of the material. (Scott J. Saunders Associates)

(B) End closures and ribs of this boron slat for aircraft are bonded to the inner skin with epoxy adhesives.

Fig. 8-20. High strength makes epoxy resins very useful.

winding processes (Fig. 8-19). Epoxy has a very good chemical and fatigue resistance, thus epoxy resins replace the less costly unsaturated polyester resins in many applications. A saving of one-third of the resin mass is realized when epoxy resins are used in place of polyesters.

Epoxy-glass laminates find many uses because they have high strength-to-mass ratio. Superior adhesion to all materials and wide compatibility make epoxy resins desirable (Fig. 8-20). Laminated circuit boards, radomes, aircraft parts, and filament-wound pipes, tanks, and containers are only a few uses for this plastics material. (Fig. 8-21)

Filled epoxy resins are commonly used for special castings. These strong compounds may be used for low-cost tooling. In dies, jigs, fixtures, and molds for short production runs, epoxies are replacing other tooling materials. Faithful reproduction of details is obtained when epoxy compounds are cast against prototypes or patterns (Fig. 8-22).

Many different fillers are used in caulking and patching compounds containing epoxy resins. The adhesive qualities and low shrinkage of epoxies during cure make them durable in caulking or patching applications.

Electrical potting is another casting use. Epoxies are outstanding in protecting electronic parts from moisture, heat, and corrosive chemicals (Fig. 8-23). Electric motor parts, high-voltage transformers, relays, coils, and many other components may be protected from severe environments by being potted in epoxy resins.

Molding compounds of epoxy resins and fibrous reinforcements can be molded by injection, compression, or transfer processes. They are molded into small electrical items and appliance parts, and have many modular uses.

Versatility is achieved by controlling the resin manufacture, the curing agents and the rate of cure. These resins may be formulated to give results ranging from soft, flexible compounds to hard, chemical-resistant products. By incorporation of a blowing agent, low-density epoxy foams may be produced. The qualities offered by the epoxy plastics are adhesion, chemical resistance, toughness, and excellent electrical characteristics.

When the epoxy resins were first introduced in the 1950's, they were recognized as outstanding coating materials (Fig. 8-24). They are more expensive than other coating materials but their adhe-

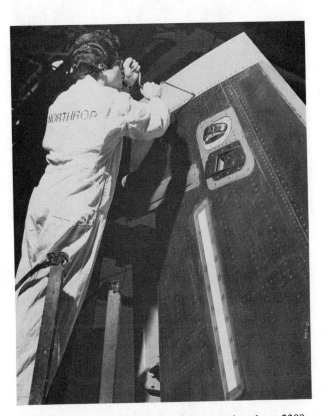

Fig. 8–21. (A) Graphite/epoxy composites are used on the vertical tail and the entire horizontal stabilizer of Northrop's multirole F-20 Tigershark tactical fighter. These workers are preparing the graphite/epoxy composite material, which is lighter than aluminum and stronger than steel.

(B) The F/A-18 Hornet strike fighter carries about 2300 pounds of graphite/epoxy structures. (Northrop Aircraft Division)

sion qualities and chemical inertness make them competitive. Five advantages and three disadvantages of epoxy resins are shown below.

Advantages of Epoxy

1. Wide range of cure conditions, from room temperature to 350 °F [178 °C]
2. No volatiles formed during cure
3. Excellent adhesion
4. Can be cross-linked with other materials

5. Suitable for all thermosetting processing methods

Disadvantages of Epoxy

1. Poor oxidative stability; some moisture sensitivity
2. Thermal stability limited to 350–450 °F [178–232 °C]
3. Many grades are expensive

Fig. 8-22. Casting mold of epoxy resin, for use in prototype or low-volume production. (Henkel Corp.)

Fig. 8-23. A polyamide-epoxy resin blend was used to encapsulate this selenium rectifier. (Henkel Corp.)

(A) Epoxy spray coating is the second can from the left.

(B) Epoxy coating on metal substrate of modern farm silo storage bins. (Bolted Tank Group, Butler Mfg. Co.)

Fig. 8-24. Epoxy coating.

Epoxy-based finishes are used on driveways, concrete floors, porches, metal appliances, and wooden furniture. Epoxy finishes on home appliances are a major application of this durable, abrasion-resistant finish. Epoxy coatings have replaced glass enamel finishes for tank car and other container linings that need to resist chemicals. Ship hulls and bulkheads may be coated with epoxy. More durable finishes mean fewer repairs and reduced surface tension between the ship and water. These factors reduce maintenance and fuel costs.

The flexibility of many epoxy coatings makes them popular for postforming coated metal parts. For example, sheets of metal are coated while flat. They are then formed or bent into shallow pans with no damage to the coating.

The ability of epoxy adhesives to bond to dissimilar materials has allowed them to replace soldering, welding, riveting, and other joining methods. The aircraft and automotive industries use these adhesives where heat or other bonding methods might

distort the surface. Honeycomb or panel structures make use of the superb adhesive and thermal properties of epoxy.

Table 8-6 gives some of the properties of various epoxies. Epoxy copolymers are made by crosslinking with phenolics, melamines, polyamide, urea, polyester, and some elastomers.

8-5 FURAN

Furan resins are derivatives of furfurylaldehyde and furfuryl alcohol (Fig. 8-25). Acid catalysts are used in polymerization therefore it is vital to apply a protective coating to substrates attacked by acids.

Furan plastics have excellent chemical resistance can can withstand temperatures as high as 265 °F [130 °C]. They are used primarily as additives, binders, or adhesives. Furfurylaldehyde has been co-reacted with phenolic plastics. Resins based on furfuryl alcohol are used with amino resins to improve wetting. Their wetting and adhesive capabilities make these resins ideal for impregnating agents. Reinforced and laminated products of furan plastics include tanks, pipes, ducting and construction panels. Furan resins are also used as sand binders in foundries (Table 8-7). Two advantages and disadvantages of furans follow.

Advantages of Furans

1. Produced from nonpetrochemical sources
2. Excellent chemical resistance

Disadvantages of Furans

1. Hard to process, limited to fiber-reinforced plastics
2. Subject to attack by halogens

8-6 PHENOLICS (PF)

Phenolics (phenol-aldehyde) were among the first true synthetic resins produced. They are known chemically as phenol-formaldehyde (PF). Their history reaches back to the work of Adolph Baeyer in 1872. In 1909, the chemist Baekeland invented and patented a technique for combining phenol (C_6H_5OH, also called carbolic acid) and gaseous formaldehyde (H_2CO).

Table 8-6. Properties of Epoxies

Property	Epoxy Molding Compounds		
	Glass-Filled	Mineral-Filled	Microballoon-Filled
Molding qualities	Excellent	Excellent	Good
Relative density	1.6–2.0	1.6–2.0	0.75–1.00
Tensile strength, MPa	69–207	34–103	17–28
(psi)	(10 000–30 000)	(5 000–15 000)	(2 500–4 000)
Compressive strength, MPa	172–276	124–276	69–103
(psi)	(25 000–40 000)	(18 000–40 000)	(10 000–15 000)
Impact strength, Izod, J/mm	0.5–1.5	0.015–0.02	0.008–0.013
(ft lb/in)	(10–30)	(0.3–0.4)	(0.15–0.25)
Hardness, Rockwell	M100–M110	M100–M110	
Thermal expansion, $10^{-4}/°C$	2.8–8.9	5.1–12.7	
Resistance to heat, °C	150–260	150–260	
(°F)	(300–500)	(300–500)	
Dielectric strength, V/mm	11 810–15 750	11 810–15 750	14 960–16 535
Dielectric constant, (at 60 Hz)	3.5–5	3.5–5	
Dissipation factor (at 60 Hz)	0.01	0.01	
Arc resistance, s	120–180	150–190	120–150
Water absorption, (24 h), %	0.05–0.20	0.04	0.10–0.20
Burning rate	Self-extinguishing	Self-extinguishing	Self-extinguishing
Effect of sunlight	Slight	Slight	Slight
Effect of acids	Negligible	None	Slight
Effect of alkalies	None	Slight	Slight
Effect of solvents	None	None	Slight
Machining qualities	Good	Fair	Good
Optical qualities	Opaque	Opaque	Opaque

Furfurylaldehyde   Furfuryl Alcohol   Furon

Fig. 8–25. Acid catalysts will cause condensation of furfurylaldehyde or furfuryl alcohol. Crosslinking occurs between furan rings.

Table 8-7. Properties of Furan

Property	Furan (Asbestos-Filled)
Molding qualities	Good
Relative density	1.75
Tensile strength, MPa	20–31
(psi)	(2 900–4 500)
Compressive strength, MPa	68–72
(psi)	(9 900–10 450)
Hardness, Rockwell	R110
Resistance to heat, °C	130
(°F)	(266)
Water absorption (24 h), %	0.01–2.0
Burning rate	Slow
Effect of sunlight	None
Effect of acids	Attacked
Effect of alkalies	Little
Effect of solvents	Resistant
Machining qualities	Fair
Optical qualities	Opaque

The success of the phenol-formaldehyde resins later stimulated research into urea- and melamine-formaldehyde resins.

The resin formed from the reaction of phenol with formaldehyde (an aldehyde) is known as a *phenolic*. Figure 8-26 shows the reaction of a phenol

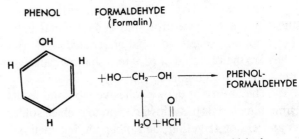

Fig. 8-26. Reaction of phenol and formaldehyde.

with formaldehyde. This involves a condensation reaction in which water is formed as a byproduct (A-stage). The first phenol formaldehyde reactions produce a low-molecular-mass resin that is compounded with fillers and other ingredients (B-stage). During the molding process, the resin is transformed into a highly cross-linked thermosetting plastics product by heat and pressure (C-stage).

Although the monomer solution of phenol is commercially used, cresols, xylenols, resorcinols, or synthetically produced oil-soluble phenols may be used. Furfural may replace the formaldehyde.

In *one-stage resins,* a *resol* is produced by reacting a phenol with an excess amount of aldehyde in the presence of a catalyst (not acid). Sodium and ammonium hydroxide are common catalysts. This product is soluble and low in molecular mass. It will form large molecules without addition of a hardening agent during the molding cycle.

Two-stage resins are produced when phenol is present in excess with an acid catalyst. The low-molecular mass and soluble *novolac* resin is the result. It will remain a linear thermoplastic resin unless compounds capable of forming cross-linkage on heating are added. They are called *two-stage resins* because some agent must be added before molding (Fig. 8-27).

The *A-stage* novolac resin is a fusible and soluble thermoplastic. The *B-stage* resin is produced by thermally blending the A-stage and hexamethylenetetramine. The B-stage is usually sold in a granular or powder form. Fillers, pigments, lubricants and other additives are compounded with the resin in this stage. During molding, heat and pressure convert the B-stage resin into an insoluble, infusible, *C-stage* thermosetting plastics.

Phenolics are not used as frequently as they once were because so many new plastics have been developed (Fig. 8-28). Their low cost, moldability, and physical properties make phenolics leaders in the thermoset field, however. These materials are widely used as molding powders, resin binders, coatings, and adhesives.

Molding powders or compounds of novolac resins are rarely used without a filler. The filler is not used simply to reduce cost. It improves physi-

(A) Billiard balls of molded phenolic.

(B) Molded parts for a humidifier. (Durez Division, Hooker Chemical Corp.)

(C) End panels for a broiler oven. (Durez Division, Hooker Chemical Corp.)

Fig. 8-28. Some uses of phenolics.

Fig. 8-27. Final curing or heat-hardening should be considered a further condensation process.

cal properties, increases adaptability for processing, and reduces shrinkage. Curing time, shrinkage, and molding pressures may be reduced by preheating phenol-formaldehyde compounds. Advances in equipment and techniques have kept phenolics competitive with many thermoplastics and metals (Fig. 8-29). Phenolics are used in conventional transfer and compression molding operations, and in injection and reciprocating-screw machines. Molded phenolic parts are abrasive and hard to machine. Although molded phenolics have many uses as electrical insulation, they exhibit poor tracking resistances under very humid conditions. In a few uses, they have been replaced by thermoplastics.

Phenolic resins have a major appearance drawback—they are too dark in color for use as surface layers on decorative laminates and as adhesives where glue joints may show. Phenolic-resin impregnated cotton fabric, wood, or paper are often used in the making of gear wheels, bearings,

(A) Automotive brake system parts of phenolic.

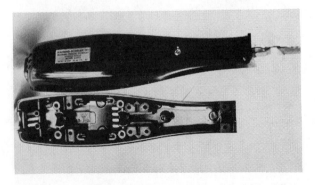

(B) High-impact phenolic used for handle of electric carving knife.

Fig. 8-29. Phenolics remain competitive with thermoplastics. (Durez Division, Hooker Chemical Corp.)

(A) Paper and cloth.

(B) Electrical circuit board.

Fig. 8-30. Materials impregnated with phenolic resin.

substrates for electrical circuit boards, and melamine decorative laminates (Fig. 8-30). These laminates are usually made in large presses under controlled heat and pressure. Many methods of impregnation are used including dipping, coating, and spreading.

Phenol formaldehyde resins may be cast into many profile shapes, such as billiard balls, cutlery handles, and novelty items.

Phenolic-based resins are available in liquid, powder, flake, and film forms. The ability of these resins to impregnate and bond with wood and other materials is the reason for their success as adhesives. They improve adhesion and heat resistance, and are widely used in the making of plywood and as binder adhesives in wood-particle moldings. Wood-particle boards are used in many building applications, such as sheathing, subflooring, and core stocks.

Phenolic resins are used as binders for abrasive grinding wheels. The abrasive grit and resin are simply molded into the desired shape and cured. Resin binders are an important ingredient in the shell molds and cores used in foundries (Fig. 8-31). These molds and cores produce very

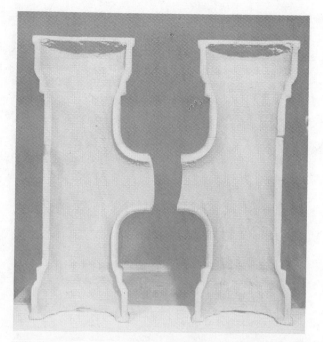

(A) Phenolic resin was used as a binder for this sand core.

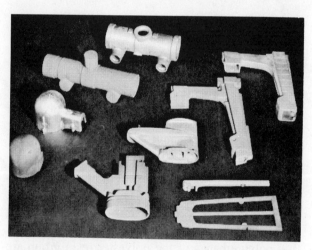

(B) Resin-bonded sand cores and resulting metal castings.

Fig. 8-31. Resin binders used in foundry work. (Acme Resin Co.)

smooth metal castings. As heat-resistant binders, phenolic resins are used in making brake linings and clutch facings.

Because of their high resistance to water, alkalies, chemicals, heat, and abrasion, phenolics are sometimes chosen for use in finishes. They are used for coating appliances, machinery, or other devices requiring maximum heat resistance.

A high-strength, heat- and fire-resistant foam may be produced using phenolic resins. The foam may be produced in the plant or on the site by rapidly mixing a blowing agent and catalyst with the resin. As the chemical reaction generates heat and

begins the polymerization process, the blowing agent vaporizes. This causes the resin to expand into a multicellular, semipermeable structure. These foams may be used as fill for honeycomb structures in aircraft, flotation materials, acoustic and thermal insulation, and as packing materials for fragile objects.

Microballons (small hollow spheres) may be produced of phenolic plastics filled with nitrogen. These spheres vary from 0.0002 to 0.0032 in. [0.005 to 0.08 mm] in diameter. They may be mixed with other resins to produce syntactic foams. These foams find uses as insulative fillers. They serve as vapor barriers when placed on volatile liquids such as petroleum.

Table 8-8 gives properties of phenolic materials, and eight advantages and four disadvantages of phenolic resins are shown below.

Advantages of Phenolics

1. Comparatively low cost
2. Suitable for use at temperatures to 400 °F [205 °C]
3. Excellent solvent resistance
4. Rigid
5. Good compressive strength
6. High resistivity
7. Self-extinguishing
8. Very good electrical characteristics

Disadvantages of Phenolics

1. Need fillers for moldings
2. Poor resistance to bases and oxidizers
3. Volatiles released during cure (a condensation polymer)
4. Dark color (due to oxidation discoloration)

Phenol-Aralkyl

In 1976, the Ciba-Geigy Corporation introduced a group of resins based on aralkyl ethers and phenols. Two basic prepolymer grades are available. Both are sold as 100 percent prepolymer resin. One grade cures by a condensation reaction. The other undergoes an addition reaction similar to epoxy. Condensation grades are blended with phenolic novolac resins to improve phenolic properties. Addition polymerization grades are finding uses in the fabrication of laminates. These resins are used as binders for the making of cutting wheels, printed circuit boards, bearings, appliance parts, and engine components. Because of excellent mechanical properties, processing advantages, and

Table 8-8.   Properties of Phenolics

Property	Phenol-Formaldehyde (Unfilled)	Phenol-Formaldehyde (Macerated Fabric)	Phenolic Casting Resin (Unfilled)
Molding qualities	Fair	Fair to good	
Relative density	1.25–1.30	1.36–1.43	1.236–1.320
Tensile strength, MPa	48–55	21–62	34–62
(psi)	(7 000–8 000)	(3 000–9 000)	(5 000–9 000)
Compressive strength, MPa	69–207	103–207	83–103
(psi)	(10 000–30 000)	(15 000–30 000)	(12 000–15 000)
Impact strength, Izod, J/mm	0.01–0.018	0.038–0.4	0.012–0.02
(ft lb/in)	(0.20–0.36)	(0.75–8)	(0.24–0.40)
Hardness, Rockwell	M124–M128	E79–E82	M93–M120
Thermal expansion, 10^{-4}/°C	6.4–15.2	2.5–10	17.3
Resistance to heat, °C	120	105–120	70
(°F)	(250)	(220–250)	(160)
Dielectric strength, V/mm	11 810–15 750	7 875–15 750	9 845–15 750
Dielectric constant (at 60 Hz)	5–6.5	5.2–21	6.5–17.5
Dissipation factor (at 60 Hz)	0.06–0.10	0.08–0.64	0.10–0.15
Arc resistance, s	Tracks	Tracks	
Water absorption (24 h), %	0.1–0.2	0.40–0.75	0.2–0.4
Burning rate	Very slow	Very slow	Very slow
Effect of sunlight	Darkens	Darkens	Darkens
Effect of acids	Decomposed by oxidizing acids	Decomposed by oxidizing acids	None
Effect of alkalies	Decomposes	Attacked	Attacked
Effect of solvents	Resistant	Resistant	Resistant
Machining qualities	Fair to good	Good	Excellent
Optical qualities	Transparent to translucent	Opaque	Transparent to opaque

Table 8-9.   Properties of Phenol-Aralkyl

Property	Phenol-Aralkyl (Glass-Filled)
Molding qualities	Good
Relative density	1.70–1.80
Tensile strength, MPa	48–62
(psi)	(6 900–9 000)
Compressive strength, MPa	206–241
(psi)	(3 000–3 500)
Impact strength, Izod, J/mm	0.02–0.03
(ft lb/in)	(0.4–0.6)
Hardness, Rockwell	
Resistance to heat, °C	250
(°F)	(480)
Dielectric strength, V/mm	
Dielectric constant (at 1 MHz)	2.5–4.0
Dissipation factor (at 1 MHz)	0.02–0.03
Water absorption (24 h), %	0.05
Effect of acids	None to slight
Effect of alkalies	Attacked
Effect of solvents	Resistant
Machining qualities	Fair
Optical qualities	Opaque

thermal capabilities, these prepolymers will find other uses. Table 8-9 lists phenol-aralkyl properties.

8-7 UNSATURATED POLYESTERS

The term *polyester resin* encompasses a variety of materials. It is often confused with other polyester classifications. A polyester is formed by the reaction of a polybasic acid and a polyhydric alcohol. Changes with acids, with acids and bases, and with some unsaturated reactants permit cross-linking, forming thermosetting plastics.

The term polyester resin should refer to unsaturated resins based on dibasic acids and dihydric alcohols. These resins are capable of cross-linking with unsaturated monomers (often styrene). Alkyds and polyurethanes of the polyester resin group are discussed individually.

Sometimes the term *fiber glass* has been used to indicate unsaturated polyester plastics. This term should refer only to fibrous pieces of glass. Various resins may be used with glass fiber acting as a reinforcing agent. The main use for unsaturated polyester resin is in the making of reinforced plastics. Glass fiber is the most-used reinforcement.

Credit for the first preparation of polyester resins (alkyd type) is usually attributed to the Swedish chemist Jons Jacob Berzelius in 1847 and to Gay-Lussac and Pelouze in 1833. Further development was conducted by W. H. Carothers and by R. H. Kienle. Through the 1930s most of the work on polyesters was aimed at developing and improving paint and varnish applications. Further interest in the resin was stimulated by Carleton Ellis in 1937. He found that by adding unsaturated monomers to.

unsaturated polyesters, cross-linking and polymerization time was greatly reduced. Ellis has been called the father of unsaturated polyesters.

Large-scale industrial use of unsaturated polyesters developed quickly as wartime shortages spurred development of many resin uses. Reinforced polyester structures and parts were widely used during World War II.

The word *polyester* is derived from two chemical processing terms, *poly*merization and *ester*ification. In esterification, an organic acid is combined with an alcohol to form an ester and water. A simple esterification reaction is shown in Figure 8-32. (See Alkyds, above.)

The reverse of the esterification reaction is called *saponification*. In order to obtain a good yield of ester in a condensation reaction, water must be removed to prevent saponification (Fig. 8-33). If a polybasic acid (such as maleic acid) is caused to react, and the water is removed as it is formed, the result will be an *unsaturated polyester*. *Unsaturated* means that the double-bonded carbon atoms are reactive or possess unused valance bonds. These can be attached to another atom or molecule, thus such a polyester is capable of cross-linkage. There are many other reactive or unsaturated monomers that can be used to change or tailor the resin to meet a special purpose or use. Vinyl toluene, chlorostyrene, methyl methacrylate, and diallyl phthalate are commonly used monomers. Unsaturated styrene is an ideal, low-cost monomer most often used with polyesters (Fig. 8-34).

The four main functions of a monomer are as follows:

1. To act as a solvent carrier for the unsaturated polyester
2. To lower viscosity (thin)
3. To enhance selected properties for specific uses

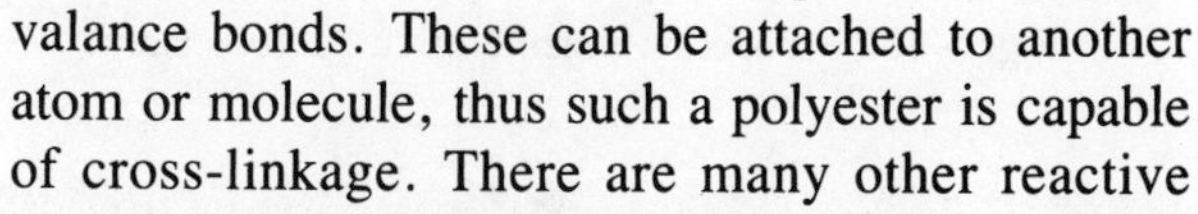

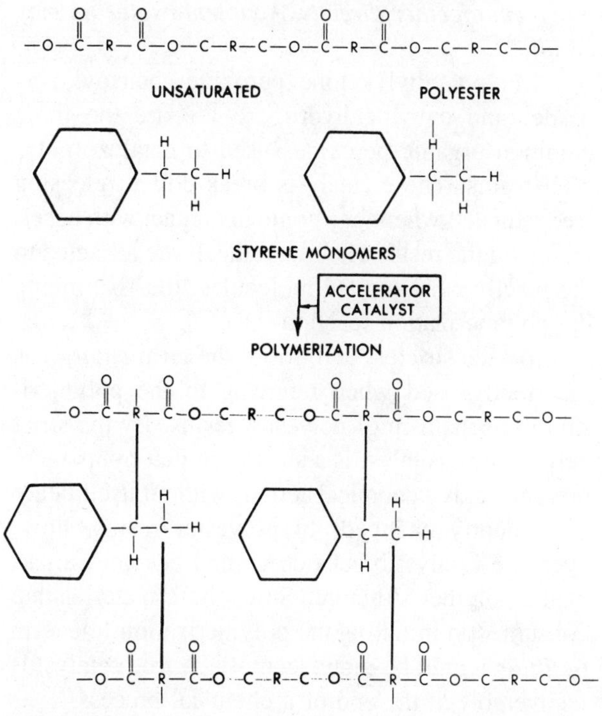

Fig. 8-34. Polymerization reaction with unsaturated polyester and styrene monomers.

4. To provide a rapid means of reacting (cross-linking) with the unsaturated linkages in the polyester

As the molecules randomly collide and occasional bonds are completed, a very slow polymerization (cross-linking) process will occur. This process may take days or weeks in simple mixtures of polyesters and monomers.

To speed up polymerization at room temperature, accelerators (promoters) and catalysts (initiators) are added. The accelerators commonly used are cobalt naphthenate, diethyl aniline, and dimethyl aniline. Polyester resins will usually have the accelerator added by the manufacturer unless otherwise specified. Resins that contain an accelerator require only a catalyst to provide rapid polymerization at room temperatures. With the addition of an accelerator the shelf life of the resin is appreciably shortened. Inhibitors such as hydroquinone may be added to stabilize or retard premature polymerization. These additives do not interfere with the final polymerization to any great extent. The speed of cure can be influenced by temperature, light, and the amounts of additives.

Polyester resins may be formulated without accelerators. All resins should be kept in a cool, dark storage area until used. ***Warning:*** *If the accelerator and catalyst are supplied separately, never*

Fig. 8-32. Examples of esterification reaction.

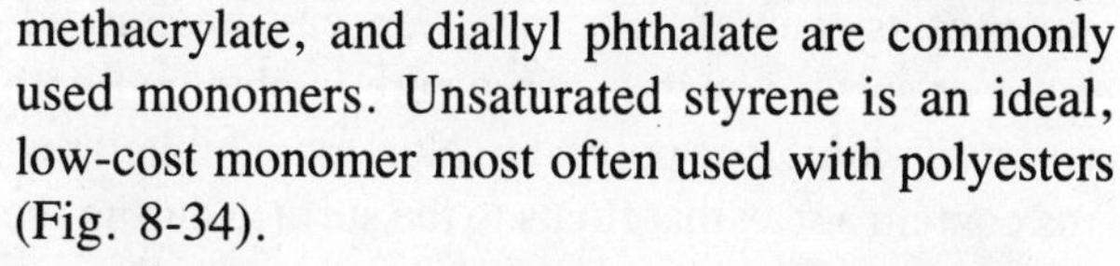

Fig. 8-33. To prevent saponification, water should be removed in an esterification reaction.

mix them together directly. A violent explosion may result.

Methyl ethyl ketone peroxide, benzoyl peroxide, and cumene hydroperoxide are the three common organic peroxides used to catalyze polyester resins. These catalysts break down, releasing free radicals, when they come in contact with accelerators in the resin. The free radicals are attracted to the reactive unsaturated molecules, thus beginning the polymerization reaction.

By the strictest definition, the term *catalyst* is incorrectly used when referring to the polymerization mechanism of polyester resins. By the strict definition, a catalyst is a substance that by its mere presence aids a chemical action, without itself being permanently changed. In polyester resins, however, the catalyst breaks down and becomes a part of the polymer structure. Since these materials are consumed in initiating the polymerization, the term *initiator* would be more accurate. A true catalyst is recoverable at the end of a chemical process.

Exposure to radiation, ultraviolet light, and heat have also been used to begin the polymerization of double-bonded molecules. If catalysts are used, the resin mix becomes correspondingly more sensitive to heat and light. On a hot day or in the sunlight, less catalyst is required for polymerization. On a cold day, more catalyst would be needed. The resin and catalyst also could be warmed to produce a rapid cure.

The final curing reaction is called *addition polymerization* because no byproducts are present as a result of the reaction. In phenol-formaldehyde reactions, the curing reaction is called *condensation polymerization* because a byproduct, water, is present. (Fig. 8-35) See polyallyl esters and allylics.

Polyester may be specially modified for a wide variety of uses by altering the chemical structure or by using additives. With higher percentages of unsaturated acid, more cross-linkage is possible. A stiffer, harder product results. The adding of saturated acids will increase toughness and flexibility. Thixotropic fillers, pigments, and lubricants also may be added to the resin.

Polyester resins that contain no wax are susceptible to *air inhibition*. When exposed directly to air, such resins remain undercured, soft, and tacky for some time after setting. This is desirable when multiple layers are to be built up. Resins purchased from the manufacturer without wax are referred to as *air-inhibited* resins. The absence of wax permits

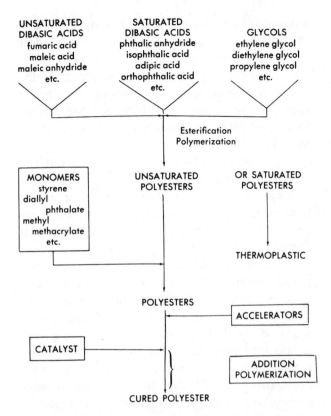

Fig. 8-35. Production scheme of cured polyester.

better bonds between multiple layers in hand-layup operations.

In some cases, a tackfree cure of air-exposed surfaces is wanted. For one-step castings, moldings, or surface coats, such a cure is obtained with a *non-air-inhibited* polyester. Non-air-inhibited resins contain a wax that floats to the surface during the curing operation blocking out the air and allowing the surface to cure tackfree. Many waxes may be used in such resins including household paraffin wax, carnauba wax, beeswax, stearic acid, and others. The use of waxes adversely affects adhesion; therefore, if more layers are to be added, all wax must be removed from the surface by sanding.

By altering the basic combination of raw materials, fillers, reinforcements, cure time, and treatment technique, a wide range of properties is possible.

Polyester finds its main use in making composite products. The primary value of reinforcement is to gain high strength-to-mass ratio (Fig. 8-36). Glass fiber is the most common reinforcing agent. Asbestos, sisal, many plastics fibers, and whisker filaments are also used. The type of reinforcement chosen depends on the end use and method of fabrication. Reinforced polyesters are

(A) The roof of this pavilion consists of 640 panels of fibrous glass-reinforced plastics. (Hooker Chemical Corp.)

(B) Glass-reinforced polyester is used for this welding helmet.

Fig. 8-36. Products of fibrous-glass-reinforced polyester have a high strength-to-mass ratio.

Fig. 8–37. Classic composite 1953 Chevrolet Corvette was first composite production body. It was composed of glass fibers in a matrix of polyester.

Fig. 8–38. This vessel is made of a reinforced corrosion resistant polyester resin. It is used to remove sulfuric acid mist from other gasses. (ICI Americas, Incorporated)

among the strongest materials known. They have been used in automobile bodies and in boat hulls and because of their high strength-to-mass ratio, they have aircraft and aerospace uses as well. (Fig. 8-37)

Other applications include radar domes, ducts, storage tanks, sports equipment, trays, furniture, luggage, sinks, and many kinds of ornaments. (Fig. 8-38)

Unreinforced casting grades of polyesters are used for embedding, potting, casting, and sealing. Resins filled with wood flour may be cast in silicone molds to produce precise copies of wood carvings and trim. Special resins may be emulsified with water, further reducing costs. These resins, referred to as *water-extended resins*, may contain up to 70 percent water. The castings undergo some shrinkage due to water loss.

Fabrication methods for reinforced polyesters include hand layup, sprayup, matched molding, premix molding, pressure-bag molding, vacuum-bag molding, casting, and continuous laminating. Other molding modifications are sometimes used. Compression molding equipment is sometimes used with doughlike premixes containing all ingredients. Table 8-10 gives some of the properties of many polyesters. The following list contains six advantages and two disadvantages of polyesters.

Table 8-10.   Properties of Thermosetting Polyesters

Property	Thermosetting Polyester (Cast)	Thermosetting Polyester (Glass Cloth)
Molding qualities	Excellent	Excellent
Relative density	1.10–1.46	1.50–2.10
Tensile strength, MPa	41–90	207–345
(psi)	(6 000–13 000)	(30 000–50 000)
Compressive strength, MPa	90–252	172–345
(psi)	(13 000–36 500)	(25 000–50 000)
Impact strength, Izod, J/mm	0.01–0.02	0.25–1.5
(ft lb/in)	(0.2–0.4)	(5.0–30.0)
Hardness, Rockwell	M70–M115	M80–M120
Thermal expansion, $10^{-4}/°C$	14–25.4	3.8–7.6
Resistance to heat, °C	120	150–180
(°F)	(250)	(300–350)
Dielectric strength, V/mm	14 960–19 685	13 780–19 690
Dielectric constant, (at 60 Hz)	3.0–4.36	4.1–5.5
Dissipation factor (at 60 Hz)	0.003–0.028	0.01–0.04
Arc resistance, s	125	60–120
Water absorption (24 h), %	0.15–0.60	0.05–0.50
Burning rate	Burns to self-extinguishing	Burns to self-extinguishing
Effect of sunlight	Yellows slightly	Slight
Effect of acids	Attacked by oxidizing acids	Attacked by oxidizing acids
Effect of alkalies	Attacked	Attacked
Effect of solvents	Attacked by some	Attacked by some
Machining qualities	Good	Good
Optical qualities	Transparent to opaque	Translucent to opaque

Advantages of Unsaturated Polyesters

1. Wide curing latitude
2. May be used for medical devices (artificial limbs)
3. Accept high filler levels
4. Thermosetting materials
5. Inexpensive tooling
6. Non-burning halogenated grades available

Disadvantages of Unsaturated Polyesters

1. Upper service temperature limited to 93 °C (200 °F)
2. Poor resistance to solvents

There are a number of thermoset interpenetrating polymer networks (IPN) involving a cross-linked unsaturated polyester, vinyl ester, or polyester-urethane copolymer in a urethane network. You will recall that an IPN (Chapter 2) is a configuration of two or more polymers, each existing in a network. There is a synergistic effect when one of the polymers is synthesized in the presence of the other. Urethane/polyester network resins have good wet-out and may be used in pultrusion, filament winding, RIM, RTM, sprayup, and other composite reinforcing methods. Thermoplastic IPNs do not form chemical crosslinks as do thermoset IPNs. There is a physical entanglement that interferes with polymer mobility. Thermoplastic

IPN has been produced using PA, PBT, POM and PP with silicone as the IPN.

8-8 THERMOSETTING POLYIMIDE

Polyimides may exist as either thermoplastic or thermosetting materials. Addition polyimides are available as thermosets. Condensation polyimides thermally decompose before they reach their melting point during processing.

Thermosetting polyimides are molded by injection, transfer, extrusion, and compression methods.

Thermosetting polyimides find uses in aircraft engine parts, automobile wheels, electrical dielectrics, and coatings. (See Thermoplastic Polyimides in Chapter 7).

Table 8-11 gives properties of thermosetting polyimides.

Both thermosetting and thermoplastic polyimides are considered high temperature polymers.

Some improvements in resin systems in recent years have resulted in a high temperature polymer that is less brittle and more easily processed. Some systems begin with a thermoplastic polymer based on an aromatic tetracarboxylic dianhydride and an aromatic diamine. A completely

Table 8-11. Properties of Thermosetting Polyimide and Bismaleimide

Property	Thermosetting Polyimide (Unfilled)	Bismaleimide 1:1	Bismaleimide 10:0.87
Molding qualities	Good	good	good
Relative density	1.43	—	—
Tensile strength, MPa	86	81	92
(psi)	(12 500)	(11 900)	(13 600)
Compressive strength, MPa	275	201	207
(psi)	(39 900)	(29 900)	(30 500)
Impact strength, Izod, J/mm	0.075	—	—
(ft lb/in)	(1.5)	—	—
Hardness, Rockwell	E50		
Thermal expansion, 10^{-4}/°C	13.71		
Resistance to heat, °C	350	272	285
(°F)	(660)	(523)	(545)
Dielectric strength, V/mm	22 050		
Dielectric constant (at 60 Hz)	3.6		
Dissipation factor (at 60 Hz)	0.0018		
Water absorption (24 h), %	0.24		
Effect of acids	Slowly attacked		
Effect of alkalies	Attacked		
Effect of solvents	Very resistant		
Matching qualities	Good		
Optical qualities	Opaque		

imidized powder is the result. Unlike addition polyimides, this product can be processed from many common organic solvents (e.g., cyclohexanone). Although the polyimide is completely imidized, additional crosslinking occurs during processing.

Bismaleimides (BMI) are a class of polyimides with a general structure containing reactive double bonds on each end of the molecule.

Various functional groups, such as vinyls, allyls, or amines, are used as co-curing agents with bismaleimides to improve the properties of the homopolymer.

In one bismaleimide system, component A (4,4′-bismaleimidodiphenylmethane) and component B (o-,o′-diallyl-bisphenol A) are reacted together to provide a BMI molecule with a tougher, flexible backbone.

Tests have shown improved strength and toughness with formulations using a higher ratio (1.0:0.87) of BMI to diallylbisphenol A. Typical properties of two bismaleimide systems are shown in Table 8-11.

8-9 POLYURETHANE (PU)

The term *polyurethane* refers to the reaction of polyisocyanates (-NCO-) and polyhydroxyl (-OH-) groups. A simple reaction of isocyanate and an alcohol is shown below. The reaction product is urethane, not a polyurethane.

$$R \cdot NCO + HOR_1 \longrightarrow R \cdot NH \cdot COOR_1$$

PolyIsocyanate PolyHydroxyl   PolyUrethane

The German chemists Wurtz, in 1848, and Hentschel, in 1884, produced the first isocyanates.

These later led to the development of polyurethanes. It was Otto Bayer and his coworkers who actually made possible the commercial development of polyurethanes in 1937. Since that time, polyurethanes have developed into many commercially available forms that include coatings, elastomers, adhesives, molding compounds, foams, and fibers.

The isocyanates and di-isocyanates are highly reactive with compounds containing reactive hydrogen atoms. For this reason, polyurethane polymers may be reproduced. The recurring link of the polyurethane chain is NHCOO or NHCO.

More complex polyurethanes have been developed based on toluene di-isocyanates (TDI) and polyester, diamine, castor oil, or polyether chains. Other isocyanates used are diphenylmethane di-isocyanate (MDI), and polymethylene polyphenyl isocyanate (PAPI).

The first polyurethanes were made in Germany to compete with other polymers produced at that time. Linear aliphatic polyurethanes were used to make fibers. Linear polyurethanes are thermoplastic. They may be processed by all normal thermoplastic techniques, including injection and extrusion. Because of cost, they find limited use as fibers or filaments.

Polyurethane coatings are noted for their high abrasion resistance, unusual toughness, hardness, good flexibility, chemical resistance, and weatherability (Fig. 8-39). ASTM has defined five distinct types of polyurethane coatings, as shown in Table 8-12.

Polyurethane resins are used as clear or pigmented finishes for home, industrial, or marine use. They improve the chemical and ozone resistance of rubber and other polymers. These coatings and finishes may be simple solutions of linear polyurethanes or complex systems of polyisocyanate and such OH groups as polyesters, polyethers, and castor oil.

Many polyurethane elastomers (PUR) (rubbers), may be prepared from di-isocyanates, linear polyesters or polyether resin, and curing agents (Fig. 8-40). If formulated into a linear thermoplastic urethane, they may be processed by normal thermoplastic processing equipment. They find uses as shock absorbers, bumpers, gears, cable covers, hose jacketing, elastic thread (Spandex), and diaphragms. Common uses of cross-linked thermosetting elastometers include industrial tires, shoe heels, gaskets, seals, O-rings, pump impellers, and tread stock for tires. Polyurethane elastomers have extreme resistance to abrasion, ozone aging, and hydrocarbon fluids. These elastometers cost more than conventional rubbers but are tough, elastic, and show a wide range of flexibility at temperature extremes. (See Appendix E)

Polyurethane foams are widely used and well known. They are available in flexible, semirigid, and rigid forms in a number of different densities. Various flexible foams are used as cushioning for furniture, automobile seating, and mattresses. They are produced by reacting toluene di-isocyanate (TDI) with polyester and water in the pres-

(A) Foams, insulation, sponges, belts and gaskets of polyurethane.

(B) This self-locking polyurethane container is shockproof, fire retardant and moistureproof. (Poly-Con Industries, Inc.)

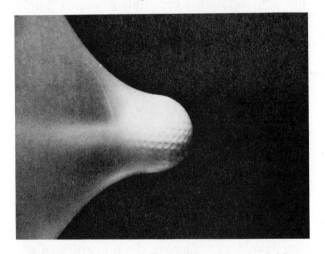

(C) This 0.003-in. (0.076-mm) film stops a hard-driven golf ball, demonstrating the puncture resistance and tensile strength of polyurethane. (B. F. Goodrich Chemical Co.)

Fig. 8-39. Applications of polyurethane.

ence of catalysts. At higher densities, they are cast or molded into drawer fronts, doors, moldings, and complete pieces of furniture. Flexible foams are open-celled structures that may be used as artificial sponges. These foams are used by the garment and textile industries for backing and insulation (Fig. 8-41).

Table 8-12. ASTM Designations for Polyurethane Coatings

ASTM Type	Components	Pot Life	Cure	Clear or Pigmented Uses
(I) Oil-modified	One	Unlimited	Air	Interior or exterior wood and marine. Industrial enamels
(II) Prepolymer	One	Extended	Moisture	Interior or exterior. Wood, rubber and leather coatings
(III) Blocked	One	Unlimited	Heat	Wire coatings and baked finishes
(IV) Prepolymer + Catalyst	Two	Limited	Amine/ catalyst Air	Industrial finishes and leather, rubber products
(V) Polyisocyanate + Polyol	Two	Limited	NCO/OH reaction	Industrial finishes and leather, rubber products

$$n\text{HOR}—(\text{OR}—)_x—\text{OH} + n\,\text{OCNR}_1\text{NCO} \rightarrow (—\text{OR}—(\text{OR})_x\text{OCONHR}_1\text{NHCO}—)_n$$

Polyester — Di-isocyanate — Polyurethane

$$n\text{HOR}(-—\text{OCOR}_2\text{CO}\cdot\text{OR}—-)_x\text{OH} + n\text{OCNR}_1\text{NCO} \rightarrow$$

Polyether — Di-isocyanate

$$—\text{OR}(—\text{OCOR}_2\text{CO}\cdot\text{OR})_x\text{OCONHR}_1\text{NHCO}—_n$$

Polyurethane

Fig. 8-40. Production of polyurethane elastomers.

Fig. 8-41. A polyurethane foam backing is used for this carpet.

Fig. 8-42. This china cabinet of polyurethane, produced by Jasper Stylemasters Plastics, has the appearance, mass, and feel of wood. (The Upjohn Company)

Semirigid foams find use as energy-absorbing materials in crash pads, arm rests, and sun visors.

The three largest uses of rigid polyurethane foam are in the making of furniture, automotive and construction moldings, and various thermal insulation uses. Replicas of wood carvings, decorative parts, and moldings are produced from high-density, self-skinning foams (Fig. 8-42). The insulation value of these foams makes them an ideal choice for insulating refrigerators and refrigerated trucks and railroad cars. They may be foamed in place for many architectural uses. They may be placed on vertical surfaces by spraying the reaction mixture through a nozzle. They have found use as flotation devices, packing, and structural reinforcement.

Rigid polyurethane is a closed cellular material produced by the reaction of TDI (prepolymer form) with polyethers and reactive blowing agents such as monofluorotrichloromethane (fluorocarbon). Diphenylmethane di-isocyanate (MDI) and polymethylene polyphenyl isocyanate (PAPI) are also used in some rigid foams. MDI foams have better dimensional stability, while PAPI foams have high temperature resistance.

Polyurethane-based caulks and sealants are inexpensive polyisocyanate materials used for encapsulation, and for construction and manufacturing uses. Various polyisocyanates are also useful adhesives. They produce strong bonds between flexible fabrics, rubbers, foams, or other materials.

Many blowing or foaming agents are explosive and toxic. When mixing or processing polyurethane foams, make certain that proper ventilation is provided.

Table 8-13 gives some properties of urethane plastics. Six advantages and four disadvantages of polyurethane plastics are listed below.

Advantages of Polyurethane

1. High abrasion resistance
2. Good low-temperature capability
3. Wide variability in molecular structure
4. Possibility of ambient curing
5. Comparatively low cost
6. Prepolymers foam readily

Disadvantages of Polyurethane

1. Poor thermal capability
2. Toxic (isocyanates are used)
3. Poor weatherability
4. Subject to attack by solvents

8-10 SILICONES (SI)

In organic chemistry, carbon is studied because of its capability of forming molecular structures with many other elements. Carbon is considered to be a reactive element. Carbon is capable of entering into more molecular combinations than is any other element. Life on earth is based on the element carbon.

The second most abundant element on earth is silicon. It has the same number of available bonding sites as carbon. Some scientists have speculated that life on other planets may be based on silicon. Others find this possibility hard to accept because silicon is an inorganic solid with a metallic appearance. Most of the earth's crust is composed of SiO_2 (silicon dioxide) in the form of sand, quartz, and flint.

The tetravalent capacity of silicon interested chemists as early as 1863. Friedrich Wohler, C. M. Crafts, Charles Friedel, F. S. Kipping, W. H. Carothers, and many others did work that led to the development of silicone polymers.

By 1943, Dow Corning Corporation was producing the first commercial silicone polymers in the United States. There are thousands of uses for these materials. The word *silicone* should be applied only to polymers containing silicon-oxygen-silicon bonding, however it is often used to denote any polymer containing silicon atoms.

In many carbon-hydrogen compounds, silicon may replace the element carbon. Methane (CH_4) may be changed to silane or silicomethane (SiH_4). Many structures similar to the aliphatic series of saturated hydrocarbons may be formed.

Table 8-13. Properties of Polyurethane

Property	Cast Urethane	Urethane Elastomer
Molding qualities	Good	Good to excellent
Relative density	1.10–1.50	1.11–1.25
Tensile strength, MPa	1–69	31–58
(psi)	(175–10 000)	(4 500–8 400)
Compressive strength, MPa	14	14
(psi)	(2 000)	(2 000)
Impact strength, Izod, J/mm	0.25 to flexible	Does not break
(ft lb/in)	(5)	
Hardness, Shore	10A–90D	30A–70D
Rockwell		M28, R60
Themal expansion, $10^{-4}/°C$	25.4–50.8	25–50
Resistance to heat, °C	90–120	90
(°F)	(190–250)	(190)
Dielectric strength, V/mm	15 750–19 690	12 990–35 435
Dielectric constant (at 60 Hz)	4–7.5	5.4–7.6
Dissipation factor (at 60 Hz)	0.015–0.017	0.015–0.048
Arc resistance, s	0.1–0.6	0.22
Water absorption (24 h), %	0.02–1.5	0.7–0.9
Burning rate	Slow to self-extinguishing	Slow to self-extinguishing
Effect of sunlight	None to yellows	None to yellows
Effect of acids	Attacked	Dissolves
Effect of alkalies	Slight to attacked	Dissolves
Effect of solvents	None to slight	Resistant
Machining qualities	Excellent	Fair to excellent
Optical qualities	Transparent to opaque	Transparent to opaque

The following general types of bonds may be of value in understanding the formation of silicone polymers:

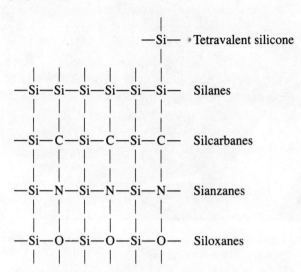

Compounds with only silicon and hydrogen atoms present are called *silanes*. When the silicon atoms are separated by carbon atoms, the structure is called a *silcarbane* (sil-CARB-ane). A *polysiloxane* is produced when more than one oxygen atom separates the silicon atoms in the chain.

$$—Si—O—O—O—Si—O—O—O—$$

A polymerized silicone molecular chain could be based on the structure shown in Figure 8-43 modified by radicals (R).

Many silicone polymers are based on chains, rings, or networks of alternating silicon and oxygen atoms. Common ones contain methyl, phenyl, or vinyl groups on the siloxane chain (Fig. 8-44). A number of polymers are formed by varying the organic radical groups on the silicon chain. Many copolymers are also available.

The amount of energy needed to produce silicone plastics makes the price high. Silicone plastics may still be economical if one considers longer product life, higher service temperatures, and flexibility at temperature extremes.

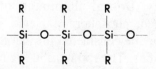

Fig. 8-43. An example of a polymerized silicone molecular chain.

(A) Based on methyl (CH_3) radical.

(B) Based on phenyl (C_6H_5) radical.

Fig. 8-44. Two siloxane polymers.

Silicones are produced in five commercially available categories: fluids, compounds, lubricants, resins, and elastomers (rubber).

Probably the best-known silicon plastics are associated with oils and ingredients for polishes. Examples are lens-cleaning tissues or water-repellent fabrics treated with a thin-film coating of silicone.

Silicone fluids are added to some liquids to prevent foaming (antifoaming), prevent transmission of vibrations (damping), and improve electrical and thermal limits of various liquids. Fluid silicones are used as additives in paints, oils and inks, as mold-release agents, as finishes for glass and fabrics, and for paper coating.

Silicone compounds are usually granular or fibrous filled materials. Because of their outstanding electrical and thermal properties, mineral- and glass-filled silicone compounds are used for encapsulation of electronic components. Figure 8-45 shows such uses of silicone coupounds.

As adhesives and sealants, silicone plastics are limited by their high cost. Their high service temperature and elastic properties make them useful for sealing, gasketing, caulking, and encapsulating, and for repairing all types of materials (Fig. 8-46).

The chemical inertness of foamed silicone is useful in breast and facial implants in plastic surgery. Its main uses include electrical and thermal insulation of electrical wires and electrical components.

As lubricants, silicones are prized because they do not deteriorate at extreme service temperatures. Silicones are used for lubricating rubber, plastics, ball bearings, valves and vacuum pumps.

Silicone resins have many uses such as releasing agents for baking dishes. Silicone resins are

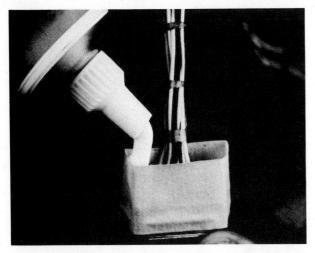

(A) Potting a small electrical component with silicone compound.

(B) Use of silicone casting resin for potting electronic components.

Fig. 8-45. Uses of silicone compounds. (Dow-Corning Corp.)

Fig. 8-46. Silicone sealant is used in the assembly of this gear box. (Dow-Corning Corp.)

also found in flexible, tough coatings used as high-temperature paints for engine manifolds and mufflers. Their waterproofing capability makes them useful in treating masonry and concrete walls.

Excellent waterproofing, thermal, and electrical properties make these resins valuable for electrical insulation in motors and generators.

Laminates reinforced with glass cloth find uses for structural parts, ducts, radomes and electronic panel boards. These silicone laminates are characterized by their excellent dielectric and thermal properties and strength-to-mass ratio.

Diatomaceous earth, glass fiber, or asbestos may be used as fillers in preparing a premix or putty for the molding of small parts from silicone resins.

Some of the best known silicones are in the form of elastomers. Few industrial rubbers or elas-

tomers can withstand long exposure to ozone (O_3) or hot mineral oils. Silicone "rubbers" are stable at elevated temperatures and remain flexible when exposed to ozone or oils.

Silicone elastomers find use as artificial organs, O-rings, gaskets, and diaphragms. They are also used as flexible molds for casting of plastics and low-melting-point metals.

Room-temperature vulcanizing (RTV) elastomers are used to copy intricate molded parts, seal joints, and adhere parts (Fig. 8-47).

Silly Putty and *Crazy Clay* are novelty silicone products (Fig. 8-48). This bouncing putty is a silicone elastomer that is also used for damping noise and as a sealing and filling compound. A hard bouncing putty will rebound to 80 percent of the height from which it is dropped. Other super-rebounding novelty items are produced from this compound.

Silicone molding compounds may be processed in the same way as other thermosetting organic plastics. Silicones are often used in resin form as castings, coatings, adhesives, or laminating compounds.

Research and development is being carried out on various other elements with covalent bonding capacity. Boron, aluminum, titanium, tin, lead, nitrogen, phosphorus, arsenic, sulphur, and selenium may be considered for inorganic or semi-organic plastics. The formulas in Figure 8-49 show a few of many possible chemical structures of inorganic or semi-organic plastics.

(A) Intricate details, such as this picture-frame decoration, may be reproduced with silicone RTV.

(B) Silicone reproduces fine detail and allows severe undercuts.

Fig. 8-47. Use of room-temperature-vulcanizing (RTV) silicone in molding. (Dow-Corning Corp.)

Fig. 8-48. A novelty toy made of a silicone compound.

Table 8-14 gives properties of silicone plastics. Seven advantages and three disadvantages of silicone are shown below.

Advantages of Silicones

1. Wide range of thermal capability, from -100 °F to 600 °F [-73 °C to 315 °C]
2. Good electrical traits
3. Wide variation in molecular structure (flexible or rigid forms)
4. Available in transparent grades

Table 8-14.   Properties of Silicones

Property	Cast Resin (Including RTV)	Molding Compounds (Mineral-Filled)	Molding Compounds (Glass-Filled)
Molding qualities	Excellent	Excellent	Good
Relative density	0.99–1.50	1.7–2	1.68–2
Tensile strength, MPa	2–7	28–41	28–45
(psi)	(350–1 000)	(4 000–6 000)	(4 000–6 500)
Compressive strength, MPa	0.7	90–124	69–103
(psi)	(100)	(13 000–18 000)	(10 000–15 000)
Impact strength, Izod, J/mm		0.013–0.018	0.15–0.75
(ft lb/in)		(0.26–0.36)	(3–15)
Hardness	Shore A15–A65	Rockwell M71–M95	Rockwell M84
Thermal expansion, 10^{-4}/°C	20–79	5–10	0.61–0.76
Resistance to heat, °C	260	315	315
(°F)	(500)	(600)	(600)
Dielectric strength, V/mm	21 665	7 875–15 750	7 875–15 750
Dielectric constant (at 60 Hz)	2.75–4.20	3.5–3.6	3.3–5.2
Dissipation factor (at 60 Hz)	0.001–0.025	0.004–0.005	0.004–0.030
Arc resistance, s	115–130	250–420	150–250
Water absorption, %	0.12 (7 days)	0.08–0.13 (24 h)	0.1–0.2 (24 h)
Burning rate	Self-extinguishing	None to slow	None to slow
Effect of sunlight	None	None to slight	None to slight
Effect of acids	Slight to severe	Slight	Slight
Effect of alkalies	Moderate to severe	Slight to marked	Slight to marked
Effect of solvents	Swells in some	Attacked by some	Attacked by some
Machining qualities	None	Fair	Fair
Optical qualities	Transparent to opaque	Opaque	Opaque

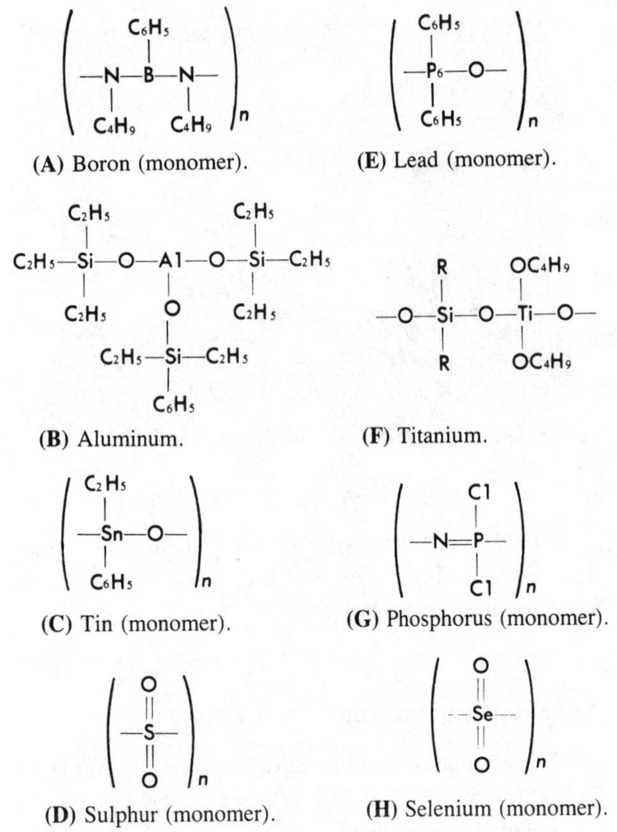

(A) Boron (monomer).

(E) Lead (monomer).

(B) Aluminum.

(F) Titanium.

(C) Tin (monomer).

(G) Phosphorus (monomer).

(D) Sulphur (monomer).

(H) Selenium (monomer).

Fig. 8-49. Possible chemical structures of inorganic or semi-inorganic plastics.

5. Low water absorption
6. Available in flame-retardant grades
7. Good chemical resistance

Disadvantages of Silicones

1. Low strength
2. Subject to attack by halogenated solvents
3. Comparatively high cost

VOCABULARY

The following vocabulary words are found in this chapter. Look up the definition of any of these words you do not understand as they apply to plastics, in the glosssary, Appendix A.

A-stage
Alkyd
Allylics
Amino
B-stage
C-stage
Casein
Condensation
Epoxy
Formalin
Isocyanate resins
Novolac
Phenolic
Polyester
Polyurethane
Potting
Rennin
RTV
Saturated compounds

8-1. Name nine thermosetting plastics.

8-2. Name the early stage in the reaction of a thermosetting resin, when the material is soluble and fusible.

8-3. Name two thermosetting plastics used extensively for surface coatings and finishes.

8-4. Name the plastics that may be produced from milk.

8-5. Which plastics are known for their adhesive quality and durable finishes?

8-6. Identify the term for a chemical reaction in which two or more molecules combine, with the separation of water.

8-7. Polyester resins without wax additives are referred to as being ____?____ resins.

8-8. What thermosetting plastics has a backbone of silicon?

8-9. Dacron is made from ____?____ .

8-10. The leading thermosetting material made into dinnerware is ____?____ .

8-11. Name two popular resins used with fiber glass in the manufacture of boats.

8-12. Identify a monomer used to thin polyester resin.

8-13. Identify a common catalyst or initiator used to cure polyester resin.

8-14. What solvent will clean polyester and epoxy resin from tools?

8-15. Name the thermosetting resin you would select to make optically clear embedments.

8-16. Name the property of thermosets that generally increases service temperature, chemical resistance, hardness, and tensile strength.

8-17. List the hazards of using resins, catalysts, promoters, etc.

8-18. Which thermosetting plastics would you select for each of the following product applications?
coatings on machinery
circuit breakers and appliance knobs
tub-shower combination composite
dinnerware
composite adhesives

8-19. Would you specify melamine-formaldehyde or urea-formaldehyde when heat and moisture resistance is required?

8-20. Which has the longer shelf life before being converted to a thermoset plastics, a resol or a novalac resin?

8-21. What atoms along with silicon are present in the backbone of the silicone molecule?

8-22. When light colored products are required, would you specify urea-formaldehyde or phenol-formaldehyde?

8-23. What does RTV mean?

8-24. Name the three broad processes that make use of solid forms of epoxy resins.

8-25. Rubber-like polymers are called ____?____ .

8-26. The catalyst used to cure RTV silicone rubber is ____?____ .

8-27. Biological implants are one use of ____?____ foams.

8-28. Name three reasons for adding styrene monomers to polyester resin.

8-29. Something used to initiate or accelerate a reaction is called a ____?____ .

8-30. Would you select urea-formaldehyde or phenolics for the widest choice of color?

8-31. Bisphenol A and epichlorohydrin are reacted to produce ____?____ .

8-32. Free radical initiators such as benzoyl peroxide are agents used to cure unsaturated ____?____ .

8-33. Hexamethylenetetramine, which yields formaldehyde when heated, is used as a curing agent to convert A-stage ____?____ resins to the C-stage.

8-34. Chemicals that greatly increase the activity of a given catalyst are called ____?____ .

8-35. Carbon dioxide reacts with ammonia to produce ____?____ resins.

8-36. Of urea, phenolic, and melamine, which has the highest heat resistance?

8-37. By forming cross-linking between fiber and acting as an elastic spring, ___?___ resin imparts crush resistance to fabrics.

8-38. It is during the final or ___?___ that an infusable thermoset plastics occurs.

8-39. Resins that harden (cure) when exposed to air are called ___?___ resins.

8-40. The two most significant amino plastics are ___?___ and ___?___ .

8-41. What is the most important product application for each of these plastics?
alkyd
amino
epoxy
polyurethane

8-42. Name three thermosetting plastics used extensively in the wood and plywood industries.

8-43. Can you select a solvent to dissolve melamine plastics?

8-44. Thermosetting plastics have ___?___ melting points but do decompose.

8-45. Thermosetting plastics may not be ___?___ by heat and/or pressure.

8-46. Three applications of silicon liquids are:

8-47. Phenol-aldehyde plastics are also known as ___?___ .

8-48. The A-stage resin or one-step PF resin made by the alkaline condensation of phenol with an adequate amount of formaldehyde is called a ___?___ .

8-49. As a rule, ___?___ resins have no filler added to impede flow or wetting of substrates.

8-50. Why is a dark resin, such as phenolformaldehyde, used in the interior layers of a decorative laminate?

8-51. For pastel colors, would you select phenolformaldehyde or polyurethane materials?

8-52. Several additives are added to ___?___ compounds to improve moldability and plastics properties.

Chapter 9

Machining and Finishing

In this chapter, you will learn how plastics and composites are machined and finished. Molded or formed plastics parts often require further processing, including such common operations as flash removal, slot cutting, polishing, and annealing. Many of the operations are similar to those used for the machining and finishing of metal or wood products.

The vast numbers of machines and processes used in the shaping and finishing of plastics do not allow discussion in detail, although certain basics apply to all machining and finishing processes. Additives, fillers, and the plastics of each family require different shaping and finishing techniques. Few plastics pieces are made solely by machining, though many molded parts must be finished or fabricated into useful items.

All machining and finishing operations present potential physical hazards. Fine dusts or particles are produced when sawing, laser cutting, or water jet cutting. Eye and face mask protection should be worn by the operator to prevent injury or inhalation of particles. (See Chapter 3 on Health and Safety)

Processing techniques for plastics are based on those used for wood and metal. Nearly all plastics may be machined (Fig. 9-1). As a rule, thermosets are more abrasive to cutting tools than thermoplastics.

Machining techniques for composite materials such as high-pressure laminates, filament wound parts, and reinforced plastics, attempt to prevent fraying and delamination of the composite. The reinforcing agents used with various matrix compounds are abrasive. Most cutting tools must be made from tungsten carbide or coated (with titanium diboride, for example). High speed steel

(M2) or diamond tipped cutters are also used. Boron/epoxy composites are generally cut using diamond tipped tools. The lower thermal conductivity and the lower modulus of elasticity (softness, flexibility) of most thermoplastics mean that tools should be kept properly sharp to allow them to cut cleanly, without burning, clogging, or causing frictional heat.

Elastic recovery causes drilled or tapped holes to become smaller than the diameter of the drills used, and turned diameters often become larger. The low melting points of some thermoplastic materials tend to make them gum, melt, or craze when machined. Plastics expand more than most materials when heated. The coefficient of thermal expansion for plastics is roughly ten times greater than that of metals. Cooling agents (liquids or air) may be needed to keep the cutting tool clean and free of chips. The benefits of cooling include increased cutting speed, smoother cuts, longer tool life, and elimination of dust. Because the polymer matrix has a high coefficient of expansion, even small variations of temperature may cause dimensional control problems.

Finishing operations include sawing, filing, drilling, tapping, turning, planing, milling, shaping, routing, sanding, shearing, punching, laser cutting, tumbling, grinding, ashing, buffing, polishing, transparent coating, polishing by solvents, annealing, and postcuring.

9-1 SAWING

Nearly all types of saws have been adapted to cutting plastics. Backsaws, coping saws, hacksaws, saber saws, hand saws, and jeweler's saws may

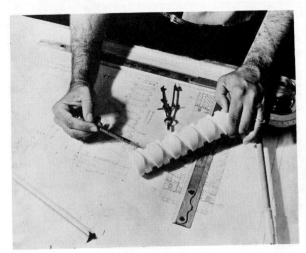

Fig. 9-1. This part was machined from bar stock. (The Polymer Corp.)

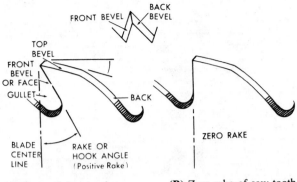

(A) Parts of the circular saw tooth.

(B) Zero rake of saw tooth.

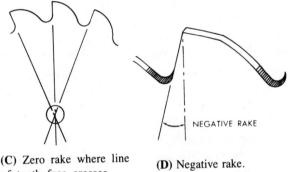

(C) Zero rake where line of tooth face crosses center of blade.

(D) Negative rake.

(E) Positive rake.

Fig. 9-2. Tooth characteristics of a circular saw blade.

be used for hobbycraft or short-run cutting. The shape of the tooth is important for proper cutting of plastics.

Circular blades should have plenty of *set* or be *hollow ground*. Blades should have a deep, well-rounded *gullet* (Fig. 9-2A). The *rake* (or hook) angle should be zero (or slightly negative). The *back clearance* should be about thirty degrees. The preferred number of teeth per centimetre varies with the thickness of the material to be cut. Four or more teeth per centimetre should be used for cutting thin materials. Fewer than four teeth per centimetre are needed for plastics over 25 mm [1 in.] thick.

A *skip-tooth* bandsaw blade is preferred (Fig. 9-3A). The wide gullet in this blade provides ample space for plastics chips to be carried out of the *kerf* (cut made by the saw). For best results, the teeth should have zero rake and some set.

Bandsaw blades may be reversed to accomplish a zero or negative rake. Abrasive carbide or diamond grit blades may be used to cut graphite and boron/epoxy composites. In all cutting operations, it is best to back up the work with a solid material to reduce chipping, fraying, and delamination of composites. Table 9-1 suggests the number of teeth per centimetre for various speeds and thicknesses of material.

NOTE: Fewer teeth per centimetre are needed for cutting plastics over 0.23 in. [6 mm] thick. Thin or flexible plastics may be cut with shears or blanking dies. Foams require cutting speeds above 8 000 fpm [40 m/s]

For cutting reinforced or filled plastics, and for many thermosetting plastics, carbide-tipped blades are recommended (Fig. 9-4). They provide accurate cuts and long blade life. Abrasive- or diamond-tipped blades may be used also. A liquid coolant is advised to prevent clogging or overheating, although jets of CO_2 are often used as a coolant while thermoplastics are being machined. All cutting tools must have protective shields and safety devices.

The feeds and speeds for cutting composites vary greatly with thickness and material, but are similar to those for nonferrous materials. See Table 9-2.

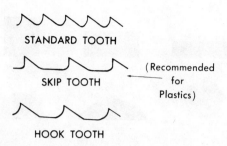

(A) Skip tooth provides large gullet and good chip clearance. The hook tooth is sometimes preferred for glass-filled thermosets.

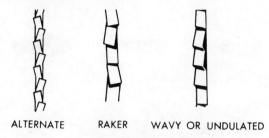

(B) Common bandsaw blade teeth (cutting edge view).

Fig. 9-3. Bandsaw blade teeth.

9-2 FILING

Thermosetting plastics are quite hard and brittle, therefore filing removes material in the form of a light powder. Aluminum type A, shear-tooth, or other files that have coarse, single-cut teeth with an angle of 45 degrees are preferred (Fig. 9-5). The deep-angled teeth help the file to clear itself of plastics chips. Many thermoplastics tend to clog files. Curved-tooth files like those used in auto body shops are good because they clear themselves of plastics chips. Special files designed for plastics should be kept clean and not used for filing metals.

9-3 DRILLING

Thermoplastic and thermosetting materials may be drilled with any standard twist drill; however, special drills designed for plastics produce better results. Carbide-tipped drills will give long life. Holes drilled in most thermoplastics and some thermosets are usually 0.002 to 0.004 in. [0.05 to 0.10 mm] undersize. Thus, a 0.23 in. [6 mm] drill will not produce a hole wide enough for a 6 mm rod. Thermoplastics may need a coolant to reduce frictional heat and gumming during drilling.

For most plastics, drills should be ground with a 70 to 120 degree point angle and a 10 to 25 degree lip-clearance angle (Fig. 9-6). The rake angle on the cutting edge should be zero or several degrees negative. Table 9-3 gives rake and point angles for many plastics materials. Highly pol-

Table 9-1. Power Saws for Cutting Plastics

Plastics	Circular Saws			Band Saws		
	Teeth per cm		Speed, m/s	Teeth per cm		Speed, m/s (>6 mm)
	(<6 mm)	(>6 mm)		(<6 mm)	(>6 mm)	
Acetal	4	3	40	8	5	7.5–9
Acrylic	3	2	15	6	3	10–20
ABS	4	3	20	4	3	5–15
Cellulose acetate	4	3	15	4	2	7.5–15
Diallyl phthalate	6	4	12.5	10	5	10–12.5
Epoxy	6	4	15	10	5	7.5–10
Ionomer	6	4	30	4	3	7.5–10
Melamine-formaldehyde	6	4	25	10	5	12.5–22.5
Phenol-formaldehyde	6	4	15	10	5	7.5–15
Polyallomer	4	3	45	3	2	5–7.5
Polyamide	6	4	25	3	2	5–7.5
Polycarbonate	4	3	40	3	2	7.5–10
Polyester	6	4	25	10	5	15–20
Polyethylene	6	4	45	3	2	7.5–10
Polyphenylene oxide	6	4	25	3	2	10–15
Polypropylene	6	4	45	3	2	7.5–10
Polystyrene	4	3	10	10	5	10–12.5
Polysulfone	4	3	15	5	3	10–15
Polyurethane	4	3	20	3	2	7.5–10
Polyvinyl chloride	4	3	15	5	3	10–15
Tetrafluoroethylene	4	3	40	4	3	7.5–10

See Appendix F, Useful Tables.

(A) Diamond cutoff used to cut boron-epoxy reinforced tube. (Advanced Structures Division, TRE Corp.)

(B) This machine skives (pares) sheets of plastics from slab stock. (McNeil Akron Corp.)

(C) Space-age composites like Kevlar are quickly and cleanly cut by water jet. (Flow Systems Inc.)

Fig. 9-4. Various methods of cutting plastics.

(A) Various file profiles.

(B) Rotary files may be used on plastics.

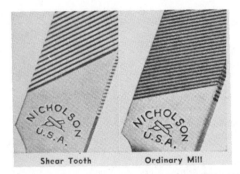

(C) Comparison of shear-tooth and ordinary mill files.

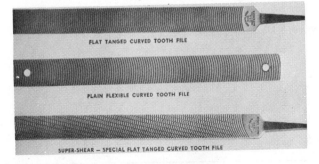

(D) Curved-tooth files used on plastics.

Fig. 9-5. Files used on plastics.

Table 9-2. Machining Composites

Operation	Material	Cutting Tool	Speeds	Feeds (<0.250 thick)
Drilling	Gl-Pe	0.250-Diamond	20 000 RPM	0.002/rev
	B-Ep	0.250-Diamond core (60–120 grit)	100 SFPM	0.002/rev
	B-Ep	0.250 2–4 flute (HSS)	25 SFPM	0.002/rev
	Kv-Ep	Spade-carbide	>25 000 RPM	0.002/rev
	Kv-Ep	Brad point-carbide	>6 000 RPM	0.002/rev
	Gl-Ep	Tungsten Carbide	<2 000 RPM	<0.5 ipm
	Gr-Ep	Tungsten Carbide	>5 000 RPM	<0.5 ipm
Band Saw	Kv-Ep	14 teeth, honed saber	3000–6000 SFPM	<30 ipm
	B-Ep	Carbide or 60 gut Diamond	2000–5000 SFPM	<30 ipm
	Hybrids		3000–6000 SFPM	<30 ipm
	Gl-Pe	14 teeth, honed saber	3000–6000 SFPM	<30 ipm
Milling	most	Four-flute carbide	300– 800 SFPM	<10 ipm
Circular Saw	Gr-Ep, B-Ep	60 grit diamond	6000 SFPM	<30 ipm
	Gl-Pe	60 tooth carbide or	5000 SFPM	<30 ipm
		60 grit diamond	5000 SFPM	<30 ipm
Lathe	Kv-Ep	carbide	250– 300 SFPM	0.002/rev
	Gl-Pe	carbide	300– 600 SFPM	0.002/rev
Shears	Kv-Ep	HSS or Carbide	—	<30 ipm
Countersink or counterbore	most	Diamond grit or carbide	20 000 RPM 6 000 RPM	<0.5 ipm
Laser (10kW)	most <0.250 thickness	CO_2 cooling	—	<30 ipm, depending upon material
Water jet	most <0.250 thickness	60 000 psi–0.10 in oriface	—	<30 ipm, depending upon material
Abrasive (sanding-grinding)	most	Silicone carbide or alumina grit (wet)	4 000 SFPM	—
Router	most	Carbide or Diamond grit	20 000 RPM	—

ished, large, slowly twisting flutes (large helix or helix rake angle) are desirable for good chip removal.

Four factors that affect cutting speeds are as follows:

1. type of plastics
2. tool geometry
3. lubricant or coolant
4. feed and depth of cut

The cutting speed of plastics is given in surface feet per minute (fpm) or metres per second (m/s). Metres per second refers to the distance the cutting edge of the drill travels in one second when measured on the circumference of the cutting tool. The following formula is used to determine surface metres per second. This information may be obtained from handbooks:

$$m/s = \pi D \times rpm$$
$$r/s = (m/s) \div (\pi D)$$

where   rpm = revolutions per minute

r/s = revolutions per second

m/s = surface metres per second

D = diameter of cutting tool in metres

π = 3.14

As a rule of thumb, plastics have a cutting speed of 200 fpm [1 m/s]. A guide for drilling thermoplastics and thermosets is shown in Table 9-4.

The rate at which the drill or cutting tool moves into the plastics is crucial. The distance the tool is fed into the work each revolution is called the *feed*. Feed is measured in inches or millimetres. Drill feed ranges from 0.001 to 0.0031 in. [0.25 to 0.8 mm] for most plastics, depending on thickness of the material. (Table 9-5)

Many of these same principles of speed and feed apply to reaming, countersinking, spotfacing, and counterboring (Fig. 9-7). In many thermoplastics, holes may be made with hollow punches. Warming the stock may help the punching operation.

Sintered-diamond core drills, countersinks, reamers, or counterbores may be used with ultrasonic energy to bore into some composites. Boron/

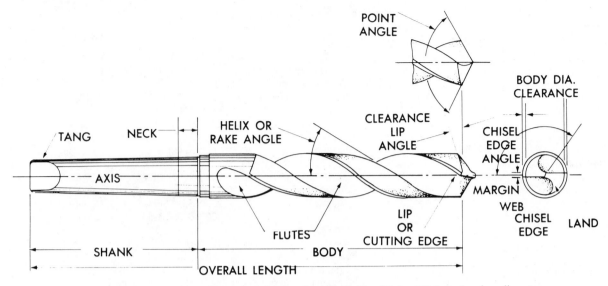

(A) Selected nomenclature for tapered-shank twist drill. Drills of ½ in. [12.5 mm] or less diameter usually have straight shanks. (Morse Twist Drill Machine Co.)

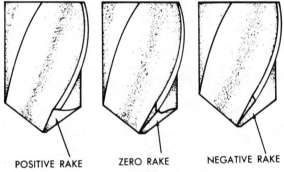

POSITIVE RAKE   ZERO RAKE   NEGATIVE RAKE

(B) A zero rake is usually preferred for plastics, but a negative rake is sometimes used with polystyrene.

Fig. 9-6. Drill nomenclature.

epoxy, graphite/boron/epoxy, and other hybrid composite materials may require ultrasonic techniques.

9-4 STAMPING, BLANKING, AND DIE CUTTING

Many thermoplastics and thin pieces of thermosets may be cut using rule, blanking, piercing, or matched molding dies (Fig. 9-8). It is done on flat parts less than 0.23 in. [6 mm] thick. Holes may be either drilled or die cut, and heating the plastics stock may aid in these operations.

Table 9-3. Drill Geometry

Material	Rake Angle	Point Angle	Clearance	Rake
Thermoplastic				
Polyethylene	10°–20°	70°–90°	9°–15°	0°
Rigid polyvinyl chloride	25°	120°	9°–15°	0°
Acrylic (polymethyl methacrylate)	25°	120°	12°–20°	0°
Polystyrene	40°–50°	60°–90°	12°–15°	0° to neg. 5°
Polyamide resin	17°	70°–90°	9°–15°	0°
Polycarbonate	25°	80°–90°	9°–15°	0°
Acetal resin	10°–20°	60°–90°	10°–15°	0°
Fluorocarbon TFE	10°–20°	70°–90°	9°–15°	0°
Thermosetting				
Paper or cotton base	25°	90°–120°	10°–15°	0°
Fibrous glass or other fillers	25°	90°–120°	10°–15°	0°

Table 9-4. Guide to Speeds for Drilling Plastics

Drill Size	Speed for Thermoplastics, r/s	Speed for Thermosets, r/s
No. 33 and smaller	85	85
No. 17 through 32	50	40
No. 1 through 16	40	28
1.5 mm.	85	85
3 mm	50	50
5 mm	40	40
6 mm	28	28
8 mm	28	20
9.5 mm.	20	16
11 mm	16	10
12.5 mm.	16	10
A–C	40	28
D–O	20	20
P–Z	20	16

Table 9-5. Drilling Feeds of Plastics

Material	Speed, m/s	Feed, mm/revolution					
		1.5	3	6	12.5	19	25
Thermoplastics							
Polyethylene	0.75–1.0	0.05	0.08	0.13	0.25	0.38	0.5
Polypropylene							
TFE fluorocarbon							
Butyrate							
High impact styrene	0.75–1.0	0.05	0.1	0.13	0.15	0.15	0.2
Acrylonitrile-butadiene-styrene							
Modified acrylic							
Nylon	0.75–1.0	0.05	0.08	0.13	0.2	0.25	0.3
Acetals							
Polycarbonate							
Acrylics	0.75–1.0	0.02	0.05	0.1	0.2	0.25	0.3
Polystyrenes	0.75–1.0	0.02	0.05	0.08	0.1	0.13	0.15
Thermosets							
Paper or cotton base	1.0–2.0	0.05	0.08	0.13	0.15	0.25	0.3
Homopolymers	0.75–1.5	0.05	0.08	0.1	0.15	0.25	0.3
Fiber glass, graphitized, and asbestos base	1.0–1.25	0.05	0.08	0.13	0.2	0.25	0.3

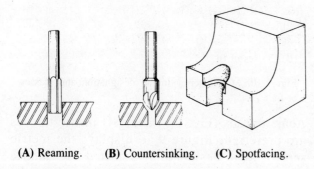

(A) Reaming.   **(B)** Countersinking.   **(C)** Spotfacing.

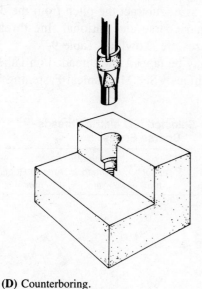

(D) Counterboring.

Fig. 9-7. Drilling operations in plastics.

Punching and shearing of laminar composite materials usually results in some delamination, edge raging, or fiber tearing. It is recommended that the part be abrasively ground to the final dimension.

9-5 TAPPING AND THREADING

Standard machine shop tools and methods may be used for tapping and threading. To prevent overheating, taps should be finish ground and have polished flutes. Lubricants may also be used to help clear chips from the hole. If transparency is needed, a wax stick may be inserted in the drilled hole before tapping. The wax lubricates, helps expel chips, and makes a more transparent thread.

Because of the elastic recovery of most plastics, oversized taps should be used. Oversized taps are designated as follows:

H1:   Basic size to basic + 0.012 mm
H2:   Basic + 0.012 mm to basic + 0.025 mm
H3:   Basic + 0.025 mm to basic + 0.038 mm
H4:   Basic + 0.038 mm to basic + 0.050 mm

The cutting speed for machine tapping should be less than 9.842 in./s [0.25 m/s]. The tap should be backed out often to clear chips. Usually, not more than 75 percent of the full thread is cut into the plastics. Sharp V-threads are not advised.

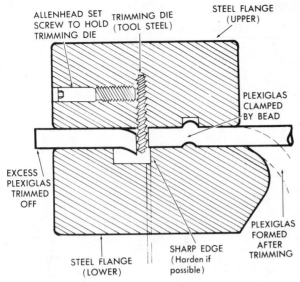

(A) Blank die and clamp for cutting plexiglas.

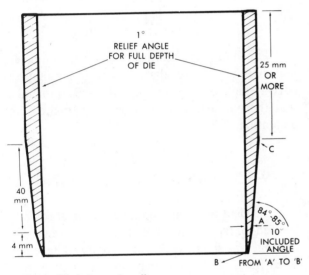

(B) Modified shoemaker die.

Fig. 9-8. Dies used for cutting plastics. (Rohm & Haas Co.)

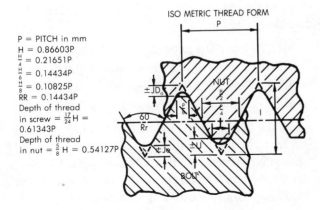

P = PITCH in mm
H = 0.86603P
$\frac{H}{4}$ = 0.21651P
$\frac{H}{6}$ = 0.14434P
$\frac{H}{8}$ = 0.10825P
RR = 0.14434P
Depth of thread
in screw = $\frac{17}{24}H$ =
0.61343P
Depth of thread
in nut = $\frac{5}{8}H$ = 0.54127P

Fig. 9-9. Simplified ISO metric thread forms.

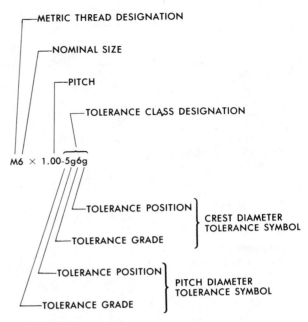

(A) Metric screw thread designation.

CLASSES OF FIT	INTERNAL THREADS (NUTS)	EXTERNAL THREADS (BOLTS)
CLOSE (close accuracy required)	5H	4h
MEDIUM (general purposes)	6H	6g
FREE (easy assembly)	7H	8g

(B) Classes of fit.

Fig. 9-10. Metric screw thread designation and classes of fit.

Acme threads (Fig. 9-9) and ISO metric threads are preferred. Figure 9-10 displays the metric screw thread designation. Selected ISO metric threads are shown in Table 9-6. (ISO stands for the International Organization for Standardization.) To obtain tap drill size, subtract the pitch from the diameter. National coarse and national fine thread and tap drill sizes are shown on Table 9-7.

Plastics may be tapped and threaded on lathes and screw machines. (See Mechanical Fastening in Chapter 18).

Table 9-6.   Selected ISO Metric Threads— Coarse Series

Diameter, mm	Pitch, mm	Tap Drill, mm	Depth of Thread, mm	Area of Root, mm²
M 2	0.40	1.60	0.25	1.79
M 2.5	0.45	2.05	0.28	2.98
M 3	0.50	2.50	0.31	4.47
M 4	0.70	3.30	0.43	7.75
M 5	0.80	4.20	0.49	12.7
M 6	1.00	5.00	0.61	17.9
M 8	1.25	6.75	0.77	32.8
M 10	1.50	8.50	0.92	52.3
M 12	1.75	10.25	1.07	76.2
M 16	2.00	14.00	1.23	144

Table 9-7. National Coarse, and National Fine Threads and Tap Drills

Size	Threads Per Inch	Major Dia.	Minor Dia.	Pitch Dia.	Tap Drill 75% Thread	Decimal Equivalent	Clearance Drill	Decimal Equivalent
2	56	.0860	.0628	.0744	50	.0700	42	.0935
	64	.0860	.0657	.0759	50	.0700	42	.0935
3	48	.099	.0719	.0855	47	.0785	36	.1065
	56	.099	.0758	.0874	45	.0820	36	.1065
4	40	.112	.0795	.0958	43	.0890	31	.1200
	48	.112	.0849	.0985	42	.0935	31	.1200
6	32	.138	.0974	.1177	36	.1065	26	.1470
	40	.138	.1055	.1218	33	.1130	26	.1470
8	32	.164	.1234	.1437	29	.1360	17	.1730
	36	.164	.1279	.1460	29	.1360	17	.1730
10	24	.190	.1359	.1629	25	.1495	8	.1990
	32	.190	.1494	.1697	21	.1590	8	.1990
12	24	.216	.1619	.1889	16	.1770	1	.2280
	28	.216	.1696	.1928	14	.1820	2	.2210
1/4	20	.250	.1850	.2175	7	.2010	G	.2610
	28	.250	.2036	.2268	3	.2130	G	.2610
5/16	18	.3125	.2403	.2764	F	.2570	21/64	.3281
	24	.3125	.2584	.2854	I	.2720	21/64	.3281
3/8	16	.3750	.2938	.3344	5/16	.3125	25/64	.3906
	24	.3750	.3209	.3479	Q	.3320	25/64	.3906
7/16	14	.4375	.3447	.3911	U	.3680	15/32	.4687
	20	.4375	.3725	.4050	25/64	.3906	29/64	.4531
1/2	13	.5000	.4001	.4500	27/64	.4219	17/32	.5312
	20	.5000	.4350	.4675	29/64	.4531	33/64	.5156
9/16	12	.5625	.4542	.5084	31/64	.4844	19/32	.5937
	18	.5625	.4903	.5264	33/64	.5156	37/64	.5781
5/8	11	.6250	.5069	.5660	17/32	.5312	21/32	.6562
	18	.6250	.5528	.5889	37/64	.5781	41/64	.6406
3/4	10	.7500	.6201	.6850	21/32	.6562	25/32	.7812
	16	.7500	.6688	.7094	11/16	.6875	49/64	.7656
7/8	9	.8750	.7307	.8028	49/64	.7656	29/32	.9062
	14	.8750	.7822	.8286	13/16	.8125	57/64	.8906
1	8	1.0000	.8376	.9188	7/8	.8750	1-1/32	1.0312
	14	1.0000	.9072	.9536	15/16	.9375	1-1/64	1.0156
1-1/8	7	1.1250	.9394	1.0322	63/64	.9844	1-5/32	1.1562
	12	1.1250	1.0167	1.0709	1-3/64	1.0469	1-5/32	1.1562
1-1/4	7	1.2500	1.0644	1.1572	1-7/64	1.1094	1-9/32	1.2812
	12	1.2500	1.1417	1.1959	1-11/64	1.1719	1-9/32	1.2812
1-1/2	6	1.5000	1.2835	1.3917	1-11/32	1.3437	1-17/32	1.5312
	12	1.5000	1.3917	1.4459	1-27/64	1.4219	1-17/32	1.5312

9-6 TURNING, MILLING, PLANING, SHAPING, AND ROUTING

High-speed steel or carbide cutting tools used for machining brass and aluminum are advised for machining plastics (Fig. 9-11A). The feeds and speeds are similar. For many plastics, a surface speed of 492 ft/min. [2.5 m/s] with feeds (depth of cut) of 0.02 to 005 in./r [0.5 to 0.12 mm] per revolution will produce good results. On cylindrical stock, a 0.049 in. [1.25 mm] cut will reduce the diameter by 0.098 in. [2.5 mm].

Climb-cutting (or *down-cutting*) a milling operation using lubrication, gives a good machined finish on plastics (Fig. 9-11B). In climb milling, the work moves in the same direction as the rotating cutter. The feed rate on multiple-edged milling cutters is expressed in millimetres of cut per cutting edge per second. The feed of a milling machine is expressed in millimetres of table movement per second rather than millimetres per spindle rotation. The formula below is used to determine the amount of feed in inches per minute or millimetres per second:

$$mm/s = t \times fpt \times r/s$$

where

t = number of teeth

mm/s = feed in millimetres per second

fpt = feed per tooth (chip load)

r/s = revolutions per second (spindle or work)

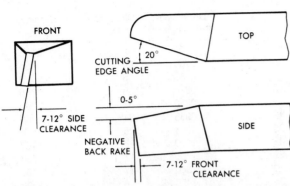

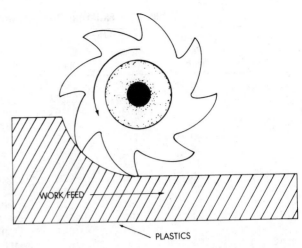

(A) Cutting tool rake and clearance angles for general-purpose turning of plastics. Note the 0-5° negative back rake angle. A sharp-pointed tool with a +20° rake is used in turning polyamides.

(B) Climb milling, or down milling, a technique in which the work moves in the same direction as the rotating cutter.

Fig. 9-11. Machining plastics.

Table 9-8.   Turning and Milling Plastics

	Turning Single Point (H-S Steel)			Milling Tool Per Tooth (H-S Steel)		
Material	Depth of cut, mm	Speed, m/s	Feed, mm/r	Depth of cut, mm	Speed, m/s	Feed, mm per tooth
Thermoplastics						
Polyethylene	3.8	0.8–1.8	0.25	3.8	2.5–3.8	0.4
Polypropylene	0.6	1.5–2	0.05	3.8	2.5–3.8	0.4
TFE-fluorocarbon				1.5	3.8–5	0.1
Butyrates				3.8	2.5–3.8	0.4
ABS	3.8	1.2–1.8	0.38	3.8	2.5–3.8	0.4
Polyamides	3.8	1.5–2	0.25	3.8	2.5–3.8	0.4
Polycarbonate	0.6	2–2.5	0.05	1.5	3.8–5	0.1
Acrylics	3.8	1.2–1.5	0.05	1.5	3.8–5	0.1
Polystyrenes, low	3.8	0.4–0.5	0.19	3.8	2.5–3.8	0.4
and medium impact	0.6	0.8–1	0.02	3.8	2.5–3.8	0.4
Thermosets						
Paper	3.8	2.5–5	0.3	1.5	2.0–2.5	0.12
and cotton base	0.6	5–10	0.13	1.5	2.0–2.5	0.12
Fiber glass	3.8	1–2.5	0.3	1.5	2.0–2.5	0.12
and graphite base	0.6	2.5–5	0.13	1.5	2.0–2.5	0.12
Asbestos base	3.8	3.2–3.8	0.3	1.5	2.0–2.5	0.12

Table 9-8 gives turning and milling data for various plastics materials. Table 9-9 gives side and end relief angles and back rake angles for cutting tools used with different plastics.

For all milling, planing, shaping, and routing work, carbide-tipped cutters are advised. Conventional high-speed steel shapers, planers, and routers used for wood-working may be employed with plastics if tools are sharpened in a proper manner. Routers and shapers are useful for cutting beads, rabbets, and flutes and for trimming edges. Carbide or diamond-tipped tools are vital for long runs, uniformity of finish, and accuracy.

Table 9-9.   Design of Turning Cutting Tool

Work Material	Side Relief Angle, Deg.	End Relief Angle, Deg.	Back Rake Angle, Deg.
Polycarbonate	3	3	0–5
Acetal	4–6	4–6	0–5
Polyamide	5–20	15–25	neg. 5–0
TFE	5–20	0.5–10	0–10
Polyethylene	5–20	0.5–10	0–10
Polypropylene	5–20	0.5–10	0–10
Acrylic	5–10	5–10	10–20
Styrene	0–5	0–5	0
Thermosets:			
Paper or Cloth	13	30–60	neg. 5–0
Glass	13	33	0

9-7 LASER CUTTING

A CO$_2$ laser (light amplification by stimulated emission of radiation) can deliver powerful radiation at a wavelength of 10.6 μm (microns). A laser may be used to make intricate holes and complex patterns in plastics (Fig. 9-12). The laser power can be controlled to merely etch the plastics surface or actually vaporize and melt it. Holes and cuts made by a laser have a slight taper, but the cuts are clean with a finished appearance. Cuts made by a laser are more precise, and tolerances are held more closely than those made with conventional machining operations. There is no physical contact between the plastics and the laser equipment, therefore no chips are produced. Laser cutting does produce a residue of fine dust; how-ever, this is easily removed by vacuum systems. Most polymers and composites may be laser machined. Some laminar composites tend to heat up, bubble and char.

9-8 INDUCED FRACTURE CUTTING

Acrylics and several other plastics including some composites, may be cut to shape by *induced fracture* methods. The methods are similar to cutting glass. A sharp tool or cutting blade is used to score, or scratch, the plastics surface. On thick pieces, both sides are scored. Pressure is applied along the scratch line, and the plastics fractures. The fracture will follow the score line (Fig. 9-13).

9-9 THERMAL CUTTING

Heated wires or dies are used to cut solid and expanded, or foamed, plastics. Hot dies are used to cut fabrics and silhouette-shaped products, while a heated wire or ribbon is commonly used to cut expanded plastics (Fig. 9-14). Thermal cutting produces a smooth edge with no chips or dust.

9-10 HYDRODYNAMIC CUTTING

High-velocity fluids may be used to cut many plastics and composites (Fig. 9-4C). Pressures of 320 MPa [46 417 psi] are used. Foamed or cellular plastics are reinforced and filled plastics have been successfully machined by this method.

9-11 SMOOTHING AND POLISHING

Smoothing and polishing techniques for plastics are like those used on woods, metals, and glass.

Because of the elastic and thermal properties of thermoplastics, many are difficult to grind. Abrasive grinding is more easily accomplished on thermosetting materials, reinforced plastics, and most composites. Grinding is not advised unless open-grit wheels are used with a coolant. Hand and machine sanding is an important operation. *Open-grit sandpaper* is used on machines to prevent clogging (loading). A number 80-grit silicon-carbide abrasive is advised for rough sanding. In any machine sanding, light pressure is used to prevent overheating the plastics.

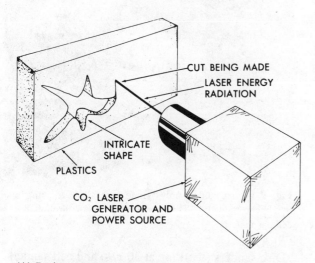

(A) Basic concept.

(B) A five axis robotic laser is used to trim the three-dimensional molded Kevlar part. (Russel Plastics Technology)

Fig. 9–12. Light energy from a laser can be used to cut intricate shapes in plastics, or to trim to final shape.

(A) Score plastics with tool.

(D) Completed cutting.

Fig. 9-13. The induced-fracture cutting method.

(B) Align score line with edge of table.

Fig. 9-14. Expanded polystyrene is easily cut using a hot Nichrome wire. The wire melts a path through the cellular material.

Disc sanders (working at 30 r/s) and belt sanders (working at a surface speed of 18 m/s [59 ft/s]) are used for dry sanding. If water coolants are used, the abrasive lasts longer and cutting action is increased. Progressively finer abrasives are used; that is, the first rough sanding using 80-grit paper should be followed by 280-grit silicon-carbide wet or dry sandpaper. The final sanding may be with 400- or 600-grit sandpaper. After the sanding is finished and the abrasives removed, further finishing operations are used.

Ashing, buffing, and polishing are done on abrasive-charged wheels. These wheels may be made of cloth, leather, or bristles. A different wheel is used for each abrasive grit. Finishing wheel speeds should not exceed 10 surface m/s [32.8 ft/s]. With the use of coolants, surface speed may be increased.

Never finish plastics on wheels used for metals. Small metal particles may be left in the wheel

(C) Press down on plastics piece to induce fracture.

and these will damage the plastics surface. Machines should be grounded. Static electricity is generated by the movement of the wheels over the plastics. Remove coarse tool marks before the finishing wheels are used.

Ashing is a finishing step in which a wet abrasive is applied to a loose muslin wheel. Number 00 pumice is commonly used (Fig. 9-15). A hood or shield is used over the wheel because the operation is wet. Surface speeds of over 20 m/s [65.6 ft/s] may be used. Overheating is avoided in this process and the loose muslin wheel is fast-cutting on irregular surfaces.

Buffing is an operation in which grease- or wax-filled abrasive bars or sticks are applied to a loose or sewn muslin wheel.

Loose buffs are used for more irregular shapes or for entering crevices. Hard buffing wheels should be avoided.

The buffing wheels are charged by holding the bars or sticks against them as they revolve, producing frictional heat that leaves the wax-filled abrasive on the wheel. The most common buffing abrasives are tripoli, rouge, or other fine silica.

Polishing, sometimes called luster buffing or burnishing, employs wax compounds containing the finest abrasives such as levigated alumina or whiting. Polishing wheels are generally made of loose flannel or chamois. A final polishing is sometimes done with clean, abrasive-free waxes on a flannel or chamois wheel. The wax fills many imperfections and protects the polished surface.

Never let the finishing wheel rotate to the edge of a part because it may be jerked from your hands. The wheel may pass over an edge but never to it. Always keep the work piece below the center of the buffing wheel. The preferred procedure is to do about half of the surface and then turn it around and finish the final portion. The part should be moved or pulled toward the operator in rapid, even strokes (Fig. 9-16). Do not spend very much time at the finishing wheels. Move stock around. If you hold a piece in one place, the heat generated by friction between the wheel and the workpiece will melt many thermoplastics.

Solvent-dip polishing of cellulosic and acrylic plastics may be used to dissolve minor surface defects (Fig. 9-17A). The parts are either dipped into or sprayed with solvents for about one minute. Solvents are sometimes used to polish edges or drilled

(A) Charging flannel wheel with whiting.

(B) Abrasive wheel passing over an edge. About half of each side is done by pulling work toward the operator.

Fig. 9-16. Buffing plastics materials. Note: Guards have been removed from equipment for this illustration. Protective devices and guards should *always* be used when working with such equipment.

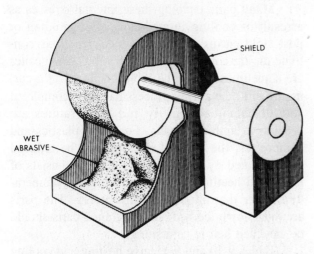

Fig. 9-15. Ashing with wet pumice is a faster cutting method than the use of grease- or wax-based compounds. Cooling action is also better.

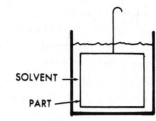

(A) Solvent-dip polishing.

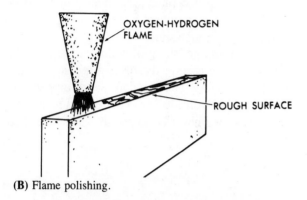

(B) Flame polishing.

Fig. 9-17. Two methods of polishing.

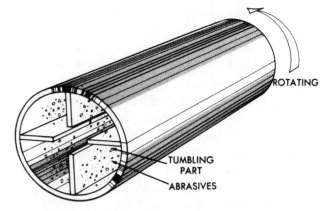

(A) Parts being tumbled in a revolving drum.

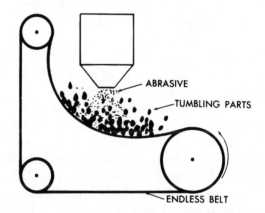

(B) Tumbling of parts on an endless revolving belt.

Fig. 9-18. Two tumbling methods.

holes. All solvent-polished parts should be annealed to prevent crazing.

Surface coatings may be used on most plastics to produce a high surface gloss, which may cost less than other finishing operations.

Flame polishing with an oxygen-hydrogen flame may be used to polish some plastics (Fig. 9-17B).

9-12 TUMBLING

The tumbling-barrel process is one of the least costly ways to finish plastics molded parts rapidly. It produces a smooth finish on plastics parts by rotating them in a drum with abrasives and lubricants, causing the parts and abrasives to rub against each other with a smoothing effect (Fig. 9-18A). The amount of material removed depends on the speed of the tumbling barrel, the abrasive grit size, and the length of the tumbling cycle.

In another tumbling process, abrasive grit is sprayed over the parts as they tumble on an endless rubber belt. Figure 9-18B shows parts being tumbled while being doused with abrasive grit.

Dry ice is sometimes used in tumbling to remove molding flash. The dry ice chills the thin flash and makes it brittle; then tumbling breaks it free in a very short time.

9-13 ANNEALING AND POSTCURING

During the molding, finishing, and fabrication processes, the plastics or composite part may develop internal stresses. Chemicals may sensitize the plastics and cause crazing.

Many parts develop these internal stresses as a result of cooling immediately after molding or post-mold curing, as the chemical reactions continue during complete polymerization. Composites are sometimes left in the mold or placed in a curing jig until the curing process has been completed and all chemical activity and temperatures are brought to ambient levels. For some plastics and composites, the internal stresses may be reduced or eliminated by annealing. Annealing consists of prolonged heating of the plastics part at temperature lower than molding temperatures. The parts are then slowly cooled. All machined parts should be annealed before cementing.

Tables 9-10 and 9-11 give heating and cooling times for the annealing of Plexiglas. Figure 9-19 shows a large oven that may be used in the process.

VOCABULARY

The following vocabulary words are found in this chapter. Look up the definition of any of these words you do not understand as they apply to plastics in the glossary, Appendix A.

Annealing
Ashing
Blanking
Charging
Die cutting
Feed
Hollow ground
Kerf
Laser cutting
m/s
Negative rake
Open-grit sandpaper
rpm
r/s
Tripoli
Tumbling
Whiting

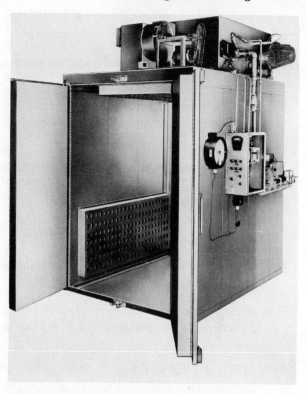

Fig. 9-19. This large oven may be used for annealing and postcuring plastics. (Precision Quincey Corp.)

Table 9-10.   Heating Times for Annealing of Plexiglas

Thickness (mm)	Time in a Forced-Circulation Oven at the Indicated Temperature, h									
	Plexiglas G, II, and 55					Plexiglas I-A				
	110 °C*	100 °C*	90 °C*	80 °C	70 °C**	90 °C*	80 °C*	70 °C*	60 °C	50 °C
1.5 to 3.8	2	3	5	10	24	2	3	5	10	24
4.8 to 9.5	2½	3½	5½	10½	24	2½	3½	5½	10½	24
12.7 to 19	3	4	6	11	24	3	4	6	11	24
22.2 to 28.5	3½	4½	6½	11½	24	3½	4½	6½	11½	24
31.8 to 38	4	5	7	12	24	4	5	7	12	24

Note: Times include period required to bring part up to annealing temperature, but not cooling time. See Table 9–9.
*Formed parts may show objectionable deformation when annealed at these temperatures.
**For Plexiglas G and Plexiglas II only. Minimum annealing temperature for Plexiglas 55 is 80 °C.
Source: *Rohm & Haas Co.*

Table 9-11.   Cooling Times for Annealing of Plexiglas

Thickness (mm)	Rate (°C)/h	Time to Cool from Annealing Temperature to Maximum Removal Temperature							
		Plexiglas G, II, and 55				Plexiglas I-A			
		230 (110 °C)	212 (100 °C)	184 (90 °C)	170 (80 °C)	194 (90 °C)	176 (80 °C)	158 (70 °C)	140 (60 °C)
1.5 to 3.8	122 (50)	¾	½	½	¼	¾	½	½	¼
4.8 to 9.5	50 10	1½	1¼	¾	½	1½	1¼	¾	½
12.7 to 19	22 − 5	3¼	2¼	1½	¾	3	2¼	1½	¾
22.2 to 28.5	18 − 8	4¼	3	2	1	4	3	2¼	1
31.8 to 38	14 −10	5¾	4½	3	1½	5¾	4½	3	1½

Note: Removal temperature is 70 °C for Plexiglas G and II, 80 °C for Plexiglas 55, and 50 °C for Plexiglas I-A.
Source: *Rohm & Haas Co.*

9-1. Name the process of slowly cooling plastics to remove internal stresses.

9-2. Identify the number of teeth per centimetre for cutting 6 mm thick polycarbonate on the bandsaw.

9-3. Name the rake angle of a circular saw when the hook angle of the tooth and the center of the blade are in line.

9-4. How many r/s are required for drilling with an 8 mm drill in thermoplastics?

9-5. The distance the cutting tool is fed into the work each revolution is called ___?___ .

9-6. What is the abbreviation or symbol for the International Organization for Standardization?

9-7. Name the slit or notch made by a saw or cutting tool.

9-8. Name the burnishing agent sometimes used to polish the edges of acrylic plastics.

9-9. Name the finishing operation in which wet abrasives are used.

9-10. Name the cutting operation that uses high velocity fluids to cut plastics.

9-11. What type of cutting edges or teeth on tools are essential for long runs, uniform finishing and accuracy?

9-12. What does the 1.00 show in M6X1.00-5g6g thread designation?

9-13. Name a popular silica abrasive used in some finishing operations.

9-14. Frictional ___?___ is a major problem in machining most plastics.

9-15. Name the operation that is preferred over sawing, for many plastics, because it produces a smoother edge.

9-16. How many teeth per centimetre should a circular saw have for cutting thin plastics materials? Thick materials?

9-17. What type of bandsaw blade and what cutting speed should you use to cut 0.118 in. [3 mm] thick acrylic plastics?

9-18. What is a skip-tooth blade?

9-19. Name some saws that may be used to cut plastics. Indicate the kinds of jobs for which each may be used.

9-20. What is drill feed? What is its range in millimetres for most plastics?

9-21. What factors affect the cutting speed of drills in plastics materials?

9-22. What precaution must be observed in drilling a hole of a given size in a plastic rod?

9-23. Why are oversized taps necessary for plastics materials?

9-24. Name the preferred thread forms for taps used on plastics.

9-25. What is climb milling? Why is it used?

9-26. What is laser cutting of plastics? Where is it used?

9-27. What grit number of sandpaper should be used for final finishing of plastics?

9-28. What is ashing? Buffing? Charging a wheel?

9-29. Which abrasives are used in buffing? In burnishing?

9-30. Briefly describe solvent-dip polishing and flame polishing.

9-31. What is tumbling? Why is it so named?

9-32. What does the smoothing in the tumbling operation?

9-33. What is annealing or postcuring of machined or molded parts? Why is this process done?

9-34. What machining operation would you select for shaping or finishing of the following products?
(a) 0.23 in. [6 mm] thick polycarbonate window glazings.
(b) Christmas tree decoration shapes made of thin sheet or films.
(c) Removing flash from radio cabinet housing.
(d) Making the edges of a plastics part smooth and glossy.

Chapter 10

Molding Processes

This chapter will discuss how the various plastics are molded from resins, powders, granules, or other forms into end products or components. Molding is one of the major processes used in converting plastics into products.

Keep in mind that many processes are used to form composite products. As we discuss in Chapters 11, 12, and 13, techniques such as extrusion, calendering, laminating, and reinforcing can produce a continuous or very long reinforcement. Other processes use short and mostly randomly oriented fibers, crystals, and fillers in the polymer matrix.

The Society of the Plastics Industry, Inc., defines the processor as one who converts plastics into products. The fabricator and finisher further fashion and decorate plastics products. Decorating, finishing, and fabrication of parts are highly specialized fields (see Chapters 18 and 19). Many of the processes are similar to those used in metalworking, papermaking, and glassblowing while others apply only to the processing of plastics. The low-cost mass production of quality plastics products depends on the work of machinists and toolmakers. Specially-formulated plastics and processing techniques have helped to change plastics processing from a craft to a true technology.

The basic processes of the plastics industry include, molding, reinforcing, laminating, casting, thermoforming, expanding, coating, decorating, machining and finishing, and assembly or fabrication. Radiation processing may be included as a special technique (See Chapter 20).

One basic assumption of molding all plastics is that they may be made fluid at some time during the operation, but they later solidify. For an operation to be classed as molding, force is required.

The molding processes discussed in this chapter include compression molding, transfer molding, injection molding, reaction injection molding, reinforced reaction injection molding, vacuum-injection molding, resin transfer molding, thermal expansion resin transfer molding, cold molding, and sintering.

10-1 COMPRESSION MOLDING

One of the oldest known molding processes is compression molding. The plastics material is placed in a mold cavity and formed by heat and pressure. As a rule, thermosetting compounds are used for compression molding, but thermoplastics may be used. The process is somewhat like making waffles. Heat and pressure force the materials into all areas of the mold. Then, after the heat hardens the substance, the part is removed from the mold cavity (Fig. 10-1).

To reduce pressure requirements and production (cure) time, the plastics material is usually preheated with infrared, induction, or other heating methods before it is placed in the mold cavity. A screw extruder is sometimes used to reduce cycle time and increase output. The screw extruder is often used to make preformed slugs that are loaded into the molding cavity. The screw-compression

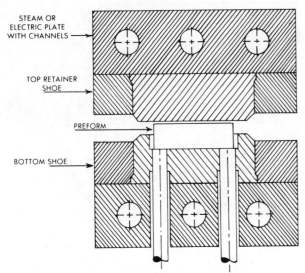

(A) Preform about to be molded.

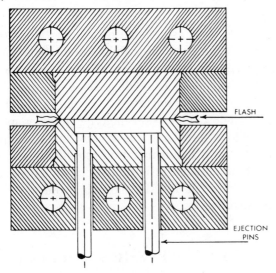

(B) Closed mold, showing flash.

Fig. 10-1. Principle of compression molding.

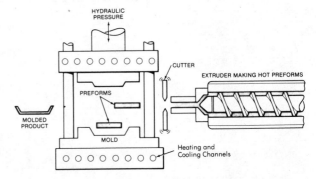

Fig. 10-2. Compression molding, showing hot preforms being fed to mold cavity.

process greatly reduces cycle time, eliminating the greatest drawback of compression molding. Compression molded parts with heavy wall thickness may be produced with up to 400 percent greater product output per mold cavity when this method is used.

Thermosetting bulk molding compounds (BMC) are used in forming polyester compounds. BMC is a mixture of fillers, resins, hardening agents, and other additives. Hot extruded preforms of this material may be loaded straight into the cold cavity or feed chute (Fig. 10-2).

Phenolic plastics, urea-formaldehyde, and melamine compounds are also popular molding materials. Like BMC, they are usually preformed for automation and speed.

Heavily filled and reinforced sheet molding compounds are used. They may be placed in alternating layers for more isotropic properties, or placed in one direction for more anisotropic properties.

Most compression molding equipment is sold by the press or platen rating. A force of 20 MPa or 2 900 psi [2 kN/cm^2] is usually required for moldings up to 1 in. [25 mm] thick. An added 725 psi [0.5 kN/cm^2] should be provided for each 1 in. [25 mm] increase. Hydraulic action provides this force (Fig. 10-3).

Steam, electricity, hot oil, and open flame are some of the means for heating the molds, platens, and related equipment. Hot oil is common because it may be heated to high temperatures with little pressure. Electricity is clean, but limited by wattage.

During preforming and the actual molding process, heat and various catalysts begin cross-linking the molecules. During the cross-linking reaction, gases, water, or other by-products may be freed. If they are trapped in the mold cavity, they may affect the plastics part and the part may be damaged, of poor quality, or marked by surface blisters. Molds are usually vented to allow the escape of these by-products.

In molding, the thermosetting plastics compound becomes cross-linked and infusible, thus these products may be removed from the molding cavity while hot. Thermoplastic materials, because they do not cross-link to any extent, must be cooled before removal. Many elastomers are molded by this process.

Long runs of moderately complex parts are often produced by compression molding. Mold maintenance and starting costs are low, there is little material waste, and large bulky parts are practical. Very complex parts are hard to mold, however. Inserts, undercuts, side draws, and small holes are

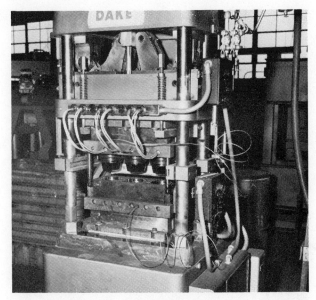

(A) Large compression molding press being tested. (Hull Corp.)

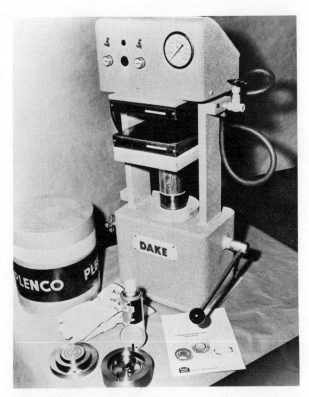

(C) Small laboratory press and mold. (Dake Corp.)

Fig. 10-3. Compression molding presses.

5. Apply heat and pressure until cure is complete (dwell).
6. Open mold and place hot part in cooling fixture.

Five advantages and disadvantages of compression molding are given here.

Advantages of Compression Molding

1. Little waste (no gates, sprues, or runners in many molds)
2. Low tooling costs
3. Process may be automated or hand-operated
4. Parts are true and round
5. Material flow is short; less chance of disturbing inserts, causing product stress, and/or eroding molds

Disadvantages of Compression Molding

1. Hard to mold complex parts
2. Inserts and fine ejector pins are easily damaged
3. Complex shapes are sometimes hard to achieve
4. Long molding cycles may be needed
5. Unacceptable parts cannot be reprocessed

Compression-molded products include dinnerware, buttons, buckles, knobs, handles, appliance housings, drawers, parts bins, radio cases, large containers, and many electrical parts (Fig. 10-4).

(B) Stokes press used to manufacture electrical plate covers.

not practical with this method when it is necessary to maintain close tolerances. (See Chapter 21.)

The sequence of compression molding may include the following six steps:

1. Clean mold and apply mold release (if required).
2. Load preform into cavity.
3. Close mold.
4. Open mold briefly to release trapped gases (breathe mold).

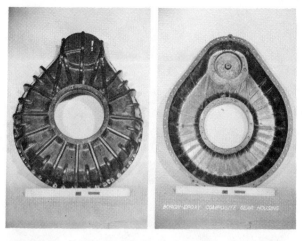

Fig. 10-4. The reinforced plastics gear housing at right is lighter and stronger than the metal housing it replaces, shown at left. Many similar parts are compression-molded. (Allison Division, Detroit Diesel)

10-2 TRANSFER MOLDING

Transfer molding has been known and practiced since World War II. The process is sometimes called plunger molding, duplex molding, reserve-plunger transfer molding, step molding, injection transfer molding, or impact molding. It is actually a variation of compression molding but differs in that the material is loaded in a chamber outside the mold cavity. An advantage of transfer molding is that the molten mass is fluid when entering the mold cavity. Fragile, complex shapes with inserts or pins may be formed with accuracy. Transfer-molding techniques are much like those of injection molding, except that thermosetting compounds are normally used.

The American Society of Tool and Manufacturing Engineers, recognizes two basic types of transfer molds:

1. Pot or sprue molds
2. Plunger molds

Plunger molds (Fig. 10-5) differ from the sprue molds (Fig. 10-6) in that the plunger or force is pushed to the parting line of the mold cavity when inserting the plastics material. In sprue molds the plastics is fed by gravity down a hole (sprue). With plunger molds, only runners and gates are left as waste on the molded part.

A third type of mold may also be included in transfer molding. In this type, molding compound is preplasticized by extruder action, then a plunger forces the melt into the mold (Fig. 10-7.)

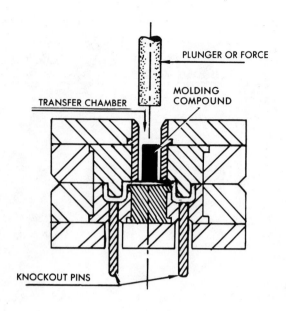

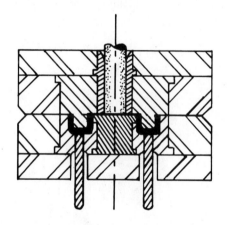

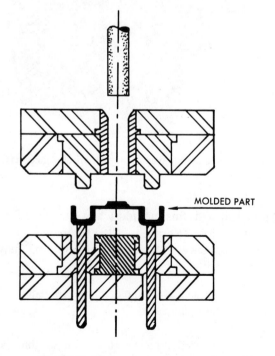

Fig. 10-5. Plunger (two-plate) transfer mold.

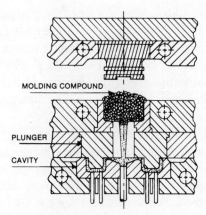

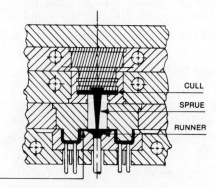

(A) Open position.

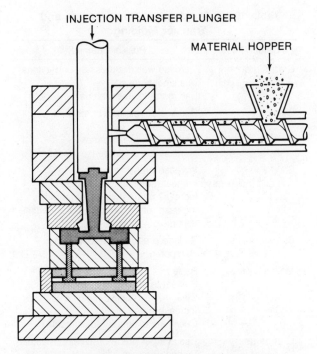

Fig. 10-7. Injection transfer molding, showing the hot extruded compound being forced into the mold cavity by the transfer plunger.

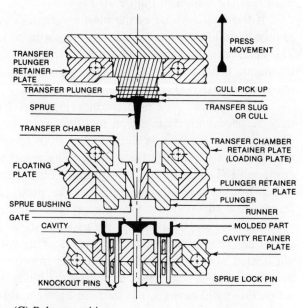

(B) Closed position.

(C) Release position.

Fig. 10-6. Integral-transfer (three-plate) mold. (American Technical Society)

The cost of detailed mold designs and high waste from culls, sprues, and flash are two major limitations of transfer molding.

Although most parts are limited in size, there are numerous applications including distributor caps, camera parts, switch parts, buttons, coil forms, terminal block insulators, and complex shapes such as cups and caps for cosmetic containers.

Problems in compression and transfer molding are found in Table 10-1, while six advantages and four disadvantages of transfer molding are shown below.

Advantages of Transfer Molding

1. Less mold erosion or wear
2. Complex parts (small diameter holes or thin wall sections) and inserts may be molded and used
3. Less flash than in compression molding
4. Densities are more even than in compression molding
5. Multiple pieces may be molded
6. Shorter mold and loading times than most compression molding processes

Disadvantages of Transfer Molding

1. More waste from runners and sprues
2. More costly equipment and molds
3. Molds must be vented
4. Runners and gates must be removed

10-3 INJECTION MOLDING

The injection molding process was in use as early as the 1860s. In 1872, the Hyatt brothers were

Table 10-1. Problems in Compression and Transfer Molding

Defect	Possible Remedy
Cracks around inserts	Increase wall thickness around inserts Use smaller inserts Use more-flexible material
Blistering	Decrease cycle and/or mold temperature Vent mold — breathe mold Increase cure — increase pressure
Short and porous moldings	Increase pressure Preheat material Increase charge weight of material Increase temperature and/or cure time Vent mold — breathe mold
Burned marks	Reduce preheating and molding temperature
Mold sticking	Raise mold temperature Preheat to eliminate moisture Clean mold — polish mold Increase cure Check knockout pin adjustments
Orange peel surface	Use a stiffer grade of molding material Preheat material Close mold slowly before applying high pressure Use finer ground materials Use lower mold temperatures
Flow marks	Use stiffer material Close mold slowly before applying high pressure Breathe mold Increase mold temperature
Warping	Cool on jig or modify design Heat mold more uniformly Use stiffer material Increase cure Lower temperature Anneal in oven
Thick flash	Reduce mold charge Reduce mold temperature Increase high pressure Close slowly — eliminate breathe Increase temperature Use softer grade material Increase clamping pressure

awarded a United States patent for injection molding. Today, the injection process is used for many thermosetting compounds and for all thermoplastics except PTFE fluoroplastics, polyimides, some aromatic polyesters and special engineering grades. The thermosets, including phenolic, urea, melamine, diallyl phthalate, epoxy, silicone, alkyd, and polyester, as well as numerous elastometers are processed by means of reciprocating-screw machines.

During injection molding, a granular plastics is heated and forced through a heated cylinder. The hot mass is then injected into the closed mold

cavity (Fig. 10-8). After cooling, the plastics part is removed. The process works somewhat like an injection glue gun; hot glue is forced into a crack or joint where it cools and solidifies. The technique is not unlike metal die casting, and in fact, most of the early injection molding mechanisms were patterned after die casting machines.

The two basic types of injection machines, with various modifications, are the plunger and the reciprocating screw. Both may have preplasticizer sections or the two techniques may be combined.

The major difference between the two methods is in the way that the molten mass is forced through the heating cylinders and injection chamber. In plunger machines, the material is forced around the torpedo and then into the mold cavity (Fig. 10-9). In the reciprocating screw machines, granular material is quickly made molten by the heated barrel and screw action. When the screw stops turning it functions as a plunger. The hot melt is then forced into the mold cavity. This machine blends colors or other materials more rapidly, because of the action of the screw (Fig. 10-10). Multiscrew machines have improved mixing and blending action. Some modern injection molding machines are shown in Figure 10-11.

If the melting time of the plastics granules is reduced, molding cycle times may be shortened. The material is therefore preplasticated (melted) in a separate heating chamber and then transferred to the main cylinder. (Fig. 10-12)

A rotary shear-cone preplasticater (Fig. 10-12E) may be used with an injection molding machine. The rotary shear machine takes up less floor space and does not use an extruder screw. The plastics compound is metered into the plunger chamber, which houses a revolving cone. As the plunger forces the compound past the opening between the spinning cone and the housing, the material shears, plasticates, and homogenizes, or blends. This hot plasticated material then passes into the injection plunger chamber where it is forced by the plunger action into mold cavities (not shown). Faster molding cycles and greater viscosity control are advantages of this technique.

Injection molding is popular because metal inserts may be used, output rates are high, surface finish can be controlled to produce any desired texture, and dimensional accuracy is good. For thermoplastics, gates, runners, and rejected parts may be ground and reused. Eight advantages and four disadvantages of injection molding are listed below.

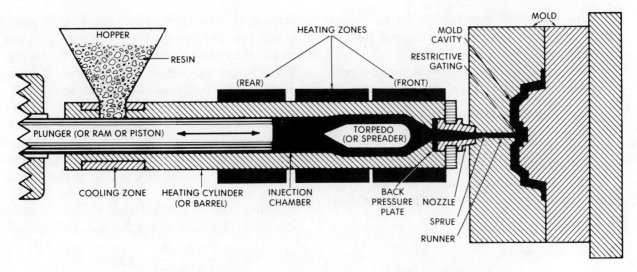

Fig. 10-8. Plunger injection molding machine, with mold in closed position.

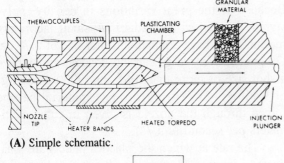

(A) Simple schematic.

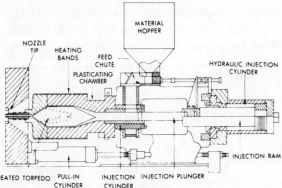

(B) Complete with ram and hydraulic cylinder. (Hydraulic Press Mfg. Co.)

Fig. 10-9. Injection molding machine (plunger).

Advantages of Injection Molding

1. High output rates
2. Fillers and inserts may be used
3. Small, complex parts with close dimensional tolerances can be molded
4. More than one material may be injected into the mold (coinjection molding)
5. Parts require little or no finishing
6. Thermoplastics scrap may be ground and reused

7. Self-skinning structural foams may be molded (reaction injection molding)
8. Process may be highly automated

Disadvantages of Injection Molding

1. High mold costs (multiple cavity molds)
2. High equipment costs
3. Quality may be affected by workmanship, process control, and other variables
4. Competition is keen

Injection molding consists of five basic steps:

1. The mold is closed and forced against the nozzle.
2. The screw or screw and plunger plasticize the material and force it forward into the mold cavity.
3. The screw or screw and plunger maintain pressure through the nozzle until the plastics is cooled or set. (Heat is continually removed from the mold by water circulation).
4. The pressure is removed as the screw or screw and plunger receive fresh material from the feed hopper.
5. The mold moves away from the nozzle and opens. The molded part is removed.

It is common to group these basic steps as a *time cycle*. All injection systems have the following four elements in a time cycle:

1. *Injection time* — the time it takes to force the plastics materials into the mold cavity
2. *Dwell time* — the time that force or pressure is kept on the material in the mold cavity

3. *Freeze time* — the time required for the material to cool or set enough for safe removal from the mold cavity

4. *Dead time* — the time required to open the mold, remove the molded part and close the mold.

A common production expression is *downtime*. This is time during which the machine is not in operation. It may be down because of maintenance, mold changes, or lack of materials.

Injection molding machines may use forces five to ten times greater than those needed on compression presses. Hydraulic, mechanical (toggle), or mechanical-hydraulic clamping units hold the mold halves closed during injection (Fig. 10-13). These forces may range from less than 2.25 to more than 4 496.40 tons-force [20 kN to 40 MN].

As a rule of thumb, 786 pounds-force [3.5 kN] of force is needed for each square centimetre of mold cavity area.

A machine with a clamp force of 337 tons-force [3 MN] should be capable of molding a polystyrene plastics part 9.8 by 12.8 in. [250 by 325 mm]. This part would have a surface area of 126 in.^{2} [812.5 cm^2]. Using the rule of thumb:

$$\frac{3000 \text{ kN}}{3.5 \text{ kN/cm}^2} = 857 \text{ cm}^2$$

Injection capacity (shot size) is the maximum amount of material the machine will eject per cycle. Because of the great variations in density, polystyrene is used as a standard. A small laboratory machine may have a maximum shot size of 0.70 oz. [20 grams]. Very large capacity machines may have a shot size of more than 19.8 lbs [9000 grams (g) or 9 kilograms (kg)]. How fast material can be forced through the nozzle is given in cubic centimeters per second (cm^3/s).

The rate at which the plastics pellets are plasticated is expressed in oz/s or lbs/h [g/s or kg/h].

The mold clamp must be capable of holding the injection pressure. These pressures may range from less than 10 159 psi to greater than 29 010 psi [70 MPa to more than 200 MPa].

Once the material has been forced into the cavity, a non-return check valve prevents loss of

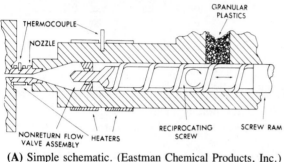

(A) Simple schematic. (Eastman Chemical Products, Inc.)

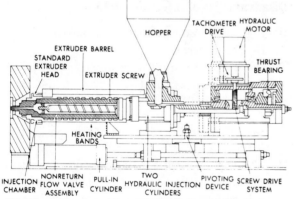

(B) Schematic of machine. (Hydraulic Press Mfg. Co.)

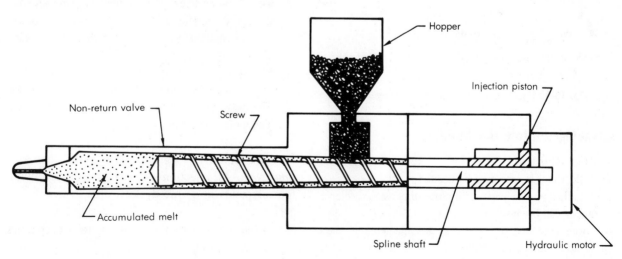

(C) Schematic of reciprocating screw injection machine.

Fig. 10-10. Injection molding machine (reciprocating screw).

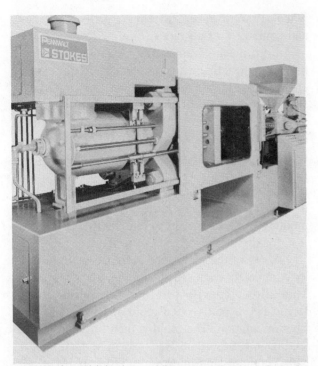

(A) Horizontal injection molding press. (Pennwalt-Stokes)

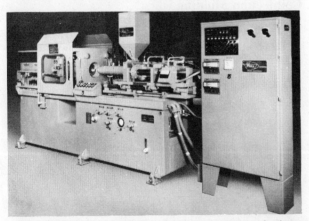

(B) Injection molding machine with two-stage preplasticator.

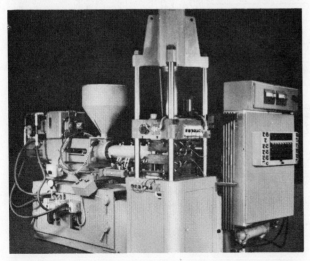

(C) This machine uses a reciprocating screw to plasticate material and inject it through a sprue bushing. (Hull Corp.)

(D) Closeup of machine shown in (C), showing sprue bushing at the parting line.

(E) Fully automatic injection press. (Pennwalt-Stokes)

Fig. 10-11. Examples of modern injection molding machines.

pressure. It also prevents materials from being forced back into the screw tip or nozzle. This valve is also designed to prevent drooling while the mold is in the open position (Fig. 10-14).

Temperature controllers, coolant conditioners, dehumidifying dryers, scrap regrinders, and other peripheral equipment are vital to all injection molding processes.

Injection molding is not practical for short production runs because of high machine, tooling, and mold costs. Problems that may occur with injection molding are listed in Table 10-2.

Common injection-molded items include toys, bathroom and kitchen wall tile, cases, housings for radio cabinets and appliances, refrigerator parts, handles, battery lights, instrument panels, steering

wheels, pump parts, fasteners, grills, bearings, and containers of all types (Fig. 10-15).

Both machine and mold designs differ when molding thermosetting materials. A non-return check valve is not required since the material is very viscous and little material is left in the barrel after injection. Screw flights are shallow with a one-to-one compression ratio. BMC or other heavily filled or reinforced materials generally use a plunger machine. The length-to-diameter (L/D) ratio in these machines generally range from 12:1 to 16:1, compared to higher ratios for injection molding of thermoplastics (Fig. 10-16). See Extrusion for discussion of L/D ratio for screws.

Many of the molds are designed with the cavity completely enclosed, including the parting line. This allows heavily reinforced materials to be forced into a partly open mold and prevents some fiber orientation. The mold is then fully closed, resulting in a fully densified part. The technique is called compression-injection molding (CIM).

Coinjection Molding

Coinjection molding is a process in which two or more materials are injected into the mold cavity (Fig. 10-17). This usually produces a skin of material on the mold surfaces and a cellular center

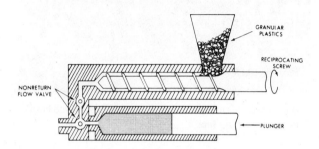

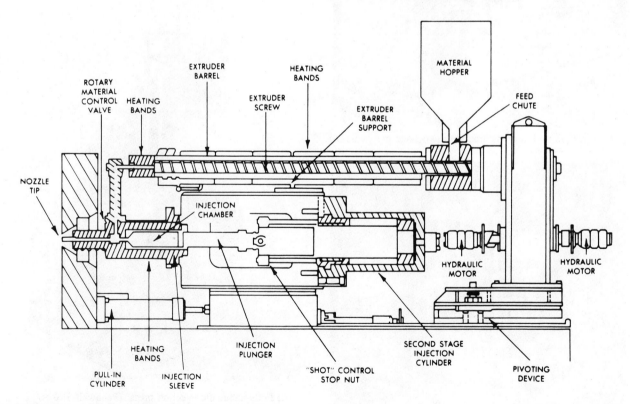

(B) Injection molding machine with two-stage preplasticator.

Fig. 10-12. Examples of injection molding equipment with preplasticators.

core. The core material includes blowing agents to produce the desired cellular densities. Injection pressures seldom exceed 15 995 psi [110 MPa].

If fiber reinforcements are used, the flow patterns may result in an orientation of fibers.

The process is sometimes incorrectly called sandwich molding, because of the composite layered effect. Only the exposed skin surfaces need to be platable or pigmented. Different families of plastics may be used for the skin or core layer.

Items that use the coinjection molding process include automotive parts, office machine housings, furniture components, and appliance housings.

10-4 REACTION INJECTION MOLDING

Reaction injection molding (RIM) is also known as liquid reaction molding or high-pressure impingement mixing.

It is a process in which several reactive chemical systems are mixed and forced into the molding cavity where the polymerization reaction occurs. Although most of the current RIM components are polyols and isocyanates, other modified polyurethanes, polyester, epoxies, and polyamide monomers are used.

The process involves impingement-atomized mixing of two or more liquids in a mixing chamber. This mixture is immediately injected into a closed mold, and a rigid, structural foamed or cellular product results (Fig. 10-18). (See Expanding Processes)

The automotive and furniture industries are the major users of RIM parts. Bumpers, belts, shock absorbing parts, fender components, and cabinet elements are familiar examples (Fig. 10-19.)

Reaction injection molding is limited only by mold and equipment size. Present machines are capable of molding 135 lbs. [300 kg] of mixture in one shot. Clamp capacity requirements are much lower than those of conventional injection molding. Clamps are often designed to be opened and closed like a book. This allows easy parts removal and operator access to the mold (Fig. 10-20). Seven advantages and four disadvantages of Reaction Injection Molding are listed on the following page.

10-5 REINFORCED REACTION INJECTION MOLDING (RRIM)

When short fibers or flakes (particulates) are used to produce a more isotropic product, the pro-

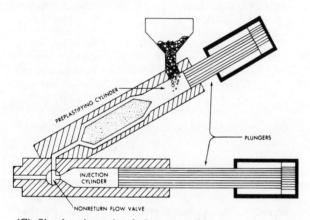

(C) Simple schematic of plunger machine.

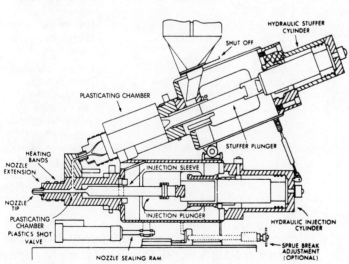

(D) Two-stage preplasticator on plunger machine.

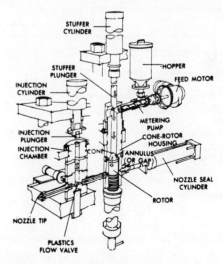

(E) Exploded view of rotary shear-cone preplasticator on plunger machine. (Borg-Warner Corp.)

Fig. 10-12. (continued)

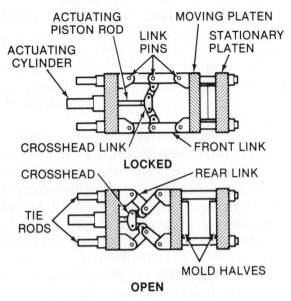

(A) Toggle clamp.

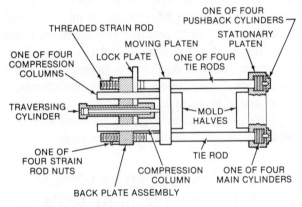

(B) Hydromechanical clamp.

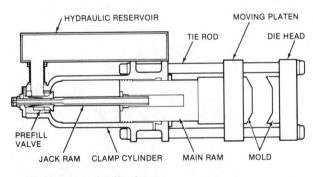

(C) Straight hydraulic clamp.

Fig. 10-13. Various types of mold clamps. (*Modern Plastics Encyclopedia*)

Table 10-2. Problems in Injection Molding

Difficulty	Cause	Possible Remedy
Black specks, spots, or streaks	Flaking off of burned plastics on cylinder walls	Purge heating cylinder
	Air trapped in mold causing burning	Vent mold properly
	Frictional burning of cold granules against cylinder walls	Use lubricated plastics
Bubbles	Moisture on granules	Dry granules before molding
Flashing	Material too hot	Reduce temperature
	Pressure too high	Lower pressure
	Poor parting line	Reface the parting line
	Insufficient clamp pressure	Increase clamp pressure
Poor finish	Mold too cold	Raise mold temperature
	Injection pressure too low	Raise injection pressure
	Water on mold face	Clean mold
	Excess mold lubricant	Clean mold
	Poor surface on mold	Polish mold
Short moldings	Cold material	Increase temperature
	Cold mold	Increase mold temperature
	Insufficient pressure	Increase pressure
	Small gates	Enlarge gates
	Entrapped air	Increase vent size
	Improper balance of plastics flow in multiple cavity molds	Correct runner system
Sink marks	Insufficient plastics in mold	Increase injection speed, check gate size
	Plastics too hot	Reduce cylinder temperature
	Injection pressure too low	Increase pressure
Warping	Part ejected too hot	Reduce plastics temperature
	Plastics too cold	Increase cylinder temperature
	Too much feed	Reduce feed
	Unbalanced gates	Change location or reduce gates
Surface marks	Cold material	Increase plastics temperature
	Cold mold	Increase mold temperature
	Slow injection	Increase injection speed
	Unbalanced flow in gates and runners	Rebalance gates or runners

cess is called *reinforced reaction injection molding (RRIM)*. Fiber loading increases monomer viscosities and abrasive wear on all flow surfaces.

Polyurethane/urea hybrid, epoxy, polyamide, polyurea, polyurethane/polyester hybrid, polydicyclopentadiene, and other resin systems have been used for RIM and RRIM. RRIM applications include automobile fenders, panels, bumpers, shields, radomes, appliance housings, and furniture components.

NON-RETURN CHECK VALVE   NON-RETURN CHECK VALVE

(A) Ring valve.   **(B)** Ball valve.

Fig. 10-14. Two check-valve assemblies used to control the flow of materials.

(A) Uppers of these shoes are polyurethane and vinyl, the heel traction bar is acetal polymer, and the swivel unit is injection-molded ionomer. (Du Pont)

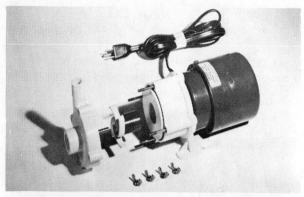

(B) Injection-molded, glass-fiber-reinforced polypropylene is used for many parts of this pump. (Fiberfil Division, Dart Industries)

Fig. 10-15. Injection molding is used for a wide variety of products.

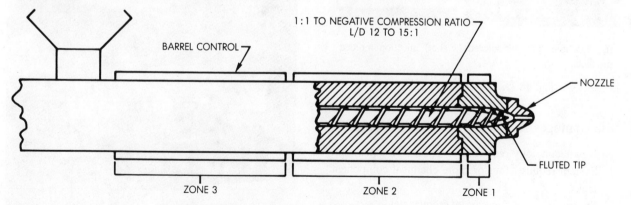

Fig. 10-16. Schematic of basic thermoset screw barrel assembly.

Advantages of Reaction Injection Molding (RIM)

1. Cellular core and integral skin for durable products
2. Fast cycle times for large products
3. Good finishes that are paintable
4. Less cost than castings
5. Polymers may be reinforced
6. Reduced tooling and energy cost (compared with injection molding)
7. Lower equipment cost, due to low pressures

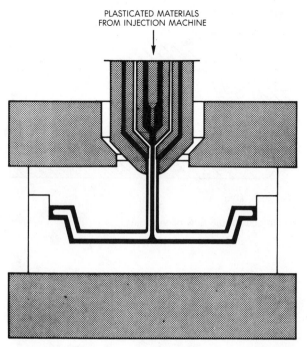

PLASTICATED MATERIALS
FROM INJECTION MACHINE

(A) Three separate channels feed this coinjection molding machine.

(B) Examples of multicolored (2 shot) injection molded products.

Fig. 10-17. Coinjection molding.

Disadvantages of Reaction Injection Molding

1. New technology requiring investment in equipment
2. System requires four or more chemical component tanks
3. System requires handling of isocyanates
4. Releasing agents required

10-6 LIQUID RESIN MOLDING

Liquid resin molding is a term used to describe products produced by a variety of low-pressure methods in which mixing is often mechanical

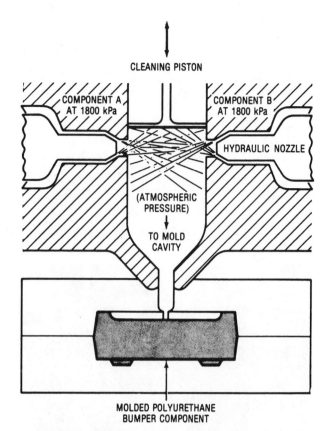

CLEANING PISTON

COMPONENT A AT 1800 kPa

COMPONENT B AT 1800 kPa

HYDRAULIC NOZZLE

(ATMOSPHERIC PRESSURE)

TO MOLD CAVITY

MOLDED POLYURETHANE BUMPER COMPONENT

Fig. 10-18. Reaction injection molding (RIM), showing impingement mixing. Components are atomized to a fine spray by a pressure drop from 2 500 psi [1 800 kPa] to atmospheric pressure.

Fig. 10-19. RIM equipment producing automobile components. (General Motors)

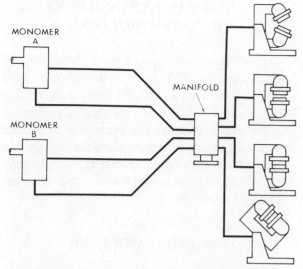

(A) The C-frame stations (clamps) open in book-fashion for fast part removal and rock back for venting.

(B) RIM molding machine with mold open.

Fig. 10-20. Molding machine system for RIM. (Cincinnati Milacron)

rather than by impingement. The term once described a very specialized process for potting and encapsulating components. LRM is used to describe a group of processing methods including resin transfer molding (RTM), vacuum injection molding (VIM), and thermal expansion resin transfer molding (TERTM) in which resins are forced under low pressure into the molding cavity and rapidly cured. Epoxies, silicones, polyesters, and polyurethanes lend themselves to use in liquid resin molding processes.

10-7 RESIN TRANSFER MOLDING

Resin transfer molding, also called resin injection molding, is a process whereby catalyzed

resin is forced into a mold in which fragile parts or reinforcements have been placed. Low pressure does not distort or move the desired fiber orientation of preforms or other materials (Fig. 10-21) Boat hulls, hatches, computer housings, fan shrouds, or other large composite structures may be pro-

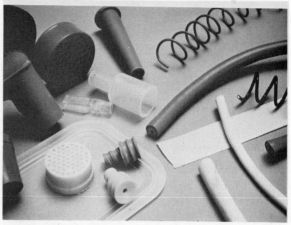

(A) RTM is recommended for such parts as electrical connector inserts, diaphragms, valves, O-rings, stoppers, and plungers as shown above. This molding technique does not distort windings or shift delicate devices in the molding cycle. (Plastics Design Forum)

(B) RTM bathtub with glossy gelcoat finish. (Molded Fiber Glass Co.)

Fig. 10-21. RTM products.

duced by this technique. The basic concept of RTM is illustrated in Figure 10-22.

Nine major advantages of RTM include:

1. Eliminates the plasticizing stage necessary with dry compounds
2. Permits encapsulation of delicate, or fragile parts
3. No hand mixing
4. Eliminates preheating and preforming
5. Lower pressures are needed
6. Minimum material waste
7. Rapid cure of resins at low temperatures
8. Improved reliability and dimensional stability
9. Reduction of material handling

10-8 VACUUM INJECTION MOLDING

In a process similar to RTM, preforms are placed on a male mold and the female mold is closed. A vacuum is drawn pulling the reactive resin system into the mold cavity. This technique is illustrated in Figure 10-23.

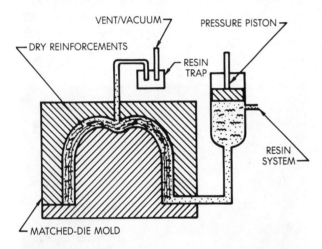

Fig. 10-22. Concept of RTM illustrated as a resin system being forced around the preform reinforcements placed in the matched-die mold.

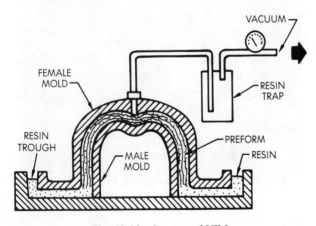

Fig. 10-23. Concept of VIM.

10-9 THERMAL EXPANSION RESIN TRANSFER MOLDING

TERTM is a variation of the RTM process. A cellular mandrel of PVC or PU is wound or wrapped with reinforcements and placed into a matched die. Epoxy or other resin systems are injected to impregnate the reinforcements. The heated die causes the cellular material to further expand, forcing the impregnated reinforcements against the mold walls. The tooling is vented to allow excess matrix or entrapped air to escape.

10-10 COLD MOLDING

Cold molding of plastics materials began in the United States in about 1908. The process was adapted from the ceramics industry. Ancient people used this basic process to form and shape clay with hand pressure. In Biblical times, ornamental clay parts were compressed in wooden male and female molds. The formed parts were removed from the molds and baked in kilns (ovens).

The process of cold molding is like compression molding, except the molds are not heated. Molding is done in conventional presses. From 14 to 2 030 to 12 039 psi [83 MPa] are used to rapidly press the molding compound into a solid form, something like a preform. After being formed in the unheated molds, the part is hardened into an infusible mass in an oven (Fig. 10-24)

Cold molding materials may be divided into two distinct groups according to the ingredients used:

1. Nonrefractory (organic)
2. Refractory (inorganic)

Compounds for organic or nonrefractory cold molded parts may include plastics (mostly phenolics), raw linseed oil, asbestos, various bitumens, sulphur, or other additives. Cold-molded products—both organic and inorganic types—as a rule lack luster. They are produced in dark colors. To help improve the surface finish of organic molded parts, molds may be heated briefly during the rapid compression cycle. The actual cure of the compound is done in ovens. Electrical insulator parts, utensil handles, battery boxes, and valve wheels are common uses.

Inorganic or refractory cold molding compounds may contain formulations of cement, asbestos fibers, clay, and lime (silica) in a water blend.

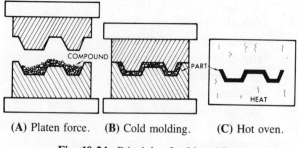

(A) Platen force.   (B) Cold molding.   (C) Hot oven.

Fig. 10-24. Principle of cold molding.

Once the materials are molded, they are placed in ovens with a steam atmosphere. The hot steam cures the cement and hardens the molded piece. These products are used where extremely high temperatures exist. They are not plastics products, but are included as examples of cold molding.

Cold molding has the five following advantages:

1. Low-cost materials may be used
2. Cooling and equipment costs are low
3. Electrical properties are good
4. Parts are cured in a baking oven and not in the mold
5. Production rates are high (especially with small, thin-walled parts)

10-11 SINTERING

Sintering is the process of compressing a powdered plastics in a mold at temperatures just below its melting point for about one-half hour (Fig. 10-25). The powdered particles are fused (sintered) together, but the mass as a whole does not melt. Bonding is done by the exchange of atoms between the individual particles. After the fusion process, the material may be postformed under heat and pressure to the needed dimensions.

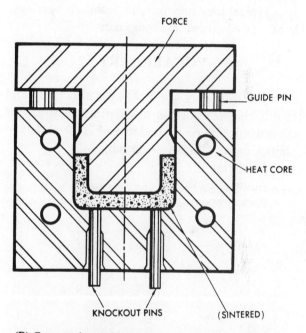

(B) Compression and heating.

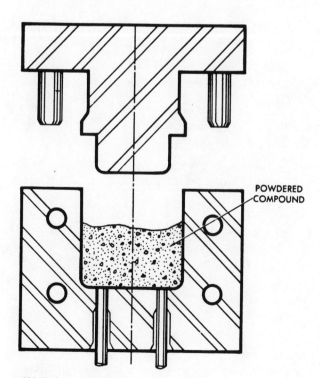

(A) Before compression.

(C) PTFE parts have been sintered to basic shape before machining. (Chempiast, Inc.)

Fig. 10-25. Sintering plastics parts.

The three major variables governing the sintering process are temperature, time, and the composition of the plastics.

The process is adapted from the sintering operations of powder metallurgy. Sintering can be used to process polytetrafluoroethylene, polyamides, and other specially filled plastics and is the primary method by which polytetrafluoroethylene is processed. Dense parts with first rate electrical and mechanical properties may be produced. The cost of tools and production is high, and parts with thin walls or variations in cross-sectional thickness are hard to form.

Typical uses may include bushings, bearings, hubs, and electrical insulating parts.

VOCABULARY

The following vocabularay words are found in this chapter. Look up the definition of any of these words you no not understand as they apply to plastics in the glossary, Appendix A.

Annealing
Automatic cycle
Back pressure
Banbury
Barrel
Barrel vent
Breathing
Cold molding
Compression mold
Compression molding
Dwell
Female mold
Flash
Hopper dryer
Impingement
Injection blow molding
Injection molding
LRM
Plasticate
Platens
Purging
RIM
RRIM
RTM
Sintering
TERTM
Transfer molding
VIM

10-1. The excess material left on a product after compression molding is called ___?___ .

10-2. The amount of material used to fill a mold during an injection molding process is called ___?___ .

10-3. The chief advantage of injection molding over other molding processes is ___?___ .

10-4. In ___?___ injection machines, the material is forced around a torpedo.

10-5. In ___?___ injection machines, the material is rapidly plasticated and compounded before being forced into the mold.

10-6. Name three ways to reduce production time in compression molding.

10-7. Name three ways to reduce the cycle time in compression molding.

10-8. Opening the compression mold to allow gasses to escape during the molding cycle is called ___?___ .

10-9. The mark on a molding where the halves of the mold meet in closing is called ___?___ .

10-10. Compression molds are usually made of ___?___ .

10-11. The indented half of a mold designed to receive the male half is called the ___?___ mold.

10-12. The process of molding or forming items from pressed powders at a temperature just below the plastics' melting point is called ___?___ .

10-13. A well-known compounding machine with contrarotating rotors which mix and blend the material is called ___?___ .

10-14. The major difference between matched-die and compression molding is that lower ___?___ and ___?___ are needed for matched-die processing.

10-15. The ___?___ is made smaller than the runner so that parts are more easily removed.

10-16. Both ___?___ plate and ___?___ pack help create back pressure pulsations from the screw of a screw injection machine.

10-17. Name three major disadvantages of transfer molding.

10-18. The time it takes to close a mold, form a part, open the mold and remove the cooled part is called the ___?___ time, in injection molding.

10-19. In ___?___ molding, the material is loaded in a chamber outside the mold cavity before it is made fluid and forced into the mold cavity.

10-20. The two factors that rate the capacity of an injection molding machine are ___?___ and ___?___ .

10-21. Thermosetting materials with fragile, intricate shapes, and with inserts or pins, may be ___?___ or ___?___ molded.

10-22. Name the two molding processes that use thermosetting and selected thermoplastic materials in preform or bulk molding compound forms.

10-23. Name three processes that are similar to transfer molding processes.

10-24. Name four commonly used polymers in liquid resin molding.

10-25. A process of holding a material at an elevated temperature below its melting point to permit stress relaxation without distortion of part shape is called ___?___ .

10-26. Which molding process would you select to make each of the following products?
tail-light lens
office machine housings
small radio cabinet
ash tray
car fender and bumper

10-27. A common expression for the time during which a molding machine is not operating is ___?___ .

10-28. Reaction injection molding is also known as ___?___ or high-pressure impingement mixing.

Chapter 11

Extrusion Processes

The word extrusion is derived from the Latin word **extrudere** meaning (ex) out and (trudere) to push. In this chapter we will discuss typical uses of the extruder. The process of extrusion is continuous. It forms thermoplastic materials into three main products:

1. profile shapes (rods, fibers, tubes, etc.)
2. films and sheets
3. covering on wire and cables.

Extruders are also used in compounding materials and in the compression, transfer, and injection-molding processes. Blow molding and calendering will also be included. The extruder is used to produce the parison for blow molding, and the rotating rolls of the calender act as an extruder, melting and forcing the melt forward.

11-1 EXTRUSION

Extrusion is like the process used for making wire and other metallic profile shapes. The process is something like operation of a sausage-stuffing machine or a hand-operated cake decorator.

Dry powder, granular, or heavily reinforced plastics is heated and forced through an orifice in a die. The heart of the process is the extruder. It plasticates (melts and mixes) the material and forces it through the die. Screw extruders are the most common, but ram or plunger types have special uses.

Gutta percha, rubber, and shellac were extruded as early as 1845 by ram machines. Screw machines began to appear in Germany and the United States in the early 1930s. Today, the extruder is probably the most widely used plastics process machine. It is used for plasticizing units in other processing techniques. Injection molding is the most widely used molding process. Extrusion is common because extrusion dies are fairly simple and low in cost, co-extrusion dies are costly. (See Co-extrusion.) Large amounts of material may be forced through these dies in a continuous form.

Screw extruders are shown in Figures 11-1 and 11-2. The speed and shape of the close-fitted screw dictate the output, milling rate, and die pressure of the extruder. The polymer is compacted, heated, degassed, compressed, and plasticated by the action of the screw. Screws are characterized by their L/D ratios. A 20:1 screw could be 1.96 in. [50 mm] in diameter and 39.37 in. [1 000 mm] long. Screws of 16:1 and 40:1 are also used. Some screw designs are shown in Figure 11-3.

The channel depth of the screw decreases beginning at the transition section. This continuous reduction forces out air and compacts the material (Fig. 11-3A). A breaker plate acts as a mechanical seal between the barrel and die and it also holds the screen pack in place. Screen packs are used to filter out pieces of foreign material and create back pressure on the molten plastics. As screens become plugged, back pressure increases (Fig. 11-4). Sometimes external screen packs are mounted between the head clamp and die. These screens are more easily changed.

Some machines are equipped with valves to control back pressure and replace screens. Most are equipped with a continuous screen changer. This continuous ribbon of screen (sometimes rotary) can be automatically controlled to maintain a steady head pressure despite varying levels of contamination in the polymer or other flow rate conditions.

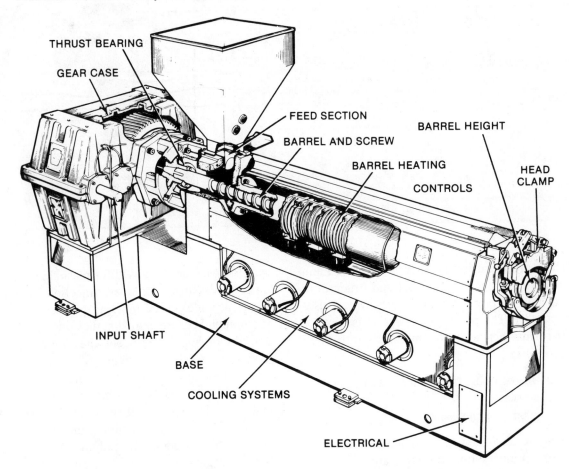

Fig. 11-1. Extruder, with parts labeled. (Davis-Standard Division, Crompton & Knowles Corp.)

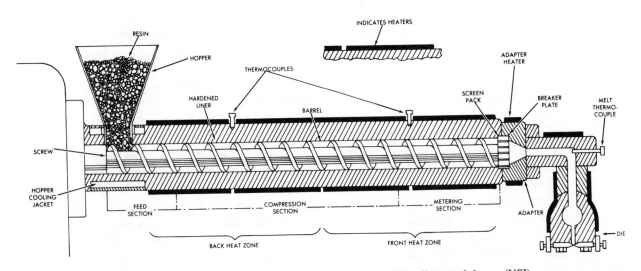

Fig. 11-2. Cross-section of a typical screw extruder, with the die turned down. (USI)

With thermosetting materials, it is imperative that the die head be easily removed to fully purge any unwanted or semicured materials. Barrels are made of high-strength steel with hardened bores. When using reinforcements, screws and barrels must be specially coated or hardened for added wear resistance. Most wear occurs at the tip and

metering section. Barrel liner inserts are used to replace worn bores.

Barrels may be vented to rid the molten polymer of unwanted vapors.

Electric heaters are used around the barrel to help melt the plastics. Once the extruder is mixing, blending, and forcing the material through the

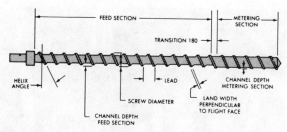

(A) Metering screw. (*Processing of Thermoplastic Materials*)

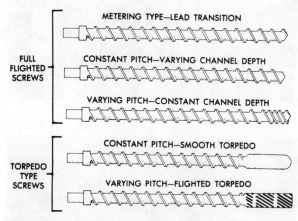

(B) Common extruder screws. (*Processing of Thermoplastic Materials*)

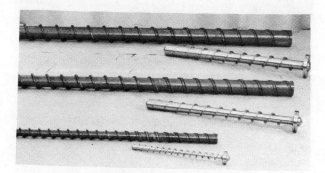

(C) Side view of extruder screws. (Cameron-Waldron Division, Midland-Ross Corp.)

(D) End view of extruder screws. (Cameron-Waldron Division, Midland-Ross Corp.)

Fig. 11-3. Common extruder screws. (Cameron-Waldron Division, Midland-Ross Corp.)

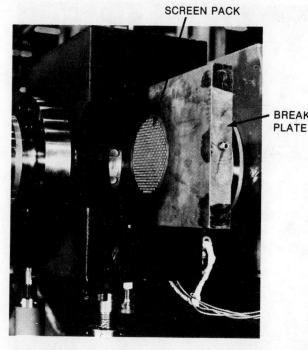

Fig. 11-4. The screen pack creates back-pressure on material in the extruder barrel. (Cameron-Waldron Division, Midland Ross Corp.)

die, frictional heat produced by the action of the screw may be enough to partly plasticate the material. External heaters are used to maintain a fixed temperature once the process is started. Some materials require special screw designs for processing. Some machines have twin, parallel screws for greater capacity and better mixing (Fig. 11-5).

Extruders are sold by the barrel (inside diameter) and the amount of material they will plasticate per minute or hour. Extruder capacity with low-density polyethylene may vary from less than 4.5 lb [2 kg] to more than 11 000 lb [5 000 kg] per hour. Screw diameters vary from 1 in. [25 mm] to more than 32 in. [800 mm]. Multiple-screw extruders are popular for heat-sensitive materials.

In most extruder processes, the plastics go through the following five steps:

1. *The extruder* — the material is plasticated and forced out through the die or orifice
2. *The die* — the hot molten or soft plastics takes shape
3. *Forming* — the hot material is further stretched or shaped
4. *Post-forming* — the material is trimmed, cut, or further shaped
5. *Secondary processing* — the material is further cut or fabricated into other shapes or becomes part of other assemblies

Fig. 11-5. Twin parallel extruder screws. (North American Bitruder Co.)

(A) Die and cutter inside pelletizing head.

(B) Hot pelletizing from extruder (center).

Fig. 11-6. Extrusion of material through dies. (North American Bitruder Co.)

Polyvinyl chloride tends to decompose or degrade if subjected to prolonged exposure to high temperatures and the mechanical actions of the extruder screw. Each polymer or polymer modification may need a different or special screw design. With new advances in polymer development, improved technology in extruder design is needed as well.

Extruders are also used to compound and blend basic plastics with plasticizers, fillers, colorants, and other ingredients. Granular molding materials are made by extruding these materials through dies. The spaghetti-like extrusion is chilled, then chopped into the granular molding forms that are used widely in extruders and injection molding machines (Fig. 11-6).

The die actually forms the molten plastics as it comes out of the extruder. Dies may be made of mild steel but should be of chromium-molybdenum steel for long runs. Stainless alloys are used with corrosive materials. Die design and construction is a very broad and complex topic. (See Chapters 20 and 21).

To produce exact cross-sectional dimensions after the extrudate has cooled, allowances must be made in the orifice design. In complex cross sections where thin sections or sharp edges are formed, cooling occurs more quickly at such portions. These areas shrink first, therefore they are smaller than the rest of the section. This means that the die and the shape of the plastics (extrudate) may be different. To correct this problem, the orifice in the die is made larger at these points (Fig. 11-7).

Extruder products may be divided into six basic areas of use:

1. Rod and profiles
2. Pipe
3. Film and sheet
4. Monofilaments
5. Extrusion coatings
6. Wire and cable covering

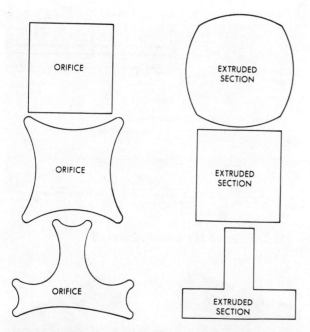

Fig. 11-7. Relationships between die orifices and extruded sections.

(A) Note the fingers that help shape the material, and water jets that cool the extruded shape. (Alma Plastics Co.)

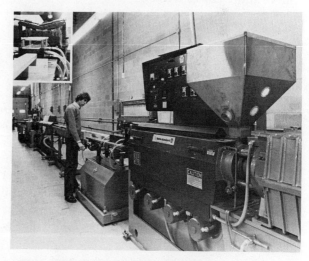

(B) Forming dies help shape the extrudate.

Fig. 11-8. Extrusion production aids.

Profile Extrusion

Production of rod shapes from a round orifice in a die is probably the simplest of the extrusion processes. Many extrusions are produced horizontally through dies and cooled by air jets, water troughs, or cooling sleeves. (Fig. 11-8) Many profile shapes are produced for a variety of uses. Vinyl siding and tracks or channels for sliding doors and windows are well-known examples.

Extrudates are sometimes postformed into different shapes by the use of sizing plates, shoes, or rollers. A flat-tape shape may be postformed into a corrugated form. Round rods may be postformed into oval or other new shapes while the extrudate is still hot (Fig. 11-9).

Pipe Extrusion

Pipes (tubular forms) are shaped by the exterior dimensions of the orifice and by the mandrel (sometimes called a *pin*), which shapes inside dimensions (Fig. 11-10). The mandrel is held in place by thin pieces of metal called spiders.

The diameter of the pipe or tube is also controlled by the tension of the take-up mechanism. If the tube is pulled faster than speed of the extrudate melt, the product will be smaller and thinner than the die.

To prevent the tube from collapsing before cooling, it is pinched shut on the end and air is forced in through the die. This air pressure ex-

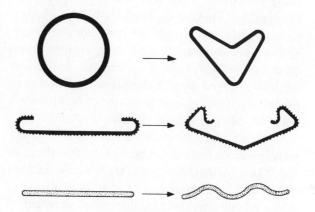

Fig. 11-9. The extrudates at left are postformed into the shapes at right by being passed through rollers.

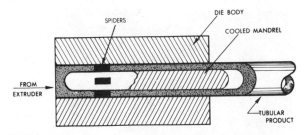

Fig. 11-10. In this pipe-forming operation, hot material is extruded around a cold mandrel or pin. (Du Pont Co.)

pands the pipe slightly. The hot tube may be pulled through a sizing ring or vacuum ring to hold the outside diameter to a close tolerance. The thickness of the pipe wall is controlled by the mandrel and die size.

Sheet and Film Extrusion

There are two basic methods of producing film:

1. Slot or cast film extrusion
2. Blow (lay-flat or tubing) film extrusion

Both sheet and film forms are produced by extruding molten thermoplastic materials through dies with a long horizontal slot.

In both *T* and *coat hanger-shaped* die construction, the molten material is fed to the center of the die. It is then formed by the die lands and adjustable jaw (Fig. 11-11). The width may be controlled by the external deckle bars or by the actual die width.

In Figure 11-12A an adjustable choke or restrictor bar is used in the extrusion of sheeting. Sheeting dies are, as a rule, more heavily constructed and possess longer die lands than film dies.

In Figure 11-13A the sheet is extruded into a quench tank, while in Figure 11-13B the sheet is being drawn over chill rollers in a process sometimes called film casting. Both chill-roller and water-tank sheet extrusions are used commercially. Temperature, vibration, and water currents must be controlled with care when using the water-tank method to allow clear, defect-free sheeting to be produced. Heavy-gauge films are currently being made by this high-speed method.

With chill-roller methods there is a great improvement in optical properties of the sheet of film. This method also lends itself to the making of stiffer film extrudate. The slots of both extrusion methods provide close thickness tolerances of the film and also orient the molecular structure in one direction.

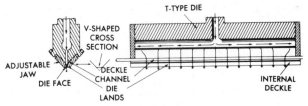

(A) Cross section of T-shaped die.

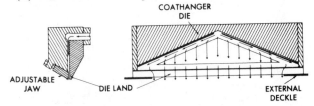

(B) Cross section of coathanger-shaped die.

(C) T-shaped sheet dies. (Cameron-Waldron Division, Midland-Ross Corp.)

Fig. 11-11. Two types of extrusion dies.

Table 11-1 gives some common sheet-extrusion problems and their remedies.

Several resins may be extruded at one time into a multimanifold die to produce a sheet or film composed of different resins. (See Composites and Laminating)

Films may be cold extruded (sometimes called process casting) in solution form on polished conveyors (Fig. 11-14). The solvents are removed by evaporation and collected for reuse. Polyvinyls, polycarbonates, acetates, and polysulfones are cold-extruded plastics. (See Chapter 11, Casting Processes)

11-2 BLOWN-FILM EXTRUSION

Blown or tubular-extruded film is distinctly different from slot-extruded film. Most large or less costly films are produced by the blown-film extrusion method (Fig. 11-15).

In blown-film extrusion, the film is produced by forcing molten material through a die and around a mandrel. It emerges from the orifice in tube form (Fig. 11-16). This process is like that

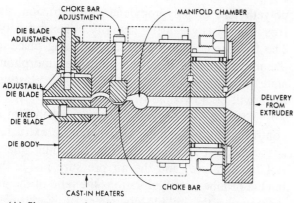

(A) Sheet extrusion die (Phillips Petroleum Co.)

Fig. 11-12. Dies with adjustable choke bars.

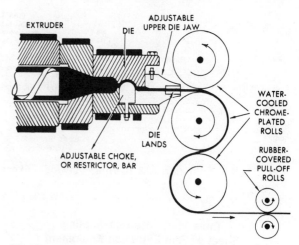

(B) Sheeting die with takeoff unit. (USI)

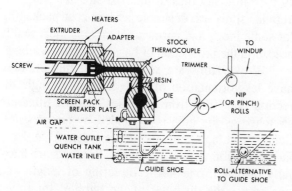

(A) Cross section of the front part of a flat-film extruder, the quench tank, and takeoff equipment. (USI)

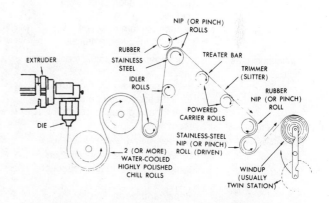

(B) Schematic of chill-roll film extrusion equipment.

(C) Sheet extrusion from die into nip of chill rollers. (North American Bitruder Co.)

Fig. 11-13. Examples of sheet and film extrusion.

(D) Sheet extrusion, looking toward extruder. Finished sheet will be sheared to desired lengths. (North American Bitruder Co.)

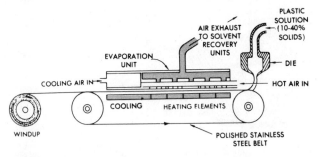

Fig. 11-14. Cold extrusion of film, showing solvent recovery.

Table 11-1.   Troubleshooting Sheet or Film Extrusion Equipment

Defect	Possible Remedy
Continuous lines in direction of extrusion	Repair or clean out die Die contamination or polish of rollers scored Reduce die temperatures Use properly dried materials
Continuous lines across sheet	Jerky operation—adjust tension on sheet Reduce polishing roll temperatures or increase roll temperatures Check back-pressure gauge for surging
Discoloration	Use proper die and screw design Minimize material contamination Temperature too high—too much regrind Repair and clean out die
Dimensional variation across sheet	Adjust bead at polishing rolls Balance die heats Reduce polishing roll temperatures Check temperature controllers Repair or clean out die
Voids in sheet	Balance extruder line conditions Use proper screw design Minimize material contamination Reduce stock temperature
Dull strip	Die set too narrow at this point Minimize material contamination Increase die temperature Repair or clean out die
Pits, craters	Balance extruder line conditions Minimize material contamination Use properly dried materials Control stock feed Reduce stock temperature

used to make pipes or tubes. This tube, or bubble, is expanded by blowing air through the center of the mandrel until the desired film thickness is reached. This process is something like blowing up a balloon. The tube is usually cooled by air from a cooling ring around the die. The *frost line* is the zone where the temperature of the tube has fallen below the softening point of the plastics. In polyethylene or polypropylene film extrusion, the frost zone is evident; it actually appears frosty. The frost zone shows the change taking place as the plastics cools from the melt (an amorphous state) to a crystalline state. With some plastics, there will be no visible frost line.

The size and thickness of the finished film is controlled by several factors including extrusion speed, takeoff speed, die (orifice) opening, material temperature, and the air pressure inside the bubble or tube. Blow-up ratio is the ratio of the die diameter to the bubble diameter. Blow-extruded film is sold as seamless tubing, as flat film, or as film folded in a number of ways. Film producers may slit the tubing on one edge during windup. If the tube is blown to a diameter of 6.5 ft [2 m] the flat film will have a width (slit and opened) of over 19 ft [6 m]. Slot dies of this size are not practical. Tubular films are desirable as low-cost packaging for some foods and garments. Only one heat seal is needed in the production of bags from blown tubing.

Blown films are semioriented; that is, they have less orientation of molecules in a single direction than film from slot dies. Blown films are stretched as the tube is expanded by air pressure. Such stretching results in a more balanced molecular orientation in two directions. Products are biaxially oriented; one in the direction of length and one across the diameter of the bubble. Improved physical properties are an asset of blown film. However, clarity, surface defects, and film thicknesses are harder to regulate than with slot extrusion. Table 11-2 gives troubleshooting information for blown extruded film.

11-3 FILAMENT EXTRUSION

Monofilaments are produced much as are profile shapes, except that a multiorifice die is used. These dies contain many small openings from which the molten material emerges. Such dies are used to produce granular pellets, monofilaments, and multifilament strands.

Filament shapes are made by forcing plastics through small orifices in a process referred to as spinning. The plastics is shaped by the opening in the die or spinneret. This process may have been named "spinning" from the method of spinning natural fibers. The small opening under the jaw of the silkworm is also called the "spinneret."

Spinnerets are often made of such metals as platinum, which will resist acids and orifice wear,

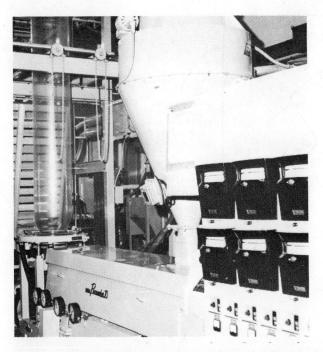

(A) Extruder, with blown film being taken off. (Chemplex Co.)

(C) Closeup view, looking down at extruder die and gauge bars. (Chemplex Co.)

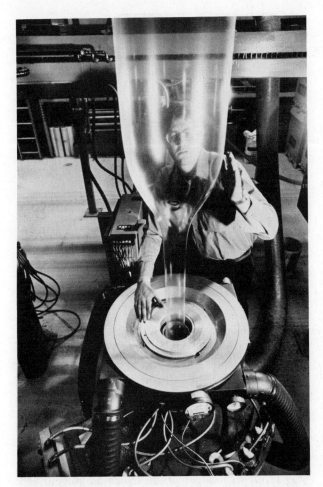

(B) Closeup view of extruded blown film. (Chemplex Co.)

Fig. 11-15. Blown film extrusion.

(D) Note size of blow-extruded film after blowings. (BASF)

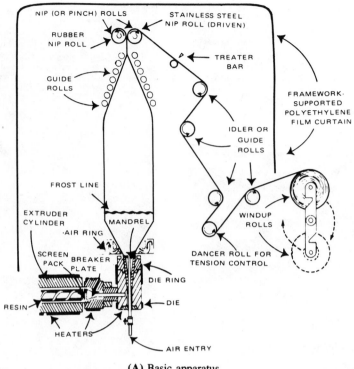

(A) Basic apparatus.

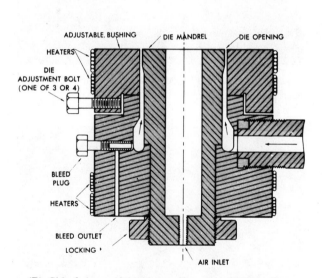

(B) Side-fed manifold blown film die. (Phillips Petroleum Co.)

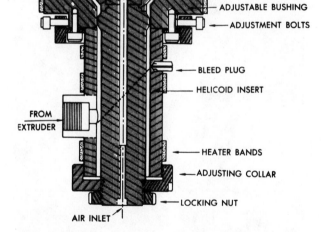

(C) Adjustable-opening blown film die. (Phillips Petroleum Co.)

Fig. 11-16. Schematic drawings of blown-film extrusion procedures. (U.S. Industrial Chemicals)

because these orifices are often much finer than the diameter of human hair. To be forced or extruded out of these small openings, the plastics must be made fluid.

Acrylic fiber is produced in the spinning process shown in Figure 11-17. A thick chemical solution is extruded into a coagulation bath through the tiny holes of the spinneret. In the bath, the solution coagulates (becomes a solid) and becomes Acrilan acrylic fiber. The fiber is washed, dried, crimped,

cut into staple lengths, and baled for shipment to textile mills where it is converted into carpeting, wearing apparel, and many other products.

There are three basic methods of spinning fibers:

1. In *melt spinning,* plastics such as polyethylene, polypropylene, polyvinyl, polyamide, or thermoplastic polyesters are melted and forced out the spin-

Table 11-2. Troubleshooting Blown Film

Defect	Possible Remedy
Black specks in film	Clean die and extruder Change screen pack Check resin for contamination
Die lines in film	Lower die pressure Increase melt temperature Polish all rough edges in film path Check nip rolls—make smooth
Bubble bounces	Increase screw rpm and nip roll speed Enclose tower or stop drafts Adjust cooling ring to obtain constant air velocity around ring
Poor optical and physical properties	Raise melt temperature Increase blow-up ratio Increase frost line height Clean die lips, extruder, and rollers
Failures at fold	Decrease nip roll pressure
Failure at weld lines	If possible, bleed die at weld line Heat die spiders—insulate air lines there Increase melt temperature Check for contamination
Film won't run continuously	Clean die and extruder Lower melt temperature Increase film thickness

neret. As the filaments hit the air, they solidify and are passed through other conditioners (Fig. 11-18A).

2. In *solvent spinning,* plastics such as acrylics, cellulose acetate, and polyvinyl chloride are dissolved by certain solvents. The solution is forced through the spinneret (Fig. 11-18B), and the filament then passes through a stream of hot air. The air aids in evaporating the solvents from the slender fibers. For economy, these solvents must be recovered for use again.

3. The first step of *wet spinning* (Fig. 11-18C) is like solvent spinning. The plastics is dissolved in chemical solvents. This fluid solution is forced out through the spinneret into a coagulating bath that makes the plastics gel into a solid filament form. Some members of the cellulosic, acrylic, and polyvinyl plastics families may be processed by wet spinning.

All three processes begin by forcing fluid plastics through a spinneret. They end by solidifying the filament through cooling, evaporation, or coagulation (Table 11-3).

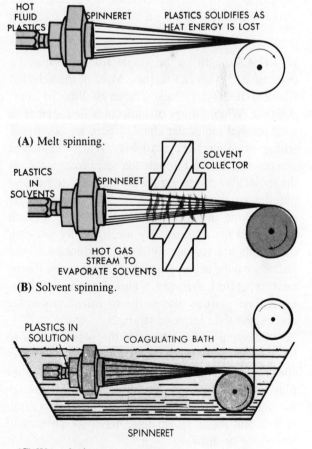

(A) Melt spinning.

(B) Solvent spinning.

(C) Wet spinning.

Fig. 11-18. Three basic methods of spinning plastics fibers.

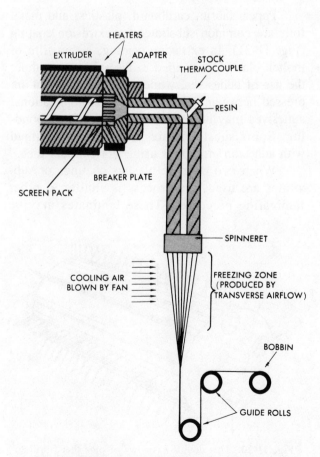

Fig. 11-17. Production of acrylic fiber by spinning.

Table 11-3. Selected Fibers and Production Processes

Fibers	Extrusion Spinning Process
Acrylics and modacrylics	
Acrilan	Wet
Creslan	Wet
Dynel (vinyl-acrylic)	Solvent
Orlon	Solvent
Verel	Solvent
Cellulose esters	
Acetate (Acele, Estron)	Solvent
Triacetate (Arnel)	Solvent
Cellulose, regenerated	
Rayon (viscose, cuprammonium)	Wet
Olefins	
Polyethylene	Melt
Polypropylene (Avisun, Herculon)	Melt
Polyamides	
Nylon 6,6, Nylon 6, Qiana	Melt
Polyesters	
Dacron, Trevira, Kodel, Fortrel	Melt
Polyurethanes	
Glospan	Wet
Lycra	Solvent
Numa	Wet
Vinyls and vinylidines	
Saran	Melt
Vinyon N	Solvent

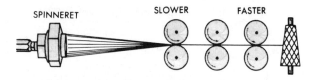

Fig. 11-19. Drawing plastics filaments.

High-bulk fibers and yarns may be produced from films. In this process, a composite film that has been coextruded or laminated is used. It is mechanically drawn and cut (fibrillated) into fine strands. The fibrillation is done during the stretching or drawing process as the film passes between serrating rollers rotating at different speeds. The teeth of these rollers cut the film into fibrous form (Fig. 11-20). The composite film develops internal stresses as it is extruded, drawn and fibrillated. This unequal stress orientation in the film layers makes the fibers curl and exhibit properties much like those of natural fibers.

11-4 EXTRUSION COATING

Paper, fabric, cardboard, plastics, and metal foils are common substrates for extrusion coating (Fig. 11-21). In extrusion coating, a thin film of molten plastics is applied to the substrate without the use of adhesives, while substrate and film are pressed between rollers. For special applications, adhesives may be needed to ensure proper bonding. Some substrates are preheated and primed with adhesion promoters using slot extruder dies.

When two or more plastics coatings or substrates are used, the process is similar to other laminating processes. These laminates may be

The strength of the single filaments may be determined by several factors. Most of the selected filament fibers are linear and crystalline in composition. When groups of molecules lie together in long parallel molecular chains, there are additional strong bonding sites available. When the liquid plastics are forced through the spinneret, many of the molecular chains are forced closer together and parallel to the filament axis. This packing, arranging, and drawing provides increased strength throughout the filament. By mechanically working the filament, further molecular orientation and packing can be accomplished. This mechanical process is called *drawing*. The drawing of non-crystalline plastics also helps to orient molecular chains and thus improve strength.

Drawing or stretching of the plastics is done by running the filaments through a series of variable-speed rollers (Fig. 11-19). The drawing of crystalline plastics is continuous. As the fiber goes through the drawing process, each roller is rotated at a faster speed, and roller speed determines the amount of stretching or drawing.

Monofilaments range from 0.0047 to 0.0059 in. [0.12 mm to 1.50 mm] in diameter. They may be handled individually or by takeoff machines.

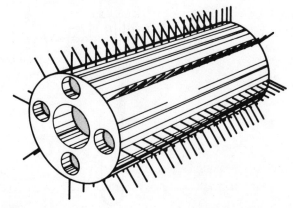

Fig. 11-20. This device has rows of pins that fibrillate yarn of very fine denier.

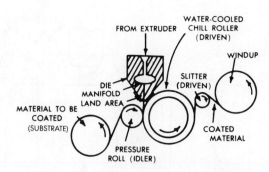

Fig. 11-21. Extrusion coating of substrates.

composed of several layers of plastics or layers of other materials. (See Coating Processes)

Wire and Cable Extrusion Covering

Extrusion coating of wire and cable is shown in Figure 11-22. During this process a molten plastics is forced around the wire or cable as it passes through the die. The die actually controls and forms the coating on the wire. Wires and cables are usually heated before coating to remove moisture and ensure adhesion. As the coated wire emerges from the crosshead die, it is cooled in a water bath. Two or more wires may be coated at

one time. Television and appliance cords are common examples. Wooden strips, cotton rope, and plastics filaments may be coated by this process. (See Coating Processes)

11-5 BLOW MOLDING

This process is sometimes listed as a molding technique because force is used to press the hot, soft tubular material against the mold walls.

Blow molding is a technique adopted and modified from the glass industry for making one-piece containers and other articles. The process has been used for centuries for making glass bottles. Blow molding of thermoplastics did not develop until the late 1950s. In 1880, blow molding was accomplished by heating and clamping two sheets of celluloid in a mold. Air was then forced in to form a blow-molded baby rattle. This may have been the first blow-molded thermoplastic article produced in the United States.

The basic principle of blow-molding is simple (Fig. 11-23). A hollow tube (Parison) of molten thermoplastic is placed in a female mold and the mold closed. It is then forced (blown) by air pres-

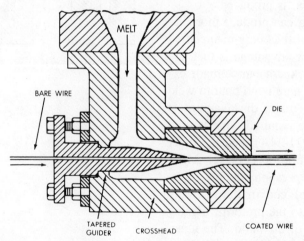

(A) A crosshead holds the wire-coating die and the tapered guide as the soft plastics flows around the moving wire.

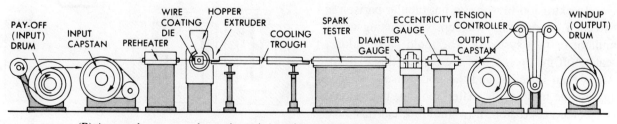

(B) A general component layout in a wire-coating extrusion plant. (U. S. Industrial Chemicals Co.)

Fig. 11-22. Extrusion coating of wire and cable.

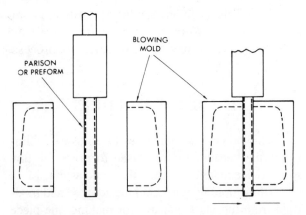

(A) Molded hollow tube (parison) is placed between mold halves, which then close.

(B) The still-molten parison is pinched off and inflated by an air blast. The blast forces the plastics against the cold walls of the mold. Once the product has cooled, the mold opens and the object is ejected.

Fig. 11-23. Blow molding sequence. (USI)

sure against the walls of the mold. After a cooling cycle, the mold opens and ejects the finished product. The process is used to produce many containers, toys, packaging units, automobile parts, and appliance housings.

There are two basic blow molding methods:

1. Injection blowing
2. Extrusion blowing

The major difference is in the way that the hot, hollow tube, or *parison,* is produced.

Injection blow molding can produce more accurately the desired material thickness in specified areas of the part. The major advantage is that any shape with varying wall thickness can be made exactly the same each time. There is no bottom weld or scrap to reprocess. The major disadvantage is that two different molds are required. One is used to mold the preform (Fig. 11-24A), and the other for the actual blowing operation (Fig. 11-24B). During the blow-molding operation, the hot injection-molded preform is placed in the blowing mold. Air is then forced into the preform, making it expand against the walls of the mold. The injection-blow process has been called *transfer blow* because the injected preform must be transferred to the blowing mold (Fig. 11-25).

In extrusion blowing, a hot tubular parison is extruded continuously (except when using accumulator or ram systems). The mold halves close, sealing off the open end of the parison (Fig. 11-26). Air is then injected, and the hot parison expands against the mold walls. After cooling, the product is ejected. Extrusion blow molding can produce articles large enough to hold 2 646 gal. [10 000 L] of water. Preforms of this size are too costly, how-

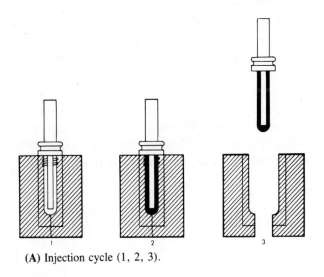

(A) Injection cycle (1, 2, 3).

(B) Blowing cycle (4, 5, 6).

Fig. 11-24. The injection blow molding process.

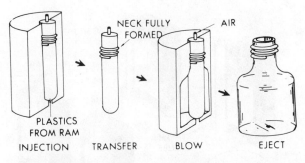

Fig. 11-25. Injection blow process. (Monsanto Co.)

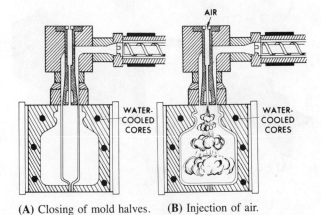

(A) Closing of mold halves. **(B)** Injection of air.

Fig. 11-26. Extrusion blow molding.

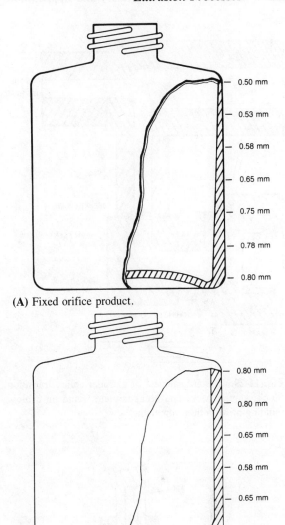

(A) Fixed orifice product.

(B) Programmed orifice product,

Fig. 11-27. Parison programming with variable die orifice. Note wall thicknesses.

ever. Blow extrusion offers strain-free articles at high production rate but scrap reprocessing is required. Controlling wall thickness is the largest disadvantage. By controlling (sometimes called *programming*) the wall thickness of the extruded parison, thinning is reduced. For a part requiring an extremely large body but needing strength at the corners, a parison could be produced with the corner areas much thicker than the walls (Fig. 11-27).

In Figure 11-28, the arrangement of the extruder and die parts is shown. By this method, one or more continuous parisons may be extruded. In Figure 11-29, hot plastics is fed into an accumulator and then forced through the die. A controlled length of parison is produced when the ram or plunger operates. The extruder fills the accumulator and the cycle begins again.

The wall thickness of the tube or parison may be controlled (programmed) to suit the container configuration. This is done by using a die with a variable orifice, as shown in Figure 11-30.

Many different ways of forming the blow-molded product have been developed (Fig. 11-31), and each process may have an advantage in molding a given product. One manufacturer forms the container and fills it in a single operation. The

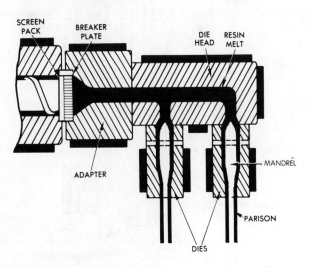

Fig. 11-28. Parts found in most extrusion blow molders.

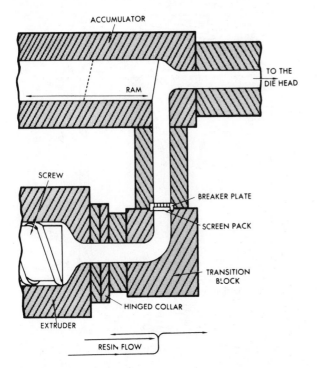

Fig. 11-29. The arrangement of extruder collar, transition block, screen pack, and breaker plate found on a blow molding press with accumulator.

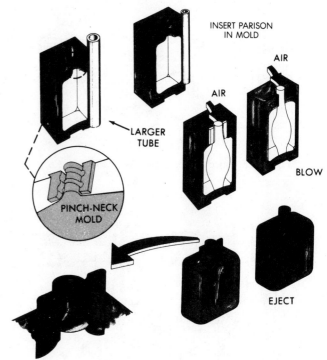

(A) Pinch-neck and regular processes.

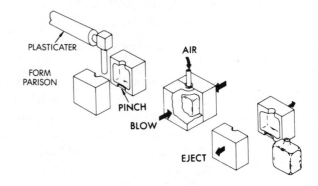

(B) Basic pinch-parison process.

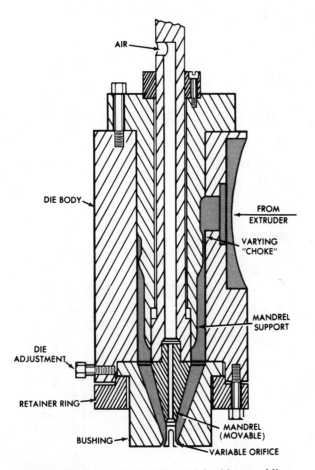

Fig. 11-30. Programming die used for blow molding. (Phillips Petroleum Co.)

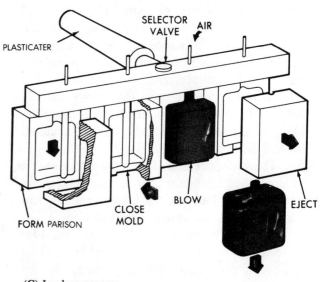

(C) In-place process.

Fig. 11-31. Various blow molding processes. (Monsanto Co.)

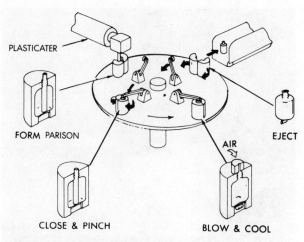

(D) Pinch-parison rotary process

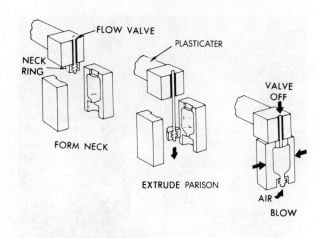

(E) Neck-ring process.

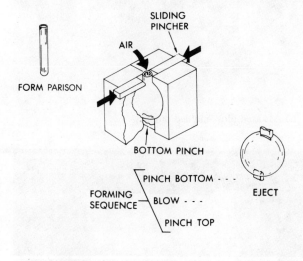

(F) Trapped-air process.

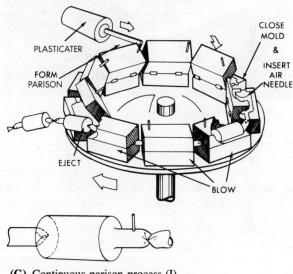

(G) Continuous-parison process (I).

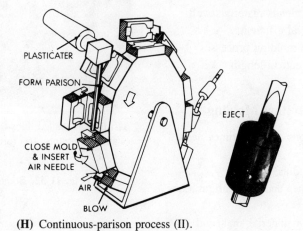

(H) Continuous-parison process (II).

Fig. 11-31. (Continued)

product, rather than compressed air, is forced into the parison.

Extrusion blow molding may be used to produce all forms of composites, including fibrous, particulate, and laminar. Short fibers are used to produce a variety of reinforced blow-molded products. (See Pultrusion)

Figure 11-32 shows some blow-molded products and a mold. Note the pinched parison on the product in Figure 11-33A. Table 11-4 gives some

(A) Assorted blow-molded products.

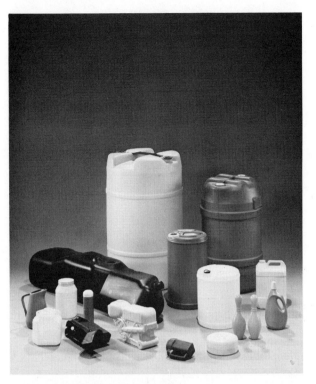

(B) Assorted blow-molded products.

common problems in blow molding and their remedies, while six advantages and five disadvantages of extrusion blow molding are shown below.

Advantages of Extrusion Blow Molding

1. Most thermoplastics and many thermosets may be used
2. Die costs are lower than those for injection molding
3. Extruder compounds and blends materials well
4. Extruder plasticates material efficiently
5. Extruder is basic to many molding processes
6. Extrudates may be any practical length

Disadvantages of Extrusion Blow Molding

1. Costly secondary operations are sometimes needed
2. Machine cost is high
3. Purging and trimming produce waste
4. Limited programmed shapes and die configurations are available
5. Screw design must match material melt and flow characteristics for efficient operation

Many blow-molded products are flame-treated to enhance antistatic properties. This makes the surface more receptive to ink or other decorative media. (See Chapter 16, Decorating Processes)

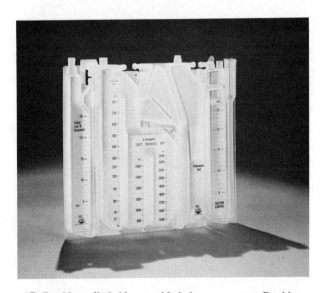

(C) Double-walled, blow-molded chest evacuator. Double-walled products are rigid, yet have some flexibility. (Geauga)

Fig. 11-32. Some blow-molded products.

11-6 BLOW MOLDING VARIATIONS

Four blow molding variations should be mentioned:

1. Cold parison
2. Sheet

(A) Note the pinched parison and untrimmed, or un-separated, lids. (Hoover Universal)

(B) Note the pinchoff on the milk containers from this six-head blow molder.

(C) A spin-off trimmer for wide-mouth jars.

Fig. 11-33. These containers require trimming.

3. Stretched or biaxial
4. Multilayered (co-extrusion or co-injection)

In the *cold parison* process, the parison is extruded by normal means (either injection or extrusion), then cooled and stored. The parison is later heated and blown to shape. The major advantage is that the parison can be shipped to other locations or stored in case of a breakdown or materials shortage.

Multilayer bottles may be produced by co-injection blow-molding or co-extrusion methods. The three-layered product generally contains a barrier layer sandwiched between two body layers.

In the *sheet blow molding* process, hot extruded sheets are blow-formed as they are pinched between mold halves. The edges are fused together by the pinching action of the mold. Two different-colored sheets may be extruded and formed into a product with two separate colors (Fig. 11-34). The pinch-welded seams are the biggest disadvantage; in addition, two extruders are usually needed, and there is much scrap to be reprocessed.

In the *stretch or biaxial blow molding* process, molded preforms and extruded tubes may be stretched before blowing. This produces a blow-molded product with better clarity, reduced creep, higher impact strength, improved gas and water vapor barrier properties, and lower mass. Homopolymers may be used instead of more costly copolymers.

In injection-molded preforms, a rod stretches the hot preform during the blow cycle (Fig. 11-35).

Table 11-4.   Troubleshooting Blow Molding

Defect	Possible Remedy
Excess parison stretch	Reduce stock temperature Increase extrusion rate Reduce die tip heat
Die lines	Die surface poorly finished or dirty Blowing air orifice too small—more air Extrusion rate too slow—parison cooling
Uneven parison thickness	Center mandrel and die Check heater bands for uneven heating Increase extrusion rate Reduce melt temperature Program parison
Parison curls up	Excessive temperature difference between mandrel and die body Increase heating period Uneven wall thickness or die temperature
Bubbles (fisheyes) in parison	Check resin for moisture Reduce extruder temperatures for better melt control Tighten die tip bolts Reduce feed-section temperature Check resin for contamination
Streaks in parison	Check die for damage Check melt for contamination Increase back pressure on extruder Clean and repair die
Poor surface	Extrusion temperature too low Die temperature too low Poor tool finish or dirty tools Blowing air pressure too low Mold temperature too low Blowing speed too slow
Parison blowout	Reduce melt temperature Reduce air pressure or orifice size Align parison and check for contamination Check for hot spots in mold and parison
Poor weld at pinch-off	Parison temperature too high Mold temperature too high Mold closing speed too fast Pinch-off land too short or improperly designed
Container breaks on weld lines	Increase melt temperature Decrease melt temperature Check pinch-off areas Check mold temperature and decrease cycle time
Container sticks in mold	Check mold design—eliminate undercuts Reduce mold temperature and melt temperature Increase cycle time
Part weight too heavy	Parison temperature too low Melt index of resin too low Annular opening too large

Table 11-4.   Troubleshooting Blow Molding (Continued)

Defect	Possible Remedy
Warpage of container	Check mold cooling Check for proper resin distribution Lower melt temperature Reduce cycle time for cooling
Flashing around container	Lower melt temperature Check blowing pressure and air start time Check for molds closing on parison Check air start time and pressure

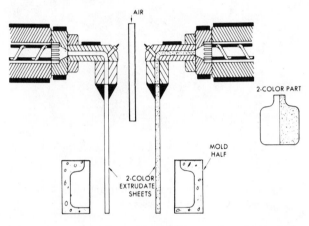

Fig. 11-34. Extrudates of two different colors are pinched between mold halves to form a sheet-blown two-color part.

In parison or tube methods, the hot tube is stretched prior to the blow cycle (Fig. 11-36).

Co-extrusion blow molding actually produces a laminar bottle product. (See Laminating) Several extruders are used to extrude the material into the manifold. The multilayered container is then blow molded from the emerging parison. For long shelf life, some food containers may have seven layers. A common food container consists of PP/adhesive/EVOH/adhesive/PP (Fig. 11-37). See multilayered sheet products discussed under laminating processes.

11-7 CALENDERING

In calendering, thermoplastic materials are squeezed to final thickness by heated rollers (Fig. 11-38). Films and sheet forms with a glossy or embossed finish may be produced by this method (Fig. 11-39). Much calendered film is used in the textile industry. Embossed or textured film is used to produce leatherlike apparel, handbags, shoes, and luggage.

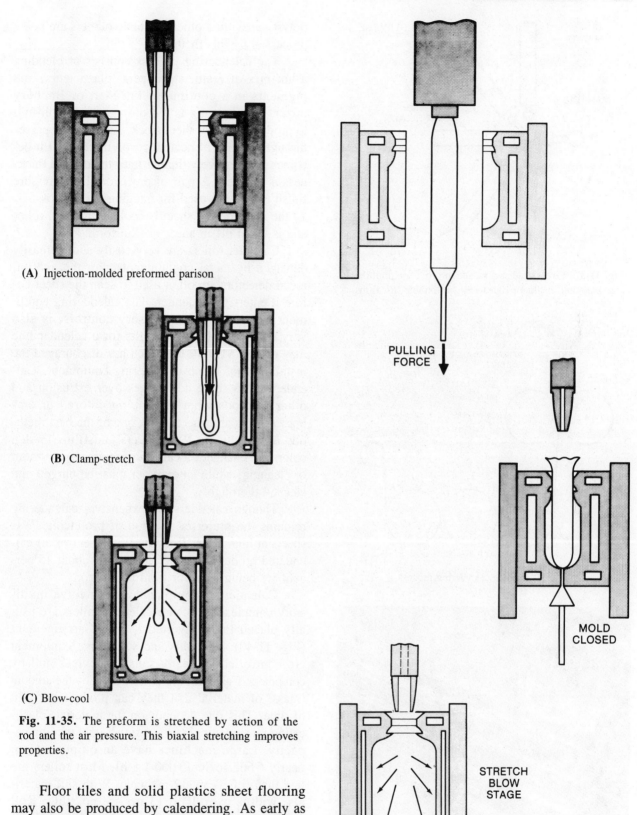

(A) Injection-molded preformed parison

(B) Clamp-stretch

(C) Blow-cool

Fig. 11-35. The preform is stretched by action of the rod and the air pressure. This biaxial stretching improves properties.

PULLING FORCE

MOLD CLOSED

STRETCH BLOW STAGE

Fig. 11-36. Biaxially oriented product is produced by pulling the extruded tube, then blowing.

Floor tiles and solid plastics sheet flooring may also be produced by calendering. As early as 1836, two-roller calenders were being used to process sheet rubber. Today, vinyls and a variety of polyurethanes, polyethylene-polypropylene copolymers, polyethylene-vinyl acetate copolymers, polyethylene-ethyl acrylate copolymers, modified

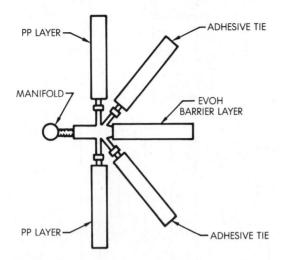

Fig. 11-37. Five extruders are used to produce a multilayered parison (for blow molded parts) or bubble (for films).

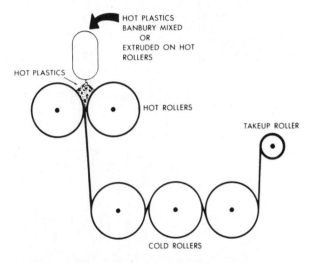

Fig. 11-38. The calendering process.

Fig. 11-39. Calendering a thermoplastic material. (Monsanto Co.)

polystyrenes, and other synthetic rubbers are being processed by this method.

The calendering process consists of blending a hot mix of resin, stabilizers, plasticizers, and pigments in a continuous kneader, or Banbury mixer. This hot mix is fed through a two-roll mill to produce a heavy sheet stock. As the sheet passes through a series of heated, revolving rollers, it becomes progressively thinner until the desired thickness is reached. A pair of precision, high-pressure finish rollers are used for gauging and embossing. At the end, the hot sheet is cooled on a chill roller and is taken off in sheet or film form.

Calender rollers are very costly and are easily damaged by metal contaminants. For this reason, metal detectors are often used to scan the sheet before it enters the calender. The calendering equipment, together with accessory controls, is also very costly. Replacement costs for a calender line may exceed $1 million, which has discouraged the installation of much calendering equipment. Calendering has some advantages over extrusion and other methods when producing colored or embossed films and sheets. It is one of the best methods for producing PVC sheets and films. When colors are changed, a calender requires a minimum of cleaning, while an extruder must be purged and cleaned thoroughly.

Though calenders are expensive, calendering remains the preferred method of producing PVC sheets at high rates. Roughly 95 percent of all calendered products are PVC, and only about 15 percent are being used for rigid production.

Calendering equipment may involve multistory complexes, because calender rollers are usually placed in an inverted *L* or a *Z* arrangement (Fig. 11-40). The rolls and supportive equipment are controlled by many sensing devices and by computers. Calenders are usually rated by the amount (mass) of material that they can produce per unit of time. This rate depends on the material, plasticating rate, required surface finish, and take-up capacity. Large machines have an output rate of nearly 6662 lb/hr [3000 kg/h]. Most rollers are less than 6 ft. 7 in. [2 m] wide. With soft materials, widths over 10 ft [3 m] are possible. Roll forces may approach 39.34 tons-force [350 kN] for thin, rigid materials.

Multilayered sheet structures of plastics or composites may be made using multiple calenders. Line speeds are relatively slow, around 30 cm/s

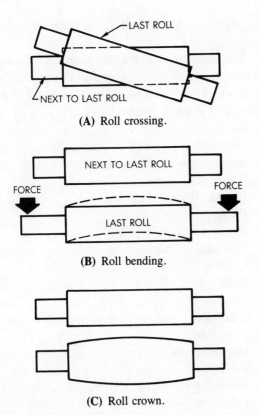

(A) Roll crossing.

(B) Roll bending.

(C) Roll crown.

Fig. 11-40. Methods used for correcting sheet profile.

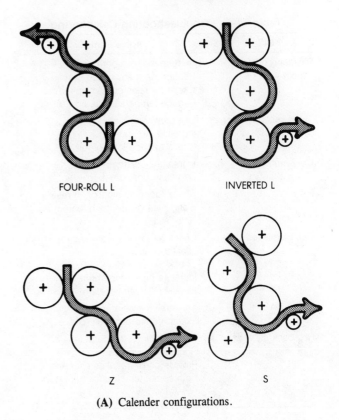

FOUR-ROLL L

INVERTED L

Z

S

(A) Calender configurations.

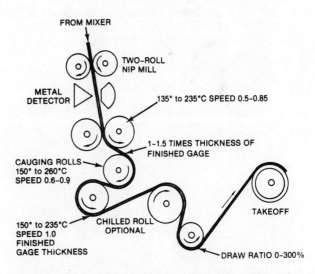

(B) A series of rollers arranged in a Z-form is used to calender thermoplastic sheet material.

Fig. 11-41. Common calender roll configurations.

on 1.5 m wide rolls, because of the variety of materials and temperatures.

Several methods are used to compensate for roll deflection (bending or bowing): 1) force is applied to the outboard or inboard main bearings, 2) rolls are produced with slight crowns, or 3) one roll is skewed with respect to the other (roll crossing). (See Figure 11-41)

The term calendering is also used, but incorrectly, in discussing the application of a plastics film to one or both sides of fabric or paper. These operations should be classed as either coating or laminating. As coating machines, calenders may press paper, fabric, or other substrates together as they pass through the rollers.

Table 11-5 gives some common calendering problems and remedies. Below are listed three advantages and four disadvantages of calendering.

Advantages of Calendering

1. Offers a wide range of surface finishes and textures
2. May be used to laminate fabric or other materials
3. In continuous operation, will produce any practical length
4. High rate; versatility

Disadvantages of Calendering

1. Machines are large and very costly
2. Costly support equipment is needed
3. Product width and thickness are limited
4. Roller deflection must be compensated for

Table 11-5.   Troubleshooting Calendering

Defect	Possible Remedy
Blistering of film or sheet	Reduce melt temperature Reduce program speed of rolls Check for resin contamination Reduce temperature of chill rolls
Thick section in center, thin edges	Use crowned rolls Increase nip opening Check bearing load of rolls
Cold marks or crow's feet	Increase stock temperature Decrease program feed
Pin holes	Check for contamination in resin Blend plasticizers more thoroughly into resin
Dull blemishes	Check for lubricant or resin contamination Check roll surfaces Increase melt temperature Increase roll temperature
Rough finish	Raise roll temperature Excessive nip between rolls
Roll bank on sheet	Incorrect stock temperature—increase/decrease Provide constant takeoff speeds Reduce roll temperatures and nip clearance

VOCABULARY

The following vocabulary words are found in this chapter. Look up the definition of any of these words you do not understand as they apply to plastics in the glossary, Appendix A.

Back pressure
Banbury
Barrel
Barrel vent
Breaker plate
Blow molding
Blowup ratio
Calender or calendering
Coating
Drawing
Extrudate
Extrusion
Frost line
Hopper dryer
Parison
Pinch-off
Spinneret

11-1. The hollow plastics tube used in blow molding is called a ___?___ .

11-2. The process that forces hot plastics through machine dies to form continuous shapes is called ___?___ .

11-3. Which molding process would be used to make long sections of plastics pipe?

11-4. A well-known compounding machine with contrarotating rotors to mix and blend the material is called ___?___ .

11-5. Name three extrusion spinning processes.

11-6. The process of stretching a thermoplastic sheet, rod, or filament to reduce its cross-sectional area and change its physical traits is called ___?___ .

11-7. Because of molecular orientation, ___?___ film extrusion film is generally stronger than ___?___ film.

11-8. The extruder screw moves the plastics materials through the ___?___ of the machine.

11-9. The ___?___ in the extruder helps support the screen pack and creates additional mixing of the plastics before it leaves the die.

11-10. The general term for the product or material delivered by an extruder is ___?___ .

11-11. Name the two methods used to make films by the extrusion process.

11-12. The ___?___ line is the name of the zone that appears frosty in the extrusion blow film process.

11-13. In cold extrusion, the screw action mixes and forces the unheated material through the die. Solvents are removed or a chemical bath causes the ___?___ to solidify.

11-14. Blow molding is a refinement of ancient ___?___ processes.

11-15. The two basic blow-molding methods are ___?___ blowing and ___?___ blowing.

11-16. In blow molding, a die with a variable orifice may be used to control ___?___ and ___?___ thickness.

11-17. In the ___?___ process, hot material is pressed between two or more rotating rollers.

11-18. Patterns or textures may be placed on calendered films or sheets by passing the soft, hot plastics between ___?___ rolls.

11-19. Spinnerets are used to produce ___?___ and ___?___ plastics forms.

11-20. Is the extruder or the calender equipment more versatile for making plastics film and sheet?

Chapter 12

Laminating Processes

In this chapter, you will discover that laminating processes are used to produce structural shapes, plates, sheets, angles, channels, rods, tubes, and other shapes by combining layers of materials. We will discuss reinforcing processes in Chapter 13.

You should recall that the terms *laminate, laminar,* and *laminated* describe a composite form or product made by the process of laminating.

A laminated plastics may be defined as, "a plastics material consisting of superimposed layers of a filler, impregnated or coated by a synthetic resin, that have been bonded together, usually by means of heat and pressure, to form a single piece." See Reinforced and Laminated Plastics in Chapter 5.

Simply stated, lamination is the process of combining two or more composite layers into one composite piece. The components may be superimposed layers of filler, reinforcement, polymers, impregnated or coated paper, fabric, foil, or other materials.

Composite materials offer the five following advantages:

1. Better strength-to-mass ratio
2. Better chemical and electrical resistance
3. Greater dimensional stability
4. Improved mechanical and physical properties
5. Reduced product costs for many designs

In some laminates, the plastics is used as an adhesive holding various substrates together. Examples are plywood, some metal laminates, and honeycomb construction panels and other sandwich composites.

Laminates are usually flat sheets while reinforced plastics products may take many shapes, although high-pressure laminates may be molded

into some shapes. The result of the laminating process should not be confused with reinforced plastics in which the reinforcement is not in a laminated form. Reinforcing is done through many processing operations, including compression, transfer, and injection molding, casting, calendering, and rotational molding.

12-1 LAMINAR COMPOSITES

Traditionally, laminated plastics were products made by molding pressures exceeding 1 015 psi [7 000 kPa]. Today's laminated products may include some materials used with processing pressures of less than 7 000 kPa. Various reinforcements may be used in both high- and low-pressure laminations.

Cams, pulleys, gears, fan blades, decorative tops, printed circuit boards, and nameplate stock are typical laminated products.

In high-pressure lamination, thermosetting resins are widely used to impregnate the base materials. Urea, melamine, phenolic, polyester, epoxy, and other resins are used (Table 12-1).

For some products, molds are used to form the impregnated stock into various shapes, such as rods, tubes, cups, plates, and cones. In this case, matched metal molds must be used in fusing and compressing the mass into a laminated structure.

High-pressure laminates (Fig. 12-1) possess a wide range of properties. One of the earliest phenolic laminates was produced in 1905 by J. P. Wright, the founder of the Continental Fiber Company. This was five years before Baekeland had patented his ideas of using sheets impregnated with phenolic resins as laminates. Many high-pressure industrial laminates were made from a layup

Table 12-1.   Selected Resins/Plastics and Materials Used in Lamination

Resin/Plastics	Paper	Cotton Fabric	Asbestos Fabric	Fibrous Glass Fabric/Mat	Metallic Foils	Composites, Honeycomb, etc.
Acrylic	LP	—	—	LP		—
Polyamide	LP	—	—	LP	LP	—
Polyethylene	LP	—	—	LP	LP	—
Polypropylene	LP	—	—	LP	LP	—
Polystyrene	—	—	—	LP		
Polyvinyl chloride	LP	—	—	—	LP	LP
Polyester	LP-HP	LP	LP	LP	—	LP
Phenolic	LP-HP	LP-HP	HP	HP-LP	HP	LP-HP
Epoxy	LP	LP-HP	—	LP-HP	LP	LP
Melamine	HP	HP	—		HP	
Silicone	—	—	—	HP	—	LP

LP —Low-pressure (laminate).
HP—High-pressure (laminate).
——— —Only limited amounts manufactured in this category.

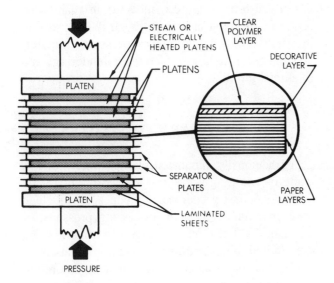

Fig. 12-1. Concept of multiple stacking of laminates in press.

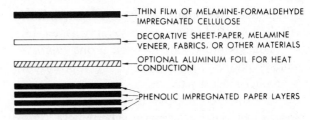

Fig. 12-2. Typical arrangement of layers in a decorative high-pressure laminate.

of paper, cloth, asbestos, synthetic fiber, or fibrous glass using this basic patent. Today, there are over fifty standard industrial grades of laminates for electrical, chemical, and mechanical uses. At first, high-pressure industrial laminates took the place of mica as a quality electrical insulation material. Around 1913, the Formica Corporation emerged, making high-pressure laminates to replace mica (For Mica) in many electrical and mechanical uses. By 1930, the National Electrical Manufacturers Association (NEMA) saw the potential of decorative laminates, and in 1947, a separate NEMA section was established to work with government agencies and associations interested in setting up product standards for decorative laminates (Fig. 12-2).

The major disadvantage of high-pressure lamination is slow production rates compared to high-speed injection molding. Many flat laminates may be competitive with the rates of other molding processes, however.

The actual impregnation of the reinforcing layer is done by a number of methods, including premix, dipping, coating, or spreading (Fig. 12-3). After a drying period, the impregnated laminating stock is cut to the desired size and placed in multi-sandwich form between the metal plates (separation plates) of the press. Press platens are not smooth enough for the desired finish. These plates may be glossy, matte, or embossed. Metal foils are sometimes used between surface layers to produce a decorative finish. In decorative laminates, a printed pattern layer and a protective overlay sheet are superimposed on the base material. The prepared stock is subjected to heat and high pressure. The combined heat and pressure cause the resin to flow and the layers to compact into one polymerized mass. Polymerization may also be done by chemical or radiation sources. When the thermosetting resins have cured or the thermoplastic resins have cooled, the laminate is removed from the press.

The use of thermoplastic materials in lamination has some limitations; however, demand for

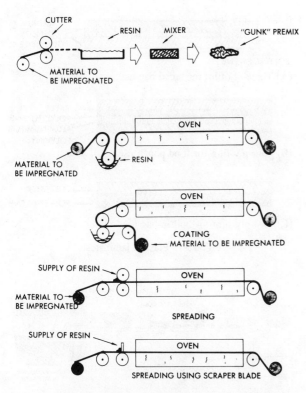

Fig. 12-3. Various impregnation methods.

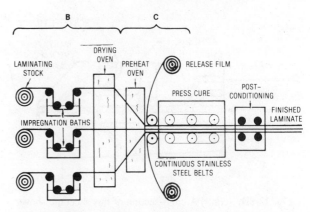

(A) Continuous lamination, in which the (B-stage) resin is changed into an infusible (C-state) plastics.

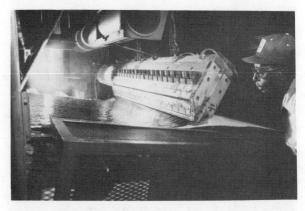

(B) Manufacturing a fiberglass reinforced thermoplastic composite sheet. (AZDEL, Inc.)

Fig. 12-4. Continuous lamination of thermosetting and thermoplastic matrix composite sheets.

such products is growing. Hot melts are used in producing composite-layered laminates, and acrylic monomers are used where color or clarity is important. Certain thermoplastic powders and films may be heated by various energy sources and used in fusing and compressing the mass in a laminate structure. Many of these thermoplastic laminates may be postformed, hot or cold, on dies or molds with varying cross-sectional thicknesses.

In *continuous laminating* (Fig. 12-4), fabrics or other reinforcements are saturated with resin and passed between two plastics film layers, cellophane, ethylene, or vinyl, for example. The thickness of the laminated composite is controlled by the number of layers and by a set of squeeze rollers. The laminate is then drawn through a heating zone to speed polymerization. Corrugated awnings, skylights, and structural panels are products of continuous laminating.

Extruders are an important part of many laminating operations. (Fig. 12-5) Films, foils, paper, fabric, reinforcements, or several different polymers may be made into a composite laminate. Continuous extrusion laminating of different materials, colors, or compositions are made into engraving stock, refrigerator liners, and thermoformed containers.

Coextruded composite films of polyethylene and vinyl acetate produce a tough, durable two-

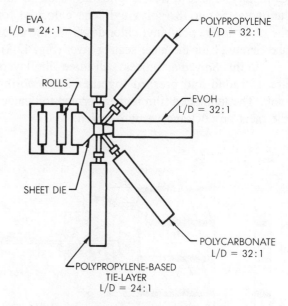

Fig. 12-5A. Five extruders are used to produce a true composite laminate.

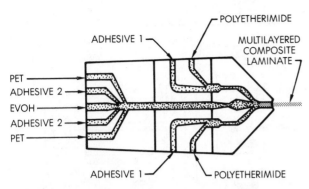

Fig. 12-5B. (con't) A cross-section of this sheet die shows that at least two extruders were used for extruding the adhesive (1&2), one for the outside polyetherimide layers, one for the two PET layers and one for the EVOH center barrier layer.

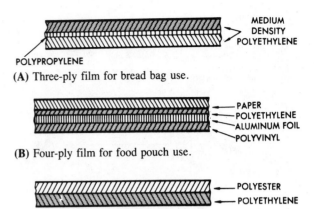

(A) Three-ply film for bread bag use.

(B) Four-ply film for food pouch use.

(C) Two-ply film for boil-in-bag use.

Fig. 12-7. Plastics laminates.

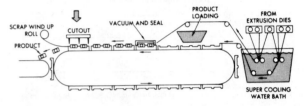

(A) Extrusion and packing.

layer film that may be heat-sealed. The plastics layers are brought together in the molten state and extruded through a single die opening to make the multilayered laminate film (Fig. 12-6). This laminated material made possible a new processing technique in which many composites can be post-formed on metal-stamping presses. Some of these thermoplastic laminates may be draped or thermo-formed. See Chapter 15 and cold stamping.

A three-ply laminate film is used to wrap bread products (Fig. 12-7). This film laminate is composed of an inner core (ply) or polypropylene with two outer layers (plies) of polyethylene.

A vinyl plastics film laminate was developed specifically for packaging meat products by the Dow Chemical Company and the Oscar Mayer Company. Three different plies are used in this laminate; Saran 18 (polyvinylidene chloride) for the outer layer, polyvinyl chloride 88 for the core, and Saran 22 for the inner sealing layer (Fig. 12-8).

In this *Saranpac* process, all three film layers are extruded and pressed together in a cooling tank. The laminated film is then formed to contain the meat product and vacuum-sealed.

(B) Three-ply laminate film is used to wrap bread loaves.

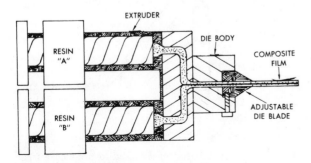

Fig. 12-6. Production of coextruded film.

(C) Saranpac process is used for meat products.

(D) Nuts packaged in acetate-polyethylene film.

(E) Foil-paper-polyethylene packages for dry meats.

Fig. 12-8. Extrusion and packing technique, with some product examples.

Table 12-2.   Selected Extruded Film Laminates and Applications

Laminate Material	Application
Paper-polyethylene-vinyl	Sealable pouches for dried milk, soups, etc.
Acetate-polyethylene	Tough, heat-sealable packing for nuts
Foil-paper-polyethylene	Moisture barrier pouch for soup mixes, dry milk, etc.
Polycarbonate-polyethylene	Tough, puncture-resistant skin packages
Paper-polyethylene-foil-polyethylene	Strong heat seals for dehydrated soups
Paper-polyethylene-foil-vinyl	Heat-sealable pouches for instant coffee
Polyester-polyethylene	Tough, moistureproof boil-in-bag pouches for foods
Cellulose-polyethylene-foil-polyethylene	Gas and moisture barrier for pouches of ketchup, mustard, jam, etc.
Acetate-foil-vinyl	Opaque, heat-sealing pouches for pharmaceuticals
Paper-acetate	Glossy, scratch-resistant material for record covers, paperback books

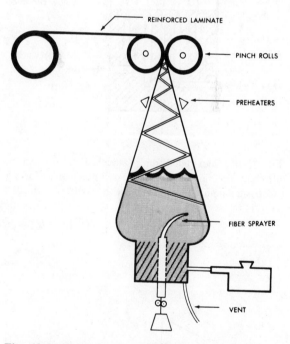

Fig. 12-9. Making filament-reinforced sheet by blown film process.

There are numerous applications of extruded film laminates. Only a few examples are listed in Table 12-2.

In one proprietary blow film process, a composite sheet is produced by filament reinforcing the inside of the hot blown film and pressing the fibrous layered film between pinch rollers. This concept is illustrated in Figure 12-9. In another variation, two different polymers are blow film coextruded with molecular orientation and then squeezed together into a laminate sheet. (Fig. 12-10)

12-3 SANDWICH COMPOSITES

Sandwich construction consists of two relatively dense, thin, strong outer layers or skins separated by a lightweight core material. The outer layers must be strong to carry the axial and in-plane shear loading. Most of the tensile and compressive forces are transferred to these layers. These facing materials may be made of impreg-nated paper, fibrous glass, metals, woods, polymers, or other composite materials. The core material may be solid, cellular, honeycomb, or any other configuration. The core material transfers the loads from one facing to the other. Cellular materials are commonly used in sandwich panels to make refrigerator liners for truck boxes, railcars, food

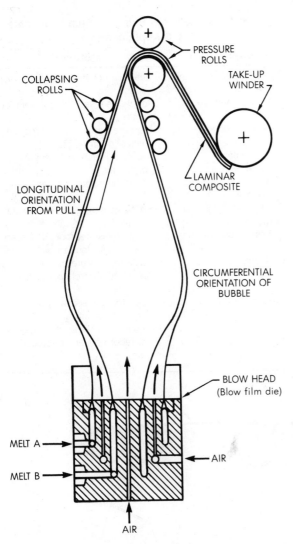

PRESSURE ROLLS

COLLAPSING ROLLS

TAKE-UP WINDER

LAMINAR COMPOSITE

LONGITUDINAL ORIENTATION FROM PULL

CIRCUMFERENTIAL ORIENTATION OF BUBBLE

BLOW HEAD (Blow film die)

MELT A

MELT B

AIR

AIR

Fig. 12-10. Coextruded film is stretched and oriented in different directions and then pressed together to form a composite laminar sheet. One melt (B) could be formed.

coolers, and exterior panels for mobile homes. Solid wooden cores have been used in doors, snow skis, boat hulls, pallets, and furniture.

Honeycomb core materials made of resin-impregnated kraft paper, aluminum, glass-reinforced polymers, titanium, or other materials are among the strongest core structures for their mass. Aluminum is the most commonly used honeycomb core material. All honeycombs are anisotropic, and properties depend on the composition, cell size, and geometry. Two major methods of producing honeycomb core materials are illustrated in Figures 12-11 and 12-12. The properties of selected aluminum honeycomb materials are shown in Table 12-3. Properties of several glass-reinforced plastics honeycombs are shown in Table 12-4.

All properties, including thermal and electrical, depend on the selection of facings, core, and bonding agent. Adhesive bonding is critical if shear and axial loads are to be transmitted to and from the core material. Polyimide, epoxy, and phenolics are commonly used. Adhesive films are resin-impregnated fiber matting, cloth, or paper may be used in the core-to-face bond.

Sandwich composites are manufactured through several techniques, from simple hand lay-up operations to vacuum bag and autoclave techniques. Extruders are commonly used to deposit a thin layer of plastics material on both sides of a continuous extruded, or roller-fed, core. In Figure 12-13, three extruders are shown producing a sandwich laminate, possibly made from more than one polymerized material. It may have either a solid-center core or one composed of low-density materials.

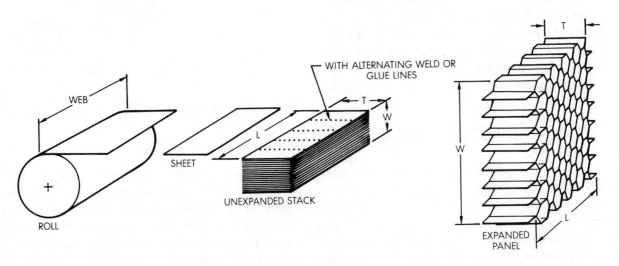

WEB

SHEET

WITH ALTERNATING WELD OR GLUE LINES

UNEXPANDED STACK

ROLL

EXPANDED PANEL

Fig. 12-11. Honeycomb manufacture by the expansion process.

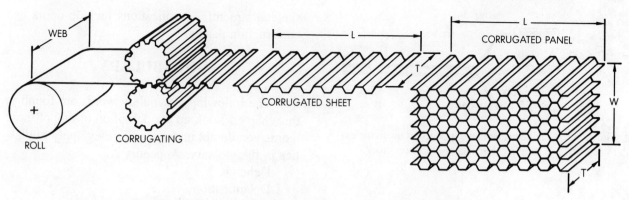

Fig. 12-12. Honeycomb manufacture by the corrugation process.

Table 12-3. Properties of 5056, 5052, and 2024 Hexagonal Aluminum Honeycomb

Honeycomb Cell-Material-Gauge	Nominal Density, kg/m³	Compressive			Plate Shear			
		Bare	Stabilized		"L" Direction		"W" Direction	
		Strength, kPa	Strength, kPa	Modulus, MPa	Strength, kPa	Modulus, MPa	Strength, kPa	Modulus, MPa
5056 Hexagonal Aluminum Honeycomb:								
1/16–5056–0.0007	101	6894	7584	2275	4447	655	2551	262
1/16–5056–0.001	144	11721	12410	3447	6756	758	4136	344
1/8 –5056–0.0007	50	2344	2482	668	1723	310	1068	137
1/8 –5056–0.001	72	4343	4619	1275	2930	482	1758	262
5/32–5056–0.001	61	3275	3447	965	2310	393	1413	165
3/16–5056–0.001	50	2344	2482	669	1758	310	1069	138
1/4 –5056–0.001	37	1413	1448	400	1172	221	724	103
5052 Alloy Hexagonal Aluminum Honeycomb:								
1/16–5052–0.0007	101	5998	6274	1896	3516	621	2206	276
1/8 –5052–0.0007	50	1862	1999	517	1448	310	896	152
1/8 –5052–0.001	72	3585	3758	1034	2344	483	1517	214
5/32–5052–0.0007	42	1379	1482	379	1138	255	689	131
5/32–5052–0.001	61	2723	2827	758	1862	386	1207	182
3/16–5052–0.001	50	1862	1999	517	1448	310	896	152
3/16–5052–0.002	91	5309	5585	1517	3172	621	2068	265
1/4 –5052–0.001	37	1138	1207	310	965	221	586	113
1/4 –5052–0.004	127	9377	9791	2344	4826	896	3034	364
3/8 –5052–0.001	26	586	655	138	586	145	345	76

Table 12-4. Properties of Several Glass-Reinforced Plastics Honeycombs

Honeycomb Material-Cell-Density	Compressive			Plate Shear			
	Bare	Stabilized		"L" Direction		"W" Direction	
	Strength, kPa	Strength, kPa	Modulus, MPa	Strength, kPa	Modulus, MPa	Strength, kPa	Modulus, MPa
Glass-Reinforced Polyimide Honeycomb:							
HRH 327–3/16–4.0		3033	344	1930	199	896	68
HRH 327–3/16–6.0		5377	599	3171	310	1585	103
HRH 327–3/8 –4.0		3033	344	1930	199	1034	82
Glass-Reinforced Phenolic Honeycomb (Bias Weave Reinforcement):							
HFT–1/8 –4.0	2688	3964	310	2068	220	1034	82
HFT–1/8 –8.0	9997	11203	689	3964	331	2344	172
HFT–3/16–3.0	1896	2585	220	1378	165	689	62
Glass-Reinforced Polyester Honeycomb:							
HRP–3/16–4.0	3447	4137	393	1793	79	965	34
HRP–3/16–8.0	9653	11032	1131	4551	234	2758	103
HRP–1/4 –4.5	4344	4826	483	2068	97	1172	41
HRP–1/4 –6.5	7076	8136	827	3103	172	1793	76
HRP–3/8 –4.5	4205	4757	448	2068	97	1172	41
HRP–3/8 –6.0	6205	6895	689	2758	155	1793	69

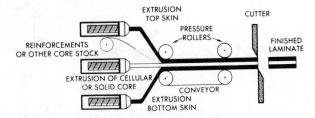

Fig. 12-13. Continuous laminate production involving extruders and use of reinforcements or other core stock.

SMC and other skin materials may be used in press laminating in matched-die operations requiring ribs or more complicated shapes. In a process called *foam reservoir molding* or *elastic reservoir molding*, open-celled polyurethane foam is impregnated with epoxy. The two skin layers are then pressed against the spongy core forcing some of the epoxy adhesive to adhere to the two face skins. The foam and matrix become a catacomb-like, skeletal structure. Applications include doors and construction panels.

VOCABULARY

The following vocabulary words are found in this chapter. Look up the definition of any of these words you do not understand as they apply to plastics in the glossary, Appendix A.

Debond
Delamination
High-pressure laminates
Honeycomb
Impregnate
Interlaminar shear
Laminate
Laminated plastics
Lamination
Low-pressure laminates
Matrix
Sandwich construction

12-1 Two major disadvantages of high-press lamination are low ___?___ rates and high ___?___ pressures.

12-2. The process in which two or more layers of materials are bonded together is called ___?___ .

12-3. If there are unfavorable interlaminar stresses, what may occur? How can this be effectively prevented?

12-4. What are the major applications for high-pressure laminates?

12-5. How are extruders used to produce laminates? Are calenders used?

12-6. What properties are favorable for applications using honeycomb laminated components?

12-7. Name four honeycomb core materials and describe the merits of each in a particular application.

12-8. Define a laminated plastics and describe how several products may be formed.

12-9. Defend the selection of sandwich construction in house doors, airplane components and cargo cantainers.

12-10. Describe the process of continuous laminating and include the type of materials used and typical product applications.

12-11. Describe the differences and similarities between the processes and products of laminating and reinforcing.

The term *reinforced plastics* is not very descriptive. It simply implies that an agent has been added to improve or "reinforce" the product. The SPE defines reinforced plastics as "a plastics composition in which reinforcements are embedded with strength properties greatly superior to those of the base resin." Specific terms such as *advanced, high-strength, engineered,* or *structural* composites came into use in the 1960s. With them, a stiffer, higher modulus material of "exotic" reinforcements in new matrices was used.

Today, *reinforced plastics* is used to describe several forms of composite materials, produced by any one of ten reinforcing processes. Someday, we may classify all laminating and reinforcing processes as *composite processing.*

Some composite reinforcing techniques are variations of laminating because in them, two or more different materials are combined in layers. Other techniques are simply modifications of processing methods that produce a new material with specific or unique properties.

The following composite reinforcing processes will be discussed in this chapter:

1. Matched die
2. Hand layup
3. Sprayup
4. Rigidized vacuum forming
5. Vacuum-bag
6. Pressure-bag
7. Filament winding
8. Centrifugal reinforcing
9. Pultrusion
10. Cold stamping/forming

In each process, the molds, dies, or rollers must be made with care to ensure proper release of the finished product. Film, wax, and silicone releasing agents are most often used on mold surfaces. Reinforced molding compounds should not be confused with laminates, although they occasionally are (Fig. 13-1). See Chapter 5, Reinforced and Laminated Plastics.

In the past, only thermosetting plastics were commercially reinforced in large quantity; today, the demand for reinforced thermoplastics (RTP) is increasing. Because the thermoplastic materials may be processed in many different ways, many innovative uses have resulted.

Reinforced molding compounds may be molded by injection, matched-die, transfer, compression, or extrusion methods to produce products with complex shapes and a broad range of physical properties. There is some difficulty in blow molding small, thin-walled items. Injection molding is the most common method of processing reinforced thermoplastics compounds. (See Injection, Extrusion, and Compression Molding in index.)

Short fibers of milled or chopped glass are most often used to reinforce molding compounds. See Table 13-1 for a list of properties of fibrous-glass reinforced plastics. Plastics fibers and exotic metallic and crystalline whiskers are used as well.

13-1 MATCHED DIE

Matched male and female molds are used to shape reinforced resins (Fig. 13-2). BMC, SMC, TMC, XMC, and preformed shapes are used. Bulk molding compounds are a putty-like mixture of resin, catalysts, fillers, and short fiber reinforcements. BMC are isotropic with fiber lengths usually less than 0.015 inches. They are commonly

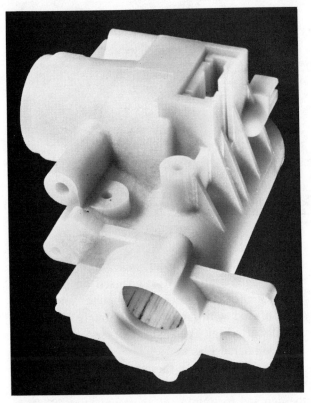

(A) A steering column lock housing made of injection molded, glass reinforced nylon instead of the traditional aluminum die-cast and machined housing.

(C) Reinforced parts and housing for a chain saw designed for rugged use.

Fig. 13-1. Some examples of reinforced plastics parts. (Dow Chemical Co.)

(B) Fan and other parts for hair dryer made of reinforced plastics.

Fig. 13-2. Matched mold for making reinforced plastics chairs. (Cincinnati Milicron)

molded by compression, transfer, and injection molding processes. BMC generally come in rope-like preforms for automated equipment. See Chapter 5, Molding Compounds.

Bulk molding compounds (sometimes called premix or dough molding compounds, DMC) have

Table 13-1. Typical Properties of Fibrous-Glass Reinforced Plastics

Plastics	Relative Density	Tensile Strength, 1 000 MPa	Compressive Strength, 1 000 MPa	Thermal Expansion $10^{-4}/°C$	Deflection Temperature (at 264 MPa), °C
Acetal	1.54–1.69	62–124	83–86	4.8–4.9	1 062–1 599
Epoxy	1.8–2.0	10–207	207–262	2.8–8.9	834–1 599
Melamine	1.8–2.0	34–69	138–241	3.8–4.3	1 406
Phenolic	1.75–1.95	34–69	117–179	2–5.1	1 027–2 178
Phenylene oxide	1.21–1.36	97–117	124–207	2.8–5.6	910–986
Polycarbonate	1.34–1.58	90–145	117–124	3.6–5.1	965–1 000
Polyester (thermoplastic)	1.48–1.63	69–117	124–134	3.6	1 379–1 586
Polyester (thermosetting)	1.35–2.3	172–207	103–207	3.8–6.4	1 406–1 792
Polyethylene	1.09–1.28	48–76	34–41	4.3–6.9	800–876
Polypropylene	1.04–1.22	41–62	45–48	4.1–6.1	910–1 027
Polystyrene	1.20–1.34	69–103	90–131	4.3–5.6	683–717
Polysulfone	1.31–1.47	76–117	131–145	4.3	1 179–1 220
Silicone	1.87	28–41	83–138	None	<3 323

some advantages over preform processing. This fibrous putty may be extruded into H-beam or other profile shapes and automatically fed into the matched-die. These operations are sometimes further divided into open- and closed-mold processing. Injection, compression, transfer, stamping, and matched-die, are formed in closed molds. Open-mold processing includes hand layup, sprayup, and filament winding. Open- and closed-composite molding processes often overlap. Sometimes open-mold techniques are used to preform the part and closed operations are used to complete the forming or molding. For example, filament wound shapes (classed as preforms) are sometimes further processed in matched-die, compression molds, or laminating presses.

Sheet molding compounds are leather-like mixtures of resins, catalysts, fillers, and reinforcements. They have replaced many preform and mat techniques. Since they are made in sheet form, fibers may be much longer than in BMC. SMC provide greater glass loadings (70 percent glass) and lighter products. The longer fibers provide improved mechanical properties. SMC require processing pressures in the range of 507 to 2 030 psi [3.5 to 14 MPa]. Processing temperatures vary with product design and polymer formulation. SMC are fed to molds and passed through pressing, curing, and demolding in one continuous cycle. This technique eliminates waiting through full cure cycle times at the press.

Designs that require thick sections and heavy rib details call for thick molding compounds (TMC). XMC provide strength in a dominant fiber direction. Some are filament wound on a mandrel. A protective film is wrapped over the final layer and the preform is removed and stored for later molding.

Major markets for BMC and SMC molded parts are in the transportation and appliance industry. Shower floors, heater housings, and appliance cases are made of BMC. As the name SMC implies, large parts such as automotive body panels, hoods, small boat hulls, furniture, and appliance components are made from this material in matched-die molds.

A variation of this process is *macerated* reinforced processing. Macerated parts are produced by chopping the reinforcing materials into pieces 0.003 to 0.155 in. [2 to 100 mm] long to be processed in the matched molds. Reinforced resin products produced from matched-die molds are strong and may have a superb surface finish, both inside and out. Mold and equipment costs are high, however. A list of five advantages and disadvantages of matched-die processing is shown below.

Advantages of Matched-die

1. Both interior and exterior surfaces are finished
2. Complex shapes (including ribs and thin details) are possible
3. Minimum trimming of parts is needed
4. Products have good mechanical properties, close part tolerances, and corrosion resistance
5. Cost and reject rate are relatively low

Disadvantages of Matched-die

1. Preform, BMC, TMC, XMC, and SMC require more equipment, handling, and storage
2. Press guides must have good parallelism for close tolerances

3. Molds and tooling are costly, compared to open molds
4. Surfaces may be porous or wavy
5. There are no transparent products

13-2 HAND LAYUP OR CONTACT PROCESSING

Thermosetting resins are used in *hand layup molding* (Figs. 13-3 and 13-4). Because only atmospheric pressure is used in applying the saturated reinforcing material to the mold, this process is not, by strict definition, a molding process. *Contact molding* or *open molding* are more descriptive terms. After the mold (either male or female) is coated with a releasing agent, a layer of catalyzed

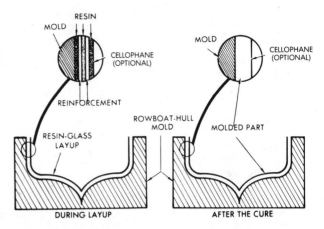

Fig. 13-4. In hand layup, reinforcing material in mat or fabric form is applied to the mold, then saturated with a selected thermosetting resin.

(A) Hand layup or contact processing being used to apply fibrous glass to a titanium-honeycomb structure. (Bell Helicopter Co.)

(B) Contact mold for hand layup of honeycomb reinforcements for radome. (McMillan Radiation Labs., Inc.)

Fig. 13-3. Hand layup processing.

resin is applied which is allowed to polymerize to the gel (tacky) state.

This first layer is a specially formulated gelcoat resin used in industry to improve flexibility, blister resistance, surface finish or color, stain resistance, and weatherability. Gel coats based on neopentyl glycol, trimethylpentanediol glycol, and propylene glycol provide a major advantage as surface treatments for reinforced polyester products.

The gel coat forms a protective surface layer through which fibrous reinforcements do not penetrate. A prime cause of deterioration of fibrous reinforced plastics is penetration of water, which takes place when fibers protrude at the surface. Once the gel coat has partially set, reinforcement is applied. Then, more catalyzed resin is poured, brushed, or sprayed over the reinforcement. This sequence is repeated until the desired thickness is reached. In each layer, the mixture is worked to the mold shape with hand rollers, then the reinforced composite laminate is allowed to harden or cure (Fig. 13-5). External heating is sometimes used to speed polymerization (Fig. 13-6).

Hand layup and sprayed operations are often used alternately. In order to obtain a resin-rich, superior surface finish, a lightweight veil or surface mat is sometimes placed next to the gel layer. The coarser reinforcements are then placed over this layer. Some operations and designs use preforms, cloth, mat, and roving materials for directional or additional strength in selected areas of the part.

Among the major advantages of hand layup are low-cost tooling, minimal equipment required, and the ability to mold large components. As for disadvantages, the process is labor intensive and dependent upon the skill of the operator, while the

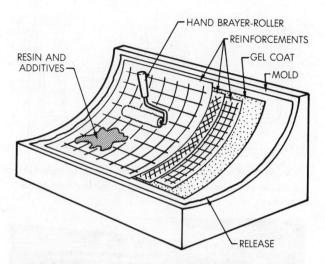

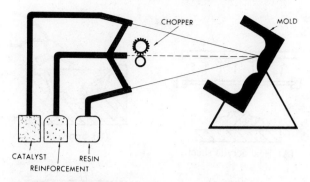

Fig. 13-5. Hand layup implies that little equipment is required. In all layup operations, it is important to bray out (remove) any bubbles trapped between the layers.

Fig. 13-6. Intrared ovens are used to cure resin in about 18 minutes. These glass-reinforced guitar bodies are produced by the hand layup method. (Fostoria Industries, Inc.)

Fig. 13-7. The sprayup method can cover either simple or complex shapes easily, an advantage over the hand layup method.

production rate is low. Also the messy process exposes workers to hazardous chemicals.

13-3 SPRAYUP

In sprayup, catalyst, resin, and chopped roving may be sprayed simultaneously onto mold shapes (Fig. 13-7). While it is considered a variation of hand layup, this process can be accomplished by hand or machine. After the gel coat has been applied, sprayup of resin and chopped fibers begins. Careful roll out is important, to avoid damaging the gel coat. Roll out aids in densifying (eliminating air pockets and aiding in wetting action) the composite. Poor roll out can induce structural weakness by leaving air bubbles, dislocating the fibers, or causing poor wet out (coating of reinforcement). Heat may be applied to speed cure

and to increase production. This low-cost method allows production of very complex shapes. Production rates are high compared to hand layup methods. Care must be taken to apply uniform layers of materials; otherwise, mechanical properties may not be consistent throughout the product. Highly contoured or stressed areas can be given additional thickness, or metallic tapping plates, stiffeners, or other reinforcing components can be placed in desired areas and over-sprayed.

13-4 RIGIDIZED VACUUM-FORMING

In a process sometimes called *rigidized shell sprayup*, a thermoplastic sheet is thermoformed into the desired shape, eliminating the gel coat. PVC, PMMA, ABS, and PC are commonly used. This shell is reinforced (by spray or hand layup) on the back side to produce a strong composite bathtub, sink, bathtub-shower combination, small boat, exterior signs, car top carrier, or other similar products. This method is illustrated in Figure 13-8. The list below gives four advantages and three disadvantages of rigidized vacuum-forming.

Advantages of Rigidized Vacuum-Forming

1. Thermoplastic skin gives smooth finish
2. Eliminates surface gel coat voids and gel time
3. Requires only a thermoforming mold
4. Output rates are faster than those obtained by sprayup methods

Disadvantages of Rigidized Vacuum-Forming

1. Needs thermoforming equipment and sheet storage space
2. Costly materials (thermoplastic surface sheets)
3. Repair of damaged surface sheets is difficult

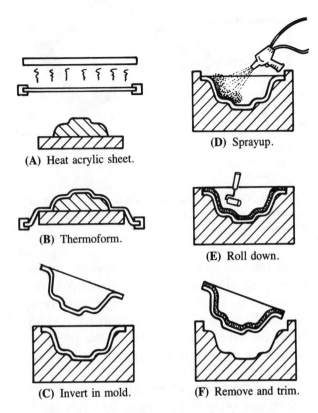

(A) Heat acrylic sheet.

(B) Thermoform.

(C) Invert in mold.

(D) Sprayup.

(E) Roll down.

(F) Remove and trim.

Fig. 13-8. The rigidized vacuum-forming process.

Cold-Mold Thermoforming

This process is like rigidized vacuum-forming, but the major differences are that two surfaces are finished, and that tolerances are more closely controlled. The process involves thermoforming a sheet, then reinforcing the back side by preform, mat or sprayup methods. Next, the composite is pressed between matched dies until cured. Polymerization is done by chemical means at room temperature. A major disadvantage is the additional die cost and the curing time in the mold. See Chapter 10, Cold Molding.

13-5 VACUUM-BAG

During vacuum-bag processing, a plastics film (usually polyvinyl alcohol, neoprene, polyethylene, or polyester) is placed over the layup. About 25 in. of mercury [85 kPa of vacuum] is drawn between the film and the mold (Fig. 13-9).

Vacuum is usually measured in millimetres of mercury drawn in a graduated tube, or in pascals of pressure. Vacuum in millimetres of mercury corresponding to 85 kPa can be calculated using this formula:

$$\frac{101 \text{ kPa}}{85 \text{ kPa}} = \frac{760 \text{ mm}}{x}$$

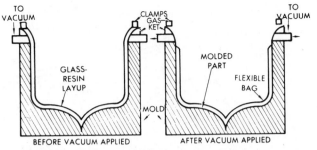

(A) Application of pressure to a layup results in improved strength and a better surface on the unfinished side of the product.

(B) Cargo carrier is vacuum-bag formed, using three layers of prepreg and one of dry glass cloth. (FMC Corp.)

(C) Cargo carrier installed on aircraft. (FMC Corp.)

Fig. 13-9. Vacuum-bag processing.

or $x = 656$ mm of mercury

where $x = $ unknown (mm of mercury)

101 kPa = known atmospheric pressure,

760 mm = mm of mercury corresponding to 101 kPa of pressure.

The plastics film forces the reinforcing material against the mold surface, producing a high-density product free of air bubbles. Tooling for vacuum bag processing is costly when large pieces are made. Output is slow compared to the high-speed production rates for injection molding.

Both male and female tooling are used. If a smooth surface is required on the exterior of a boat hull, a female mold would be selected. A male mold would probably be selected for a sink. Because heat is required in many operations, ceramic or metal tooling is used. Infrared induction, dielectric, xeone flash, or beam radiation can be used to aid or speed cure.

The mold surface must be protected to allow removal of the finished composite. Plastic films, waxes, silicone resins, PE, PTFE, PVAl, polyester (Mylar), and polyamide films are used as releasing agents.

Popular *wet layup* resins are epoxies and polyesters. SMC, TMC, and other pre-pregs of polysulfone, polyimide, phenolics, diallyl phthalate, silicones or other resin systems may be used.

Reinforcements may include honeycomb materials and mats, fabrics, paper, foils, or other pre-impregnated forms.

Wet layup vacuum-bag processing is illustrated in Figure 13-10. After the tooling has been carefully protected with a releasing wax or film (depending upon part geometry), a peel ply of finely woven polyester or polyamide fabric is carefully positioned. Sometimes a sacrificial ply (usually a fine, resin-impregnated fabric) is placed on the mold surface. The laminate layers are then placed in a specific design pattern on the peel ply. A second peel ply is placed on the laminate layers, followed by a release film or fabric. Dacron and

Teflon are commonly used as release fabrics. Because perforations allow air and excess resin to escape, this layer is sometimes called the breather ply. Bleeder plies of cloth or mat are laid on the release fabric to collect air and resin that are forced in. On some composite compositions a caul plate is used to insure a smooth surface and minimize variations in temperature during the curing process. Several vent or breather plies are then laid so air can freely pass along the surface of the part inside the bag. The bag can be made of any flexible material that is air-tight and won't dissolve in the matrix. Silicone rubber blanket, Neoprene, natural rubber, PE, PVAl, cellophane, or PA are commonly used. A vacuum of 25 inches of mercury, or about 12 psi of external pressure, is then drawn. To prevent excess liquid resin from being drawn into the vacuum lines, a resin trap is used. When additional density or difficult design requirements are needed, pressure-bag, rubber-plunger, rubber bag, autoclave, and hydroclave forming techniques are used.

Dry pre-impregnated materials are usually more difficult to form into complex shapes. Additional pressure, plug assistance, and external heat sources are used to soften and to aid in shaping the composite against the tooling.

13-6 PRESSURE-BAG

Pressure-bag processing is also costly and slow, but large, dense products with good finishes both inside and out are possible. Pressure-bag processing

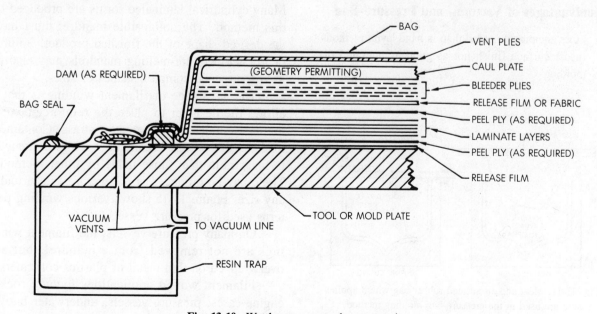

Fig. 13-10. Wet layup vacuum-bag processing.

uses a rubber bag to force the laminating compound against the contours of the mold. About 5.1 psi [35 kPa] of pressure is applied to the bag during the heating and curing cycle (Fig. 13-11). Pressures seldom exceed 50.8 psi [350 kPa].

The mold and compounds may be placed in a steam or heated gas autoclave after layup. Autoclave pressures of 50.8 to 101.5 psi [350 to 700 kPa] will achieve greater glass loading and aid in air removal.

The term *hydroclave* implies that a hot fluid is used to press the plies against the mold. In all pressure designs, the tooling (including the flexible bag) must be able to withstand the molding pressures. Pressure-bag techniques to force the layup against the mold walls may be used for long hollow pipes, tubes, tanks or, other objects with parallel walls. At least one end of the object must be open to insert and remove the bag.

Three advantages and four disadvantages of vacuum and pressure-bag processing are found in the following list.

Advantages of Vacuum- and Pressure-Bag

1. Greater glass loading and fewer voids than hand layup methods
2. Inside surface has better finish than hand layup methods
3. Better adhesion in composites

Disadvantages of Vacuum- and Pressure-Bag

1. More equipment needed than in hand layup methods
2. Inside surface finish not as good as matched die molding

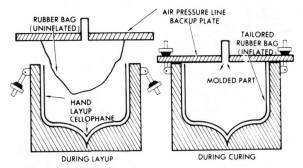

Fig. 13-11. Heat and an inflated rubber bag which applies pressure are used in the pressure-bag molding method.

3. Quality depends on skill of operator
4. Cycle times long, limiting production with single mold

13-7 FILAMENT WINDING

Filament winding produces strong parts by winding continuous fibrous reinforcements on a mold.

Long continuous filaments are able to carry more load than random, short filaments. Over 80 percent of all filament winding is accomplished with E-glass roving. Higher modulus fibers of carbon, aramid, or Kevlar may be used. For some applications, boron, wire, beryllium, polyamides, polyimides, polysulfones, bisphenol, polyesters, and other polymers are also used. Specially designed winding machines may lay down these strands in a predetermined pattern to give maximum strength in the desired direction (Fig. 13-12). During *wet winding,* excess resin matrix and entrapped air are forced (squeezed out from between strands). Filament winding tension varies from 0.25 to 1 pound per end (a group of filaments). See Figure 13-13 showing wet filament winding. Note restrictions to shape of filament wound parts.

In *dry winding* pre-impregnated, B-stage reinforcements help to insure consistency in resin-to-reinforcement content design. These pre-impregnated reinforcements may be machine- or hand-wound on the tooling (Fig. 13-14). Curing may be accelerated by heated mandrels (tooling), ambient ovens, chemical hardeners, or other energy sources. Many cylindrical laminated forms are produced by this method. The collapsible mandrel must have the desired shape of the finished product. Soluble or low-temperature-melting mandrels may also be used for special complex shapes or sizes.

The advantage of filament winding is that it allows the designer to place the reinforcement in the areas subject to the greatest stress. Containers made by this process usually have a higher strength-to-mass ratio than those made by other methods. They may be produced at a lower cost in virtually any size. Figure 13-15 shows various winding patterns used for pressure vessels.

On many pressure vessels, the filament windings are not removed from a mandrel, but are overwrapped on thin metal or plastics containers.

Filament wound applications include rocket engine cases, pressure vessels, underwater buoys,

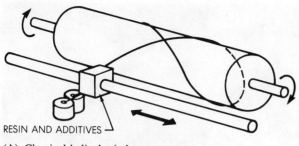

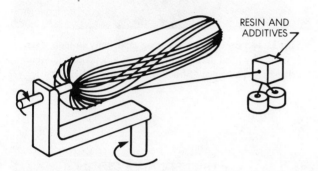

(A) Classical helical winder.

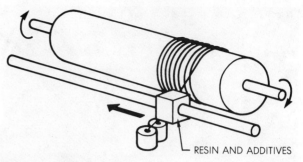

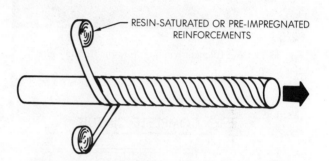

(B) Circumferential winder.

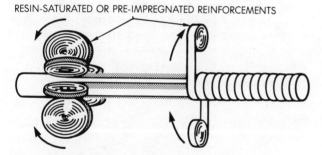

(C) Polar winder.

RESIN-SATURATED OR PRE-IMPREGNATED
REINFORCEMENTS

(D) Continuous helical winder.

RESIN-SATURATED OR PRE-IMPREGNATED REINFORCEMENTS

(E) Continuous normal-axial winder.

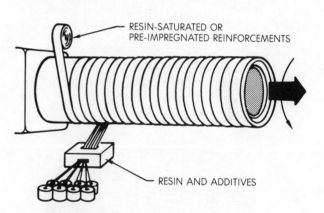

RESIN-SATURATED OR
PRE-IMPREGNATED REINFORCEMENTS

RESIN AND ADDITIVES

(F) Continuous rotating mandrel with wrap.

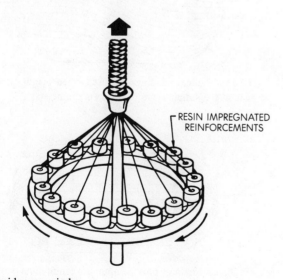

RESIN IMPREGNATED
REINFORCEMENTS

(G) Braid-wrap winder.

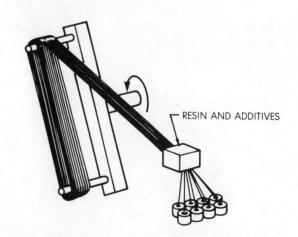

RESIN AND ADDITIVES

(H) Loop-wrap winder.

Fig. 13-12. Selected winding methods and designs.

Fig. 13-13. Wet filament winding. (Owens-Corning Fiberglas Corp.)

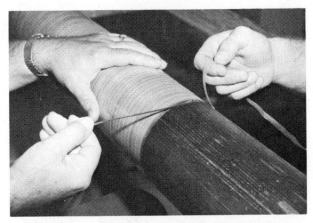

Fig. 13-14. Boron-epoxy prepreg tape, only ⅛ inch wide, is wound by hand during production of a helicopter tail-rotor driveshaft. (Advanced Structures Division, TRE Corp.)

TO COMPLETE CIRCUIT

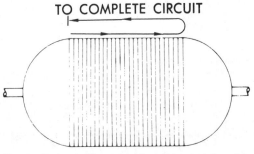

(A) Circular loop windings provide optimum girth or loop strength in a filament wound structure.

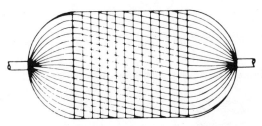

(B) Single circuit helical windings combined with circular loop windings provide high axial tensile strength.

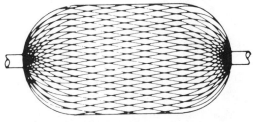

(C) Multiple circuit helical windings allow optimum use of the glass filament's strain characteristics, without the addition of loop windings.

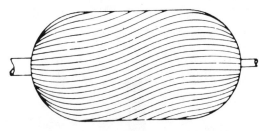

(D) Dual helical windings are used when openings at the ends of the structure are of different diameters.

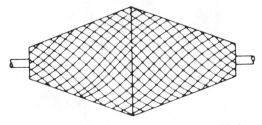

(E) Variable helical windings can produce odd-shaped structures.

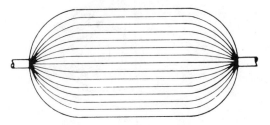

(F) Planar windings provide optimum longitudinal strength (with respect to the winding axis).

Fig. 13-15. Advantages of various types of filament winding. (CIBA-GEIGY)

radomes, nose cones, storage tanks, pipers, automotive leaf springs, helicopter blades, spacecraft spars, fuselage, and other aerospace parts.

13-8 CENTRIFUGAL REINFORCING

In centrifugal reinforcing, the resin and reinforced materials are formed against the mold surface as it rotates (Fig. 13-16). During this rotation resin is distributed uniformly through the reinforcement by centrifugal force. Heat is then applied to help polymerize the resin. Tanks and tubing may be produced in this manner. (See Rotational Casting in Chapter 11)

13-9 PULTRUSION

In pultrusion, resin-soaked matting or rovings (along with other fillers) are pulled through a long die heated to between 250 to 300 °F [120 and 150 °C]. The product is shaped and the resin polymerized as it is drawn through the die. Radiofrequency or microwave heating may also be used to speed production rates. The process appears to be like that used for extrusion. In the extrusion process the homogeneous material is *pushed* through the die opening. In pultrusion, however, resin-soaked reinforcements are *pulled* through a heated die where the resin is cured (Fig. 13-17).

Pultrusion does are generally 24 to 60 inches in length and heated to aid in the polymerization process. Cure must be carefully controlled to prevent cracking, delamination, incomplete cure, or sticking to the die surfaces.

Output varies from a few millimetres to over 3 m/min. The many resins in use include vinyl esters, polyesters, and epoxies. Fibrous glass is the

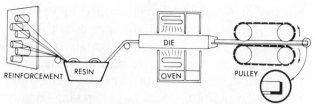

(A) Basic scheme of continuous pultrusion. (Morrison Molded Fiber Glass Co.)

(B) Structural supporting members of fiber glass reinforced polyester for the operating floor of this chemical mixing plant were produced by pultrusion. (Morrison Molded Fiber Glass Co.)

Fig. 13-17. Pultrusion method and a product application.

most widely used reinforcement, although asbestos, graphite, carbon, boron, polyester, and polyamide fibers may be used. Reinforcements can be positioned in the pultrusion product where needs for strength are the highest.

Hot melt thermoplastic materials and reinforcements may also be used. Parallel orientation of reinforcements produces a strong composite in the direction of the fibers. Some operations may use SMC or wound preforms in combination with other continuous reinforcements to improve omnidirectional properties.

Siding, gutters, I-beams, fishing rods, automotive springs, frames, airfoils, hammer handles, skis, tent poles, golf shafts, ladders, tennis racquets, vaulting poles, and other profile shapes are some items produced by pultrusion (Fig. 13-17).

Pulforming is a variation of pultrusion. As materials are pulled from the reinforcement creels and impregnated with resin and other compounds a

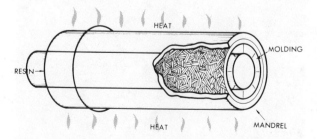

Fig. 13-16. In the centrifugal method, chopped reinforcement and resin are evenly distributed on the inner surface of a hollow mandrel. The assembly rotates inside an oven to provide heat for curing.

number of forming devices (molds) of various cross-sectional shapes form the composite part. In one method, rotary male and female dies are brought together on the pultruded material and the composite cured. Curved parts can be formed by forcing the pulform into a large circular female mold with a flexible steel belt. The mold and belt are heated to speed cure in a continuous pulformed operation. See Figure 13-18. Applications include hammer handles, bows, curved springs, and other products that do not have a continuous cross-sectional shape.

13-10 COLD STAMPING/FORMING

Fibrous-glass reinforced thermoplastics, available in sheet form, may be cold-formed much like metals (Fig. 13-19). Long reinforcements are used to improve the strength-to-mass ratio.

(A) Glass rovings feeding through wet-out tank.

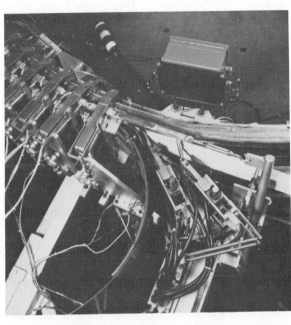

(B) Heated belt die closure on cure section.

(C) Spring stock exiting the die/belt section.

(D) Flying spring stock cut-off saw.

Fig. 13-18. Pulforming of a curved composite leaf spring. (Goldsworthy Engineering).

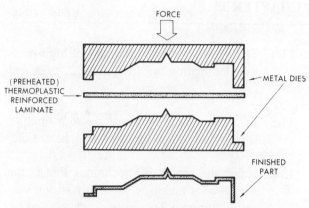

(A) Composite is preheated and formed on cold conventional metal-stamping dies and equipment.

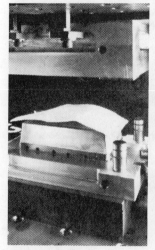 

(B) Preheated sheet is formed between cooled matched metal dies. (G. R. T. L. Co.)

(C) Since the blank is smaller than the finished part, it flows out to the periphery of the dies without forming flash or trim. (G. R. T. L. Co.)

Fig. 13-19. Cold stamping.

During the forming operation, the sheet is preheated to about 392 °F [200 °C] and then formed on normal metal stamping presses. It is possible to produce parts with complex designs and varying wall thickness with this method. Production rates may exceed 260 parts per hour. Uses include motor covers, fan guards, wheel covers, battery trays, lamp housings, seat backs, and many interior automotive trim panels.

According to one authority, stampable reinforced thermoplastic composite sheets could become a replacement for stamped steel from Detroit. Many of these sheets are paintable with a class-A finish out from the mold. Non-class-A automotive finish applications account for about 80 percent of the glass/PP sheet demand. PC/PBT, PPO/PBT and PPO/PA are alloys combined with modified glass mat or other special reinforcements in stampable, formable sheets.

Non-impregnated *commingled* blends of continuous thermoplastic filaments (PEEK, PPS) with reinforcing filaments of carbon, glass aramid, or metals may be made into yarns, fabrics, or felts. Woven, braided, or knitted three-dimensional commingled preforms are then heated under pressure in the mold. The thermoplastic filaments melt and wet out the adjacent reinforcements. These forms are a versatile material for composites.

VOCABULARY

The following vocabulary words are found in this chapter. Look up the definition of any of these words you do not understand as they apply to plastics in the glossary, Appendix A.

Autoclave
Biaxial winding
Blanket
Bleeder cloth
Breather
Caul plate
Commingled
Fiber orientation
Filament winding
Gel coat
Mandrel
Polar winding
Pressure-bag molding
Pultrusion
Reinforced molding compound
Reinforced plastics
Resin rich area
Sprayup

13-1. Plastics with strength properties increased by the addition of filler and reinforcing fibers to the base resin are called _____ .

13-2. The thin unreinforced layer of resin placed on the surface of a mold in the hand layup process is called a _____ .

13-3. The process of using reinforcements to improve selected properties of plastics parts is called _____ .

13-4. The two major disadvantages of high-pressure lamination are low _____ rates and high _____ pressures.

13-5. In _____ molding, resin soaked matting or rovings are pulled through a long heated die.

13-6. Name the processing technique that is the simplest for making a relatively strong vessel with a complex shape.

In this chapter, you will learn that in plastics casting processes the material is placed into a mold and allowed to harden. These casting processes are similar to those used for making candles, baking a cake, or casting metal, ceramic, or glass.

To fill a mold using only atmospheric pressure, the polymer must approach a liquid state. Acetals, PC, PP, and many others simply do not become liquid enough to flow into molds. Most polymer melts must be forced (molded or extruded) into mold cavities with pressures exceeding 5 000 psi. Most hot polymers have a viscosity similar to bread dough. Most monomers are more like pancake syrup.

Casting includes a number of processes in which monomers, modified monomers, powders, or solvent solutions are poured into a mold, where they become a solid plastics mass. The plastics state may be achieved by evaporation, chemical action, cooling, or external heat. No pressure is applied in casting, only the force of the mass itself and atmospheric pressures are involved. The cast material solidifies in the mold and the product is removed from the mold, or form, after curing. This process is like the casting of metals. Review casting compounds in Chapter 5.

The early castings of phenolic by Baekeland were dark colored products with limited uses. By contrast, today, both thermosetting and thermoplastic resins of many colors are cast into a great number of useful products. The major advantages of casting are that large, inexpensive molds can be used, small numbers of items can be produced economically, and parts may also be stress-free.

Casting techniques may be placed into six distinct groups: simple casting, film casting, hot-melt casting, slush casting and static casting, rotational casting (sometimes called rotational molding), and dip casting.

14-1 SIMPLE CASTING

In simple casting, liquid resins or molten plastics are poured into molds and allowed to polymerize or cool. The plastics part is then removed from the mold for further processing if necessary. Simple castings find uses as rods, tubes, cylinders, sheets, and other shapes that are easily fabricated or machined into finished products. Potting, encapsulation, and embedment are also classed as simple casting. Foams may be cast, but are discussed under the foaming processes in Chapter 16. Some cast and reinforced toolings and fixtures are included as simple castings.

In simple casting, molds may be made of wood, metal, plaster, selected plastics, selected elastomers, or glass. Silicones, for example, are often cast over patterns to make molds in which plastics or other materials may be cast. Common casting resins include acrylics, polyesters, silicones, epoxies, phenolics, ethyl cellulose, cellulose acetate butyrate, and polyurethanes. Probably the most well known is polyester resin because it is used in crafts and hobby work. Water-extended polyesters and polyurethanes are used to cast furniture and cabinet parts. (Fig. 14-1).

Examples of products made by simple casting include jewelry, billiard balls, cast sheets for windows, furniture parts, watch crystals, sunglass lenses, handles for tools, desk sets, knobs, table tops, sinks, and fancy buttons. Figure 14-2 shows the basic principle of simple casting.

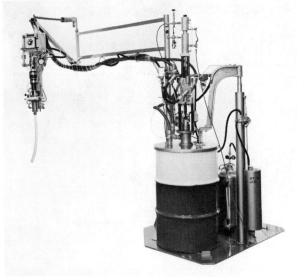

(A) This machine blends components for polyester products produced by simple casting. (Pyles Industries, Inc.)

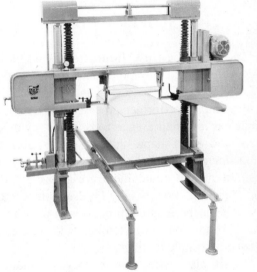

(B) A large block of cellular plastics, cast in an open mold, will be cut into slabs by this machine. (McNeil Akron Corp.)

(C) Water-extended polyester was used to make these picture and mirror frames by simple casting.

Fig. 14-1. Simple casting and cast products.

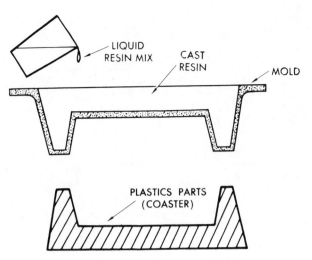

Fig. 14-2. Solid casting in an open, one piece plastics mold.

A number of thermoplastics and thermosets may be cast into sheet form. Acrylic sheets are often produced by pouring a catalyzed monomer or partially polymerized resin between two parallel plates of glass (Fig. 14-3). The glass is sealed with a gasket material to prevent leakage and to help control the thickness of the cast sheet. After the resin has fully polymerized in an oven or autoclave, the acrylic sheet is separated from the glass plates and reheated to relieve stresses that occur during the casting process. The faces are covered with masking paper to protect the sheet during shipment, handling, and fabrication. Sheets may be purchased untrimmed, with the sealing material still sticking to the edges (Fig. 14-4).

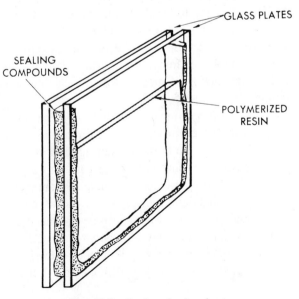

Fig. 14-3. Casting plastics sheets.

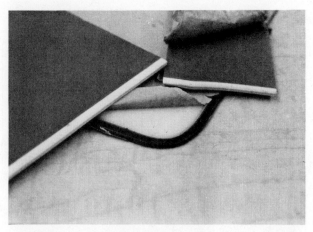

Fig. 14-4. Sealing materials still adhere to the edges of these untrimmed acrylic plastics sheets.

When heat processing or solvent casting cannot produce film or sheet stocks of required thickness, another process, called skiving, is used. Sheets of cellulose nitrate and other plastics may be sliced (skived) from blocks that have been softened by solvents. After the residual solvents have evaporated, the skived piece is pressed between polished plates to improve the surface finish (Fig. 14-5).

14-2 FILM CASTING

The casting of film involves dissolving plastics granules or powder, along with plasticizers, colorants, or other additives, in a suitable solvent. The solvent solution of plastics is then poured onto a stainless steel belt. The solvents are evaporated by the addition of heat, and the film deposit is left on the moving belt. The film is stripped or removed and wound on a takeup roller (Fig. 14-6). This film may be cast as a coating or laminate directly on fabric, paper, or other substrates.

Other cast film uses include water-soluble packaging for bleaches and detergents and skin and blister packaging.

Solvent casting of film offers the following three advantages over other heat melt processes:

1. Additives for heat stabilization and for lubrication are not needed
2. Films are uniform in thickness and optically clear
3. No orientation or stress is possible with this method

To be economically feasible, solvent casting of film requires a solvent recovery system. Plastics that may be solvent cast include cellulose acetate, cellulose butyrate, cellulose propionate, ethyl cellu-

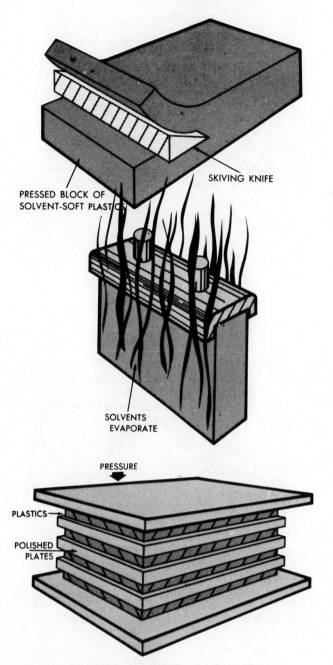

Fig. 14-5. Skiving sheets from a block of plastics.

lose, polyvinyl chloride, polymethyl methacrylate, polycarbonate, polyvinyl alcohol, and other copolymers. Casting liquid plastics latexes on Teflon-coated surfaces, rather than stainless steel, may also be used to produce special films.

Aqueous dispersions of polytetrafluoroethylene and polyvinyl fluoride are cast on heated belts at temperatures below their melting points. This method provides a handy way to make films and sheets of materials hard to process by other means. These films are used as nonstick coatings,

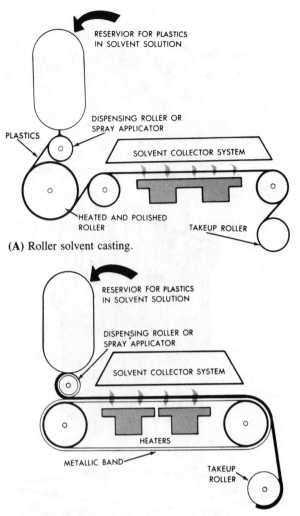

(A) Roller solvent casting.

(B) Band solvent casting.

Fig. 14-6. Film casting.

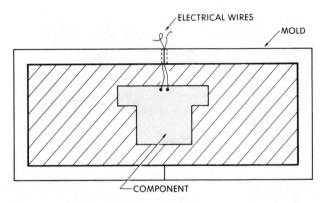

Fig. 14-7. Hot-melt encapsulation of an electronic component.

Electrical parts may be protected from hostile environments by being placed in molds and having hot resin poured over the components. When cool, the plastics provides protection for wires and vital parts. The encapsulated or potted components may then be placed with other assemblies to produce the finished product. Some encapsulations and pottings are not cast in separate molds but are produced by pouring the molten compound directly over the components inside the case of the finished product. The insulation of parts in a radio chassis or a motor is a well-known example. If the components are cast in place and not removed from a mold shape, they must be classed as coatings.

gasket material, and sealing components for pipes and joints.

14-3 HOT-MELT CASTING

Hot-melt plastics were used for casting during World War II. Today, hot-melt formulations may be based on ethyl cellulose, cellulose acetate butyrate, polyamide, butyl methacrylate, polyethylene, and other mixtures. The largest use is for strippable coatings and adhesives. Hot-melt resins may be used for making molds for casting other materials. Also, hot-melt resins are used in a casting process for potting and encapsulation (Fig. 14-7). Not all potting compounds are thermoplastic and hot-melting. Silicone is most often used for coating, sealing, and casting, but epoxy and polyester resins are also used for these purposes.

14-4 SLUSH CASTING AND STATIC CASTING

Slush casting involves pouring dispersions of polyvinyl chloride or other plastics into a heated, hollow, open mold. As the material strikes the walls of the mold, it begins to solidify (Fig. 14-8A). The wall thickness of the molded part increases as the temperature is increased or solution is left in the hot mold. When the desired wall thickness is reached, the excess material is poured from the mold and the mold is then placed in an oven until the plastics fuses together or evaporation of solvents is complete. After water cooling, the mold is opened and the product removed. Commercial molds are usually made from aluminum since this metal allows rapid cycling and lower tooling costs. Ceramic, steel, plaster, or plastics molds may be used also. Vibrating, spinning, or the use of vacuum chambers may be necessary to drive out air bubbles in the plastisol product.

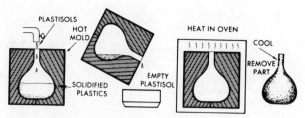

(A) Basic slush casting with plastisols.

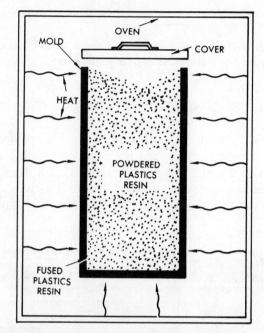

(B) Static casting with dry thermoplastic powders.

Fig. 14-8. Principle of slush casting.

Thermoplastic powders are also used in a dry process sometimes called *static casting*. The metal mold is filled with powdered plastics and placed in a hot oven (Fig. 11-8B). As the heat penetrates the mold, the powder melts and fuses to the mold wall. When the desired wall thickness is obtained, the excess powder is removed from the mold. The mold is then returned to the oven until all powder particles

have completely fused together. Huge storage tanks and containers with heavy walls are examples of products made by this casting method. Cellular polystyrene or polyurethanes may be placed in the remaining space in manufacture of tough-skinned flotation devices.

Slush-cast items are hollow but will have an opening like that found in doll parts, syringe bulbs, and special containers. Any design in the mold will be on the outside of the product.

In a related process known as *vibrational microlamination* (VIM), a combination of heat and vibration is used. Thin layers of homopolymer alternating with reinforced layers are used to produce huge storage tanks, hollow toys, syringe bulbs, or other containers ((Fig. 14-9).

14-5 ROTATIONAL CASTING

Rotational casting (sometimes called centrifugal casting) is similar to slush casting. (The process is also sometimes incorrectly called rotational molding.) Rotational casting rotates on two planes, while centrifugal casting rotates on one plane. Large pipes and tubes are sometimes centrifugal-cast. It may be used for hollow, completely closed objects such as balls, toys, containers, and industrial parts including armrests, sun-visors, fuel tanks, and floats.

Plastics powders, monomers, or dispersions are measured and placed in multi-piece aluminum molds. The mold is then placed in an oven and rotated in two planes (axes) at the same time (Fig. 14-10). This action spreads the material evenly on the walls of the hot mold. The plastics melts and fuses as it touches the hot mold surfaces, making a one-piece coating. The heating cycle is complete when all powders or dispersions

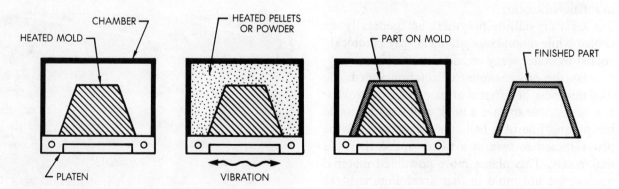

Fig. 14-9. Vibrational microlamination (VIM) Different formulations, types of plastics, and/or reinforcements may be alternately used to produce a true laminate.

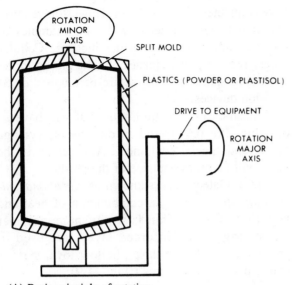

(A) Basic principle of rotation.

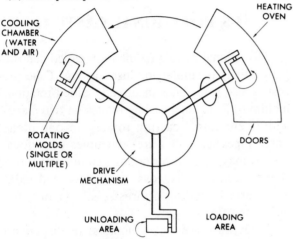

(B) Top view of three-mold rotational unit.

Fig. 14-10. Principle of rotational casting.

have melted and fused together, but the mold continues to rotate as it enters a cooling chamber. Finally, the cooled plastics product is removed.

Nearly all thermoplastic powders may be used in rotational casting.

Cast crystalline polymers are generally air cooled while amorphous polymers may be quickly cooled by water spray or bath.

By the programming of rotation speed, the wall thickness in different areas may be controlled. If it is desirable to have a thick wall section around the parting line of a ball, the minor axis can be programmed to turn at a faster speed than the major axis. This places more powdered material against the hot mold in that area. Figure 14-11 shows how this is done. Rotational casting equipment is shown in Figure 14-12.

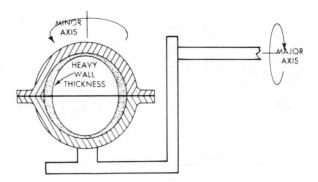

Fig. 14-11. Wall thickness in a ball can be varied by rotating the minor axis faster than the major axis.

Pipes, ducts, submarine launcher tubes, and other shapes can be made by rotating the mold on one axis (centrifugal casting). Inflatable mandrels and pre-preg, reinforced materials may be placed next to the skin layer to produce a high density composite with ribs or other geometric designs. In one operation wet layup is placed on the mold

(A) Large rotational-cast product removed from mold. (Plastics Design Forum)

(B) Rotational casting of large tank with spray water-cooling. (McNeil Femco Corp.)

(C) Rotational casting eight parts at once. (McNeil Femco Corp.)

Fig. 14-12. Rotational casting machines.

(B) This horse was rotational-cast from a three-piece mold. (McNeil Femco Corp.)

wall. Centrifugal force causes the reinforcement and matrix to take the shape of the mold.

Rotational-cast parts may be produced as large commercial containers, tubing (with ends cut open), ice chests, pans, and boxes. Luggage may be cast as one piece which is then cut at the seam to form two perfectly fitting halves. Figure 14-13 shows some rotational-cast items.

Foam filled and double walled items, including true composites, can be produced. Short fiber reinforcements are used, but care must be taken to prevent wicking of protruding reinforcements. Placing a homopolymer layer over the reinforced polymer may overcome this problem. In one opera-

(C) Rotational-cast trash container. (Phillips Petroleum Co.)

Fig. 14-13. Rotational cast products.

tion a solid outer skin layer is produced, followed by the release of a second charge of material from a "dump box" in the mold.

14-6 DIP CASTING

Dip casting should not be confused with dip coating. (This process is also sometimes incorrectly called dip molding.) Coatings are not re-

(A) Furniture made from rotational casting. (Design Forum)

moved from the substrate. In dip casting, a preheated mandrel the shape and size of the *inside* of the product is lowered into a plastisol dispersion (Fig. 14-14). As the resin hits the hot mold sur-

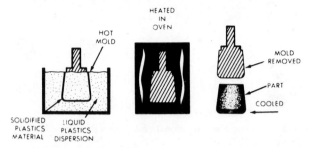

(A) Scheme of dip casting.

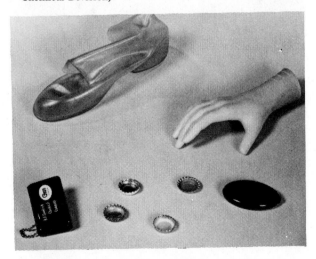

(B) Dip-cast parts for recreational uses. (BF Goodrich Chemical Division)

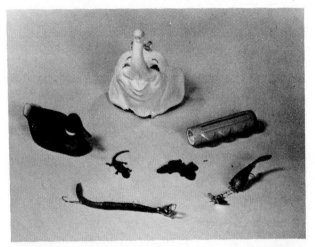

(C) Assorted dip-cast items. (BF Goodrich Chemical Division)

Fig. 14-14. Dip casting and resulting products.

face, it begins to melt and fuse. The thickness of the piece continues to increase as it remains in the solution. If additional thickness is desired, the coated piece may be reheated and dipped again. After the desired thickness is obtained, the mold is removed from the oven and cooled and the part is then stripped from the mold. Thickness of the product is determined by the temperature of the mold and the time the hot mold is in the plastisol.

Several layers or alternate colors and formulations may be applied by alternately heating and dipping. Any design on the mold will be on the inside of the product.

Plastics gloves, overshoes, coin purses, spark plug covers, and toys are examples of dip-cast products. Listed below are six advantages and four disadvantages of casting processes.

Advantages of Casting

1. Costs of equipment, tooling, and molds are low
2. Not a complex forming method
3. Wide number of treatment techniques
4. Products have little or no stress
5. Relatively low material costs
6. Rotational casting produces one-piece hollow objects

Disadvantages of Casting

1. Low output rate and high cycle time
2. Dimensional accuracy is only fair
3. Moisture and air bubbles may be problems
4. Solvents and other additives may be dangerous

VOCABULARY

The following vocabulary words are found in this chapter. Look up the definition of any of these words you do not understand as they apply to plastics in the glossary, Appendix A.

Casting
Dip casting
Encapsulating
Potting
Rotational casting
Slush casting
Vibrational microlamination

14-1. Identify the amount of pressure required for casting processes.

14-2. List three reasons for taking precautions when using polyester resins and catalyst.

14-3. The process of submerging a hot mold into a resin and removing the plastics from the molds is called ___?___ .

14-4. A major disadvantage of rotational casting is the ___?___ .

14-5. Name three methods that may drive out air bubbles in simple or plastisol castings.

14-6. What materials can be used as molds for dip casting?

14-7. Name five materials that may be used for molds in the casting process.

14-8. Identify the plastics sheets that are often cast between two polished sheets of glass. Protective paper is then applied to the surface.

14-9. Name six advantages of casting processes.

14-10. List three typical products produced by rotational casting.

14-11. Molds for static casting are normally made of ___?___ .

14-12. What determines the wall thickness of dip castings?

14-13. Name three measures that would help to reduce the problem of air bubbles in simple castings.

14-14. Name a process similar to slush casting where dry thermoplastic powders are used.

14-15. Identify the major difference between dip coating and dip casting.

14-16. Hollow, one-piece objects may be made by the ___?___ casting process.

14-17. In order to completely fuse the powders or dispersions, a second heating cycle is necessary with slush or ___?___ casting.

14-18. Because easily shaped tooling materials are used, casting molds are normally ___?___ expensive than molds for injection molding.

14-19. What is the major disadvantage with all casting operations?

14-20. What two parameters determine the wall thickness of a rotational casting?

14-21. If a plastics is not easily processed by heat methods, what process may be used to make very thin films?

14-22. Which casting processes would you select to make each of the following products?
(a) window glazing
(b) fuel tank
(c) coin purse
(d) furniture parts

14-23. Plastisol products are popular because inexpensive methods and molds are used with ___?___ casting.

14-24. Name four reasons why castings processes are less costly than molding operations:

14-25. Name the common cause of pits or pockmarks in or on castings.

In this chapter, eleven basic thermoforming techniques are discussed. In each technique, a heated thermoplastic product is formed and allowed to cool.

Thermoforming processes are possible because thermoplastic sheets can be softened and reshaped, and the new shape is retained when the material is cooled. Most thermoplastic materials may be formed by this process; however, acetals, polyamides, and fluorocarbons are not usually thermoformed. Extruded, calendered, laminated, cast, and blown films or sheet forms may be thermoformed.

Forcing a heated thermoplastic material to take the shape of a mold by mechanical, air, or vacuum pressure is common. Considerable energy, time, and space can be saved by directly thermoforming sheets as they leave the extruder.

Both male (plug) and female (cavity) molds are used. Molds require sufficient draft to assure stress-free parts removal. Tooling costs are usually low, and parts with large surface areas may be produced economically. Prototypes and short runs are also practical. Although dimensional accuracy is good, thinning is a problem in some part designs.

Tooling can run from low-cost plaster molds to expensive water-cooled steel molds, but the most common tooling material is cast aluminum. Wood, gypsum, hardboard, pressed wood, cast phenolic resins, filled or unfilled polyester or epoxy resins, sprayed metal, and steel also may be used for molds.

One source dates thermoforming back to the ancient Egyptians. They found that animal horns and tortoise shells could be heated and formed into a variety of vessels and shapes. In the United States, John Hyatt thermoformed Celluloid sheets over wooden cores for piano keys.

Today, sheets and films may be thermoformed by the basic techniques of straight vacuum forming, drape forming, matched-mold forming, pressure-bubble plug-assist vacuum forming, plug-assist vacuum forming, plug-assist pressure forming, vacuum snap-back forming, pressure-bubble vacuum snap-back forming, trapped-sheet contact-pressure forming, free forming, and mechanical forming.

Items produced by thermoforming include signs, light fixtures, ice-cube trays, ducts, drawers, instrument panels, tote trays, housewares, toys, refrigerator panels, transparent aircraft enclosures, and boat windshields (Fig. 15-1). Blister and skin packaging of products are familiar applications of thermoforming. Replacement parts and hardware are examples of items that are sometimes skin packaged. Skin packaging requires no mold; the plastics film is simply formed over the product. Cookies, pills, and other products are commonly packaged by blister packaging. Single portions of butter, jellies, and other foods are sometimes packaged in blister packs.

Expanded polystyrene is formed into meat trays and egg cartons. Crystallized PET and multilayered composites are used for retortable and dual-ovenable food trays and bowls.

True composites based on thermoplastic matrices are thermoformed. These fibrous, reinforced materials may be arranged in plies to achieve isotropic or anisotropic mechanical properties with different cross-sectional thicknesses. Although there is some local movement and slipping of the warp and weft fibers, the laminate does not thin and stretch in the mold. Because of the viscous nature of the composites, matched-mold and pressure forming techniques are used.

(A) Sports car with body made of panels that were vacuum-formed, glued together, and painted. (U.S. Gypsum)

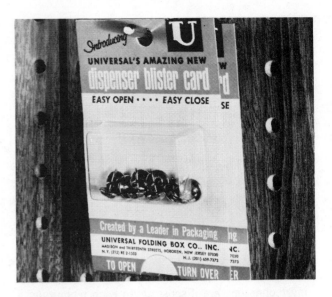

(B) Blister package with reclosable sliding door. (Celanese Plastic Materials Co.)

(C) Deep-draw, straight-walled containers are thermo-formed. (Shell)

(D) Clear plastics frames for a display package are thermo-formed on a continuous web from plastics sheet. (Celanese Plastic Materials Co.)

Fig. 15-1. Some articles fabricated by thermoforming processes.

Some modern industrial thermoforming machines are shown in Figure 15-2.

15-1 STRAIGHT VACUUM FORMING

Vacuum forming is the most versatile and widely used thermoforming process. Vacuum

(E) Vacuum-formed clear covers protect and display bakery products.

(A) High-speed pressure/vacuum former operates from either roll stock or inline with an extruder.

(B) Rotary style of unit used for large industrial components at a fairly high production rate.

(C) Twin-sheet thermoforming machine with separate, independent clamping frames.

Fig. 15-2. Modern industrial thermoforming machines. (Brown Machine Co.)

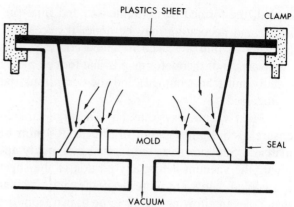

(A) A clamped and heated plastic sheet is forced down into the mold by air pressure after a vacuum is drawn in the mold. (Atlas Vac Machine Co.)

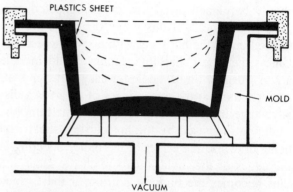

(B) Plastics sheet cools as it contacts the mold. (Atlas Vac Machine Co.)

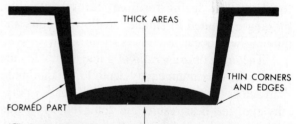

(C) Areas of the sheet that touched the mold last are the thinnest. (Atlas Vac Machine Co.)

(D) Note the frame that holds the heated thermoplastics sheet as it is drawn down over the mold. (Chemplex Co.)

Fig. 15-3. Straight vacuum forming.

equipment costs less than pressure or mechanical processing equipment.

In straight vacuum forming, a plastics sheet is clamped in a frame and heated. While the hot sheet is rubbery, or in an elastic state, it is placed over a female mold cavity. The air is removed from this cavity by vacuum (Fig. 15-3) and atmospheric pressure (10 kPa) forces the hot sheet against the walls and contours of the mold. When the plastics has cooled, the formed part is removed, and final finishing and decorating may be done, if necessary. Blowers or fans are used to speed cooling. One disadvantage of thermoforming is that formed pieces usually must be trimmed, and the scrap must be reprocessed.

Most vacuum systems have a surge tank to ensure a constant vacuum of 500 to 760 mm of mercury. Superior parts are formed by quickly applying the vacuum before any portion of the sheet has cooled. Slots are more desirable and efficient than holes in allowing the air to be drawn from the mold. Slots or holes should be smaller than 0.025 in. [0.65 mm] in diameter to avoid surface blemishes on the formed part. A hole or slot should be placed in all low or unconnected portions of the mold. If this is not done, air may be trapped under the hot sheet with no way to escape. Unless they are collapsible, molds should include a 2 to 7 degree (draft) angle for easy part removal.

Thinning at the upper edges of a part is a disadvantage in using relatively deep female molds. Thinning is caused by the hot plastics sheet first being drawn to the center of the mold. Sheeting at the edges of the mold must stretch the most and thus becomes the thinnest portion of the formed item. If preprinted flat sheets are formed, thinning must be kept in mind when trying to compensate for distortion during forming. Straight vacuum forming is limited to simple, shallow designs and thinning will occur in corners.

The *draw* or *draw ratio* of a female mold is the ratio of the maximum cavity depth to the minimum span across the top opening. For high-density polyethylene, the best results are achieved when this ratio does not exceed 0.7:1. Thermoforming equipment and dies are relatively inexpensive.

When the plastics has cooled, it is removed for trimming or postprocessing, if needed. Mark-off (marks from the mold) is on the *inside* of the product while such marks appear on the *outside* of the part in straight vacuum forming.

15-2 DRAPE FORMING

Drape forming (incorrectly called mechanical forming) is similar to straight vacuum forming except that after the plastics is framed and heated, it is mechanically stretched over a male mold. A vacuum (actually, a pressure differential) is applied that pushes the hot plastics against all portions of the mold (Fig. 15-4). The sheet touching the mold remains close to its original thickness. Side walls are formed by the material draping between the top edges of the mold and the bottom seal area at the base.

It is possible to drape-form items with a depth to diameter ratio of nearly 4:1. High draw ratios are possible with drape forming, however, this technique is also more complex. Male molds are easy to make and, as a rule, cost less than female ones, but male molds are more easily damaged.

Drape forming has also been used to form a hot plastics sheet over male or female molds by gravitational forces alone. Female molds are preferred for multicavity forming because there must be more spacing if male molds are selected.

15-3 MATCHED-MOLD FORMING

Matched-mold forming is similar to compression molding. A heated sheet is trapped and formed between male and female dies that may be

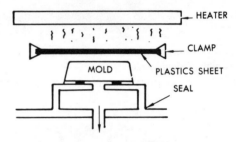

(A) Clamped heated plastics may be pulled over the mold, or the mold may be forced into the sheet.

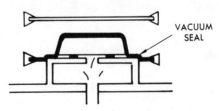

(B) Once the sheet has formed a seal around the mold, a vacuum is drawn to pull the plastics sheet tightly against the mold surface.

(C) Final wall thickness distribution in the molded part.

Fig. 15-4. Principle of drape forming plastics. (Atlas Vac Machine Co.)

made of wood, plaster, epoxy, or other materials (Fig. 15-5). Accurate parts with close-tolerances may be quickly produced in costly water-cooled molds. Very good molded detail and dimensional accuracy can be obtained with water-cooled molds, including lettering and grained surfaces. There is mark-off on both sides of the finished product; there-fore, mold dies *must* be protected from scratches or damage because such defects would be reproduced by the thermoplastic materials. A smooth-surface mold should not be used with polyolefins because air may be trapped between the hot plastics and a highly polished mold. Sandblasted mold surfaces are usually used for these materials.

Figure 15-6 shows a matched-mold operation in which the blank is smaller than the finished

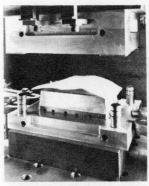 

(A) Preheated sheet is placed between cooled matched metal dies.

(B) The finished formed part is larger than the original sheet.

Fig. 15-6. Matched-mold forming being done on a conventional metal-stamping press. (G.R.T.L. Co.)

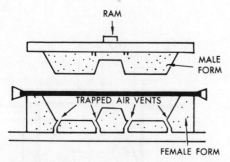

(A) The heated plastics sheet may be clamped over the female die, as shown, or draped over the mold form.

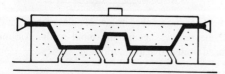

(B) Vents allow trapped air to escape as the mold closes and forms the part.

(C) Distribution of materials in the product depends on the shapes of the two dies.

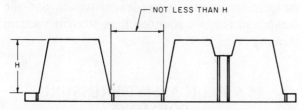

(D) Male mold forms must be spaced at a distance equal to or greater than their height or webbing may occur.

Fig. 15-5. Principle of matched-mold forming. (Atlas Vac Machine Co.)

part. Under the ram, it flows out to the periphery without forming trim or flash. The total mold cycle is 10 to 20 s.

15-4 PRESSURE-BUBBLE PLUG-ASSIST VACUUM FORMING

For deep thermoforming, pressure-bubble plug-assist vacuum forming is an important process. By this process, it is possible to control the thickness of the formed article. The item may have uniform thickness or the thickness may be varied.

Once the sheet has been placed in the frame and heated, controlled air pressure creates a bubble (Fig. 15-7). This bubble stretches the material to a predetermined height, usually controlled by a photocell. The male plug assist is then lowered forcing the stretched stock down into the cavity. The male plug is normally heated to avoid chilling the plastics prematurely. The plug is made as large as possible so the plastics is stretched close to the final shape of the finished product. Plug penetration should be from 70 to 80 percent of the mold cavity depth. Air pressure is then applied from the plug side while at the same time a vacuum is drawn on the cavity to help form the hot sheet. For many products, vacuum alone is used to complete formation of the sheet. In Figure 15-7, both vacuum and pressure are applied during the forming process. The female mold must be vented to allow trapped air to escape from between the plastics and the mold.

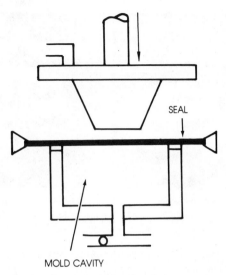

(A) The plastics sheet is heated and sealed across the mold cavity.

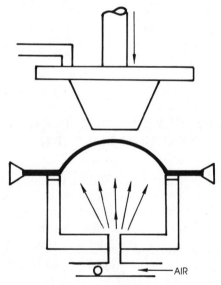

(B) Air is introduced, blowing the sheet upward into an evenly stretched bubble.

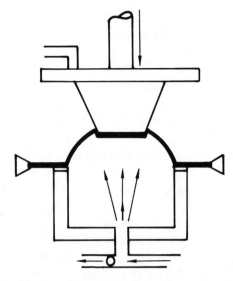

(C) A plug shaped roughly to the cavity contour presses downward into the bubble, forcing it into the mold.

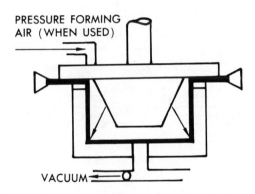

(D) When the plug reaches its lowest point, vacuum is drawn to pull the plastics against the mold walls. Air may be introduced from above to aid forming.

Fig. 15-7. Pressure-bubble plug-assist vacuum forming. (Atlas Vac Machine Co.)

15-5 PLUG-ASSIST VACUUM FORMING

To help prevent corner or periphery thinning of cup- or box-shaped articles, a plug assist is used to mechanically stretch and pull additional plastics stock into the female cavity (Fig. 15-8). The plug is normally heated to just below the forming temperature of the sheet stock. The plug should be from 10 to 20 percent smaller in length and width than the female mold. Once the plug has forced the hot sheet into the cavity, air is drawn from the mold, completing the formation of the part. The plug design or shape determines the wall thickness, as shown in cross-section in Figure 15-8D.

Plug-assist vacuum and pressure forming allows deep drawing, and permits shorter cooling cycles and better control of wall thickness. Close temperature control is needed, however, and the equipment is more complex than straight vacuum forming (Fig. 15-9).

15-6 PLUG-ASSIST PRESSURE FORMING

Plug-assist pressure forming is similar to plug-assist vacuum forming in that the plug forces the hot plastics into the female cavity. Air pressure ap-

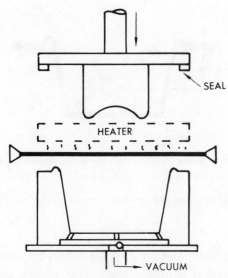

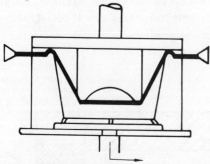

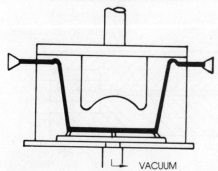

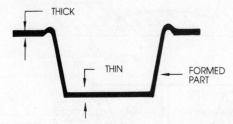

(A) Heated, clamped plastics sheet is positioned over mold cavity.

(B) A plug, shaped roughly like the mold cavity, plunges into the plastics sheet to prestretch it.

(C) When the plug reaches the limit of its travel, a vacuum is drawn in the mold cavity.

(D) Areas of the plug touching the sheet first form thickened areas due to chilling effect.

Fig. 15-8. Plug-assist vacuum forming. (Atlas Vac Machine Co.)

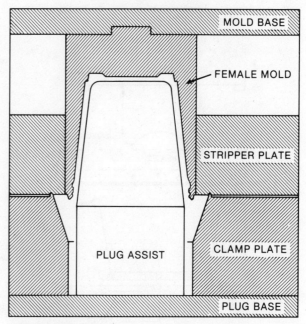

(A) Clamp layout.

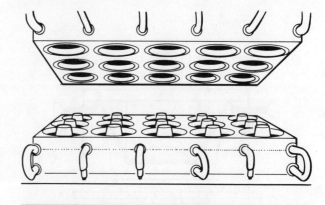

(B) RAM part clamps.

Fig. 15-9. Restricted-area molding (RAM), with individual part clamps built into the mold. This helps to control material draw and reduces draw ratio. (Brown Machine Co.)

plied from the plug forces the plastics sheet against the walls of the mold (Fig. 15-10).

15-7 SOLID PHASE PRESSURE FORMING (SPPF)

A process called *solid phase pressure forming* is similar to plug assist forming. The technique begins with a solid blank (extruded, compression molded, sintered powders) which is heated to just below its melting point. Polypropylene or other multilayered PP sheets are used. The blank is then pressed into a sheet form and transferred to the thermoforming press. A plug further stretches the

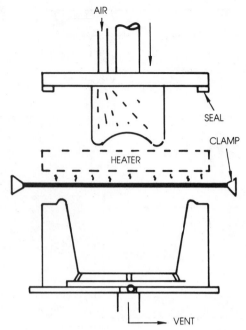

(A) Heated, clamped sheet is positioned over the mold cavity.

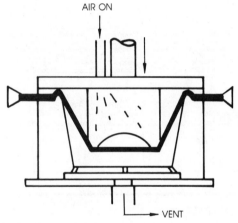

(B) As the plug touches the sheet, air is allowed to vent from beneath the sheet.

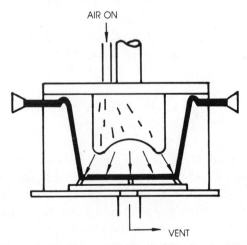

(C) As the plug completes its stroke and seals the mold, air pressure is applied from the plug side, forcing the plastics against the mold.

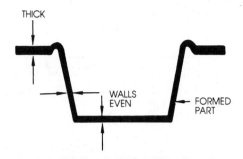

(D) Plug-assist pressure forming is capable of producing products with uniform wall thickness.

Fig. 15-10. Plug-assist pressure forming. (Atlas Vac Machine Co.)

hot material and air pressure forces the hot material against the mold sides (Fig. 15-11). The two (biaxial) stretching operations cause molecular orientation, which enhances the strength, toughness, and environmental-stress crack resistance of the thermoformed product.

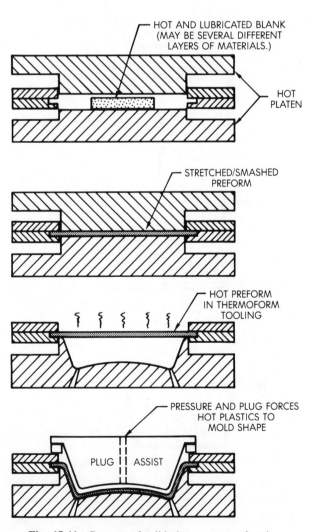

Fig. 15-11. Concept of solid phase pressure forming.

15-8 VACUUM SNAP-BACK FORMING

In vacuum snap-back forming, the hot plastics sheet is placed over a box and a vacuum is drawn that causes a bubble to be forced into the box (Fig. 15-12). A male mold is lowered and the vacuum in the box is released, causing the plastics to *snap-back* around the male mold. A vacuum may also be drawn in the male mold to help pull the plastics into place.

Vacuum snap-back forming allows complex parts with recesses to be formed.

15-9 PRESSURE-BUBBLE VACUUM SNAP-BACK FORMING

As the name implies, the sheet is heated and then stretched into a bubble shape by air pressure (Fig. 15-13). The sheet prestretches about 35 to 40 percent. The male mold is then lowered. A vacuum is applied to the male mold while air pressure is forced into the female cavity. This causes the hot sheet to snap-back around the male mold. Mark-off is on the male mold side.

Pressure-bubble vacuum snap-back forming allows deep drawing and the formation of complex parts, but the equipment is complex and costly.

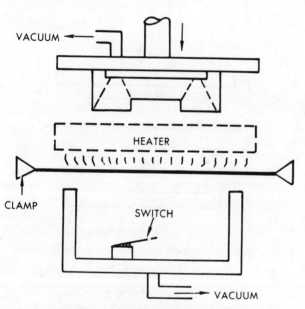

(A) Plastics sheet is heated and sealed over the top of the vacuum box. (Atlas Vac Machine Co.)

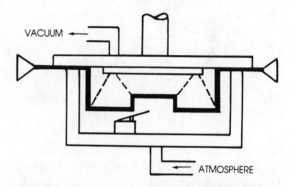

(C) The male plug is lowered and a vacuum drawn through it. At the same time, vacuum beneath the sheet is vented. (Atlas Vac Machine Co.)

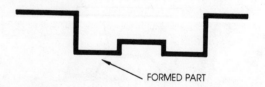

(D) External deep draws can be obtained with this process to form luggage, auto parts, and other items. (Atlas Vac Machine Co.)

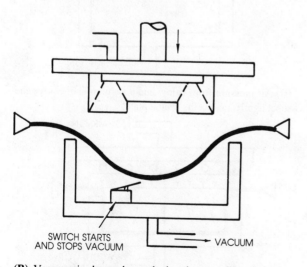

(B) Vacuum is drawn beneath the sheet, pulling it into a concave shape. (Atlas Vac Machine Co.)

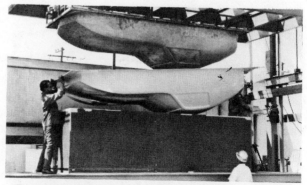

(E) A complete auto body is vacuum-formed from 6.3-mm (¼-in.) thick ABS-polycarbonate during a 20-minute cycle. (Borg-Warner)

Fig. 15-12. Vacuum snap-back forming. (Atlas Vac Machine Co.)

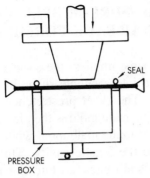

(A) Heated plastics sheet is clamped and sealed across a pressure box.

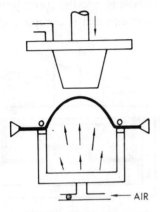

(B) Air pressure is introduced beneath the sheet, causing a large bubble to form.

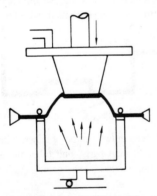

(C) A plug is forced into the bubble, while air pressure is maintained at a constant level.

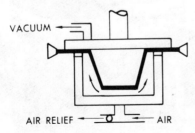

(D) Air pressure beneath the bubble and a vacuum at the plug side create a uniform draw.

Fig. 15-13. Pressure-bubble vacuum snap-back forming. (Atlas Vac Machine Co.)

15-10 TRAPPED-SHEET CONTACT-HEAT PRESSURE FORMING

This process is like straight vacuum forming except that air pressure and a vacuum assist may be used to force the hot plastics into a female mold. Figure 15-14 shows the steps of this process.

15-11 AIR-SLIP FORMING

Air-slip forming is similar to snap-back forming except for the method of creating the stretch bubble. This concept is illustrated in Figure 15-15.

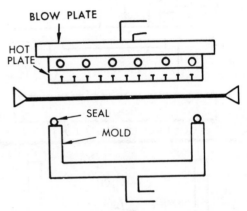

(A) A flat, porous plate allows air to be blown through its face.

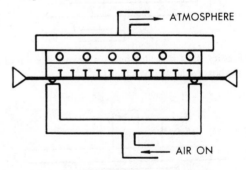

(B) Air pressure from below and a vacuum above force the sheet tightly against the heated plate.

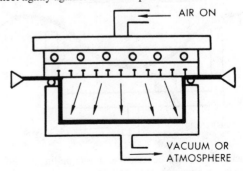

(C) Air is blown through the plate to force the plastics into the mold cavity.

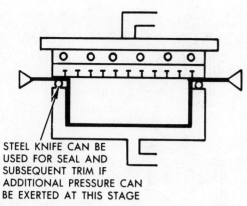

STEEL KNIFE CAN BE USED FOR SEAL AND SUBSEQUENT TRIM IF ADDITIONAL PRESSURE CAN BE EXERTED AT THIS STAGE

(D) After forming, additional pressure may be exerted.

Fig. 15-14. Trapped-sheet contact-heat pressure forming. (Atlas Vac Machine Co.)

15-12 FREE FORMING

In free forming, air pressures of over 390 psi (2.7 MPa) may be used to blow a hot plastics sheet through the silhouette of a female mold (Fig. 15-16). The air pressure causes the sheet to form a smooth bubble-shaped article. A stop may be used to form special contours in the bubble. Skylight panels and aircraft canopies are well-known examples of this technique. Unless a stop is used, there is no mark-off. Only air touches each side of the material. There will be mark-off from clamping.

15-13 MECHANICAL FORMING

In mechanical forming, no vacuum or air pressure is used to form the part. It is similar to matched molding; however, close-fitting matched male and female molds are *not* used. Only the mechanical force of bending, stretching, or holding the hot sheet is used.

This process is sometimes classed as a fabrication or postforming operation. The forming process may make use of simple wooden forming jigs to give the desired shape using ovens, a strip heater, or heat guns for the heat source. Flat stock may be heated and wrapped around cylindrical shapes or stock may be heated in a narrow strip and bent at right angles. Tubes, rods, and other profile shapes may be mechanically formed (Fig. 15-17).

Plug-and-ring-forming (Fig. 15-18) is sometimes classed as a separate forming process. No vacuum or air pressure is used, however, and it may be classed as a type of mechanical forming.

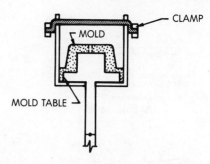

(A) Sheet is clamped to the top of a vertical walled chamber.

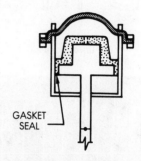

(B) Pre-bellow is achieved by a pressure buildup between the sheet and mold table.

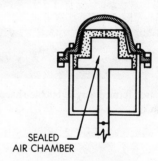

(C) Mold rises in chamber. Mold table is gasketed at edges of chamber wall.

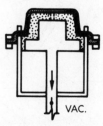

(D) Space between mold and sheet is evacuated as sheet is formed against mold by differential air pressure.

Fig. 15-15. Air-slip forming.

The process consists of a male mold shape and a similarly shaped female silhouette mold (not a matched mold). The hot plastics is forced through the *ring* (not necessarily a rough shape) of the female mold by the male. The cooling plastics

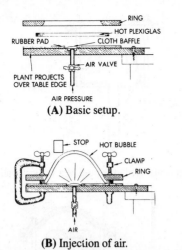

(A) Basic setup.

(B) Injection of air.

(C) Removing free-formed acrylic bubble-shaped product. (Rohm & Haas Co.)

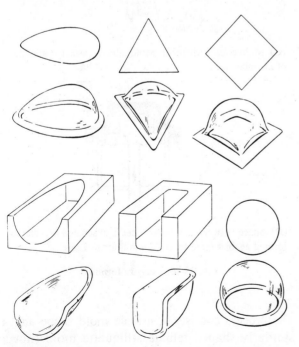

(D) Examples of free-form shapes that can be obtained with various openings. (Rohm & Haas Co.)

Fig. 15-16. Free forming of plastics bubbles.

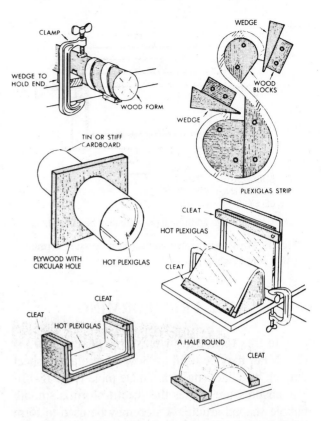

Fig. 15-17. Examples of mechanical forming. (Rohm & Hass Co.)

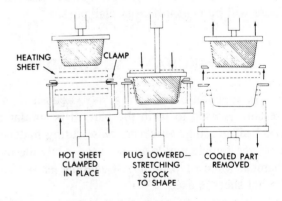

(A) Basic principle of plug and ring forming.

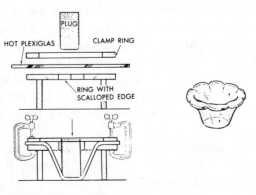

(B) Vase. (Rohm & Haas Co.)

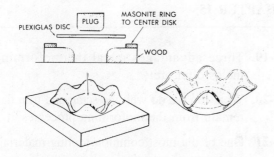

(C) Decorative bowl. (Rohm & Haas Co.)

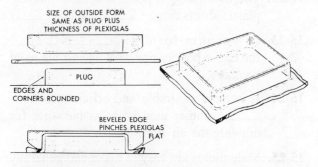

(D) Plastics pan. (Rohm & Haas Co.)

Fig. 15-18. Examples of plug and ring forming.

take the shape of the male mold that it touches. Table 15-1 gives some common problems encountered in thermoforming plastics.

VOCABULARY

The following vocabulary words are found in this chapter. Look up the definition of any of these words you do not understand as they apply to plastics in the glossary, Appendix A.

Air-slip forming
Drape forming
Free forming
Matched-mold forming
Mechanical forming
Shrink wrapping
Snap-back forming
Thermoforming
Vacuum forming

Table 15-1. Troubleshooting Thermoforming

Defect	Possible Remedy
Pinholes or ruptures	Vacuum holes too large, too much vacuum or uneven heating Attach baffles to the top clamping frame
Webbing or bridging	Sharp corners on deep draw, change design or mold layout Use mechanical drape or plug assists or add vacuum holes Check vacuum system and shorten heating cycle
Markoff	Slow draping action may trap air Clean mold or remove high surface gloss from mold Remove all tool marks or wood grain patterns from mold Mold may be chilling plastic sheet too quickly
Excessive post shrinkage	Rotate sheet in relation to mold Increase cooling time
Blisters or bubbles	Overheating sheet—lower heater temperature Ingredients of sheet formulation incorrect or hygroscopic
Sticking to mold	Smooth mold or increase taper and draft Use mechanical releasing tools, air pressure, or mold release Mold may be too warm or increase cooling cycle
Incompletely formed pieces	Lengthen heating cycle and increase vacuum Add vacuum holes
Distorted pieces	Poor mold design—check tapers and ribs Increase cooling cycle or cool molds Sheet removed too quickly while still hot
Change in color intensity	Use proper mold design and allow for thinning of piece Lengthen heating cycle and warm mold and assists Use heavier gauge sheet and add vacuum holes

15-1. Name three materials that vacuum forming molds can be made from.

15-2. Identify the packaging that makes use of the product as the mold.

15-3. Identify the process that will produce products with the greatest detail.

15-4. Mechanical forming is the name incorrectly used for ___?___ forming.

15-5. The name of the unit or tool used to heat a small section of plastics so it may be bent at a sharp angle is ___?___ .

15-6. The sides of a thermoforming mold are tapered to aid in removal of the part. This taper is called ___?___ .

15-7. Name four typical thermoformed products.

15-8. Which thermoforming technique is used to form a part with a very deep draw?

15-9. A ___?___ or ___?___ is normally placed in all low or unconnected portions of the thermoforming mold.

15-10. The ___?___ of a female mold is the ratio of the maximum cavity depth to the minimum span across the top opening.

15-11. What is a major disadvantage of straight vacuum forming using deep cavities?

15-12. In ___?___ forming no vacuum or air pressure is used to form the hot plastics sheet.

15-13. Name the term for marks left on the formed sheet if the mold is not smooth or clean.

15-14. If vacuum holes are too large, or if there is too much vacuum or uneven heating, ___?___ or ___?___ will occur.

15-15. Is a male or female mold used in free forming?

15-16. In plug-assisted methods, plug penetration would normally not exceed ___?___ percent of the mold cavity depth.

15-17. In ___?___-and-___?___ forming, a male and similarly shaped female silhouette mold shape the hot plastics.

15-18. Sharp corners on a deep draw may cause ___?___ or bridging.

15-19. Three advantages metal thermoforming molds offer are:

15-20. What can we do with the scrap and trim that remain from thermoforming processes?

15-21. One of the most common tooling materials for thermoforming is ___?___ .

15-22. Thermoforming is possible because thermoplastic sheets can be ___?___ and ___?___ .

15-23. In vacuum forming, ___?___ ___?___ forces the hot plastics sheet against the mold contours.

15-24. It is more desirable and efficient to have ___?___ rather than holes as passages for drawing the air from the mold.

15-25. Male molds are easy to make and generally cost less than ___?___ molds.

15-26. Accurate, close-tolerance parts with good detail may be made by ___?___-___?___ forming.

15-27. As a general rule, a smooth-surfaced mold should not be used when forming ___?___ .

15-28. It is easier to control the thickness of deep thermoformed products by using the ___?___ or the pressure-bubble vacuum snap-back forming process.

15-29. Give two advantages of thermoforming.

15-30. Give two disadvantages of thermoforming.

15-31. Name a process where a heated thermoplastic sheet is pulled down into or over a mold surface.

15-32. Is a vacuum used in drape forming?

15-33. What is the major disadvantage of matched-mold thermoforming?

15-34. What thermoforming techniques may be used to make each of the following products?
(1) skylight
(2) wall picture
(3) freezer-chest liner
(4) serving tray or candy dish
(5) small planter box with leather graining on exterior and instructions on inside

15-35. In vacuum forming, only ___?___ pressure is used.

15-36. The major difference between plug-assisted pressure forming and plug-assisted vacuum forming is that ___?___ pressures may be used with pressure forming.

15-37. Describe straight vacuum thermoforming.

15-38. Describe how product thickness is controlled in pressure-bubble plug-assist vaccum forming.

15-39. What determines product wall thickness in plug-assist pressure thermoforming?

15-40. What do the words *snap-back* refer to in vacuum snap-back forming?

15-41. Describe free forming.

15-42. Describe mechanical forming and plug-and-ring forming.

Chapter 16

Expansion Processes

Methods of expanding plastics are described in this chapter. An expanded plastics is something like a sponge, bread, or whipped cream, since all are cellular in structure. Expanded plastics are sometimes called frothed, cellular, blown, foamed, or bubble plastics, and they may be classified by cell structure, density, type of plastics, or degree of flexibility including rigid, semirigid and flexible forms.

These low density cellular (from Latin *Cellula,* meaning small cell or room) materials may be classified as either closed-cell or open-cell. If each cell is a discrete, separate cell, it is a closed-cell material. If the cells are interconnected, with openings between cells (sponge-like), the polymer is an open-celled material. These expanded (cellular) polymers may have densities ranging from that of the solid matrix to less than 0.56 lb/ft^3 [9 kg/m^3]. Nearly all thermoplastic and thermosetting plastics can be expanded. They may be made flame retardant; Table 16-1 lists selected properties of some expanded plastics.

Resins are made into expanded plastics by six basic methods:

1. Thermal breakdown of a chemical blowing agent, freeing a gas in a plastics particle. (A popular method)

Pentanes, hexanes, halocarbons, or mixtures of these materials are forced into the plastics particles under pressure. As the bead or granule of plastics is heated the polymer becomes soft, allowing the blowing agents to vaporize. This produces an expanded piece sometimes called the prepuff, preform, or pre-expanded bead. Cooling must be carefully controlled to prevent collapse of the cell or prepuff. Sudden cooling may crease a partial,

internal vacuum in the cell. This pre-expansion is facilitated by dry heat, radio-frequency radiation, steam, or boiling water. Pre-expanded materials must be used within a few days to prevent complete loss of all volatile expanding agents. They should be kept in a cool, airtight container until ready for molding. PS, SAN, PP, PVC, and PE are made cellular by this method.

Blowing agents vaporize quickly from polyethylene pellets; thus, their shelf life is very short. Therefore, most molders simply order pre-expanded *prepuffs* for final processing. Polyolefins are sometimes cross-linked by radiation to prevent cellular collapse before cooling. Prior to molding, the prepuffs are placed into holding tanks to diffuse air into them. Once pressurized, they are molded into a closed-cell product. Prepuffs are usually stabilized by thermal drying and annealing for a period of several hours. Pentane and butane are volatile organic compounds.

2. Dissolving in the resin a gas that expands at room temperature. (A common method)

Nitrogen and other gases may be forced directly into the polymer melt. In injection or extruder equipment, special screw shaft seals are used to prevent the escape of gas from the hot matrix. As the melt leaves the die or enters the mold cavity, the gas vaporizes and causes the polymer to expand. Once the matrix melt falls below the glass transition temperature, the expansion is stabilized.

3. Mixing in a liquid or solid component that vaporizes when heated in the melt. (Sometimes done to produce structural foams) (Fig. 16-1).

Granular, powder, or liquid blowing agents may be mixed and forced through the melt. They

Table 16-1.   Selected Properties of Expanded Plastics

	Coefficient of Linear Expansion, $10^6/°C$	Water Absorption, Vol %	Flammability, mm/min	Density Range kg/m³	Thermal Conductivity W/m · K	Max. Service Temperature, °C	Compression Strength, kPa
Cellulose acetate	6.35	13–17	Slow burning	96–128	0.043	176	862–1 034
Epoxy							
Packed in place	38	1–2	Self-extinguishing	210–400	0.028–1.15	260	13–14 × 10³
Foamed in place	102		Self-extinguishing	80–128	0.035	148	551.5–758
Phenolic							
Reactive type	5–10	15–50	Self-extinguishing	16–1 280	0.036–6.48	121	172.3–419
Polyethylene	24.1	1.0	63.5	400–480	0.05–0.058	71	68.9–275.7
Polystyrene							
Extruded	11	0.1–0.5	Self-extinguishing	20–72	0.03–0.05	79	68.9–965
Expanded-beads	10.1	1.0	Self-extinguishing	16–160	0.03–0.039	85	68.9–1 375
Self-expanded-beads and others	10.1	0.01	Self-extinguishing	80–160	0.03	85	310.2–838
Polyvinylchloride							
Open cell			Self-extinguishing	48–169		50–107	
Closed cell			Self-extinguishing	64–400		50–107	
Silicone							
Premixed powder		2.1–3.2	Does not burn	192–256	0.043	343	689–2 241
Liquid resin, rigid and semirigid		0.28	Self-extinguishing	56–72	0.04–0.43	343	55
Flexible			Self-extinguishing	112–144	0.045–0.052	315	
Urethane							
Rigid	1 370	10	Self-extinguishing	32–640	0.016–0.024	148–176	172–210
Flexible	1 650		Slow burning	22–320	0.032	107	

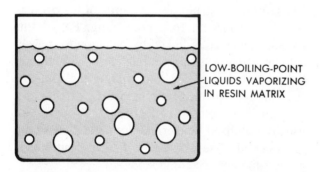

Fig. 16-1. Bubbles of gas are formed as chemicals change physical states.

Fig. 16-2. Mechanically frothed or foamed polyvinyl on thick flooring construction. Magnification 10×. (Firestone Plastics Co.)

remain compressed in the mold until forced into the atmosphere. Rapid decompression (expansion) then occurs. PS, CA, PE, PP, ABS, and PVC are expanded by this method.

4. Whipping air into the resin and then rapidly curing or cooling the resin. (Sometimes used in the production of carpet backing) (Fig. 16-2).

Vinyl esters, UF, phenolics, polyesters and some dispersion polymers are made cellular by mechanical methods.

5. Adding components that liberate gas within the resin by chemical reaction.

This is a favorite method for producing expanded materials from condensation polymers. Liquid resins, catalysts, and blowing agents are kept separate until molding. After mixing, a chemical reaction expands the matrix into a cured cellular material. As the matrix expands, many cell walls rupture, forming a catacomb structure. Polyethers, PU, UF, EP, EU, SI, isocyanurates, carbodiimide, and most elastomers use this expansion method. Polyurethanes and polystyrene account for more than 90 percent of all insulation in refrigerators, freezers, cryogenic tanks, and cellular construction insulation. Isocyanate is commonly expanded between layers of aluminum foil, felt, steel, wood, or gypsum facings. Flexible polyurethanes are open-celled, while rigid polyurethanes are closed-cell materials.

The EPA has expressed concern over the use of chlorofluorocarbons, methylene chloride, pentane, and butane as blowing agents. Chlorofluorocarbons are suspected of reducing the atmospheric ozone layer, and are expected to be phased out of use by 1990. Even pentane and butane may run afoul of clean air rules.

6. Volatizing moisture (steam left in resins by the heat generated by prior exothermic chemical reaction)

Water-blown systems may be used in some condensation polymers. In addition to the steam from the water condensate, carbon dioxide gas is liberated. Some systems combine water and halocarbon in the resin mixture.

Reinforced cellular materials may be produced with particulate and fibrous reinforcements dispersed in the polymer matrix. Fibers tend to orient themselves parallel to cell walls yielding improved rigidity. Reinforced epoxies have been mechanically mixed with pre-expanded beads (PS, PVC) and fully expanded and cured in a mold. Cellular materials are also used in the manufacture of sandwich composites. (See Sandwich and Continuous Reinforcing Chapter 13.)

Syntactic plastics are sometimes included as a separate expanded plastics group. Syntactic plastics are produced by blending microscopically small (0.03 mm) hollow balls of glass or plastics in a resin matrix binder (Fig. 16-3). This produces a light, closed-cell material. The puttylike mixture may be molded or applied by hand into spaces not easily reached by other means. The major uses of syntactic plastics include tooling, noise alleviation, thermal insulation and high-compression-strength flotation devices.

Cellular materials have been produced by placing glass spheres in the polymer matrix formulation before sintering. In a process called *leaching*, various salts or other polymers may be sintered

together. A solvent solution dissolves the crystals or selected polymer to leave a porous matrix. Alternate layers of compatible polymers, reinforcements, and soluble crystal mixtures produce a true composite component.

Expanded plastics are used for insulation, packaging, cushioning, and flotation. Some act as acoustical, as well as thermal, insulation. Others are used as moisture barriers in construction. Expanded epoxy materials are used as light tooling fixtures and models. Expanded plastics may also be used as noncorroding, light, shock-absorbing materials for automobiles, aircraft, furniture, boats, and honeycomb structures. In the textile industry, expanded plastics are used as padding and insulation to give garments a special texture or feel.

During World War II, the Dow Chemical Company introduced expanded polystyrene products in the United States and General Electric produced expanded phenolic products. The two main expanded plastics in use today are probably polystyrene and polyurethane. Polystyrene products are rigid, closed cellular structures (Fig. 16-4). Polyurethane products may be rigid or flexible. PU can also be either closed- or open-celled. Expanded products familiar to the consumer are ceiling tile, Christmas decorations, flotation materials, toys, package liners for fragile items, mattresses, pillows, carpet backing, sponges, and disposable containers (Fig. 16-5).

Fig. 16-4. Closed cellular polystyrene has excellent thermal insulating qualities. (Sinclair-Koppers Co.)

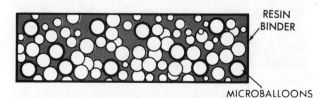

RESIN BINDER

MICROBALLOONS

Fig. 16-3. Syntactic foam.

(A) Injection-molded polyethylene bowl.

(B) Injection-molded wastebasket of polyethylene.

Fig. 16-5. Some molded products of expanded polyethylene. (Phillips Petroleum Co.)

16-1 MOLDING

Various processes have been developed for molding expandable plastics, including injection molding, compression molding, extrusion molding, dri-electric molding (a high-frequency expanding method), steam chamber, or probe molding. Integral skinned cellular, or foamed, plastics are commonly cast or molded. A solid, dense skin of plastics is formed on the mold surface, which is heated to aid in forming this skin. During the skinning process, a cellular core is formed by forcing blowing agents or gas into the melt to cause the cellular structure (Fig. 16-6).

The term *structural foam* is used to include any cellular plastics with an integral skin. Its stiffness depends largely upon the skin thickness. There are many methods of making this integral

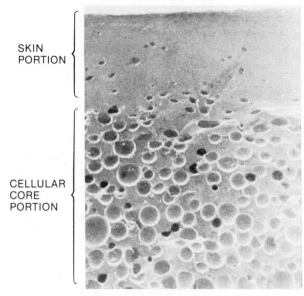

SKIN PORTION

CELLULAR CORE PORTION

Fig. 16-6. In this photograph, the transition from solid skin to cellular core may be seen.

skin; however, structural foam products are molded from the melt, by either a low-pressure, or a high-pressure process (Fig. 16-7).

Low-Pressure Processing

Low-pressure processing is the simplest, most popular and most economical method for large parts. A mixture of molten plastics and gas is injected into molds, at a pressure of from 145 to 725 psi [1 to 5 MPa]. A skin is formed as the gas cells collapse against the sides of the mold. Skin thickness is controlled by the amount of melt forced into the mold, mold temperature, and pressure. The type of blowing agent or the amount of gas forced into the melt also helps determine the skin thickness and density of the part.

Low pressure techniques produce closed-cell parts that are nearly stress free. The collapse of the cells on the mold surface sometimes produces a surface swirl pattern on parts.

High-Pressure Processing

In high-pressure processing, the heated melt is forced into the mold with pressures of from 4 351 to 20 307 psi [30 to 140 MPa]. The mold is completely filled with the melt, thus allowing the melt to become solid against the mold or die surfaces. To allow for expansion, the mold cavity is increased by having the mold open slightly or by withdrawing cores. The melt expands within this increase in volume.

In one injection and extrusion method, the expanding agent is added directly to the hot mix. The

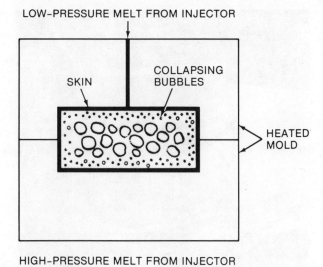

LOW-PRESSURE MELT FROM INJECTOR

SKIN

COLLAPSING BUBBLES

HEATED MOLD

HIGH-PRESSURE MELT FROM INJECTOR

SKIN FORMING

SOLID PHASE

MOLD CLOSED

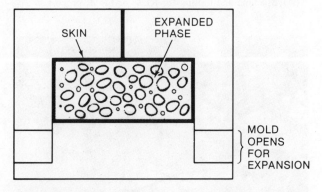

SKIN

EXPANDED PHASE

MOLD OPENS FOR EXPANSION

Fig. 16-7. Low- and high-pressure methods of expanding products with an integral skin.

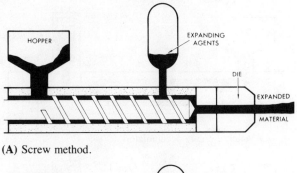

HOPPER

EXPANDING AGENTS

DIE

EXPANDED MATERIAL

(A) Screw method.

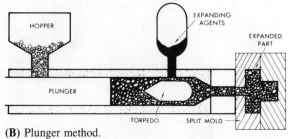

HOPPER

EXPANDING AGENTS

EXPANDED PART

PLUNGER

TORPEDO

SPLIT MOLD

(B) Plunger method.

Fig. 16-8. Adding expanding agents directly to the hot mix.

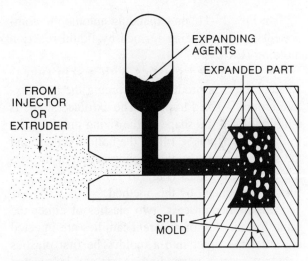

EXPANDING AGENTS

EXPANDED PART

FROM INJECTOR OR EXTRUDER

SPLIT MOLD

Fig. 16-9. A molding method in which the expanding agent is added to the molten plastics just before the plastics enters the mold.

pressure of the plunger or extruder does not allow the material to expand until it is forced into a mold (Fig. 16-8).

In other injection and extrusion methods, the blowing or expanding agent, colorants, and other additives are metered directly into the molten plastics just before it enters the mold (Fig. 16-9). The expansion then takes place in the molding cavity.

In yet another process, the extruded material with expanding agents is fed into an accumulator (Fig. 16-10). When the predetermined charge is reached, a plunger forces the material into the mold cavity, where expansion takes place. (See Reaction Injected Molding in Chapter 10)

Reaction injection molding (RIM), reinforced reaction injection molding (RRIM), and mat molding/RIM (MM/RIM) have grown rapidly after extensive use in the automotive industry and for numerous structural applications. Reinforcements greatly improve dimensional stability, impact strength, and modulus. In a modified resin transfer molding technique, long fibers or mats are placed in the mold and a mixture of reactive components are forced into the cavity. The resulting MM/RIM product is tough, light, and strong.

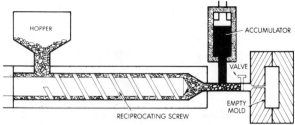

(A) Material enters accumulator.

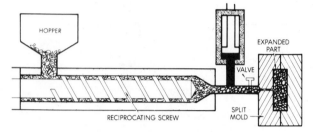

(B) Accumulator forces material into mold cavity.

Fig. 16-10. A molding method in which an accumulator is used.

In Fig. 16-11, an expanding automobile component is shown being formed by liquid reaction molding (LRM).

An expanded plastics with a skin (unexpanded layer) is produced by forcing the hot mixture around a fixed torpedo. The extruded shape is a hollow form the shape of the sizing die. The extrudate then expands, filling the hollow center, and the skin is formed by the cooling action of the sizing and cooling dies (Fig. 16-12). Structural profiles are produced by this method.

In another process, two plastics of either the same formulation or different families are injected one after the other into a mold. The first plastics does not contain expanding agents and is injected partly into the mold. The second plastics, containing the expanding agents, is then injected against the first plastics forcing it against the edges of the mold and forming a shell around the expandable plastics. To close off the shell, more of the first resin is injected into the mold, fully encapsulating the second resin. The part made has an outer skin of one plastics and inner core of expanded plastics.

In a unique process called gas counter-pressure, gas is forced into an empty, sealed mold cavity. The melt is then forced into the cavity against the gas pressure. Expansion begins as the mold is vented.

Extruded polyvinyl materials may be expanded as they emerge from the die or may be stored for future expansion. They are used in the

(A) Machine and tooling for RIM, or LRM, process.

(B) Automobile part being removed from the mold.

Fig. 16-11. Reaction injection molding (RIM) also known as liquid reaction molding (LRM) is used to produce an automobile component with a skinned foam. (Cincinnati Milacron)

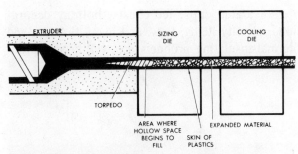

(A) A method of production.

(B) A vinyl formed mat with a skin on both sides.

Fig. 16-12. Forming a skin on expanded plastics.

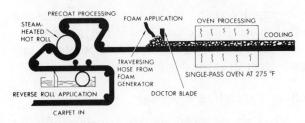

(A) Applying carpet backing (Union Carbide Corp.)

(B) Tough artificial turf of polyamide will not pull loose, and is widely used for athletic fields. (3M Co.)

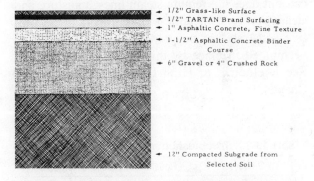

→ 1/2" Grass-like Surface
→ 1/2" TARTAN Brand Surfacing
→ 1" Asphaltic Concrete, Fine Texture
→ 1-1/2" Asphaltic Concrete Binder Course
→ 6" Gravel or 4" Crushed Rock

→ 12" Compacted Subgrade from Selected Soil

(C) Cross section of base used for artificial turf. (3M Co.)

garment industry as single components or given a cloth backing (Fig. 16-13).

Foamed materials are used as backing on carpets or other flooring, as shown in Figure 16-14.

In compression molding of expandable plastics, the resin formulation is extruded into the molding chamber and the mold closed. The molten resin quickly expands, filling the mold cavity.

One of the largest markets is for extruded polystyrene logs, planks, and sheets. They are produced by extruding from a die the molten plastics containing the expanding agent. The expansion oc-

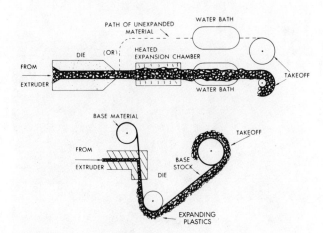

Fig. 16-13. Two methods of expanding a plastics as it emerges from the die.

(D) Workers spreading the cushioning layer that will be surfaced with the artificial grass material. (3M Co.)

Fig. 16-14. Foams are used in many flooring applications.

curs rapidly at the die orifice. Rods, tubes, or other shapes may be produced in this manner.

Table 16-2 gives solutions to problems that may occur in expansion processes.

Other Expanding Processes

Not all expanded plastics make use of a hot melt method. Polystyrene is commonly produced in the form of small beads containing an expanding agent. These beads may be pre-expanded by heat or radiation and then placed in a mold cavity where they are heated again, often by steam, causing further expansion (Fig. 16-15). These beads expand up to forty times their original size (Fig. 16-16) and the expanding pressure packs the beads into a closed cellular structure. In Figure 16-15, a typical bead-expanding system is shown. Core box vents are used to allow the steam into the mold cavity, and these vents leave marks on the expanded product (Fig. 16-15C).

Thermal excitation of molecules by high-frequency radio energy is also used to expand beads. This is sometimes called dri-electric molding because it does not require steam lines, moisture and steam vents, or metallic molds. Inlays or decorative substrates of paper, fabric, or plastics may be molded in place by this method (Fig. 16-17). See Dielectric or High-Frequency Bonding in Chapter 18.

Insulated cups, ice chests, holiday decorations, novelty items, toys, and many flotation and thermal-insulating products are well-known products of this method.

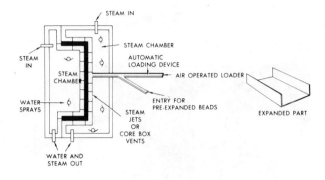

(A) Typical bead-expanding mold that uses steam.

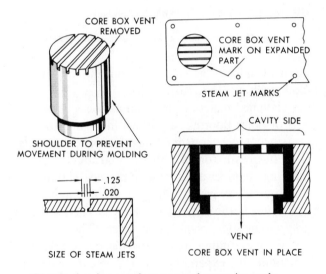

(B) Part showing core box vent and steam jet marks.

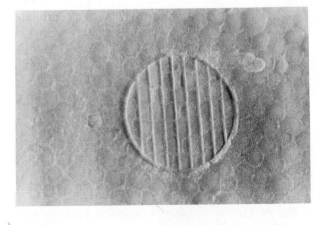

(C) Closeup of core box vent mark left on expanded product.

Table 16-2. Troubleshooting Expansion Processes

Defect	Possible Causes and Remedies
Mold not filled	Venting, short shot, trapped gas, increase pressure, increase amount of material, use fresh materials
Pits or holes in surface	Reduce amount of mold release — mold release interferring with blowing agent, increase mold temperature, not enough material in mold, melt temperature incorrect, concentration of blowing agent incorrect, polish die or mold surface
Distorted pieces	Poor mold design, increase mold time, increase cure cycle, allow thick sections to cool longer, increase mold strength
Sticking to mold	Increase mold release, select correct release, poor mold design, cool mold, polish mold surfaces
Parts too dense	Increase blowing agent or gas, use fresh pre-expanded beads, lower injection pressures, reduce melt
Part density varies	Mix compounds thoroughly, check screw design, increase mold temperature, increase melt temperature, increase dwell and cure time.

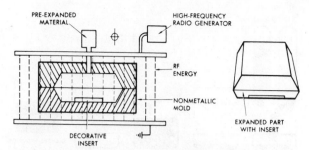

(A) Basic principle of expanding plastics using radio-frequency energy.

(D) Molding machine for producing expandable polystyrene cups.

Fig. 16-15. Expanding plastics without using a hot-melt method.

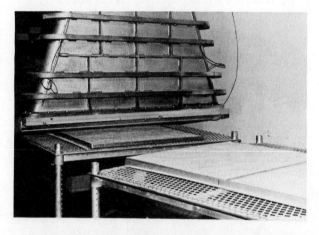

(B) Radiation may also be used to expand and cure plastics.

Fig. 16-17. Plastics may be expanded by radiation.

Fig. 16-16. Unexpanded (left) and expanded (right) beads of polystyrene.

16-2 CASTING

In casting expandable plastics materials, the resin mix containing catalysts and chemical expanding agents is placed in a mold where it expands into a cellular structure (Fig. 16-18). Polyurethanes, polyethers, urea-formaldehyde, polyvinyls, and phenolics are often cast-expanded. Flotation devices, sponges, mattresses, and safety cushioning materials are often cast. Large slabs or blocks of flexible polyurethanes are cast in open and closed molds. These slabs or blocks are cut into mattress stock or shredded for cushioning. Crash pads and pillow products may be cast in closed molds.

Slab stock is commonly produced by continuous production processes. Some are extruded, while many are simply cast or sprayed on a continuous belt. This stock is used as cores for sandwich or laminated composites.

16-3 EXPANDING-IN-PLACE

Expanding-in-place is similar to casting except that the expanded plastics and the mold, together, become the finished product. Insulation in truck trailers, rail cars, and refrigerator doors; flotation material in boats; and coatings on fabrics are examples.

In this process, the resin, catalyst, expanding agents, and other ingredients are mixed and poured into the cavity (Fig. 16-19). The expansion takes place at room temperature, but the mixture may be heated for a greater expanding reaction. This method is said to be done *in situ* (in place or position). Syntactic forms of plastics also may be placed in this category.

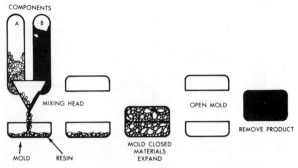

(A) Production scheme for casting plastics materials.

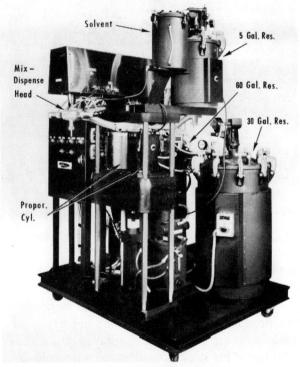

(B) A machine that blends liquid resins with curing agents to produce polyurethane foam. (Hull Corp.)

(C) Examples of shaped foam. (McNeil Akron Corp.)

Fig. 16-18. The production of cast-foamed plastics.

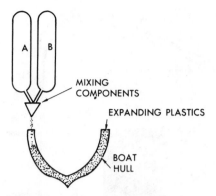

(A) Between inner and outer hulls of a boat.

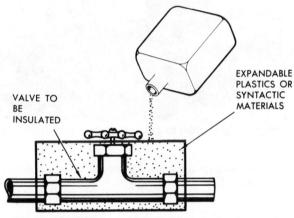

(B) Around a valve.

Fig. 16-19. Expanded-in-place forming.

16-4 SPRAYING

A special spraying device is used to place expandable plastics on mold surfaces or on walls and roofs for insulation. Figure 16-20 shows two examples of this type of forming.

Five advantages and six disadvantages of expanded plastics are listed below.

Advantages of Expanded Plastics

1. Light, less costly products with low thermal conductivity
2. Wide range of formulations from rigid to flexible
3. Wide range of processing techniques
4. Less costly molds for low-pressure and casting methods, large products are possible
5. Parts may have high strength-to-mass ratios

Disadvantages of Expanded Plastics

1. Process slow, some need cure cycle
2. Special equipment needed for hot melt methods
3. Tool and mold designs more costly for high-pressure method
4. Surface finish may be hard to control

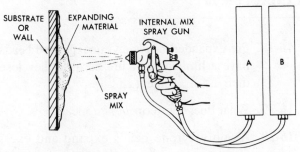

(A) Spraying on wall.

(B) Making a house shell.

Fig. 16-20. Some examples of spray forming.

5. Part size limited in high-pressure method
6. Some processes emit volatile gases or toxic fumes

VOCABULARY

The following vocabulary words are found in this chapter. Look up the definition of any of these words you do not understand as they apply to plastics in the glossary, Appendix A.

Blow systems
Closed-cell
Expanded (foamed) plastics
Expand-in-place
Foaming agents
In-situ foaming
Leaching
Liquid Reaction Molding (LRM)
MM/RIM
Open-cell
Pre-expand
Prepuffs
Structural foam
Syntactic foam

16-1. Name three terms used to describe the expanding process.

16-2. The cellular structure of expanded plastics is ___?___ or ___?___ .

16-3. Cellular plastics have ___?___ relative densities than solid plastics.

16-4. Plastics that are to be expanded in place are generally which of the two classes of plastics ___?___ .

16-5. Name six methods of forming the cellular structure in the expanding processes.

16-6. Some expanded plastics may be thermoformed. They are commonly made of ___?___ (plastics).

16-7. List four broad product uses of expanded plastics.

16-8. What is used to expand polystyrene beads in the mold?

16-9. Cellular plastics with intregral skin are called ___?___ .

16-10. What type of cell structure would a life vest have?

16-11. Supple expanded plastics are used chiefly for ___?___ .

16-12. The two main expanded plastics in use today are probably ___?___ and ___?___ .

16-13. Name four basic processes of foaming plastics.

16-14. What process would be selected to mold the front of an automobile?

16-15. Styroform is a ___?___ for polystyrene cellular plastics.

16-16. Which expanding process or processes would be used to produce each of the following products?
(a) mattress
(b) padded dashboard
(c) egg carton
(d) ice chest or cooler
(e) insulation in home walls

16-17. As polystyrene beads are heated, the ___?___ agent causes the bead to swell.

16-18. Pre-expanded beads and unexpanded beads have a limited shelf life because they lose their ___?___ agent.

16-19. What does the term *in situ* mean?

16-20. As polystyrene beads expand and exert force against the walls of other beads, they form a ___?___ structure with no ruptured cells.

16-21. Name four items that may be produced by casting of expanded plastics.

16-22. Expanding in place is appropriate when the cellular material will not be ___?___ .

16-23. In casting, the expanded plastics is always ___?___ from the mold.

16-24. Name the forming process in which the resin and expanding agent are atomized and forced out of a gun to strike a mold or substrate.

16-25. Where is a great amount of flexible polyurethane foam used?

16-26. It is the ___?___ that determines the physical properties of expanded plastics.

16-27. Name a major application for rigid polyurethane foams.

16-28. Many cellular profile shapes with a surface skin are produced by ___?___ methods.

16-29. Expanded polystyrene parts and molds must be ___?___ before removing the expanded part because latent heat in the center of the plastics part may continue to cause expansion.

16-30. Increasing foam density will ___?___ thermal conductivity but ___?___ nail-holding strength.

16-31. Small glass or plastics balls are sometimes used to make ___?___ plastics.

16-32. Name four methods of pre-expanding polystyrene beads.

16-33. In low pressure forming of structural framed products, pressures from 1 to ___?___ MPa are common.

16-34. If the product is not completely shaped or the mold not filled, ___?___ may be the cause.

16-35. On some expanded polystyrene products, you can see marks left by the ___?___ used to allow steam into the mold cavity.

16-36. Two major disadvantages of high pressure methods of expanding products are ___?___ molds and ___?___ part size.

16-37. Describe one way of producing an expanded plastics that has a skin.

16-38. Identify a use for syntatic foams and state why the foam would be advantageous for use in the application.

Chapter 17

Coating Processes

There are plastics coatings on cars, houses, machinery, and even fingernails. In this chapter, you will learn how coatings are applied. Many coatings are applied to a substrate to enhance the properties of the product by protecting, insulating, lubricating, or adding durable beauty. Coatings may have a combination of properties no other material can match such as flexibility, texture, color, and transparency.

For a process to be classed as *coating*, the plastics material must remain on the substrate. Dip casting and film casting products are not coatings since the plastics is removed from the substrate or mold. Coating may be confused with other processes because similar equipment is used and variations in processing add to the confusion.

It is important to select coating materials that are reasonably close to the thermal expansion of the substrate to be coated. This becomes especially important if reinforcements are used. Some reinforced coating processes are probably better classified as modifications of laminating and reinforcing techniques. Reinforcements can help stabilize the coating matrix. Adhesion is one of the most critical factors in any coating operation. Some substrates must be properly prepared before coating. (See Corona discharge, Plasma, and Flame treatments in Chapter 19)

In extrusion processing, wire coating is a good example showing how more than one material may be put through an extrusion die. In extrusion or calendering of films, hot films are often placed on other substrates coating them. The use of liquid dispersions or solvent solutions is a casting method if the film is removed; it is a coating process if the film remains on the substrate.

There are nine broad (sometimes overlapping) techniques by which plastics are placed on substrates: extrusion coating, calender coating, powder coating, transfer coating, knife or roller coating, dip coating, spray coating, metal coating, and brush coating.

17-1 EXTRUSION COATING

Single or coextrusion of hot melt may be placed on or around the substrate. Extrusion film coating is a technique in which a hot film of plastics is placed on a substrate and allowed to cool. For best adhesion, the hot film should strike the preheated and dried substrate before it reaches the nip of the pressure roller (Fig. 17-1). The chill roller is water-cooled to speed cooling of the hot film. It is usually chrome-plated for durability and high gloss transfer, and it may be embossed to produce special textures on the film surface. The thickness of the film is controlled by the die orifice and by the surface speed of the chill roller. Because the substrate is moving faster than the hot extrudate as it comes out of the extruder die, the extrudate is drawn out to the desired thickness just before it reaches the nip of the pressure and chill rollers.

Several coating techniques result in a thin laminated composite. Because the primary objective of the coating is to provide protection, some materials are considered more effective for coating than others. Polyolefins, EVA, PET, PVC, PA, and other polymers are commonly used in extrusion coating of various substrates to provide moisture, gas, and liquid barriers and heat-sealable

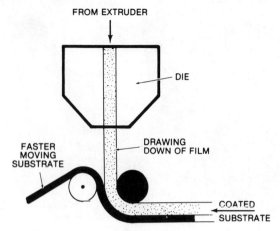

(A) Basic concept of extrusion coating.

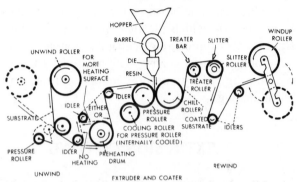

(B) Extrusion coating setup, with wind and unwind equipment. (USI)

Fig. 17-1. Extrusion film coating.

surfaces. The polyethylene coating on paperboard milk cartons is a familiar example of liquid barrier and heat-sealable coatings.

Expanded plastics are also extruded onto various substrates. The substrate may also be drawn through the extrusion die, as in the coating of wire, cable, rods, and some textiles (Fig. 17-2). Five advantages and two disadvantages of extrusion coating are shown below.

Advantages of Extrusion Coating

1. Multilayer plastics may be placed on substrate
2. No solvents are needed
3. Thickness applied to substrate may be varied
4. Uniform coating thickness on wire and cable
5. Cellular coatings may be placed on substrate

Disadvantages of Extrusion Coating

1. Extrudates are hot melts
2. Equipment is expensive

17-2 CALENDER COATING

Calendered films may be used as a coating on many substrates, in a method similar to extrusion coating. The hot film is squeezed onto the sub-

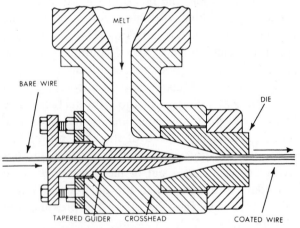

(A) Basic principles of wire coating.

(C) An extruder coats cable with polyethylene plastics. (Western Electric Co.)

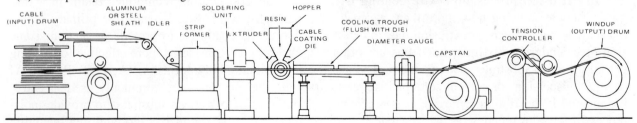

(B) Layout of cable-coating extrusion plant.

Fig. 17-2. Process and products of extrusion wire coating.

strate by the pressure of the heated gauging rollers (Fig. 17-3).

Melt roll coating is a modification of calendering. In this process, the preheated substrate is pressed into the hot melt by a rubber-covered roller, or an embossing roll may be used. The coated material is cooled and placed on windup rolls (Fig. 17-4).

Pressure-sensitive and heat-reactive hot melts that are commonly used as adhesives may be coated on a substrate. Paper, plastics, and textiles are coated by this process (Fig. 17-5).

Coating on a paper substrate may add beauty, strength, scuff resistance, moisture, and soil resistance or provide a sealing system for making a package.

Five advantages and two disadvantages of calender coating are listed below.

Fig. 17-5. Coating industrial fabrics with Zimmer coater. (Zimmer Plastics, GMbH)

Advantages of Calender Coating

1. High-speed continuous process
2. Precise thickness control
3. Pressure-sensitive and heat-reactive hot melts may be used

4. Coatings are stress-free
5. Short runs are relatively economical

Disadvantages of Calender coating

1. Equipment cost is high
2. Additional equipment needed for flat stock

17-3 POWDER COATING

Although ten techniques are known for applying plastics powder coatings, fluidized-bed, electrostatic-bed, and electrostatic powder gun techniques are the three major processes used today. The process of coating a substrate with a dry plastics powder is sometimes called *dry painting*.

PE, EP, PA, CAB, PP, PU, ACS, PVC, DAP, AN, and PMMA are made into powder (solventless) formulations for various powder coating techniques. After coating, some techniques require

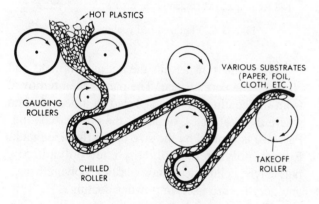

Fig. 17-3. Calender coating.

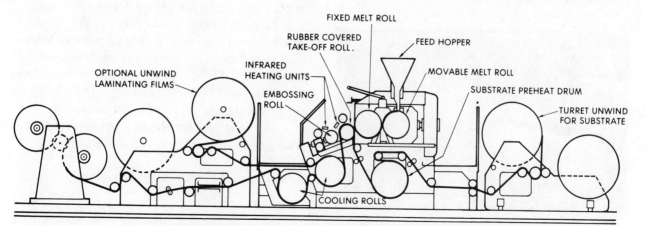

Fig. 17-4. Two-roll melt coater. (Zimmer Plastics, GMbH)

additional heating to assure complete fusion or cure.

Fluidized-Bed Coating

In fluidized-bed coating, a heated part is suspended in a tank of finely powdered plastics, usually a thermoplastic (Fig. 17-6A). The bottom of the tank has a porous base membrane to allow air (or inert gas) to atomize the powdered plastics into a cloudlike dust storm. Perhaps "fog cloud" would be more descriptive because the air velocity is carefully controlled. This air-solid phase looks and acts like a boiling liquid — hence the term fluidized bed.

(D) The parts on the conveyer are cleaned, heated, then coated by the fluidized bed method. (Michigan Oven Co.)

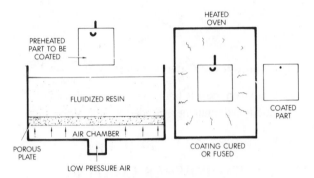

(A) Principle of operation (W. S. Rockwell Co.)

(E) Transformer can tops, at right, are being given a primer application before being heated and coated by the fluidized bed technique. (Michigan Oven Co.)

Fig. 17-6. Fluidized bed coating process.

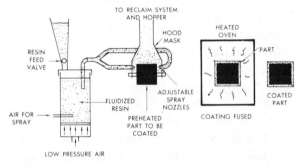

(B) Fluidized bed spray coating technique. (W. S. Rockwell Co.)

When the powder hits the hot part, it melts and clings to the part surface. The part is then removed from the coating tank and placed in a heated oven where the heat fuses, or cures, the powder coating. Part size is limited by the size of the fluidized tank. Epoxy, polyesters, polyethylene, polyamides, polyvinyls, cellulosics, fluoroplastics, polyurethanes, and acrylics are used in powder coating.

The fluidized-bed process originated in Germany in 1953 and has since grown into a useful plastics process in the United States.

In a variation of this process the fluidized powder is sprayed onto preheated parts in a separate chamber. The overspray is collected and reused (Fig. 17-6B). (This process is sometimes called *fluidized-bed spray coating*.) The coating on the part is then fused in a heated oven.

The following list shows three advantages and six disadvantages of fluidized-bed coatings.

(C) Large fluidized bed coating operation. Metal dip preparation is shown at left. (Michigan Oven Co.)

Advantages of Fluidized-Bed Coating

1. Thickness and uniformity
2. Thermoplastics and some thermosets may be used
3. No solvents needed

Disadvantages of Fluidized-Bed Coating

1. Substrate must be heated above plastics melt or fusion temperature
2. Primer may be needed
3. Thin coatings are hard to control
4. Continuous automation of line is difficult
5. Post cure is needed
6. Surface finish may be uneven (orange-peel)

Electrostatic-Bed Coating

In electrostatic bed coating, a fine cloud of negatively charged plastics powders is sprayed and deposited on a positively charged object (Fig. 17-7).

Polarity may be reversed for some operations. Over 100 000 volts with low (less than 100 mA) amperage are used to charge the particles as they are atomized by air or airless equipment. The electrostatic attraction causes the particles to cover all conductive surfaces of the substrate. These parts may or may not require preheating. If preheating is not used, the curing or fusing must take place before the plastics powder loses its charge. The curing is done in a heated oven, as shown in Fig. 17-7. Thin foils, screens, pipes, parts for dishwashers, refrigerators, washing machines, cars, and marine and farm machines are electrostatic-bed coated. Five advantages and six disadvantages of electrostatic-bed coating are:

Advantages of Electrostatic-Bed Coating

1. Thin, even coats are easily applied
2. No preheating needed
3. Process is readily automated
4. Reduced overspray
5. Improved finish quality

Disadvantages of Electrostatic-Bed Coating

1. Thick coatings need preheating of substrate
2. Small openings or tight angles are hard to coat

3. Dust recovery system may be needed
4. Only ionic resins or plastics can be used
5. Post cure is generally needed
6. Substrates may require special preparation

Electrostatic Powder Gun Coating

The electrostatic powder gun process is similar to painting with a spray gun. In this process the dry plastics powder is given a negative electrical charge as it is sprayed on the grounded object to be coated (Fig. 17-8). Fusion or curing must take place in ovens before the powder particles lose their electrical charge, otherwise they will fall from the part. It is possible to coat complex shapes with this method. The fusing oven is the limiting factor in relation to size. Automobile manufacturers may replace liquid finishing processes with powder coating methods in the future. Hundreds of products are coated using this process, including outdoor fencing, chemical tanks, plating racks, and dishwasher, refrigerator, and washer parts.

The list below shows five advantages and seven disadvantages of electrostatic gun coating.

Advantages of Electrostatic Powder Gun Coating

1. Thin, even coats are easily applied
2. No preheating necessary
3. Process is readily automated
4. Short runs and coating of odd-shaped pieces are practical
5. Lower equipment cost than electrostatic-bed coating.

Disadvantages of Electrostatic Powder Gun Coating

1. Thick coatings need preheating of substrate
2. Small openings or tight angles are hard to coat

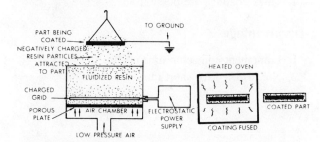

Fig. 17-7. Electrostatic bed coating process. (W. S. Rockwell Co.)

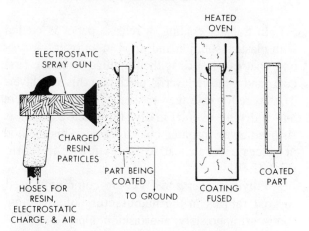

Fig. 17-8. Electrostatic powder gun coating process. (W. S. Rockwell Co.)

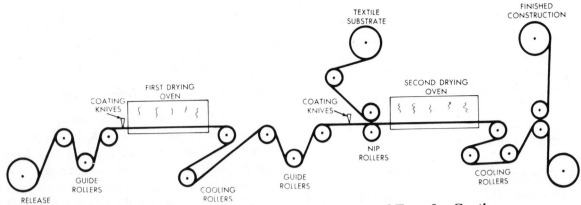

(A) Diagram of a transfer coating line.

(B) The covers of these books are examples of a plastics coating on paper substrates.

Fig. 17-9. Transfer coating process and products.

3. Dust recovery system may be needed
4. Only ionic resins or plastics can be used
5. Post cure is needed
6. High labor cost
7. Thickness harder to control

17-4 TRANSFER COATING

In transfer coating, a release paper is coated with plastics solution and dried in an oven. A second coat of plastics is then applied over the first coat, and a fabric layer is placed on this wet layer. The coated textile then passes through nip rollers and a drying oven. Finally, the release paper is stripped away from the coated fabric. This method produces a tough leatherlike skin on the fabric (Fig. 17-9).

Polyurethanes and PVC are commonly used to coat fabrics in the manufacture of awnings, footwear, upholstery, and fashion apparel.

Two advantages and one disadvantage of this process are shown below.

Advantages of Transfer Coating

1. Multicoated and colored substrates possible
2. Wide choice of substrates may be coated

Disadvantage of Transfer Coating

1. Release paper and additional equipment needed

17-5 KNIFE OR ROLLER COATING

Knife and roller coating methods are another means of spreading a dispersion or solvent mixture of plastics on a substrate. The curing or drying of the plastics coating may be done by heating ovens, evaporating systems, heated rollers, catalysts, or irradiation.

The knife method may involve a simple blade scraper or a narrow jet of air called an *air knife* (Fig. 17-10A). Both sides of the substrate may be coated by this method.

Coating may be done by a combination of rollers as shown in Figures 17-10B and 17-10F. Paper and fabric are often coated by this method. Listed below are four advantages and two disadvantages of knife or roller coating processes.

Advantages of Knife or Roller Coating

1. High-speed continuous process
2. Excellent thickness control
3. Plastisol coatings are stress- and strain-free
4. Thick coatings are possible

Disadvantages

1. Equipment and setup time costly
2. Not justified for short runs

17-6 DIP COATING

Dip coatings are applied by dipping a heated object in liquid dispersions or solvent mixtures of

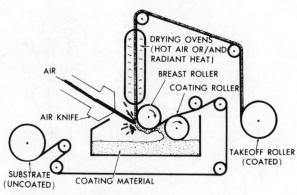

(A) Typical air-knife coating line.

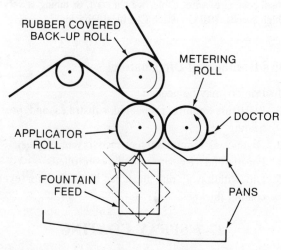

FOUNTAIN FED CONTRACOATER

(B) Contracoater method. (Black-Clawson Co., Inc., Fulton Operations)

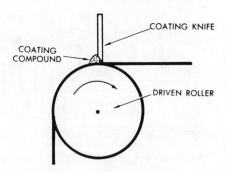

(C) Knife-over-roller coating head. (Waldron Division, Midland-Ross Corp.)

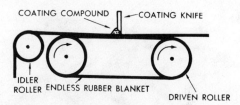

(D) Continuous blanket knife coater. (Waldron Division, Midland-Ross Corp.)

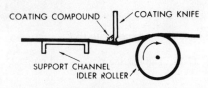

(E) Floating doctor knife. (Waldron Division, Midland-Ross Corp.)

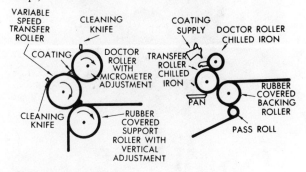

(F) Reverse roll coaters.

Fig. 17-10. Knife and roller coating processes.

plastics. The most common plastics used is polyvinyl chloride. For dispersions, a heating cycle is required to fuse or cure the plastics on the coated object although some dip coatings may harden by simple evaporation of solvents. Generally, 10 min of heating is needed for each millimeter of coating thickness. Curing temperatures run from 350° to 375 °F [175° to 190 °C]. Tool handles and dish-drainer racks are the most common dip-coated products. Objects are limited by the size of the dipping tank (Fig. 17-11).

To ensure that replacement parts arrive in good condition, and so that they may be stored under varying conditions, strippable coatings are often used. They are placed on gears, guns, and other hardware. When machined or polished surfaces need protection during fabrication or other operations, strippable coatings may be applied. These coatings have good cohesion, but relatively poor adhesion; therefore, they may be stripped or

peeled from the part. Strippable coatings are sometimes used as masking films in electroplating or in applying paints (Fig. 17-12).

Wires, cables, woven cords, and tubing may be coated in a modified dipping process wherein these substrates pass through a supply of plastisol or organosol. Preheating the substrate and postheating the product speeds fusion. The coat thickness is controlled in several ways. The strand may be passed through a die opening, fixing the size and shape of the coating. If no die is used, viscos-

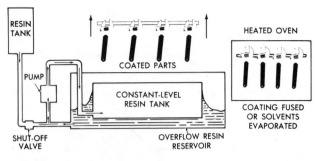

(A) Dip coating technique.

(B) Tool handles dip-coated with PVC.

Fig. 17-11. Dip coating process and products.

Fig. 17-12. Strippable coatings protect cutting tools.

ity and temperature then decide the size and shape. Several passes through the plastisol or organosol will increase thickness. This process may be done in a vertical or horizontal position (Fig. 17-13). Two advantages and five disadvantages of dip coating are shown on the following page.

Advantages of Dip Coating

1. Light or heavy coatings may be applied on complex shapes
2. Relatively inexpensive equipment is used

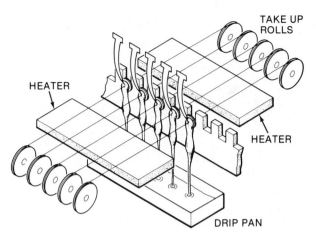

Fig. 17-13. This modified dipping process can apply a plastisol coating to wire, cable, woven cord, or tubing at very high speeds. (BF Goodrich Chemical Division)

Disadvantages of Dip Coating

1. Primers may be needed
2. Plastisols require preheated substrates and postheating
3. Organosols require solvent recovery or exhaust
4. Pot life and viscosity must be controlled
5. Dip withdrawal rate must be controlled for even coating thickness

17-7 SPRAY COATING

In spray coating, dispersions, solvent solutions, or molten powders are atomized by the action of air (or inert gas) or the pressure of the solution itself (airless) and deposited on the substrate. Spray coating of furniture, houses, and vehicles with plastics paints or varnishes are examples. Dispersions of polyvinyl chloride (plastisol) have been spray coated on railroad cars (Fig. 17-14).

In a process sometimes called *flame coating,* finely ground powders are blown through a specially designed burner nozzle of a spray gun (Fig. 17-15). The powder is quickly melted as it passes through this gas or electrically heated nozzle. The hot molten plastics quickly cools and adheres to the substrate. This process is useful for items that are too large for other coating methods. Three advantages and disadvantages of spray coating are listed below.

Advantages of Spray Coating

1. Low equipment cost
2. Short runs are economical
3. Fast and adaptable to variation in size

(A) Coated parts for the electrical industry. (Quelcor, Inc.)

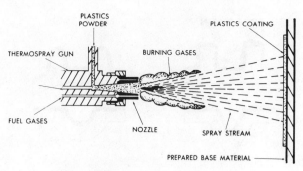

Fig. 17-15. Principle of flame coating.

(B) Coated hood. (Michigan Chrome & Chemical Co.)

(C) Aerosol spray cans of plastics materials for coatings.

Fig. 17-14. Sprays and spray-coated products.

Disadvantages of Spray Coating

1. Hard to control coating thickness
2. Labor costs may be high
3. Overspray and surface defects (runs and orange peel) may be a problem

17-8 METAL COATING

Perhaps metal coating should not be classed as a basic process of the plastics industry; however, because many plastics are associated with this process, the following information is useful. In addition to use as a decorative finish, metal coatings may provide an electrically conducting surface, a wear- and corrosion-resistant surface, or added heat deflection. The major methods for applying a metal coating on a substrate are with adhesives, through electroplating, through vacuum metalizing, and by sputter-coating techniques.

Adhesives

Adhesives are used to apply foils to many surfaces. The textile industry has used this method to adhere metal foils to special garment designs. Complex or irregular parts are difficult to coat, and polyethylene, fluoroplastics, and polyamides are difficult to adhere metals to.

Electroplating

Electroplating is done on many plastics. Both the resin and the mold design must be considered in producing a metal coating on plastics parts. Ribs, fins, slots, or indentations should be rounded or given tapers (Fig. 17-16). Phenolic, urea, acetal, ABS, polycarbonate, polyphenylene oxide, acrylics, and polysulfone are often plated.

An electrolysis preplating step is done by carefully cleaning the plastics part and etching the surface to ensure adhesion (Figs. 17-17, 17-18). The etched part is again cleaned and the surfaces *seeded* with an inactive noble-metal catalyst. An accelerator is added to activate the noble metal, and the ionic solution of metal reacts autocatalyti-

POOR DESIGN BETTER DESIGN

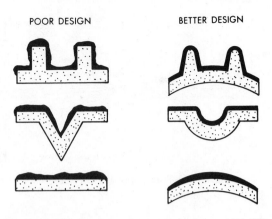

Fig. 17-16. When parts are to be plated, large-radius fillets and bends are desirable.

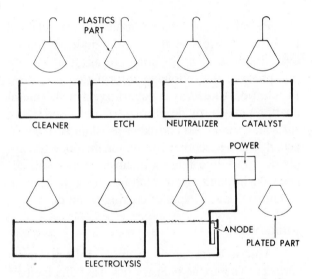

PLASTICS PART

CLEANER   ETCH   NEUTRALIZER   CATALYST

POWER

ANODE
PLATED PART

ELECTROLYSIS

Fig. 17-17. Sequence of operations in electroplating plastics.

cally in the electrolysis solution. Copper, silver, and nickel electrolysis solutions are used in preparing a deposit from 10 to 30 millionths of an inch [0.25 to 0.80 micrometres] thick. Once a conductive surface has been established, commercial plating solutions such as chrome, nickel, brass, gold, copper, and zinc, may be used. Most plated plastics acquire a chromelike finish. The following list shows two advantages and seven disadvantages of electroplating on plastics.

Advantages of Electroplating

1. Mirror-like finishes
2. Good thickness control

Disadvantages of Electroplating

1. Holes and sharp angles are hard to plate
2. Some plastics are not easily plated
3. Plastics must be cleaned and etched before plating
4. Cycle time is long
5. High initial cost
6. Costly for short runs
7. Surface finish of plastic must be near perfect in smoothness

Vacuum Metalizing

In vacuum metalizing, plastics parts or films are thoroughly cleaned and given a base coat of lacquer to fill the surface defects and seal the pores of the plastics. Polyolefins and polyamides are chemically etched to ensure good adhesion. The plastics are then placed in a vacuum chamber, and small pieces or strips of the coating metal (chro-

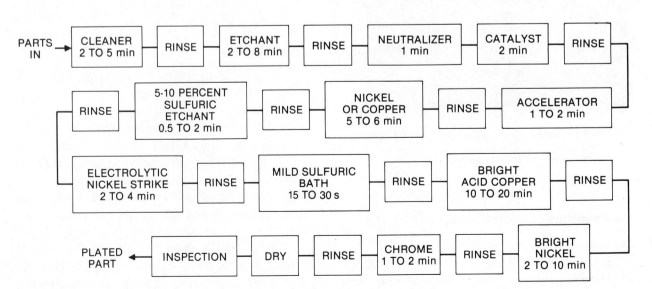

PARTS IN → CLEANER 2 TO 5 min → RINSE → ETCHANT 2 TO 8 min → RINSE → NEUTRALIZER 1 min → CATALYST 2 min → RINSE →

RINSE → 5-10 PERCENT SULFURIC ETCHANT 0.5 TO 2 min → RINSE → NICKEL OR COPPER 5 TO 6 min → RINSE → ACCELERATOR 1 TO 2 min →

ELECTROLYTIC NICKEL STRIKE 2 TO 4 min → RINSE → MILD SULFURIC BATH 15 TO 30 s → RINSE → BRIGHT ACID COPPER 10 TO 20 min → RINSE →

PLATED PART ← INSPECTION ← DRY ← RINSE ← CHROME 1 TO 2 min ← RINSE ← BRIGHT NICKEL 2 TO 10 min

Fig. 17-18. Flow chart of typical plating process.

mium, gold, silver, zinc, or aluminum) are placed on special heating filaments. The chamber is sealed and the vacuum cycle started. When the desired vacuum is reached (0.5 micrometre Hg or about 0.07 Pa), the filaments are heated. The pieces of metal melt (by high voltage) and vaporize, coating everything the vapor touches in the chamber and condensing or solidifying on cooler surfaces (Fig. 17-19). Parts must be rotated for full coverage, because the vaporized metal travels in a straight path. Once the plating is done, the vacuum

(A) Cleaning and etching part.

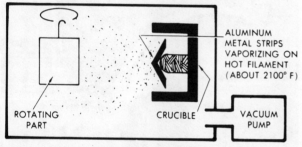

(B) Vaporizing aluminum to coat plastics.

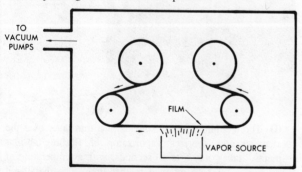

(C) Vacuum metallizing on plastics film.

(D) Type element for electric typewriter is metal-coated plastics.

Fig. 17-19. Vacuum metallizing of plastics parts.

is released and the parts are removed. To help protect the plated surface from oxidization and abrasion, a lacquer coating is applied. (Fig. 17-20). This finish is best suited for interior applications.

By alternately evaporating two or more metals, it is possible to create a chrome/copper/chrome or stainless/copper/stainless laminate. The process is sometimes called *laminated vapor plating*. The list below shows four advantages and five disadvantages of the vacuum metalizing coating process.

Advantages of Vacuum Metalizing Coating

1. Ultrathin, uniform coatings
2. Nearly all plastics may be used
3. Mirror-like finishes
4. No chemical processing

Disadvantages of Vacuum Metalizing Coating

1. Plastics must be coated with lacquer for good results
2. Vacuum chamber limits part size and output rate
3. Scratches or flaws are exaggerated
4. High initial cost
5. Costly for short runs

Sputter Coating

Metals or refractories may be deposited by sputtering systems. Magnetron electronic equipment is used to spray the metal coating. Chromium atoms fall (sputter) on the plastics surface as argon gas strikes an electrode made of the coating metal. Typical thicknesses are 0.000 19 to 0.002 75 in [0.005 mm to 0.07 mm]. A clear protective coating of PU, acrylate, or cellulosic is then applied as protection of the metallic coating. This process is used for coating knobs, films, light reflectors, car trim, and plumbing fixtures. Four advantages and three disadvantages of this process follow.

Advantages of Sputter Coating

1. Ultrathin coatings
2. Excellent adhesion
3. Automated line systems possible
4. Parts are electrically conductive

Disadvantatges of Sputter Coating

1. High technology and capital investment
2. Scratches or flaws are exaggerated
3. Protective coating needed over sputter layer

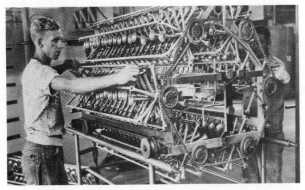

(A) Horns are loaded onto holding fixtures for placement into vacuum chamber. The fixture is rotated in the chamber during the metallizing process.

(B) A base coat of lacquer may be applied by dipping (shown here), spraying, or flow-coating, then baked in oven. The lacquer will smooth out small surface defects and provide an initial gloss.

17-9 BRUSH COATING

Solvent and solventless coatings are often brushed onto a substrate by hand. Many paints and finishes are applied in this manner. Solventless finishes are two-component systems of resins and curing agents that are mixed and applied. Polyester, epoxy, silicone, and some polyurethane resins are used in solventless formulations. Solvent-based coatings may need air-drying or heating to cure.

Finish quality depends on the skill of the applicator and the type and viscosity of the material.

Protective coatings on large metal tanks are often applied by hand or spray-up methods. If placed underground, the tanks must be protected from corrosion and electrolytic action.

Familiar examples are coatings on houses, machinery, furniture, and fingernails. Two advan-

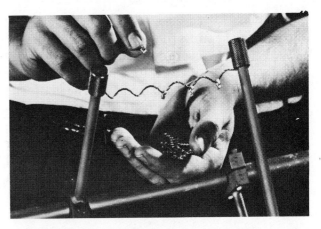

(C) Small staples of plating materials (aluminum, in this case) are placed on coils of stranded tungsten wire filaments. After the chamber is closed and proper vacuum drawn, the filaments are heated to incandescence.

(D) The aluminum melts, spreads in a thin layer over the elements, and vaporizes. Vaporizaion, or "flashing the filaments," takes only 5 to 10 seconds, with a temperature of 1 100 °C (2 100 °F) attained in that time. The metallized products are then removed from the vacuum chamber and dipped in a protective topcoat lacquer. The transparent topcoat can be dyed, allowing a wide choice of colors.

Fig. 17-20. Technique for metallizing toy horns. (Pennwalt-Stokes Corp.)

tages and three disadvantages of brush coating processes are shown below.

Advantages of Brush Coating

1. Low equipment cost
2. Short runs and prototypes are not costly

Disadvantages of Brush Coating

1. High labor costs
2. Poor thickness control
3. Finish hard to control and reproduce

VOCABULARY

The following vocabulary words are found in this chapter. Look up the definition of any of these words you do not understand as they apply to plastics in the glossary, Appendix A.

Dip coating
Extrusion coating
Flame spraying
Fluidized bed
Knife coating
Laminated vapor plating
Vacuum metalizing

17-1. For a process to be classified as a coating the plastics must remain on the ___?___ .

17-2. The technique by which hot extruded plastics is pressed on a substrate without adhesive is called ___?___ .

17-3. The coating used to protect tools from rusting and from damage to cutting edges during shipment is called ___?___ .

17-4. Name two processes used to put a coating on wire.

17-5. Melt roll coating is a modification of the ___?___ process.

17-6. What process is used to place coatings on tool handles?

17-7. What is the major advantage of electroplate coating?

17-8. What are two major advantages of brush coating?

17-9. Which coating process would be used to produce reflective window shade films?

17-10. How many minutes of heating time are required for each millimetre of dip-coating thickness.

17-11. Name four methods that may be used to spread dispersion or solvent mixtures on a substrate.

17-12. The ideal wall thickness for a vinyl dipped coin purse is ___?___ .

17-13. The temperature used to cure plastisols is ___?___ .

17-14. The main element in curing a plastisol coating is ___?___ .

17-15. The main element in curing an organisol coating is ___?___ .

17-16. Name four methods or techniques that may be used to place a coating on a fabric.

17-17. Name the two major disadvantages of extrusion coating.

17-18. What causes the powder to atomize in fluidized bed processing?

17-19. The process of coating a substrate with a dry plastics powder is sometimes called ___?___ .

17-20. Name the major disadvantage of transfer coating.

17-21. In ___?___ coating processes, the plastics and the substrate are given opposite charges.

17-22. The most commonly used plastics for dip coating is ___?___ .

17-23. To speed fusion in dip coating, ___?___ and ___?___ the substrate are common practices.

17-24. A coating of ___?___ is sometimes applied to plated surfaces to minimize oxidization and abrasion.

17-25. Metals or refractories may be deposited by ___?___ coating systems.

17-26. Name four reasons for coating a substrate.

17-27. Name the three major processes of dry-powder coating.

17-28. Name three commonly transfer-coated materials.

17-29. A narrow jet of air used to spread or disperse resins or plastics on a substrate is called an ___?___ .

17-30. Name four products that are normally spray-coated.

17-31. Name the four major methods of metalizing a substrate.

17-32. Before vacuum metalizing, parts are given a ___?___ coat to minimize surface defects, provide a reflective surface and seal the substrate.

17-33. During the vacuum metalizing cycle, the plastics parts must be ___?___ , because the vaporized metal travels in a line-of-sight path.

17-34. Textiles are coated for moisture and chemical resistance, while pots, pans and tools are coated for ___?___ and chemical resistance.

17-35. A manufacturer has the following products to be coated. Recommend coating techniques for each.
  (1) plastics grille to look like chrome
  (2) concrete wall slabs for construction
  (3) two-color jewelry piece (black and gold)
  (4) book covers
  (5) textile rainwear

17-36. Describe the fluidized-bed coating process.

17-37. In electrostatic-bed coating, how is the dry powder deposited on the part?

17-38. What is electrostatic powder gun coating? How does it differ from electrostatic-bed coating?

17-39. Briefly describe the electroplating process. Which plastics materials are well suited to this process?

17-40. How would you set up a simple process for coating tool handles with a resilient plastics?

17-41. List the coating processes requiring heat for curing. Also, list those that do not require heat for curing.

Chapter 18

Fabrication Processes

As with wood, metal, and other materials, plastics often must be assembled or fabricated. Bonding is usually the most efficient, economical, and durable method of assembly. Most fabrication processes are considered secondary operations, and are used in place of mechanical fasteners (screws, threads, clips) and welds.

In this chapter, you will learn how plastics are bonded.

Most substrates require some form of surface preparation to assure a dependable, high quality bond. Most design requirements call for careful selection of the best fabrication method for the intended application.

There are four broad methods by which plastics are bonded. Assemblies are joined by: adhesion, cohesion, mechanical fastening, or friction fitting. Each category includes several distinct assembly methods.

18-1 ADHESION

Adhesive bonding is an effective means of bonding plastics assemblies at high speeds. This bonding does not cause mingling of molecules between pieces; rather, it *adheres* the pieces together (Fig. 18-1). Gluing of wood, paper, and metals are common examples of adhesion. Hot-melt adhesives are materials that set by cooling (see Adhesives in Chapter 5).

Not all adhesive technologists agree on how adhesives function or on what solutions to classify as adhesives. In this text, no distinction will be made between permanent structural adhesives, temporary, or demountable adhesives. All will be considered simply as substances used to hold materials together by surface attachment.

Adhesives are used in the automotive, textile, aerospace, construction, packaging, and electronics industries.

18-2 COHESION

In cohesive bonds, there is mingling of the molecules between the parts (Fig. 18-2). The surfaces of the materials being joined must be converted to a state of solution (made soft or liquid) either by heat or solvents. Cohesion is the chemical adhesion that holds the materials together.

Solvent cement bonding, spin bonding, hot-gas weld bonding, heated-tool bonding, impulse bonding, dielectric bonding, ultrasonic bonding, and electromagnetic bonding are methods of cohesive assembly.

Cement Bonding

Solvent cements and dope cements are two kinds of cements in common use. The first are solvents or blends of solvents that dissolve the material, then when the solvent evaporates, the items are fused together. Dope cements are sometimes called *laminating cements* or *solvent mixes*. They are composed of solvents and a small quantity of the plastics to be joined. This cement is a viscous (syrupy) material that leaves a thin film of the parent plastics on the joint when dried.

No thermosetting, and only selected thermoplastic materials, can be solvent cemented.

Solvents with low boiling points evaporate quickly (Table 18-1); therefore, the joint must be positioned before all of the solvent evaporates. An example is methylene chloride, with a boiling point of 104 °F [40 °C].

(A) Application.

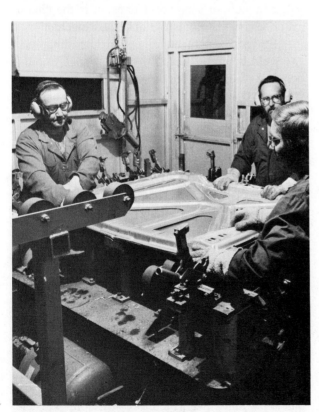

(B) Completion of bond.

Fig. 18-1. Acrylic adhesives are used to bond inner and outer panels of an automobile hood. (Reynolds Aluminum)

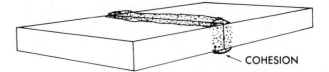

(A) Cohesive bonding.

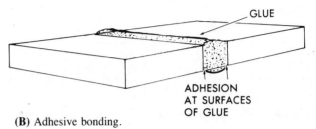

(B) Adhesive bonding.

Fig. 18-2. Cohesive and adhesive bonding compared.

Table 18-1.   Common Solvent Cements for Thermoplastics

Plastics	Solvent	Boiling Point, C°	F°
ABS	Methyl ethyl ketone	40	[104]
	Methyl isobutyl ketone		
	Methylene chloride	40	[104]
Acrylic	Ethylene dichloride	84	[183]
	Methylene chloride	40	[104]
	Vinyl trichloride	87	[189]
Cellulose Plastics:			
Acetate	Chloroform	61	[142]
	Methylene chloride	41	[106]
Butyrate, propionate	Ethylene dichloride	84	[183]
Ethyl acetate	Methyl ethyl ketone	80	[176]
Ethyl cellulose	Acetone	57	[135]
Polyamide	Aqueous phenol		
	Calcium chloride in alcohol		
Polycarbonate	Ethylene dichloride	41	[106]
	Methylene chloride	40	[104]
Polyphenylene oxide	Chloroform	61	[142]
	Ethylene dichloride	84	[183}
	Methylene chloride	40	[104]
	Toluene	110	[232]
Polysulfone	Methylene chloride	40	[104]
Polystyrene	Ethylene dichloride	84	[183]
	Methyl ethyl ketone	80	[176]
	Methylene chloride	40	[104]
	Toluene	110	[232]
Polyvinyl chloride and copolymers	Acetone	57	[135]
	Cyclohexane		
	Methyl ethyl ketone	80	[176]
	Tetrahydrofuran	65	[149]

Solvent cements may be applied to the plastics joints by any of several methods mentioned below. Regardless of method, all joints should be clean and shaved smooth. A V-joint is preferred for making butt joints by many manufacturers and fabricators (Fig. 18-3).

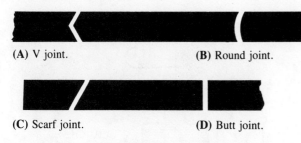

(A) V joint. **(B)** Round joint.

(C) Scarf joint. **(D)** Butt joint.

Fig. 18-3. Types of joints.

In the soaking method, joints may simply be soaked in a solvent until a soft surface is obtained. The pieces are then placed together at once under slight pressure until all solvents evaporate. If too much pressure is applied, the soft portion may be squeezed out of the joint resulting in a poor bond.

Large surfaces may be dipped into, or sprayed with, solvent cements. Cohesive bonds also may be made by allowing the solvent to flow into crack joints by capillary action. Small paint brushes and hypodermic syringes are handy cementing tools. Figure 18-4 shows a number of methods of cementing.

Spin Welding (Bonding)

Spin welding (bonding) is a friction method of joining circular thermoplastic parts. Frictional heat causes a cohesive melt when one or both parts are rotated against each other (Fig. 18-5). Depending on the diameter and the material, joints must spin at 20 ft/s [6 m/s] with less than 20 psi [138 kPa] of contact pressure. When melting takes place, the spinning is stopped and the melt solidifies under pressure.

Joints may also be spin welded by rapidly rotating a filler rod on the joint. A heavy rod of the parent material is rotated at 5 000 rpm and moved along the joints as it melts (Fig. 18-5B). The plastics weld looks like an arc weld on metal.

Vibration bonding, a variation of spin bonding, is a method by which non-circular parts may be bonded. Vibration frequencies are from 90 to

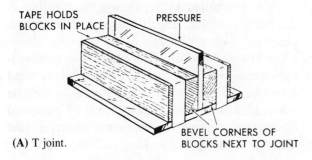

(A) T joint.

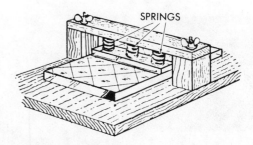

(B) Cementing a rib on a sheet.

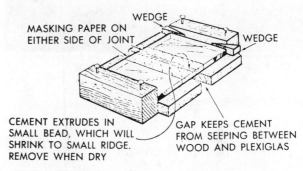

(C) Butt joint rig.

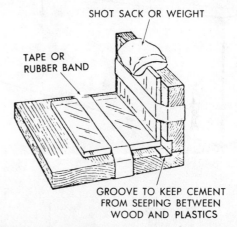

(D) Corner cementing.

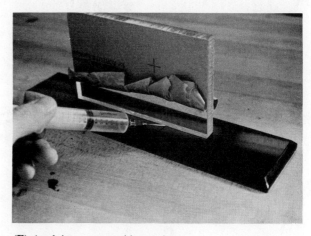

(E) Applying cement with a syringe.

Fig. 18-4. Various cementing methods. (Cadillac Plastics Co.)

(A) Plastics rod spin-welded to plastics sheet.

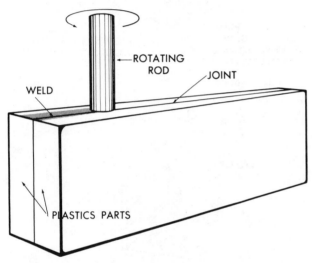

(B) A method of spin-welding joints.

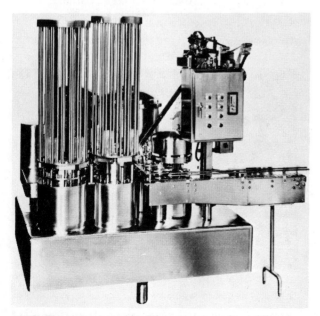

(C) This unit spin-welds, fills, and caps preformed thermo-plastic container halves. (Brown Machine Co.)

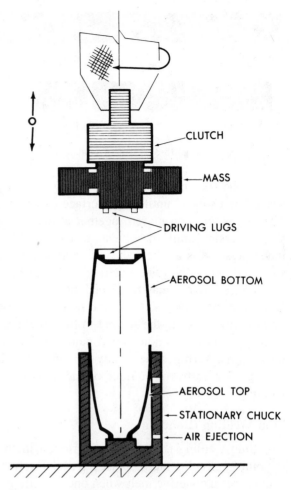

(D) Spin-welding aerosol bottle halves. (DuPont)

Fig. 18-5. Principle of spin welding (bonding).

120 Hz and joint pressures range from about 200 to 250 psi [1 300 to 1 800 kPa].

Nearly any melt-processable, thermoplastic polymer (even dissimilar polymers with compatible melt temperatures) may be assembled into bottles, tubes, and other containers.

Hot-Gas Welding (Bonding)

Hot-gas welding consists of directing a heated gas (usually nitrogen) at temperatures of 400 to 800 °F [200 to 425 °C] onto the joints to be melted together. The temperature of the flameless hot-gas torch is controlled by regulating the gas flow or the heating source. Electric heating elements are preferred with a nitrogen or air pressure of 2 to 4 psi [14 to 28 kPa]. This process is similar to open-flame welding of metals. Filler rods or materials like the parent plastics are used to build up the welded area. Welds may exceed 85 percent of the tensile strength of the parent material. (See Table 18-2) As in any welding technique, the joint

Table 18-2. Weldability of Selected Plastics

Material	Bond Strength, %	Spot Weld	Staking and Inserting	Swaging	Hot-Gas Welding	Heated-Tool Bonding	Friction Bonding	Dielectric Bonding
ABS	95–100	E	E	F	E	G	E	—
Acetal	65–70	G	E	P	G	G	G	G
Acrylics	95–100	G	E	P	E	F	G	G
Butyrates	90–100	G	G–F	G	P	G	G	E
Cellulosics	90–100	G	G–F	G	P	G	G	E
Phenoxy	90–100	G	E	G	G	G	G	G
Polyamide	90–100	E	E	F–P	G	F	G	G
Polycarbonate	95–100	E	E	G–F	E	G	G	G
Polyethylene	90–100	E	E	G	G–P	E	G	—
Polyimide	80–90	F	G	P	G	G	G	G
Polyphenylene	95–100	E	G	F–P	G	G	G	G
Polypropylene	90–100	E	E	G	G–P	G	G	—
Polysulfone	95–100	E	E	F	G	F	G	G
Polystyrene	95–100	E	E	F	E	G	E	—
Vinyls	40–100	G	G–F	F	F–P	E	E	E

Note: E — excellent, G — good, F — fair, P — poor.

area must be properly cleaned and prepared and butt joints should be beveled to 60 degrees (Fig. 18-6D).

Heated-Tool Welding (Bonding)

Heated-tool bonding, or fusion welding, is a method in which like materials are heated and the joints brought together while in the molten stage. The melted areas are then allowed to cool under pressure (Fig. 18-7A). Electric strip heaters, hot plates, soldering irons, or special heating tools are used to melt the plastics surfaces. The surfaces of the heating tools may be coated with Teflon, but the use of lubricants or other materials to prevent sticking of the plastics to the hot metal is not advised because these materials contaminate and weaken the weld.

Pipes and pipe fittings may be joined by heated-tool bonding, and one of the most common uses of heat joining is the fusing of films (Fig. 18-7B). Not all thermoplastics can be heat-sealed, but those that cannot can be coated with a layer of plastics that can be heat-sealed. Electrically heated rollers, jaws, plates, or metal bands are used to melt and fuse film layers.

Impulse Bonding

Impulse bonding may be thought of as a heated-tool method without using continuously heated tools. An impulse of electricity controlled to the proper amount is used to heat the tools (Fig. 18-8). Plastics films 0.01 in. [0.25 mm] thick are held under pressure as the tool is quickly heated and cooled.

Dielectric or High-Frequency Bonding

Dielectric bonding is used to join plastics films, fabrics, and foams. Only plastics that have a high dielectric loss characteristic (dissipation factor) may be joined by this method. Cellulose acetate, ABS, polyvinyl chloride epoxy, polyether, polyester, polyamide, and polyurethane, have sufficiently high dissipation factors to allow dielectric sealing. Polyethylene, polystyrene, and fluoroplastics have very low dissipation factors and cannot be heat-sealed electronically. The actual fusion is caused by high-frequency (radio-frequency) waves from transmitters or generators available in several kilowatt sizes. In the areas of the parts where the high frequency waves are directed, molecules try to realign themselves with the oscillations (Fig. 18-9). This rapid molecular movement causes frictional heat and the areas become molten.

The Federal Communications Commission regulates the use of high-frequency energy. The generated signals are similar to those produced by TV and FM transmitters and operate at frequencies between 20 and 40 MHz.

Ultrasonic Bonding

Ultrasonic energy is used to vibrate plastics mechanically. High-frequency mechanical vibrations in the range of 20 to 40 kHz are directed to the plastics part by a tool called a horn (Fig. 18-10A). An electronic transducer converts 60 Hz energy to the 20 to 40 kHz frequencies (Fig. 18-11). The high frequency causes the plastics molecules to vibrate, making sufficient frictional heat to melt the thermoplastic.

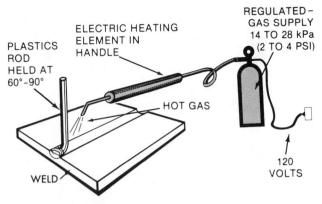

(A) Principle of hot gas welding.

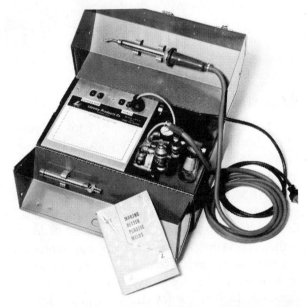

(B) Typical plastics welding unit. (Laramy Products Co.)

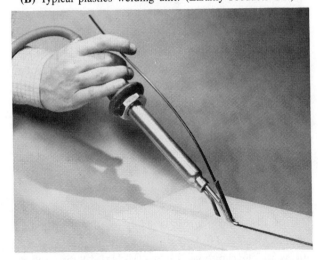

(C) Filler rod and welding tip. (Laramy Products Co.)

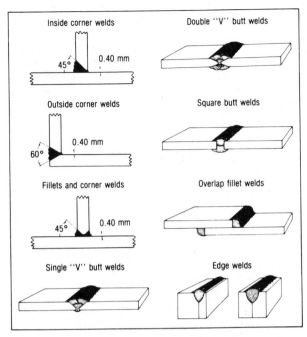

(D) Types of joints produced by hot gas welding of thermoplastics. (*Modern Plastics Magazine*)

Fig. 18-6. Hot gas welding of plastics.

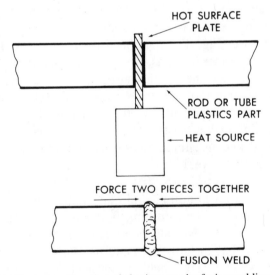

(A) Heated tool bonding of plastics parts by fusion welding.

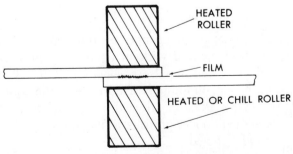

(B) Heated tool bonding of plastics film using rollers.

Ultrasonic techniques are used to activate adhesives to a molten state, to spot weld, and to sew or stitch films and fabrics together without the need for needles pland thread. Simple joints may be welded in 0.2 to 0.5 s (Fig. 18-12).

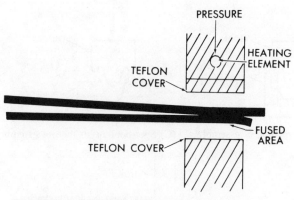

(C) Heated tool bonding of plastics film using press.

Fig. 18-7. Heated-tool bonding of plastics.

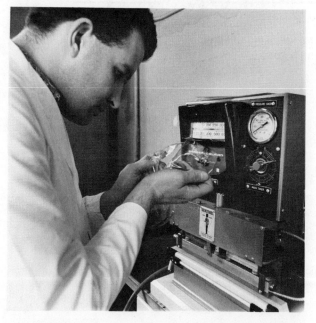

Fig. 18-8. An impulse bonding machine being used to bond film.

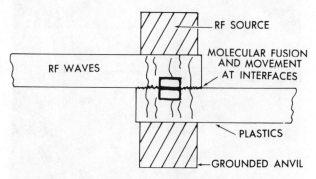

Fig. 18-9. Dielectric heat sealing with radio-frequency waves.

Staking is a term used to describe the ultrasonic or heated-tool formation of a locking head on a plastics stud that is like forming a head on a metal rivet (Fig. 18-10E). Plastics parts with studs may be assembled by this technique.

Many adhesives may be melted and caused to become cured by ultrasonic vibrations (Fig. 18-10G). Ultrasonic systems may be used to cut thermoplastic fabrics and degate parts from runner systems.

Spot or stitch bonding. Spot bonding is a process similar to metal spot-welding used on plastics up to 0.25 in. [6 mm] in thickness using specially designed horns and high-power equipment. Vibrations from the horn penetrate the first sheet and nearly half the second; then the molten material flows into the space between the sheets. Films and fabrics are stitched in a like fashion.

Insertion bonding. Metallic or plastics inserts may be placed in plastics by ultrasonic means (Fig. 18-10F). The horn is used to hold the insert and direct the high-frequency vibrations into an undersized hole. As the plastics melts, the pressure of the horn forces the insert into the hole. Upon cooling, the plastics reforms itself around the insert.

Electromagnetic Bonding

Induction bonding is an electromagnetic technique for bonding thermoplastics, although they cannot be heated directly by induction. The heat is produced by an induction generator with a power range from 1 to 5 kW and frequency range from 4 to 47 MHz. Metal powders (iron oxide, steel, ferrites, graphite) or inserts must be placed at the plastics joints. Then the metallic materials become hot when excited by the high-frequency induction source, melting the surrounding plastics. The metallic inserts or powders must remain in the final weld. Only a few seconds and slight pressure are needed for this rapid method of assembly (Fig. 18-13).

The induction coil must be as close to the joint as possible for rapid bonds. Nonmetallic tooling must be used for alignment.

18-3 MECHANICAL FASTENING

There is a wide choice of mechanical fasteners for use with plastics. Self-threading screws are used if the fastener is not to be removed very often; but when frequent disassembly is required, threaded metal inserts are placed in the plastics.

In Figure 18-14, several types of metal inserts are shown. These may be molded in, or placed in the part after molding.

Screws and metallic or plastics rivets provide permanent assembly. Standard nuts, bolts, and ma-

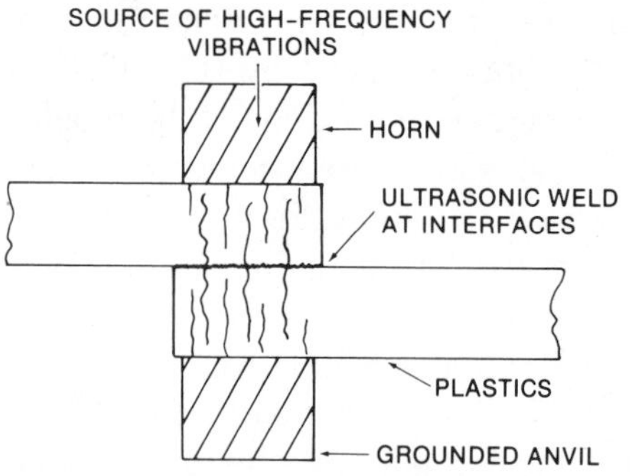

SOURCE OF HIGH-FREQUENCY
VIBRATIONS

HORN

ULTRASONIC WELD
AT INTERFACES

PLASTICS

GROUNDED ANVIL

(A) Energy directors molded into part eliminate movement
of molten plastics to edges of joint.

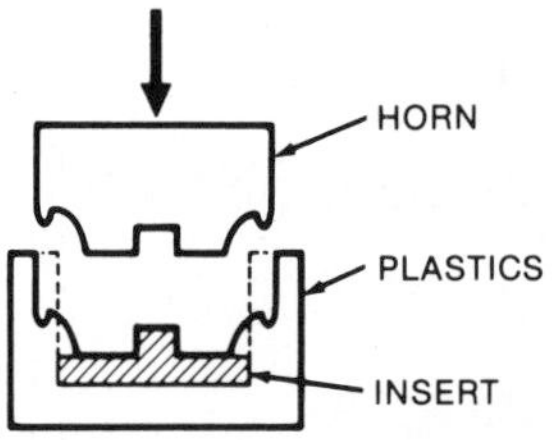

HORN

PLASTICS

INSERT

(B) Swage method of assembly.

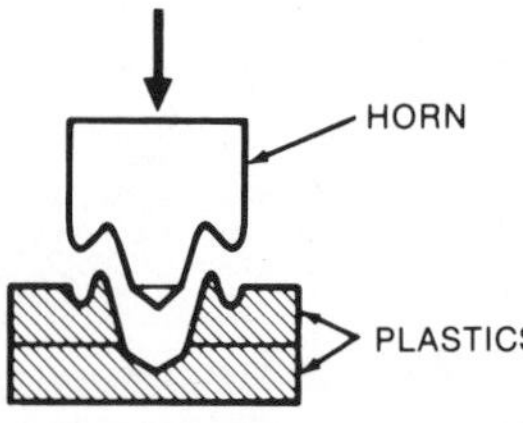

HORN

PLASTICS

(C) Spot-bonding method of assembly.

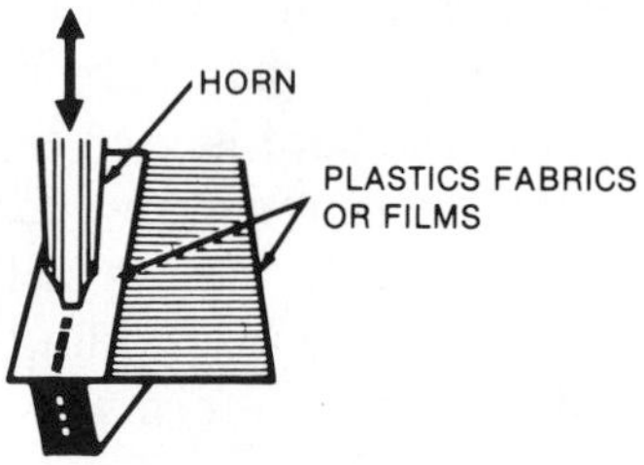

HORN

FINAL SURFACE SHAPE

MATERIAL DISPLACED FROM
HERE TO HERE

PLASTICS

METAL

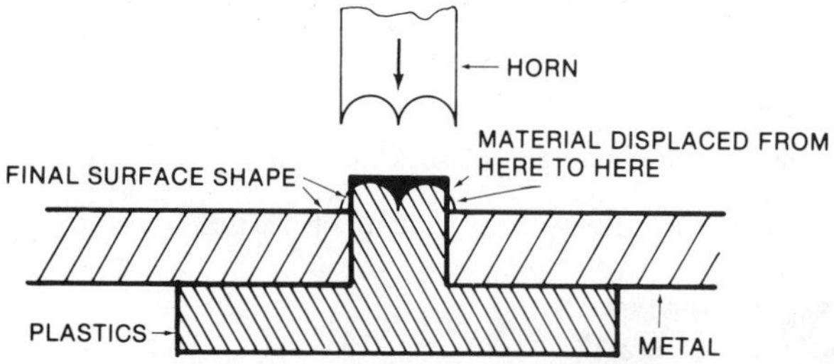

(E) Assembly of metal and plastics by stake method.
(Branson Sonic Co.)

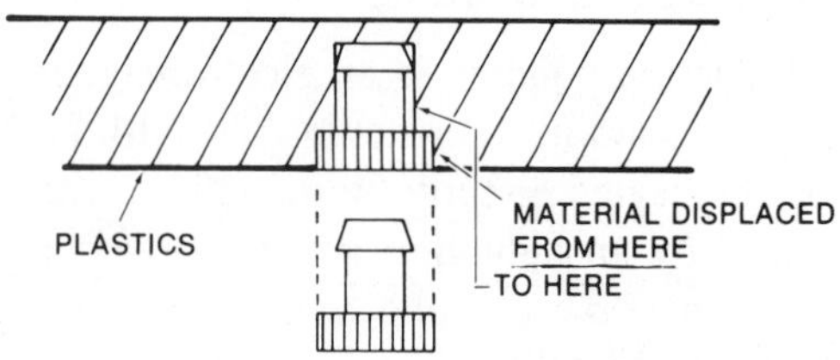

HORN

PLASTICS FABRICS
OR FILMS

(D) Stitch method of bonding.

ULTRASONIC SOURCE

ADHESIVE BECOMES
MOLTEN
AND CURES

PLASTICS

GROUNDED ANVIL

PLASTICS

MATERIAL DISPLACED
FROM HERE
TO HERE

(F) Inserting a plastics stud. (Branson Sonic Co.)

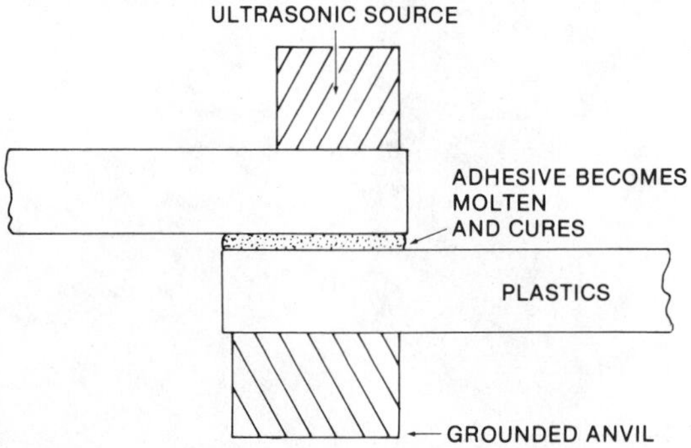

(G) Ultrasonic vibrations melt and cure adhesives.

Fig. 18-10. Ultrasonic bonding methods.

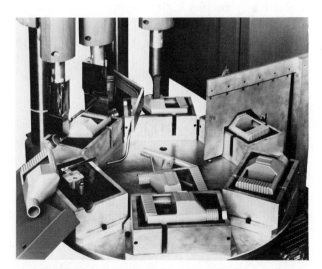

Fig. 18-11. Typical tooling for a rotary indexing ultrasonic
assembly system. Three horns were used to bond a difficult
plastics part. (Sonics & Materials, Inc.)

Fig. 18-12. A traversing ultrasonic head is used to seal
polyester film. (Sonics & Materials, Inc.)

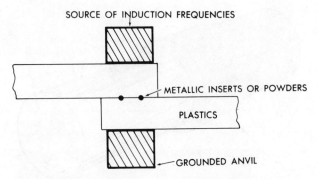

Fig. 18-13. Electromagnetic technique of induction welding.

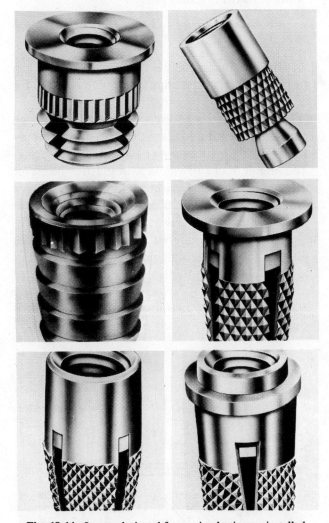

Fig. 18-14. Inserts designed for use in plastics are installed quickly and easily after molding. (Heli-Coil Products)

chine screws, made in both metals and plastics, are used in common assembly methods (Fig. 18-15).

The spring clips and nuts shown in Fig. 18-16 are low-cost, rapid mechanical fasteners. Hinges, knobs, catches, dowels, and other devices are also used for assembly of plastics.

(A) Thread-cutting (self-threading) screws designed for hard plastics. (Parker-Kalon Fasteners Co.)

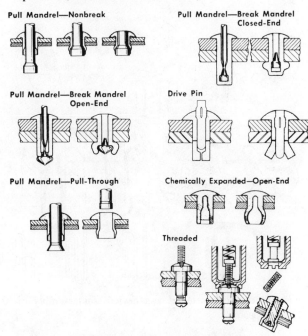

(B) Blind rivets. (Fastener Division, USM Corp.)

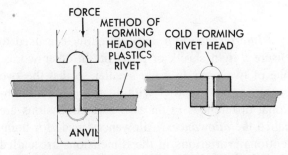

(C) Cold-forming of plastics rivets. Heads may be formed by mechanical or explosive means.

Fig. 18-15. Selected methods of assembly.

18-4 FRICTION FITTING

Friction fitting is a term used to describe a number of pressure-tight joints of permanent or temporary assemblies.

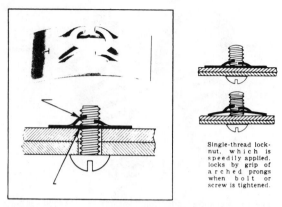

(A) Single-thread locknut. (Eaton Corporation)

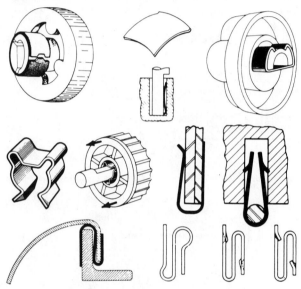

(B) Mechanical fasteners and devices for assembly of plastics parts.

Fig. 18-16. Inexpensive mechanical fasteners.

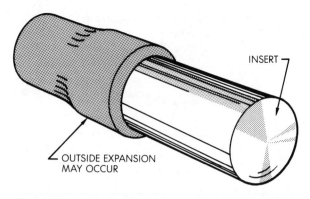

Fig. 18-17. Press-fitting, showing possible expansion of outside diameter of plastics part.

The distinction between a press fit and a snap fit is the undercut and amount of force required for assembly.

Snap fitting is a means of assembly in which parts are snapped into place. The plastics is simply forced over a lip or into an undercut retaining ring. The simple locks or catches on plastics boxes or the covers on many parts including automotive dome lenses, flashbulb covers, and instrument panels are examples (Fig. 18-18).

Shrink fitting refers to placing inserts in the plastics just after molding and allowing the plastics to cool (Fig. 18-19). It also refers to the process of placing plastics parts over substrates where they are heated until the plastics shrinks to its original shape by a property called *memory* (Fig. 18-20).

Interference is a negative allowance used to ensure a tight shrink or press fit. In other words, one of two parts is made smaller so that the two cannot be fit together without force. The intentional differences in the two part dimensions are called the *allowances*. Allowances made for unintentional variations in the dimensions are called *tolerances*. *Limits* are the maximum and the minimum dimensions which define the tolerance.

Press fits, shrink fits, and *snap fits* are terms used to indicate a pressure-tight joint of like or unlike materials without any mechanical fastener.

Press fitting may be used to insert plastics or metallic parts into other plastics components. The parts may be joined while the plastics parts are still warm. When a shaft is press-fitted into a bearing or sleeve, the outside diameter as well as the inside diameter may be expanded (Fig. 18-17).

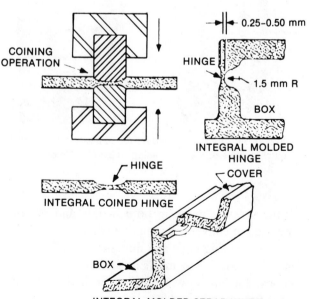

(A) Examples of integral molded and coined hinges.

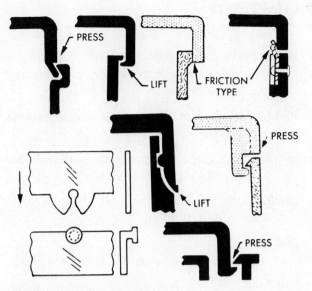

(B) Kinds of clasps for containers.

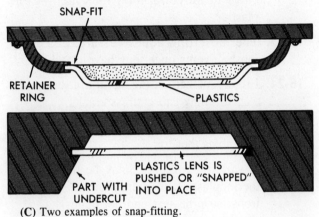

(C) Two examples of snap-fitting.

Fig. 18-18. Assembly methods using snap-fitting and integral hinge techniques.

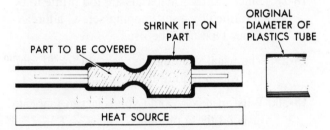

(A) Shrink-fit of plastics tube over electronic part. Note size of tube before heating.

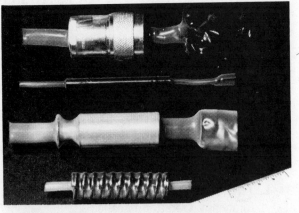

(B) Shrink-fitting plastics over parts for protection or for antistick properties. (Chemplast, Inc.)

Fig. 18-19. Shrink-fitting technique and products.

Fig. 18-20. A shrink-fitting application demonstrating the strength of this packaging. (BASF)

VOCABULARY

The following vocabulary words are found in this chapter. Look up the definition of any of these words you do not understand as they apply to plastics in the glossary, Appendix A.

Adhesion
Allowances
Cohesion
Electromagnetic bonding

Heated-tool bonding
Hot-gas weld bonding
Induction bonding
Interference
Limits
Mechanical fastening
Tolerances
Shrink fit
Spin bonding or welding
Ultrasonic bonding

18-1. Identify the action involved when materials are joined together and there is an intermingling of molecules.

18-2. Name a bonding process similar to spot welding of metals.

18-3. A ___?___ dissipation factor is essential for dielectric or high frequency bonding.

18-4. The bonding process that uses high-frequency vibration is called ___?___ .

18-5. The percent of bond strength for hot-gas weld bonding of polyethylene is ___?___ .

18-6. Name the bonding that is a method of frictionally joining circular thermoplastics together.

18-7. List three common solvents for solvent-bonding acrylic plastics.

18-8. Which solvent from Table 15-1 will evaporate most rapidly?

18-9. Cements composed of solvents and a small quantity of the plastics to be joined are called ___?___ .

18-10. Name the term used to describe the ultra-sonic forming of a locking head on plastics studs.

18-11. *Insertion* is a technique of which bonding method?

18-12. Which bonding method would be used in the meat packaging department of a grocery supermarket?

18-13. Name two plastics often bonded by hot-gas welding.

18-14. Name the bonding process in which the plastics film is quickly heated and cooled by the die.

18-15. List two major advantages of impulse bonding.

18-16. Name the four basic ways plastics are joined.

18-17. Solvents or blends of solvents that melt selected plastics joints together are known as ___?___ cements.

18-18. Into which of the four basic joining methods does spin welding fit?

18-19. In ___?___ , thermoplastic materials are softened by a jet of hot gas.

18-20. Identify the bonding method usually performed on films and fabrics.

18-21. Thermoplastics with ___?___ factors such as ABS can be sealed by dielectric joining.

18-22. In ___?___ , high-frequency causes molecules to move rapidly, thus melting the plastics.

18-23. Thermosetting adhesives are ___?___ in most solvents once they have cured.

18-24. In solvent cementing, evaporation of the ___?___ may cause stress cracks.

18-25. Joint preparation for welding thermoplastics resembles that for ___?___ .

18-26. A variety of ___?___ may be used to join thermoplastics, including knives, soldering irons, strip heaters and hot plates.

18-27. A special type of nut serving the function of a tapped hole is called an ___?___ .

18-28. Since ___?___ materials are too brittle to be deformed by a self-tapping screw, a thread-cutter type must be used.

18-29. The usual welding gas, (except for PE and PP, which require nitrogen) is ___?___ .

18-30. What methods would you select to fabricate or bond these products?
   (a) lid on a plastics butter container
   (b) knob on a TV volume control
   (c) multilayered plastics fabric bedspread
   (d) PVC pipe and exhaust hood
   (e) film wrapper for meat package
   (f) top on a plastics oil can

Chapter 19

Decoration Processes

In this chapter, you will learn that plastics may be decorated for some of the same reasons that cloth, ceramics, metals, and other materials are decorated for, and that many of the processes are similar.

A number of decorating processes are used to produce plastics parts. Decorating may be done during molding, directly afterward, or before final assembly and packaging. The least costly way to produce decorative designs on the product is to include the desired design in the cavity of the mold. These designs may be textures, raised or depressed contours, or informative messages such as trademarks, patent numbers, symbols, letters, numbers, or directions.

Embossing or *texturing* of hot melts may be called a type of rotary molding. The majority of thermoplastics sheets or films are embossed against a composition roll or matched male and female rolls. The desired pattern will be retained if there is a proper balance between the pressure of the embossing roll, heat input, and subsequent cooling. Some polyvinyls and polyurethanes are embossed with textured casting paper. After the polymer has cooled or cured the paper is removed, leaving behind the pattern of the release paper.

In decorating plastics items, in the mold or out, surface treatment and cleanliness are of chief importance. Not only must the molds remain clean and mark-free, but the molded items must be properly prepared to ensure good decorating results. *Blushing* is the result of applying coatings over items that have not been properly dried to eliminate surface moisture. *Crazing* is due to solvent cutting along lines of stress in the molded plastics. The fine cracks (crazing) may be on or under the surface or extend through a layer of the plastics material.

A change in mold design may be needed to produce a stress-free molding, which will eliminate the problem.

Prior to decoration, the surface of the plastics must be cleaned. All traces of mold release, internal plastics lubricants, and plasticizers must be removed. Plastics parts become electrostatically charged, attracting dust and disrupting the even flow of the coating. Solvent or electronic static eliminators may be used to clean and prepare plastics articles for decorating.

Polyolefins, polyacetals, and polyamides must be treated by one of the methods described below to ensure satisfactory adhesion of the decorating media.

Flame treatment consists of passing the part through a hot oxiding flame of 2 012° to 5 072 °F [1 100° to 2 800 °C]. This momentary flame exposure does not cause distortion of the plastics but makes the surface receptive to decorating methods.

Chemical treatment consists of submerging the part (or portions of the part) in an acid bath. On polyacetals and polymethylpentene polymers, the bath etches the surface making the surface receptive to decorating. For many thermoplastics, solvent vapors or baths may be used for the etching treatment.

Corona discharge is a process in which the surface of the plastics is oxidized by an electron discharge (corona). The part or film is oxidized when passed between two discharging electrodes.

Plasma treating subjects plastics to an electrical discharge in a closed vacuum chamber. Atoms on the surface of the plastics are physically changed and rearranged, making excellent adhesion possible.

The nine most widely used decorating treatments for plastics are the following: coloring, painting, hot-leaf stamping, plating, engraving, printing, in-mold decorating, heat transfer, and miscellaneous methods including pressure sensitive labels, decalcomanias (decals), and flocking.

19-1 COLORING

The way to color plastics is to blend the least costly pigments into the base resin. Color matching may be a problem, therefore, successive batches of the same color of plastics may vary slightly. Most producers of colored resins and plastics encourage the use of stock or standard colors. Plastics parts of an assembly may be produced in different plant locations and at different times; therefore, it becomes necessary that color standards be carefully considered. Plasticizers, fillers, and the molding process may affect the final product color.

Colorants in the form of dry powders, paste concentrates, organic chemicals, and metallic flakes are usually blended with a given resin mix. Banbury, two-roll, and continuous mixers are used to disperse thoroughly the pigments in the resin. The colored resin may then be cast or extruded. Water and chemical-solvent dyes have been used with success on many plastics. The procedure consists of dipping the parts in the dye bath and air drying. Three advantages and disadvantages of coloring decorating processes are given below.

Advantages of Coloring Plastics

1. Colored resin control is better in mass production
2. Dyeing is less costly for short runs
3. Surface dyeing is better for lenses

Disadvantages of Coloring Plastics

1. Some colors are hard to produce and match
2. There may be color migration and varying coloration in pieces with uneven thickness
3. Pigment mixing in resin is more costly for short runs

19-2 PAINTING

Painting plastics is a popular, low-cost way to decorate parts and provide flexibility in product color design. Transparent, clear, or colored plastics may be painted on the back surface for a striking contrast, variety, or appearance, an effect not possible by other methods. The six painting methods used in decorating plastics include:

1. spray painting
2. electrostatic spraying
3. dip painting
4. screen painting
5. fill-in marking
6. roller coating

The solvents or curing systems used in the paint must be chosen and controlled with care. As a rule, thermosetting plastics are less subject to swelling, etching, crazing, and deterioration from solvents. Temperature may be a limiting factor for the curing or baking of paints on many plastics. Radiation curing is one method used to cure coatings on plastics (See Chapter 20, Radiation Processes.)

Spray Painting

The most versatile and most often used method of decorating all sizes of plastics articles is spray painting. It is a less costly, rapid method of applying coatings. The spray guns may use air pressure (or hydraulic pressure of the paint itself) to atomize the paint.

Masking, needed when areas of the part are not to be painted, may be done with paper-backed masking tape or durable, form-fitting metal masks. Polyvinyl alcohol masks may be sprayed over areas and later removed by stripping or the use of solvents. Electroformed metal masks are preferred since they conform to the contour of the item and are durable. Four basic types of electroformed masks are shown in Figure 19-1.

Electrostatic Spraying

In electrostatic painting, the plastics surface must be treated to take an electrical charge; then the surface passes through atomized paint that has an opposite charge. The paint may be atomized by air, hydraulic pressure, or centrifugal force (Fig. 19-2). Nearly 95 percent of the atomized paint is attracted to the charged surface, thus this is a highly efficient way of applying paint. Narrow recesses are hard to coat, however, and metal masks are not practical. If dry plastics powders are used, the coated substrate must be placed in an oven to fuse the powder to the product. There are no solvents released during application or curing; however, the product must be able to withstand curing temperatures.

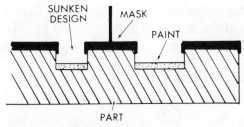

(A) Lip mask on sunken design.

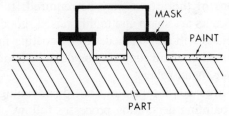

(B) Cap mask on raised design.

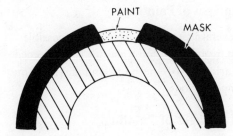

(C) Surface cutout mask.

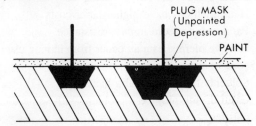

(D) Plug mask for unpainted depressions.

Fig. 19-1. Basic types of electroformed masks.

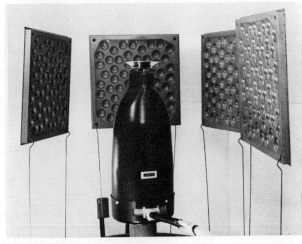

(A) Electrostatic atomization.

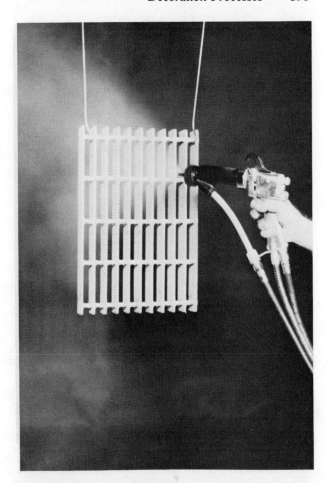

(B) Compressed air atomization.

(C) Hydraulic atomization.

Fig. 19-2. Methods of atomizing paint in electrostatic painting. (Ransburg Corp.)

Dip Painting

Dip painting (Fig. 19-3) is useful when a single color or a base color is needed. A uniform coating may be applied if the part is withdrawn from the paint very slowly. Enough time must be allowed for drainage. Excess paint may be removed by spinning the part, hand wiping, or electrostatic methods.

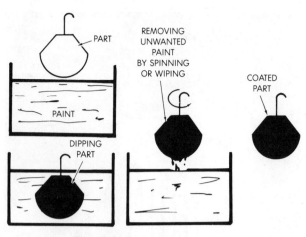

Fig. 19-3. Dip painting.

Screen Painting

Screen painting is a versatile and attractive method of decorating plastics items. It consists of forcing a special ink or paint through the small openings of a stenciled screen onto the product surface. The process is sometimes referred to as *silk screen* painting, since early screens were made of silk. Screens may be made of metal mesh or finely woven polyamide, polyester, or other plastics. A simple screen stencil is prepared by blocking out the areas where no paint is wanted. For intricate designs or lettering, photographic stencils are applied to the screen. When exposed and immersed in a developer bath, the exposed areas wash away. It is through these openings that paint will be applied to the plastics surface beneath the screen.

Fill-In Marking

In the fill-in marking process, paint is placed in low or indented portions of the article (Fig. 19-4). Letters, figures, or designs on a plastics part are produced as depressions in the molded part. These recesses are filled by spraying or wiping paint into the depressions. To ensure a sharp image, the depression should be deep and narrow; otherwise, if the depression or design is too wide, the buffing or

wiping action may remove the paint. Excess paint around the design may be removed by wiping or buffing operations.

Roller Coating

Raised portions, letters, figures, or other designs may be painted by passing a coating roller over them (Fig. 19-5). In some cases, masking out portions of the article may be required. If edges and corners are sharp and highly raised, good coating details will be obtained. Roller coating may be automated or small runs may be done by hand with a brayer (hand roller).

Three advantages and six disadvantages of using painting decorating processes follow.

Advantages of Painting

1. Several inexpensive methods are possible
2. Pretreatment of most plastics is not needed
3. Variation of methods and designs may hide imperfections

Disadvantages of Painting

1. Some plastics are solvent-sensitive
2. Hand methods have higher labor costs
3. Paint reduces cold impact resistance
4. Fisheye blemishes may occur from having used silicone or other releases
5. Solvents may be a health hazard
6. Oven drying may be a problem with some thermoplastics.

19-3 HOT-LEAF STAMPING

Hot-leaf stamping is sometimes called *roll-leaf stamping* or simply *hot stamping*. It offers a simple, economical method for producing a durable decoration on plastics. Letters, designs, trademarks, or messages may be hot-leaf stamped. The process involves a film of metal or paint on a

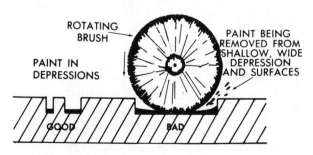

Fig. 19-4. Fill-in method of painting.

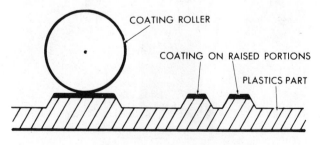

Fig. 19-5. Roller coating of raised portions.

thin carrier (usually in roll form) and a hot-stamping die. The hot die strikes the surface of the plastics part through the carrier. The paint or metallic film is fused into the impression made by the stamp, providing a durable, clear decoration. The hot-stamping dies may be made of machine-engraved or chemically etched metal. Some are made of heat-resistant, flexible silicone (Fig. 19-6). For textured, uneven, or large surfaces, silicone dies may be preferred. Roller dies are used to transfer the design to large areas.

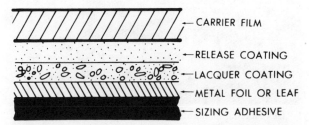

(A) Cross sectional diagram of a typical metallized hot-stamping foil.

(B) Textured silicone hot-stamping rolls produce continuous designs on flat products. (Gladen Division, Hayes-Albion Corp.)

(C) A ring hot-stamped by the roll shown in (B) above. (Gladen Division, Hayes-Albion Corp.)

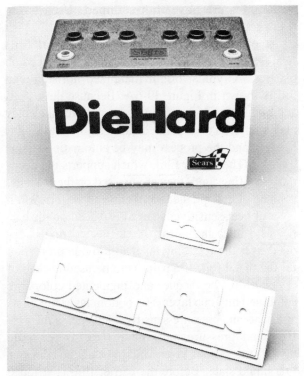

(D) Example of a product hot-stamped with a silicone die. (Gladen Division, Hayes-Albion Corp.)

(E) Molded silicone rubber die used to stamp glass. The glass bottom of the stein was coated with silk screening epoxy and forced-air dried before stamping. (Galden Division, Hayes-Albion Corp.)

Fig. 19-6. Hot stamping and examples of products.

Hot-leaf stamping can be done on all thermoplastics and some thermosets, although thermosetting materials are not easily hot-stamped because of the high heat and pressure required. On thermosets, the process is similar to branding.

Melamines are never hot-stamped. Urea-based resins are rarely decorated by this method.

Genuine gold, silver, or other metal foils (leaf), as well as paint pigments, may be placed on plastics. A typical bright metalized hot-stamping foil is shown in Figure 19-6A. Because these foils and pigments are dry, they are easy to handle, they may be placed over painted surfaces, no masking is needed, and the process may be automatic or done by hand. The carrier film, which supports the decorative coatings until they are pressed on the plastics, is made of cellophane, acetate, or polyester. A thin layer of heat-sensitive material is placed on the carrier as a releasing agent. A lacquer coating is applied over the releasing layer to provide protection for the metal foil. If a paint is to be used instead of a metal foil, the lacquer and pigmented colors are combined into one layer. The bottom layer functions as a heat- and pressure-sensitive, hot-melt adhesive. Heat and pressure must have time to penetrate (dwell) the various layers and to bring the adhesive to a liquid state. Before the carrier film is stripped away, a short cooling time is allowed, permitting the adhesive to solidify. Four advantages and two disadvantages of the hot-leaf stamping decorating process are given below.

Advantages of Hot-Leaf Stamping

1. High-speed, automated operation
2. Foils or other patterns may hide flaws or gate marks
3. No solvents are used
4. Patterns may be changed in short runs

Disadvantages of Hot-Leaf Stamping

1. Foils and release patterns are relatively costly
2. Secondary functions and equipment are costly

Figure 19-7A shows a hot-stamping press that applies multiple color decorations to a molded plastics canister in one operation. The designs are preprinted on the carrier and then are transferred and fused to the part with use of heat, pressure, and dwell. This process is dry, as is the case with all hot-stamping transfers; therefore, newly decorated parts can be handled, assembled, or packaged. This particular press setup can also use regular transfer dies and hot-stamping foil. Flat and shaped decorating areas can be accommodated and, with a special attachment, the complete circumference of cylindrical parts can be marked by this machine. Figure 19-7C shows a machine for hot-stamping a squeezable plastics tube with a

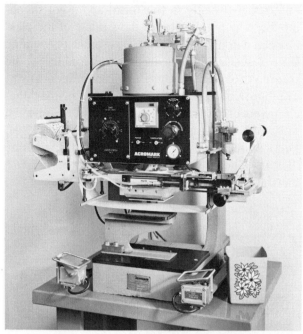

(A) Hot-stamping machine that applies multiple colors. (The Acromark Co.)

(B) Hot-stamping on four sides of a polyethylene beverage case. (Howmet Corp.)

(C) Machine for hot-stamping squeezable plastics tubes. (The Acromark Co.)

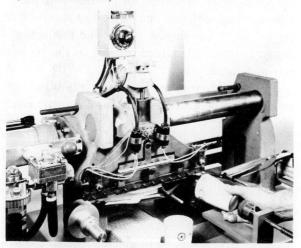

(D) Decorative hot stamping of a plastic drinking tumbler. (The Acromark Co.)

(E) Roll-on method of hot stamping being used to apply wood grain finish to a television cabinet. (Howmet Corp.)

Fig. 19-7. Hot stamping machines.

highly decorative design. The tooling consists of a rotary dial table assembly that permits continuous operation of the press, requiring only the loading of the part to the nests. Ejection after marking is automatic.

19-4 PLATING

Plating and vacuum metalizing have been discussed under the topic of metal coatings in Chapter 17. There are many functional applications of coating plastics with metal, but decorative applications outnumber functional ones. Metalized foils for dielectrics, electronics items such as semiconductors, and resistors are functional applications, as are flexible mirrors and plating for corrosion resistance. The mirror-like finish on automotive items, appliances, jewelry, and toy parts are examples of decorative applications. Four advantages and five disadvantages of plating are listed below.

Advantages of Plating

1. Metallic finish has a mirror-like quality
2. Many plastics parts need little or no polishing before plating
3. Electroplate thickness range from 0.000 38 to 0.025 mm
4. Plating is more durable than metalizing

Disadvantages of Plating

1. Mold finish and design must be considered
2. Not all plastics are easily plated
3. Setup is expensive and includes many steps
4. Many variables to control for proper adhesion, performance, and finish
5. Plating is more expensive than metalizing

19-5 ENGRAVING

Engraving is seldom used on a production scale, however, it does provide a durable means of marking and decorating plastics and is often used in engraved tool and die work. Pantographic engraving machines may be automatic or manual and are often used to engrave laminated nametags, door signs, directories, and equipment, and to place identifying names and marks on bowling balls, golf clubs, and other items. Laminated engraving sheets contain two or more layers of colored plastics. Engraving cuts through the top layer, exposing the contrasting second-color layer (Fig. 19-8).

19-6 PRINTING

There are over eleven distinct methods, and many combinations of these methods for printing on plastics.

(A) Engraving laminated plastics.

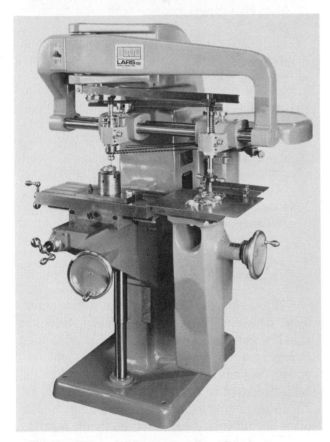

(B) Three-dimensional engraving. (Lars Corp.)

Fig. 19-8. Pantographic engraving machines.

Letterpress is a method in which raised, rigid printing plates are inked and pressed against the plastics part. The raised portion of the plate transfers the image.

Letterflex is similar to letterpress, except that flexible printing plates are used. Flexible plates may transfer their designs to irregular surfaces (Fig. 19-9A).

Flexographic printing is like letterflex, except that a liquid ink rather than a paste ink is used. The plate is often a rotary type, transferring inks that set or dry rapidly by solvent evaporation.

Dry offset is a method in which a raised, rigid printing plate transfers a paste ink image onto a special roller called an offset blanket. This roller then places the ink image on the plastics part, thus the name *offset*. If multicolor printing is required, a series of offset heads can be used to apply different colors to the blanket roller. The multicolored image is then transferred (offset) to the plastics part in a single printing step (Fig. 19-9B).

Offset lithography is similar to dry offset, except that the impression on the printing plate is not raised or sunken. The process is based on the principle that oil and water do not mix. The image or message to be printed is placed on the plate by a photographic-chemical process. Images may be placed directly on the plate by special *grease* typewriter ribbons or pencils. The greasy or treated images will be receptive to the type of ink used. Those areas not treated will be receptive to water, but will repel ink. A water roller must first pass over the offset plate. Then, the ink roller will deposit ink on the receptive areas. The image is transferred from the printing plate to a rubber offset cylinder (blanket roller) that places the image on the plastics part.

Rotogravure or *intaglio* printing involves an image that is depressed or sunken into the printing plate. Ink is applied to the entire surface of the plate and a device called a doctor blade is used to scrape the plate and remove all excess ink. The ink left in the sunken areas is transferred directly to the product.

Silk screen printing is a process in which ink or paint is forced through a fine metallic or fabric screen onto the product. A rubber squeegee is used to force paint through the screen. The screen is blank or blocked off in areas where no ink is wanted.

Stenciling is similar to silk screen printing except that the open areas (those to be printed) do not have a connecting mesh. Stencils may be positive or negative. In positive stencil printing, the image is

(A) A one-color direct-printing letterflex press. The rubber plate allows printing on irregular surfaces.

(B) This machine will print one or more colors by either dry offset or letterflex methods.

Fig. 19-9. Letterflex and dry offset printing presses. (Apex Machine Co.)

open and spray or rollers transfer the ink through these open areas onto the product. In negative stencil printing, the image is blocked out and the background is inked, leaving no ink in the stencil area. Stencil printing may be considered a masking operation.

Electrostatic printing has been adapted to several well known printing techniques. In the process, dry inks are attracted to the areas to be printed by a difference in electrical potential. There is not direct contact between the printing plate or screen and the product. There are several methods by which a screen is made conductive in the image areas and nonconductive in other areas. Dry, charged particles are held in these open areas until discharged toward an oppositely charged back plate. The object to be printed is placed between the screen and the back plate. When the ink is discharged, it strikes the substrate surface. A fixing agent is then applied to provide a permanent image. The image is faithfully reproduced, regardless of the surface configuration of the substrate. Images can be printed on the yolk of an uncooked egg or similar products by this method. Edible inks are used to identify, decorate, and supply messages on fruits and vegetables.

Heat-transfer printing is used as a decorating process and as an important printing method. The process is similar to hot-leaf stamping in that a carrier film (or paper) supports the release layer and the ink image. The thermoplastic ink is heated and transferred to the product by a heated rubber roll.

Hot-leaf stamping is the process of transferring a colorant or a decorative material from a dry carrier film to a product by heat and pressure. It is sometimes used as a printing method.

19-7 IN-MOLD DECORATING

During in-mold decorating, an overlay or coated film called a *foil* becomes part of the molded product. The decorative image and, if possible, the film carrier are made of the same material as the part to be molded. With thermosetting products, the film may be a clear cellulose sheet covered with a partially cured resinlike molding material. The in-mold overlay is placed in the mold cavity while the thermosetting material is only partially cured. The molding cycle is then completed and the decoration becomes an integral part of the product. With thermoplastic material, the overlay may be placed in the mold cavity before any molten material is forced in. The overlay should be the same material as the product. As the molten thermoplastic flows into the mold cavity, the overlay becomes fully bonded and part of the final product. It should be obvious that the placement of gates is important to prevent wrinkled or washed overlays. In both thermosetting and thermoplastic molding the overlay may be held in place in the mold by physically cutting it, so that it fits snugly in the cavity. Where physical methods are not desirable, electrostatic means of holding the overlay may be used.

In a process similar to heat-transfer decorating, blow-molded parts may be decorated in-mold. The ink or paint image is placed on a carrier film or paper. As the hot plastics expands, filling the mold cavity, the image is transferred from the carrier to the molded item. Three advantages and two disadvantages of the in-mold decorating process are shown below.

Advantages of In-Mold Decorating

1. Full color images, halftones, or combination⁓ be used
2. Very strong bonding is achieved
3. Designs and short runs are economical

Disadvantages of In-Mold Decorating

1. Hand loading, overlays, and automated machines are costly
2. Mold must be accessible and designed to minimize washing and turbulence

19-8 HEAT-TRANSFER DECORATING

In heat-transfer decorating, the image is transferred from a carrier film onto the plastics part. The structure of heat-transfer decorating stock is shown in Figure 19-10A. The preheated carrier stock is transferred to the product by a heated rubber roller (Fig. 19-10C).

A decorating or printing process that resembles a combination of engraving and offset printing is the *Tampo-Print*. In this process, a flexible transfer pad picks up the impression from the inked engraving plate (Fig. 19-11A) and transfers it to the item to be printed (Fig. 19-11B). The entire ink supply carried by the transfer pad is deposited on the part, leaving the pad clean. The flexible pad adapts to rough and uneven surfaces while maintaining ab-

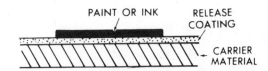

(A) Structure of heat transfer decorating stock.

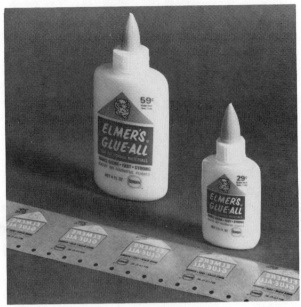

(B) Container decoration by heat transfer process. (Therimage® Products Group, Dennison Mfg. Co.)

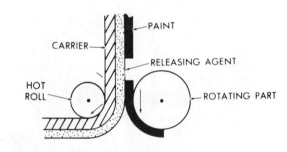

(C) Roller method of transferring design.

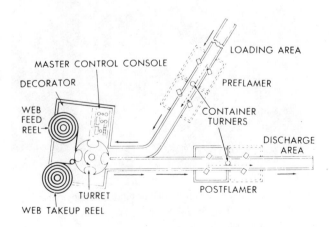

(D) Scheme for heat transfer of decorations to containers. (Therimage® Products Group, Dennison Mfg. Co.)

(E) Heat transfer of decorations, following the scheme shown in (D). (Therimage® Products Group, Dennison Mfg. Co.)

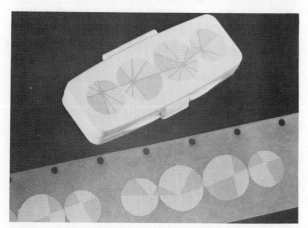

(F) Flat silicone rubber pad, 3 mm thick, [⅛ in] used to apply transfer to cosmetics container. (Gladen Division, Hayes-Albion Corp.)

(G) Various containers decorated by heat transfer process. (Therimage® Products Group, Dennison Mfg. Co.)

(H) Examples of heat-transfer decorations. Note rolls of decoration on carrier film. (Color-Dec Inc.)

Fig. 19-10. Heat transfer decorating and examples of decorated products.

solute reproduction sharpness. Printing heads of various shapes can accommodate a wide variety of objects and textures, and multicolor wet-on-wet printing, including halftones, can be accomplished. Nearly any type of printing ink or paint may be used by this simple process. Depending on the type of product, up to 20 000 parts per hour can be automatically decorated. Two advantages and disadvantages of the heat-transfer decorating process are listed below.

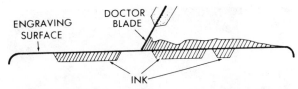

(A) Etched plate with ink held in recessed areas. Surface of the engraving is wiped clean by doctor blade.

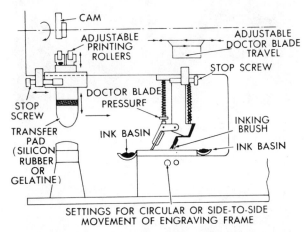

(B) Movement of the transfer pad and inking brush or squeegee.

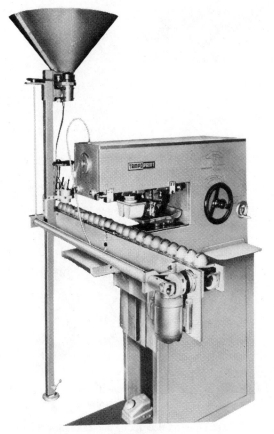

(C) Entire printing machine setup.

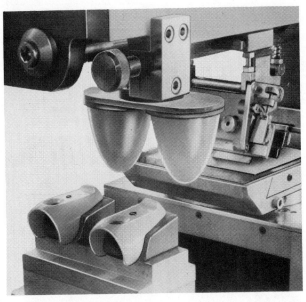

(D) Closeup of transfer pad and inking mechanism.

Fig. 19-11. The Tampo-Print process uses a transfer pad to apply ink to products. (Dependable Machine Co.)

Advantages of Heat-Transfer Decorating

1. Similar to hot-leaf stamping, except multicolored designs are possible
2. Many heat-transfer systems are available

Disadvantages of Heat-Transfer Decorating

1. Carrier film and designs are costly
2. Secondary operation and equipment are needed

19-9 MISCELLANEOUS DECORATING METHODS

There are many other decorating methods, including pressure-sensitive labels, decalcomanias, flocking, and decorative coatings or clads.

Pressure-sensitive labels are easy to apply. The designs or messages are usually printed on the adhesive-backed foil or film label and the labels are placed on the finished product by hand or mechanical means.

Decalcomanias, commonly known as decals, are a means of transferring a picture or design to plastics. They are generally a decorative film on a paper backing. The decalcomania is moistened in water and the adhesive-backed film is slipped off the paper and onto the plastics surface. This process is not widely used, mainly because decalcomanias are hard to place accurately and quickly on the plastics surface.

Flocking by mechanical or electrostatic means is an important method of placing a velvetlike finish on nearly any surface. The process consists of coating the product with an adhesive and placing plastics fibers on the adhesive areas. The velvetlike coatings or designs on wallpapers, toys, and furniture are examples.

There are a number of woodgraining decorative processes. Some are done by rolling engraved or etched woodgraining plates over a contrasting background color. This is actually a printing process adaptation. Some decorative laminates and clad coatings are also used to decorate substrates. Polyvinyl clad metal products are used in store fixtures, partitions, room dividers, furniture, automobiles, kitchen equipment, and bus interiors. These are durable as well as decorative.

Many foils and patterns are thermoformable, permitting the decoration of three-dimensional thermoformed parts. The thermoformed shell may be filled by casting or injection molding. Plastics thinning and pattern distortion must be controlled (Fig. 19-12).

Four advantages and two disadvantages of the pressure sensitive decorating process are shown below.

Advantages of Pressure Sensitive Decorating

1. Variable application rates (relatively high-speed to hand-dispensed)
2. Multi-colored patterns and designs
3. May be used on all plastics
4. Short runs and changes in patterns are economical

Disadvantages of Pressure Sensitive Decorating

1. Secondary operation and equipment are needed
2. Surface labels may wear or be removed

VOCABULARY

The following vocabulary words are found in this chapter. Look up the definition of any of these words you do not understand as they apply to plastics in the glossary, Appendix A.

Corona discharge
Electrostatic printing
Engraving
Heat-transfer decorating
Hot-leaf stamping
In-mold decorating
Offset printing
Tampo-Print

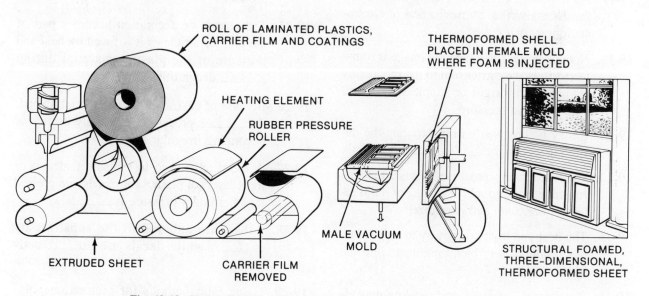

ROLL OF LAMINATED PLASTICS, CARRIER FILM AND COATINGS

THERMOFORMED SHELL PLACED IN FEMALE MOLD WHERE FOAM IS INJECTED

HEATING ELEMENT

RUBBER PRESSURE ROLLER

EXTRUDED SHEET

CARRIER FILM REMOVED

MALE VACUUM MOLD

STRUCTURAL FOAMED, THREE-DIMENSIONAL, THERMOFORMED SHEET

Fig. 19-12. Thermoforming a decorative cover for an appliance. (Dri Print Foils)

19-1. A ___?___ may be used to prevent silicone paint from depositing where it is not wanted.

19-2. Name two advantages of silicone hot-stamping dies.

19-3. Name the painting process in which surface wear does not easily remove the design.

19-4. What additive may lower the electrical resistance in a plastics?

19-5. Name the result of applying a coating over items that have not been properly dried of surface moisture.

19-6. The name of a well-known mixer used to blend plastics ingredients is ___?___ .

19-7. Hot-leaf stamping is sometimes called roll-leaf stamping or simply ___?___ .

19-8. Name three functional uses of plating.

19-9. Name three decorative uses of plating.

19-10. Electroplate thickness varies from ___?___ to ___?___ mm.

19-11. An important method of placing a velvetlike finish on a surface by mechanical or electrostatic means is called ___?___ .

19-12. Name a process in which the image is transferred from the carrier film to the product by stamping with rigid or supple shapes and with heat and pressure.

19-13. It may be less costly to incorporate the desired design in the ___?___ , dies or rollers.

19-14. Most decorating processes require that the substrate be thoroughly ___?___ of mold release, lubricants and plasticizers.

19-15. The best and least costly way to color plastics products is to blend pigments into the basic ___?___ .

19-16. Plasticizers, ___?___ , and molding may affect color.

19-17. An inexpensive and popular method of painting plastics is ___?___ .

19-18. A variety of shapes and different sizes of products may be rapidly painted by ___?___ methods.

19-19. Name the three ways of atomizing paint.

19-20. Excess paint may be removed in the dip coating process by spinning the part, hand wiping, or by ___?___ methods.

19-21. The process of forcing special inks or paint through small openings of a stenciled screen into a product surface is called ___?___ or ___?___ .

19-22. In ___?___ , the image depression should be deep and narrow for sharp detail.

19-23. Raised portions, letters, figures, or other designs are easily decorated by ___?___ methods.

19-24. Name five methods of printing on plastics products.

19-25. The pattern or decoration becomes part of the plastics article as it is fused by heat and pressure of the plastics material during ___?___ decorating.

19-26. Because of the use of a flexible printing pad, the ___?___ process is especially useful for decorating irregular surfaces.

19-27. A metal term used to indicate that two or more layers of plastics (or metals) are pressed together under pressure is ___?___ .

19-28. Labels are adhered to the substrate by ___?___ , while decals are ___?___ activated.

19-29. Name four methods for pretreating polyethylene to receive paint or ink.

Energy savings, low pollution effects, and economic advantages are offered by radiation processing. You will learn how this technology offers new product and manufacturing ideas.

Radiation processing is a growing area of technology in which ionizing and nonionizing systems are used to change and improve the physical properties of materials or components. Radiation processes may find acceptance along with chemical and thermodynamic processes.

20-1 RADIATION METHODS

The term *radiation* can refer to energy carried by either waves or particles. The carrier of wave energy is called the *photon*. In radiant energy, the photon is wavelike when in motion. It is particle-like when absorbed or emitted by an atom or molecule.

The ordinary electric light bulb with an element temperature of 4 172 °F [2 300 °C] emits radiation waves that are visible. The sun with a surface temperature of 10 832 °F [6 000 °C] emits both visible and invisible radiation. Human eyes can see radiation with wavelength as short as 400 μm [15.7 × 10^{-6} in.] and as long as 700 μm [27.5 × 10^{-6} in.] *Ultraviolet* radiations are waves of energy that can burn or tan the exposed parts of the human body yet are invisible to the human eye (Fig. 20-1). Ultraviolet radiation has wavelengths shorter than 15.7 × 10^{-6} in. [400 μm]. Photon wavelengths are measured in micrometres. Radiation from the sun, burning fuels, or radioactive elements are considered natural sources of radiation. A few of the more important radioactive elements that occur naturally are uranium, radium, thorium, and acti-

nium. These radioactive materials emit photons of energy and/or particles as the nuclei disintegrate and decrease in mass. The earth contains small traces of radioactive materials while the sun is intensely radioactive.

Radiation may be produced by nuclear reactors, by accelerators, or from natural or artificial radioisotopes. The most important source of controlled radiation is artificial radioisotopes. Scientists have brought the use and control of induced radiation to a point where it may serve human needs.

When the number of *protons* in the nucleus of an atom changes, a different element is formed. If the number of *neutrons* in the nucleus is changed, a new element is not formed; only the mass of the element is different. Different forms (masses) of the same element are called *isotopes*. Most ele-

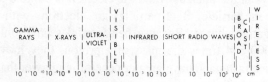

(A) Wavelengths of radiation.

 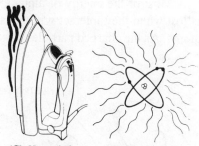

(B) Light (visible) radiation makes things visible.

(C) Heat (infrared) radiation can be felt.

(D) Radioactive radiation can't be seen or felt.

Fig. 20-1. Types of radiation.

ments have several isotopes. For example the simple element hydrogen occurs as three distinct isotopes (Fig. 20-2). Most hydrogen atoms have a mass number (the number of protons and neutrons) of 1, meaning they have no neutrons. A very small number of naturally occuring hydrogen atoms have one neutron and one proton and thus have a mass number of 2. Only when hydrogen has two neutrons and one proton (a mass number of 3) is it radioactive.

In 1900, the German physicist Max Planck advanced the idea that photons are bundles or packets of electromagnetic energy. This energy is either absorbed or emitted by atoms or molecules. The unit of energy carried by a single photon is called a *quantum* (*quanta* is the plural). Photons of energy radiation may be classified into two basic groups, electrically neutral and charged radiation.

Alpha particles are heavy, slow moving masses with a double positive charge (two protons and two neutrons). When alpha particles strike other atoms, their double positive charge removes one or more electrons, leaving the atom or molecule in a dissociated or ionized state. Ionization, you will remember from Chapter 2, is the process of changing uncharged atoms or molecules into ions. Atoms in the ionized state have either a positive or a negative charge.

Electrons ejected from the nuclei of atoms at a very high speed and high energy are called *beta particles*. When a neutron disintegrates, it becomes a proton and an electron. The proton often stays in the nucleus while the electron is emitted as a beta particle. Beta particles are electrons with a negative charge. Because beta particles have only 0.000 544 times the mass of a proton, they move much faster and have greater penetrating power than alpha particles.

Most of the energy of alpha and beta particles is lost when they interact with electrons from other atoms. As the charged particles pass through matter, they lose or transfer all their excess energy to the nuclei or orbital electrons of the atoms they encounter. Since beta particles are negative, they can push or repel electrons leaving the atom with a positive charge, or beta particles may become attached to the atom giving it a negative charge.

Gamma radiations are short, very high-frequency electromagnetic waves with no electrical charge. *Gamma rays* and *X-rays* are alike except for origin and penetrating ability. Gamma photons can penetrate even the most dense materials. More than one metre of concrete is required to stop the radiation effect of gamma rays (Fig. 20-3).

The energy of gamma photons is absorbed or lost in matter in three major ways (Fig. 20-4):

1. Energy may be lost or transferred to an electron it strikes, forcing the electron out of orbit
2. The gamma photon may strike an orbiting electron a glancing blow using only a part of its energy while the rest of the energy continues in a new direction
3. The gamma ray or photon is annihilated when it passes near the powerful electrical field of a nucleus

In the last method of gamma ray energy loss, the powerful electrical field of an atomic nucleus breaks the gamma photon into two particles of opposite charge, an electron and a positron. The positron quickly loses its energy by colliding with orbital electrons. The net effect of gamma radiation is like the effect of alpha and beta radiation. Electrons are knocked from orbit causing ionization and excitation effects in materials.

Neutrons are uncharged particles that may collide with atomic nuclei resulting in alpha and gamma radiation as energy is transferred or lost.

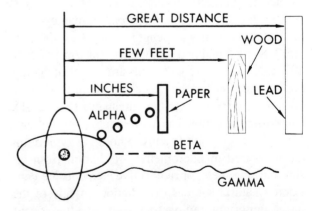

Fig. 20-3. Three types of radiation emitted by unstable atoms or radioisotopes: *alpha* (stopped by paper), *beta* (stopped by wood), and *gamma* (stopped by lead).

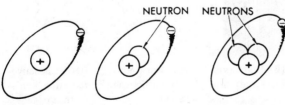

(A) Stable form: hydrogen $_1H^1$.   **(B)** Rare stable form: deuterium $_1H^2$.   **(C)** Rare radioactive form: tritium $_1H^3$.

Fig. 20-2. Isotopes of hydrogen.

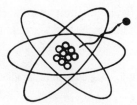

(A) Radioisotope with photon of energy being emitted.

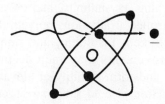

(B) Gamma energy being completely absorbed by forcing an electron from orbit and transferring energy.

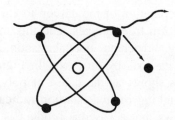

(C) Part of energy continues in new direction and part is used in ejecting electron from orbit.

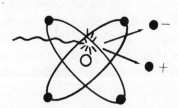

(D) Gamma radiation is annihilated. Electron and positron are created and share energy.

Fig. 20-4. Interaction of gamma radiation with matter.

20-2 RADIATION SOURCES

There are two basic types of radiation sources, those that produce ionizing radiation, and those that produce non-ionizing radiation.

Ionizing Radiation

Cobalt-60, strontium-90, and cesium-137 are three commercially available *radioisotope* sources that produce ionizing radiation. They are used because of their availability, useful characteristics, reasonably long half-life, and reasonable cost. Another source of ionizing radiation is used, or burnt, uranium slugs from fission reactor waste.

Gamma radiation, while very penetrating, is not a major source of ionizing radiation for several reasons. It is slow and may require several hours for treatment; its isotope sources are hard to control; the source cannot be turned off; experienced, well-trained workers are needed.

Electron beam accelerators are the primary ionizing source for radiation processing. Irradiation processing implies that a controlled or directed treatment of energy is being used on the polymer.

Nonionizing Radiation

Electron accelerators, such as Van de Graaff generators, cyclotrons, synchrotrons, and resonant transformers, may be used for producing non-ionizing radiation.

Electrons from machines are less penetrating than radioisotope radiation; however, they may be easily controlled and turned off when not required. These machines are capable of delivering 200 kW of power. The dose rating may be expressed in a unit called a *gray* (Gy). One gray indicates a dose absorption of one joule of energy in one kilogram of plastics (1 J/kg). As a rule, 1 kW of power is required to deliver a dose of 793 lb [10 kGy to 360 kg] of plastics per hour.

Electron radiation penetrates only a few millimetres into the plastics, but it can irradiate products at a very fast rate. Ultraviolet, infrared, induction, dielectric, and microwave are the most familiar sources of nonionizing radiation. They are generally used to speed up processing by heating, drying, and curing.

Ultraviolet radiation sources such as plasma arcs, tungsten filaments, and carbon arcs produce radiation with enough penetrating power for film and surface treatment of plastics. Prepregs have been produced that cure by exposure to sunlight.

Infrared radiation is often used in thermoforming, film extrusion, orientation, embossing, coating, laminating, drying, and curing processes.

Induction sources (electromagnetic energy) have been used to produce welds, preheat metal-filled plastics, and cure selected adhesives.

Dielectric sources (radio-frequency energy) have been used to preheat plastics, cure resins, expand polystyrene beads, melt or heat-seal plastics, and dry coatings.

Microwave sources are used to speed curing and to preheat, melt, and dry compounds.

Radiation Safety

All forms of natural radiation must be considered harmful to polymers and dangerous to work with, since they are not easily controlled; however, radiation processes are time-proven production methods used by a wide variety of industries today.

A disadvantage frequently discussed with radiation processing is safety, since an element of emotionalism often is associated with the word *radiation*.

Stray electrons and rays, high voltages, and ozone exposure are all potential hazards of radiation processing. However, safety in radiation processing is possible with an understanding of the hazards involved and with the adherence to a sound safety program. Safety standards have been published and maximum safe exposure levels for different kinds of radiation have been set by several governmental agencies.

For any company planning radiation processing, the wise manager will hire a qualified consultant to plan, design and implement a radiation safety program for the plant and its personnel.

20-3 IRRADIATION OF POLYMERS

The transfer of energy from the radiation source to the material assists in breaking bonds and thus is used for rearranging atoms into new structures. The many changes in covalent substances directly affect important physical properties. The effects of radiation on plastics may be divided into four categories:

1. Damage by radiation
2. Improvements by radiation
3. Polymerization by radiation
4. Grafting by radiation

Damage by Radiation

Breaking covalent bonds by nuclear radiation is called scission. This separation of the carbon-to-carbon bonds may lower the molecular mass of the polymer. Figure 20-5 shows that irradiation of polytetrafluoroethylene causes the long linear plastics to break into short segments. As a result of this breaking, the plastics loses strength.

Degradative symptoms include cracking, crazing, discoloration, hardening, embrittlement, softening, and other undesirable physical properties

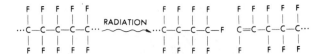

Fig. 20-5. Degradation by irradiation.

associated with molecular mass, molecular mass distribution, branching, crystallinity, and crosslinking.

With controlled irradiation, polyethylene becomes a crosslinked, insoluble, and nonmelting material. Improvements may include increased heat resistance and form stability at elevated temperatures, reduction in cold flow, stress cracking, and thermal cracking.

Effects of radiation on selected polymers are shown in Table 20-1.

The separation of the carbon-to-carbon bonds may also form free radicals, and this may lead to crosslinking, branching, polymerization, or the formation of gaseous by-products. In Figure 20-6A, radiation is shown as the cause of polymerization and crosslinking of hydrocarbon radicals. Figure 20-6B shows that a gaseous product is formed in the irradiation process. The radical (R) may be H, F, Cl, etc., as the gaseous product. The irradiation may result in atoms being knocked from the solid material, and this disassociation or displacement of an atom results in a defect in the basic structure of the polymer (Fig. 20-7). These vacancies in crystalline structures and other molecular changes re-

Table 20-1. Effects of Radiation on Selected Polymers

Polymer	Radiation Resistance	Radiation Dose for Significant Damage (Mrads)
ABS	Good	100
EP	Excellent	100–10,000
FEP	Fair	20
PC	Good	100+
PCTFE	Fair	10–20
PE	Good	100
PFV, PFV$_2$, PETFE, PECTFE	Good	100
PI	Excellent	100–10,000
PMMA	Fair	5
Polyesters (aromatic)	Good	100
Polyesters (unsaturated)	Good	1,000
Polymethylpentene	Good	30–50
PP	Fair	10
PS	Excellent	1,000
PSO	Excellent	1,000
PTFE	Poor	1
PU	Excellent	1,000+
PVC	Good	50–100
UF	Good	500

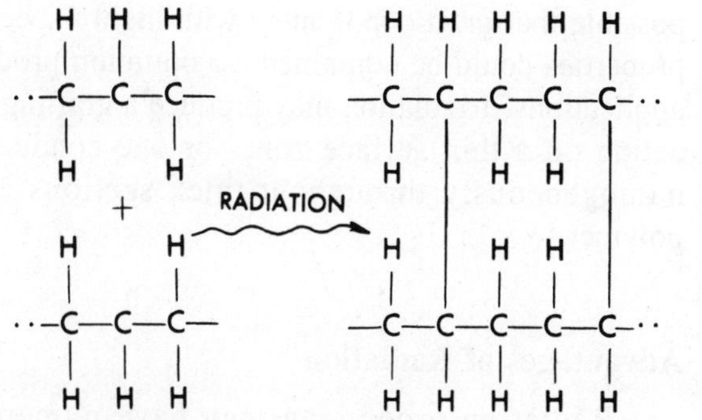

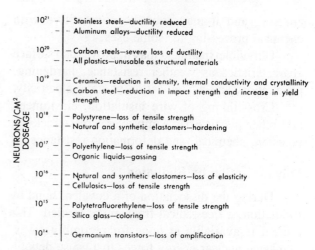

(A) Recombination leading to polymerization or cross-linking of hydrocarbon radicals.

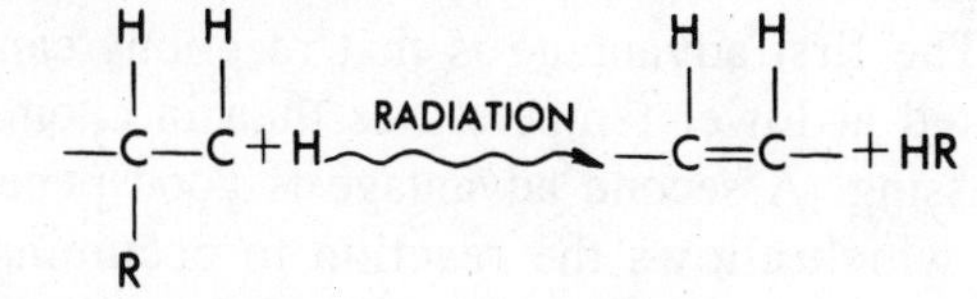

$$-\overset{\overset{\displaystyle H}{|}}{C}-\overset{\overset{\displaystyle H}{|}}{\underset{\underset{\displaystyle R}{|}}{C}}+H \xrightarrow{\text{RADIATION}} -\overset{\overset{\displaystyle H}{|}}{C}=\overset{\overset{\displaystyle H}{|}}{C}-+HR$$

(B) Gaseous product formed by radiation.

Fig. 20-6. Formation of free radicals through irradiation.

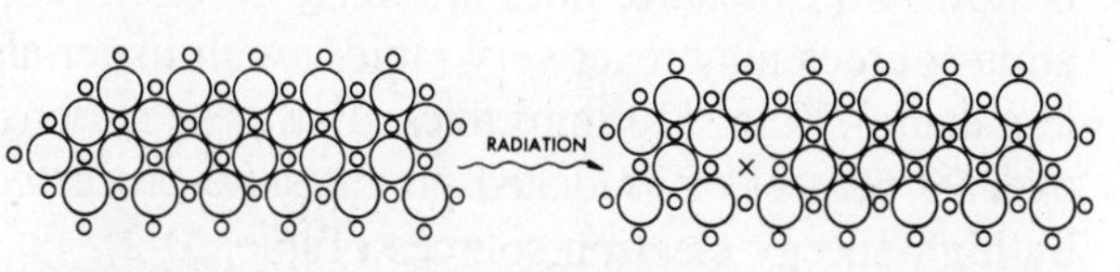

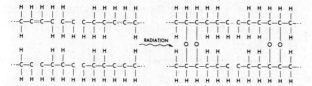

Fig. 20-7. Linear structure of plastics with missing atom. The vacancy in the crystalline structure is a potential site for radical attachment.

sult in changes in the mechanical, chemical, and electrical properties of polymers.

Cross-linkage in elastomers may be considered a form of degradation. For example, natural and synthetic rubbers become hard and brittle with more cross-linkage or branching (Figs. 20-8 and 20-9).

Mineral-, glass-, and asbestos-filled phenolics, epoxies, polyurethane, polystyrene, polyester, silicones, and furan plastics have superior radiation resistance. Unfilled methyl methacrylate, vinylidene chloride, polyesters, cellulosics, polyamides and polytetrafluoroethylene have poor radiation resistance. These plastics become brittle and their desirable optical properties are affected by discoloring and crazing. Fillers and chemical additi-

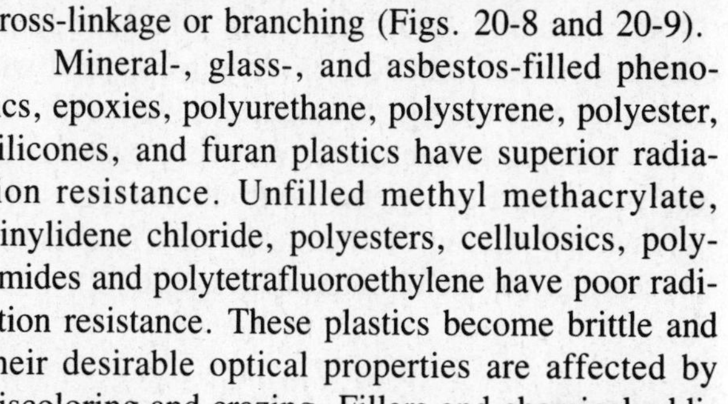

Fig. 20-8. Oxidation (radical attachment of oxygen) of polybutadiene. This crosslinking results in a rapid aging effect with loss of elastic strain.

Fig. 20-9. Changes in materials properties caused by radiation. Controlled use of radiation can be beneficial.

tives may help to absorb much radiation energy, while heavy pigmentation of the plastics may stop deep penetration of damaging radiation.

Improvements by Radiation

While some polymers are damaged by radiation, others may actually benefit from controlled amounts. Crosslinkage, grafting, and branching of thermoplastic materials may produce many of the desirable physical properties of the thermosetting plastics.

Polyethylene is one plastics that benefits from controlled, limited irradiation. Such radiation causes existing bonds to be broken and a rearrangement of the atoms into a branched structure. Branching of the PE chain elevates the softening temperature to above that of boiling water. (Excessive radiation may reverse the effect, however, by rupturing main links in the chains). Effects of radiation on selected polymers are shown in Table 20-1.

Radiation Processing. Radiation processing today is most often done with electron machines or radioisotope sources such as cobalt-60. This radiation may increase molecular mass by linking molecules of some polymers together, or it may decrease molecular mass by degrading others. It is this crosslinking and degradation that accounts for most of the property changes in plastics.

The ability of radiation to begin ionization and free-radical formation may prove superior to the ability of other agents, such as heat or chemicals.

The main industrial disadvantage of radiation-induced chemical reaction is high cost. With radiation processing integrated directly into processing lines, the cost of radiation systems has been decreasing and it may soon be competitive with chemical processing for some uses.

Ultraviolet treatment may improve such surface characteristics as weather resistance, hardening, penetration, and neutralization of static electricity.

Crosslinking of wire insulation, elastomers, and other plastics parts improves stress-cracking, abrasion, chemical, and deformation resistance.

Polymerization by Radiation

During the dissociation of a covalent bond by irradiation, a free radical fragment is formed. This radical is available at once for recombinations. The same nuclear energy forces that cause depolymerization of plastics may begin crosslinkage and polymerization of monomer resins (Fig. 20-10).

Polymerization and crosslinking are used to cure polymer coating, adhesives, or monomer layers. Typical doses (Mrads) for crosslinkable polymers range from 20-30 for PE, 5-8 for PVC, 8-16 for PVDF, 10-15 for EVA, and 6-10 for ECTFE.

Grafting by Radiation

When a given kind of monomer is polymerized and another kind of monomer is polymerized onto the primary backbone chain, a graft copolymer results. By irradiating a polymer and adding a different monomer and irradiating again, a graft copolymer is formed. The schematic structure of a graft copolymer is shown in Fig. 20-11. The recombining or structuring of two different monomer units (A and B) often yields unique properties. It is

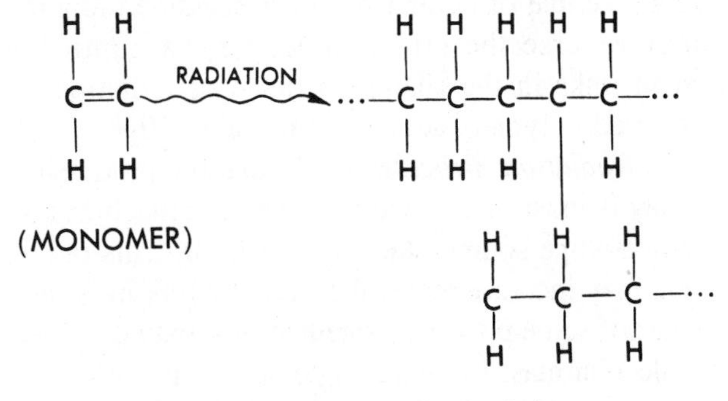

Fig. 20-10. Branching of polyethylene.

```
AAAAAAAAAAAAAAAAAAAAAAAAAAAAAA
          |
          BBBBBBBBBB
```

Fig. 20-11. In graft polymerization, a monomer of one type (B) is grafted onto a polymer of a different type (A). Because graft copolymers contain long sequences of two different monomer units some unique properties result.

possible that graft copolymers with highly specific properties could be combined for optimum product applications. Irradiation may produce a grafting reaction on a thin surface zone, or one conducted homogeneously throughout thick sections of a polymer.

Advantages of Radiation

Radiation processing may have numerous broad advantages to offset the main disadvantage of high cost.

The first advantage is that reactions can be initiated at lower temperatures than in chemical processing. A second advantage is good penetration, which allows the reaction to occur inside ordinary equipment at a uniform rate. Although gamma radiation from cobalt-60 sources can penetrate more than 12 in. [300 mm], the treatment rate is slow and exposure times are long. Electron radiation sources may react very rapidly with materials less than 0.39 in. [10 mm] thick. For these reasons, over 90 percent of irradiated products are processed by high-energy electron sources (Table 20-2).

A third advantage is that monomers can be polymerized without chemical catalysts, accelerators, and other components that may leave impurities in the polymer. A fourth advantage is that radiation-induced reactions are little affected by the presence of pigments, fillers, antioxidants, and other ingredients in the resin or polymer. A fifth advantage is that crosslinking and grafting may be done on previously shaped parts such as films, tubing, coating, moldings, and other products. Coatings in monomeric form may be applied with radiation processing, thus doing away with solvents and the collection or recovery systems for the solvents. Finally, the sixth advantage is that mixing and storing of chemicals used in chemical processing may be eliminated. (Fig. 20-12).

Applications

In addition to the processing advantages just mentioned, radiation processing may provide other marketable features not possible by other means (Fig. 20-13).

Graft and homopolymerization of various monomers on paper and fabrics improves bulkiness, resilience, acid resistance, and tensile strength. Irradiation of some cellulosic textiles has aided in the development of "dura-press" fabrics. Grafting

Table 20-2. Industrial Applications for Electron Beam Processing

Product	Product Improvements and Process Advantages	Process
Wire and cable insulation, plastic insulating tubing, plastic packaging film	Shrinkability; impact strength; cut-through, heat, solvent, stress-cracking resistance; low dielectric losses.	Crosslinking, vulcanization
Foamed polyethylene	Compression and tensile strength; reduced elongation.	Crosslinking, vulcanization
Natural and synthetic rubber	High-temperature stability; abrasion resistance; cold vulcanization; elimination of vulcanizing agents.	Crosslinking, vulcanization
Adhesives: Pressure Sensitive Flock Laminate	Increased bonding; chemical, chipping, abrasion, weathering resistance; elimination of solvent;	Curing, polymerization
Coatings, paints, and inks on: Woods Metals Plastics	100% convertibility of coating; high-speed cure, flexibility in handling techniques; low energy consumption; room-temperature cure; no limitation on colors.	Curing, polymerization
Wood and organic impregnates	Mar, scratch, abrasion, warping, swelling, weathering resistance; dimensional stability; surface uniformity; upgrading of softwoods.	Curing, polymerization
Cellulose	Enhanced chemical combination	Depolymerization
Textiles and textile fibers	Soil-release; crease, shrink, weathering resistance; improved dyeability; static dissipation; thermal stability.	Grafting
Film and paper	Surface adhesion; improved wettability.	Grafting
Medical disposables	Cold sterilization of packages and supplies.	Irridiation
Packages and containers	Reduction or elimination of residual monomer.	Polymerization
Polymer	Controlled degradation or modification of melt index.	Irridiation, Depolymerization, Crosslinking.

selected monomers to polyurethane foam, natural fibers, and plastic textiles improves weather resistance, and eases ironing, bonding, dyeing, and printing. Small radiation dosages, which degrade the surface of some plastics, improve ink adhesion to their surface.

Impregnation of monomers in wood, paper, concrete, and certain composites has increased their hardness, strength, and dimensional stability after irradiation. For example, the hardness of pine has been increased 700 percent by this method. Novolacs and resols are soluble and fusible low-molecular-mass resins used in the production of prepregs (reinforcement-impregnated resins) and impregs (resin-impregnated materials). The term *A-stage* is used to refer to novolac and resol resins. Wood, fabric, glass fibers, and paper may be saturated with A-stage resins while under a high vacuum. This supersaturated prepreg or impreg may then be exposed to cobalt radiation causing the thermosetting, A-stage material to pass through a rubbery stage referred to as the *B-stage*. Further

reaction leads to a rigid, insoluble, infusible, hard product. This last stage of polymerization is known as the *C-stage*. The terms A-, B-, and C-stage resins are also used to describe analogous states in other thermosetting resins. (See Phenolics in Chapter 8.)

A commercially available *shrinkable* polyethylene film is often used for wrapping food items. This irradiated film is crosslinked by radiation for increased strength. The film can be stretched more than 200 percent and is usually sold prestressed. When heated to 180 °F [82 °C] or higher, the film attempts to shrink back to its original dimensions, thus making a tight package. Radiation is also being used as a heatless sterilization system for packaged food and surgical supplies.

Radioisotopes are used in many measuring applications. Monomer resins, paints, or other coatings can be measured for thickness without contacting or marking the material's surface, as can extruded, or blown, films. Using this measuring method may reduce raw material consumption,

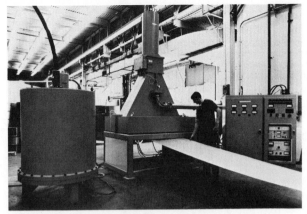

(A) When the coating is passed under the electron beam, free radicals are produced by ionization. These free radicals start a rapid buildup of long-chain molecules that become the cured resin. This curing mechanism does not require heat or catalysts.

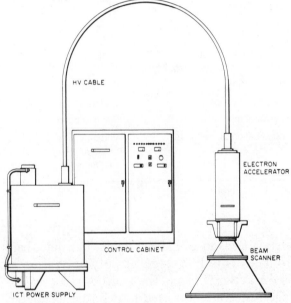

(B) Diagram of the major components of the electron-beam processing system.

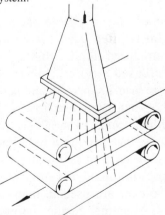

(C) Festooning is a method used for processing continuous, flexible sheets or webs of material.

Fig. 20-12. Principles of electron-beam radiation processing. (High Voltage Engineering Corp.)

Fig. 20-13. The polyethylene container in the center was exposed to controlled radiation to improve heat resistance. Containers at left and right were not treated. They lost shape at 350 °F [175 °C].

reduce or eliminate scrap, ensure more uniform thickness, and speed up output (Fig. 20-14).

Four advantages of radiation processing and three of its disadvantages are listed below.

Advantages of Radiation Processing

1. Improves many important plastics properties
2. Many non-ionizing radiation processes speed production by heating or initiating polymerization
3. No physical contact needed
4. Machine sources may be controlled with ease and require less shielding

Disadvantages of Radiation Processing

1. Gamma radiation equipment somewhat costly; some is specialized
2. Demands careful handling and trained personnel (especially ionizing radiation)

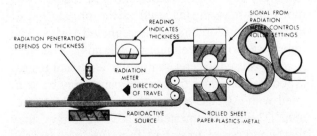

Fig. 20-14. Radioisotopes are used to continuously gauge thickness of a material without being in physical contact with it.

3. Potential danger to operator due to ionizing radiation and radioisotopes

Alpha particle
Beta particle
Gamma ray
Gray
Irradiation
Isotope
Photon
Radiation

VOCABULARY

The following vocabulary words are found in this chapter. Look up the definition of any of these words you do not understand as they apply to plastics in the glossary, Appendix A.

20-1. Name the term that refers to bombardment of plastics with a variety of subatomic particles. It may be done to polymerize and change physical properties of plastics.

20-2. The most important source of controlled radiation is ___?___ .

20-3. When one or more different kinds of monomer are attached to the primary backbone of the polymer chain, a ___?___ results.

20-4. Name five plastics that have poor radiation resistance.

20-5. Name two additives for plastics that may help stop penetration of damaging radiation.

20-6. The carrier of wave energy is called ___?___ .

20-7. Electrons traveling at very high speed and with high energy are called ___?___ particles.

20-8. Different forms of the same element with different atomic masses are called ___?___ .

20-9. Breaking of covalent bonds by nuclear radiation is called ___?___ .

20-10. The two types of radiation systems or sources are ___?___ and ___?___ .

20-11. Controlled amounts of irradiation may cause ___?___ of bonds for free radical formation and crosslinkage.

20-12. Uncontrolled irradiation may break bonds, lowering ___?___ and ___?___ .

20-13. Name four adverse effects of irradiation.

20-14. Name three possible sources of ionizing radiation for irradiating plastics.

20-15. Name four possible sources of non-ionizing radiation for irradiating plastics.

20-16. The ___?___ is the term often used to describe the dose given in irradiating polymers.

20-17. Radiation sources must be carefully handled only by trained ___?___ .

20-18. Accumulated dosages or exposure to ___?___ and ___?___ radiation may cause permanent cell damage.

20-19. Name four major advantages of irradiation processing.

20-20. Energy sources that may be used to preheat molding compounds, heat seal films, polymerize resins and expand polystyrene beads are called ___?___ .

20-21. Radiations that sunburn or tan the human body are ___?___ rays.

20-22. Identify the particles that are heavy, slow-moving masses with a double positive charge.

20-23. Burnt ___?___ slugs from reactors or fission waste may be a source of radiation.

20-24. As a rule, all forms of natural radiation must be considered ___?___ to polymers.

20-25. Over ___?___ percent of irradiated products are processed by high-energy electron sources.

20-26. For the following products or applications, would you select non-ionizing or ionizing processing?
(a) polymerization of selected resins
(b) surface treatment of films
(c) drying plastics granules or preforms

Chapter 21
Design Considerations

This chapter summarizes basic rules for designing products. Because of the diversity of materials, processes, and product uses, designing with plastics demands more experience than designing with other materials. The information presented herein should serve as a fundamental guide and a useful starting point in understanding the complexity of designing plastics products. See Appendix G for additional sources of information.

There are many sources for studying specific design problems and some of them are included in the discussions of individual materials and processes.

In the early years of their development, plastics were chosen mostly as a substitute for other materials. Some of those early products were very successful because of the consideration and thought given to the choice of materials. On the other hand, some of these products failed because the designers did not know enough about the properties of the plastics used or were motivated by costs rather than the practical use of the material. The products simply could not stand up to daily wear and tear. As the plastics industry has grown, so has the designers' knowledge of the properties of plastics. Because plastics have combinations of properties that no other materials possess, i.e., strength, lightness, flexibility, and transparency (Table 21-1), plastics are now chosen as primary materials rather than as substitutes.

The design considerations for polymer composites are more complex than those for homopolymers. Most composites vary with time under load, rate of loading, small changes in temperature, matrix composition, material form, reinforcement configuration, and fabrication method. They may be designed to be isotropic, quasi-isotropic, or anisotropic depending upon design requirements.

Three industrial tools that are revolutionizing the design process and becoming a fact of life in the 1980s are computer assisted design (CAD), computer assisted manufacturing (CAM), and computer assisted moldmaking (CAMM). Most manual drafting and hand calculation will be eliminated resulting in designers, fabricators, materials manufacturers, and toolmakers making fewer errors in part designs, material selection, and tooling configurations. Figure 21-1 shows a designer using a CAD/CAM system to improve productivity and smooth the path from design to production.

A designer is now able to use a computer in the design, engineering, and manufacture of all plastics products. During interactive process design, the designer uses the computer to rapidly draw and make changes that improve the appearance and function of the part. The graphic model appearing on the CRT screen may be rotated and

Fig. 21-1. Computer graphics (CAD/CAM) systems are designed to automate and integrate the many phases of the product development cycle. (Applicon Inc.)

Table 21-1. Plastics vs. Metals

Properties of Plastics Which May Be...

Favorable

1. Light
2. Better chemical and moisture resistance
3. Better resistance to shock and vibration
4. Transparent or translucent
5. Tend to absorb vibration and sound
6. Higher abrasion and wear resistance
7. Self-lubricating
8. Often easier to fabricate
9. Can have integral color
10. Cost trend is downward. Today's composite plastics price is approximately 11% lower than five years ago. However, the long-established, high-volume plastics—phenolics, styrenes, vinyls, for example—appear to have reached a price plateau, and prices change only when demand is out of phase with supply
11. Often cost less per finished part
12. Consolidation of parts

Unfavorable

1. Lower strength
2. Much higher thermal expansion
3. More susceptible to creep, cold flow, and deformation under load
4. Lower heat resistance—both to thermal degradation and heat distortion
5. More subject to embrittlement at low temperature
6. Softer
7. Less ductile
8. Change dimensions through absorption of moisture or solvents
9. Flammable
10. Some varieties degraded by ultraviolet radiation
11. Most cost more (per cubic millimetre) than competing metals. Nearly all cost more per kilogram.

Either Favorable or Unfavorable

1. Flexible—even rigid varieties more resilient than metals
2. Electrical nonconductors
3. Thermal insulators
4. Formed through the application of heat and pressure

Exceptions

1. Some reinforced plastics (glass-reinforced epoxies, polyesters, and phenolics) are nearly as rigid and strong (particularly in relation to mass) as most steels. May be even more dimensionally stable.
2. Some oriented films and sheets (oriented polyesters) have greater strength-to-mass ratios than cold rolled steels
3. Some plastics are now cheaper than competing metals (Nylons vs. brass, acetal vs. zinc, acrylic vs. stainless steel)
4. Some plastics are tougher at low than at normal temperatures (acrylic has no known brittle point)
5. Many plastics-metal combinations extend the range of useful applications of both (metal-vinyl laminates, leaded vinyls, metallized polyesters, and copper-filled TFE)
6. Plastics and metal components may be combined to produce a desired balance of properties (plastics parts with molded-in, threaded-metal inserts; gears with cast-iron hubs and Nylon teeth; gear trains with alternate steel and phenolic gears; rotating bearings with metal shaft and housing and Nylon or TFE bearing liner)
7. Metallic fillers in plastics make them electrically or thermally conductive or magnetic

Source: *Machine Design*: Plastics Reference Issue

viewed from different angles. Certain sections or complex shapes can be further detailed. The most important benefit of computer use in engineering is the improvement in design, labor productivity, market share, capital productivity, innovation, quality, and profitability. With computer-assisted engineering (CAE) the designer can use finite-element and other design analysis, system modeling, and simulated structural testing. CAM is useful for programming, robotic interfacing, quality control, and other operations associated with the manufacture of the product. CAD/CAE/CAM systems can evaluate and review such high-cost elements as material handling, tooling, tool maintenance, raw material costs, and scrap losses before production begins.

In computer-integrated manufacturing (CIM) all design, engineering, and manufacturing processes are intermixed to allow designers, engineers, technicians, accountants, and others access to the same database. Because these and business opera-

tions are integrated, there is a saving in terms of reduced down time, labor costs, just-in-time (JIT) manufacturing or zero-inventory, quick changeover for batch manufacturing, and rapid design changes. (See Tooling and Moldmaking in Chapter 22)

In most designs, a compromise must be made between highest performance, good appearance, efficient production, and lowest cost. Unfortunately, human needs are often given less importance than factors of cost, processes, or materials used. There are three (sometimes overlapping) major design considerations: 1) material, 2) design, and 3) production.

21-1 MATERIAL CONSIDERATIONS

Materials must be selected with the right properties to meet design, economic, and service conditions. In the past, the design was commonly changed to compensate for the material limitations.

Caution must be exercized when using information obtained from data sheets or the manufacturer about the matrix performance. Much of this data is based upon laboratory-controlled evaluation. It is also difficult to compare proprietary data from several different suppliers. This does not imply, however, that this data cannot be utilized in the screening of candidate materials.

Customers often use *specifications* in a document to state all requirements to be satisfied by the proposed product, materials, or other standards. There are several different types of *standards,* including *physical* standards as kept by the National Bureau of Standards (NBS), *regulatory* standards such as those from the EPA, *voluntary* standards recommended by technical societies, producers, trade associations, or other groups such as Underwriters Laboratories (UL), mil (military) and *public* standards promoted by professional organizations such as the ASTM.

Metrication and internationalization of standards can reduce costs associated with materials, production, inventory, design, testing, engineering, documentation, and quality control. There is substantial evidence that metrication and standardization lower costs. We must change our attitude about international measurement and standards if we are to significantly lower costs and increase our international trade.

With the computer, systematic methods of screening and selection of materials are easier. Computer models can predict and anticipate most of the ways a material can fail. In some computer models each property is assigned a value according to importance. Then each property that the part is expected to tolerate is entered. The computer will select the best combination of materials and processes.

Plastics materials must be chosen with care, keeping the final product use in mind. The properties of plastics depend more on temperature than do other materials. Plastics are more sensitive to changes in environment; therefore, many families of plastics may be limited in use. There is no one material that will possess all the qualities desired; however, undesirable characteristics may be compensated for in the product design.

The final material choice for a product is based on the most favorable balance of design, fabrication, and total cost or selling price of the finished item. Either simple or complicated designs may need fewer processing and fabrication operations using plastics, which may combine with the characteristics of the plastics material to make plastics cost competitive with other materials for specific parts.

Environment

When designing a plastics product, the physical, chemical, and thermal environments are very important considerations. The useful temperature range of most plastics seldom exceeds 392 °F [200 °C]. Many plastics parts exposed to radiant and ultraviolet energy soon suffer surface breakdown, become brittle, and lose mechanical strength. For products operating above 450 °F [230 °C], fluorocarbons, silicones, polyimides, and filled plastics must be used. The exotic environments of outer space and the human body are becoming commonplace for plastics materials. The insulation and ablative materials used for space vehicles, the artery reinforcements, monofilament sutures, heart regulators, and valves are only a small portion of these new uses.

Some plastics retain their properties at cryogenic (extremely low) temperatures. Containers, self-lubricating bearings, and flexible tubing must function properly in below-zero temperatures. The cold, hostile environments of space and earth are only two examples. Any time that refrigeration and food packaging is considered or where taste and odors are a problem, plastics may be chosen. The United States Food and Drug Administration lists acceptable plastics for packaging of foods.

The Child Protection and Toy Safety Acts of 1969 and 1976 govern the manufacture and distribution of children's toys. Should toys present an electrical, mechanical, toxic, or thermal hazard, they may be banned from sale.

In addition to extremes of temperature, humidity, radiation, abrasives, and other environmental factors, the designer must consider fire resistance. There are no fully fire resistant plastics.

Polyimide-boron fiber composites have high temperature resistance and high strength. Polyimide-graphite fiber composites can compete with metals in strength and achieve a significant weight saving at service temperatures up to 600 °F. Boron powder is sometimes added to the matrix to help stabilize the char that forms in thermal oxidation. Other flame-resistant additives or an ablative matrix may also be used. The danger from open flames is the most serious objection to the use of plastics in fabrics and architectural structures.

Remember, thermal degradation and cross-linking are not reversible phenomena. Glass transition, melting, and crystallization of most matrices are reversible.

Moisture may cause deterioration and weaken the reinforcement and matrix bond in composites. Any holes, exposed edges, or machined areas on composite designs should be given a protective coating to prevent moisture infiltration or wicking.

Electrical Characteristics

All plastics have useful electrical insulation characteristics. The selection of plastics is usually based on mechanical, thermal, and chemical properties; however, much of the pioneering in plastics was for electrical uses. The electrical insulating problems of high altitude, space, undersea, and underground environments are solved by the use of plastics. All-weather radar and underwater sonar would not be possible without the use of plastics. They are used to insulate, coat, and protect electronic components.

Particulate composites using carbon, graphite, metal, or metal-coated reinforcements provide EMI shielding for many products.

Chemical Characteristics

The chemical and electrical natures of plastics are closely related because of molecular makeup. There is no general rule for chemical resistance. Plastics must be tested in the chemical environment of their actual use. Fluorocarbons, chlorinated polyethers and polyolefins are among the most chemical-resistant materials. Some plastics react as semipermeable membranes. They allow selected chemicals or gases to pass while blocking others. The permeability of polyethylene plastics is an asset in packaging fresh fruits and meats. Silicones and other plastics allow oxygen and gases to pass through a thin membrane while at the same time they stop water molecules and many chemical ions. Selective filtration of minerals from water may be done with semipermeable plastics membranes.

Mechanical Factors

Material consideration includes the mechanical factors of fatigue, tensile, flexural, impact, and compressive strengths, hardness, damping, cold flow, thermal expansion, and dimensional stability. All of these properties were discussed in Chapter 4. Products that need dimensional stability call for careful choice of materials, although fillers will improve the dimensional stability of all plas-

tics. A factor sometimes used for evaluation and selection is strength-to-mass ratio; the ratio of tensile strength to the density of the material. Plastics can surpass steels in strength-to-mass ratio.

Example: Divide the tensile strength of the material by its density.

$$\text{Selected plastics} - \frac{0.70 \text{ GPa}}{2 \text{ g/cm}^3} = 0.350$$

$$\text{Selected steel} - \frac{1.665 \text{ GPa}}{7.7 \text{ g/cm}^3} = 0.214$$

The type and orientation of reinforcements greatly influences the properties of composite products. Some specifications in critical designs will specify a *safety factor* (SF). A safety factor (sometimes called design factor) is defined as the ratio of the ultimate strength of the material to the allowable working stress.

$$SF = \frac{\text{Ultimate Strength}}{\text{Allowable Working Stress}}$$

The SF for a composite aircraft landing gear might be 10.0, while for an automobile spring it will be only 3.0. With accurate, reliable data some designs use safety factors of 1.5 to 2.0.

Economics

The final phase of material selection is economic considerations. It is best not to include material costs in the preliminary screening of candidate materials. Materials with marginal performance properties or expensive materials need not be eliminated at this point. Either may continue to be a possible candidate depending on the processing parameters, assembly, finishing, and service conditions. A polymer with minimal performance characteristics may not be the best choice if reliability and quality are important.

Cost is always a major factor in design considerations or material selection. Strength-to-mass ratio, or chemical, electrical, and moisture resistance, may overcome price disadvantage. Some plastics cost more per kilogram than metals or other materials, but plastics often cost less per finished part. The most meaningful comparison between different plastics is cost per cubic centimetre. The apparent density and bulk factors are important in cost analysis in any molding operation.

Apparent density, sometimes called *bulk density,* is the mass per unit volume of a material. It is calculated by placing the test sample in a graduated cylinder and taking measurements. The volume (V) of the sample is the product of its height (H) and cross-sectional area (A). Thus $V = HA$.

$$\text{Apparent density} = \frac{W}{V} \text{ where}$$

V = volume in cubic centimetres occupied by the material in the measuring cylinder

H = height in centimetres of the material in the cylinder

A = cross-sectional area in square centimetres of the measuring cylinder

W = mass in grams of the material in the cylinder

Many plastics and polymer composite parts cost ten times more than steel. On a volume basis, some are lower in cost than metals.

Bulk factor is the ratio of the volume of loose molding powder to the volume of the same mass of resin after molding. Bulk factors may be calculated as follows:

$$\text{bulk factor} = \frac{D_2}{D_1}$$

where

D_2 = average density of the molded or formed specimen

D_1 = average apparent density of the plastics material prior to forming.

Economics must also include the method of production and design limitations of the product. One-piece seamless gasoline tanks may be rotationally cast or blow-molded. The latter process uses more costly equipment but can produce the products more quickly, thus reducing costs. Conversely, large storage tanks may be produced at less cost by rotational casting than by blow-molding.

Capital investment for new tooling, equipment, or physical space could result in consideration of different materials and/or processes. The labor intensive operations often associated with wet- or open-molding cannot hope to compete with the automated facilities of some companies. The number of parts to be produced and the initial production costs may be the decisive factor.

The three overriding factors in plastics design are service, production, and cost. The use and performance of the part or product must be a concern in defining some design factors. For each design, there may be several production or process options. Costs often appear to override most other concerns of design and development, and cost is often based on the production method.

Volume of sales is very important. If a mold costs $10 000 and only 10 000 parts are to be made, the mold cost would be $1.00 per part. If 1 000 000 parts are to be made, the mold cost would be $0.01 per part.

Many composite components may be more cost-effective in the long term. The products may simply outlast metals in many applications. One piece composite components could reduce the number of molds used, tooling, and assembly time in the production of a boat hull, fuselage, or automobile floor pan. Light, corrosion resistant composites or plastics components could lower energy costs of fuel for the lifetime of the transporation vehicle. Less energy is consumed (including the energy content of the raw material) in the production of polymer parts than for metal parts. Plastics are mostly petroleum derived and must continue to compete for depleting resources.

21-2 DESIGN CONSIDERATIONS

When considering the overall design conditions, the intended application or function, environment, reliability requirements, and specifications must be reviewed. The database in computer systems may alert the designer that a design is outside the parameters of the material or process selected. See Appendix G for additional sources of information.

Appearance

The consumer is probably most aware of a product's physical appearance and utility. This includes the design, color, optical properties, and surface finish. Elements of design and appearance encompass several properties at once. Color, texture, shape, and material may influence consumer appeal. The smooth, graceful lines of Danish-style furniture with dark woods and satin finish are one example. Changing any one of these elements or properties would drastically change the design and appearance of the furniture.

A few outstanding characteristics of plastics are that they may be transparent or colored, as smooth as glass, or as supple and soft as fur. For many uses, plastics may be the only materials with the desired combination of properties to fulfill service needs.

To ensure proper design, there must be close cooperation between mold makers, manufacturers, processors, and fabricators.

Thought must be given to the design of the plastics part before it is molded in order to ensure that the best combination of mechanical, electrical, chemical, and thermal properties will be obtained.

Residual stresses develop as a result of forcing the material to conform to a mold shape. These stresses are locked in during cooling or curing and matrix shrinkage. They could cause warpage in flat surfaces. Warpage is somewhat proportional to the amount of shrinkage of the matrix and is generally the result of differential shrinkage.

There are no hard and fast rules to determine the most practical wall thickness of a molded part. Ribs, bosses, flanges, and beads are common methods of adding strength without increasing wall thickness. Large flat areas should be slightly convex or crowned for greater strength and to prevent warpage from stress (Fig. 21-2).

In Table 21-2 the complexity of parts is shown for several processes. In molding, it is im-

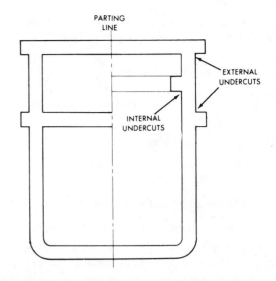

(D) A simple molding with internal and external undercuts.

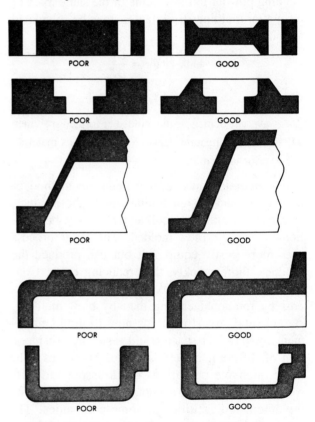

(E) The importance of uniform sectional thickness.

Fig. 21-2. Precautions to observe in production of plastics products.

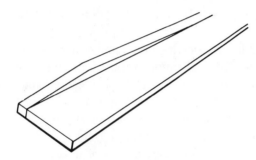

(A) Long, flat strips will warp. Ribs should be added, or the piece crowned in a convex shape.

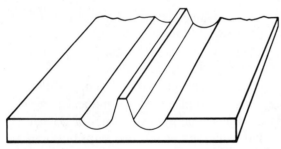

(B) Uneven sections will cause distortion, warpage, cracks, sinks, or other problems because of the difference in shrinkage from section to section.

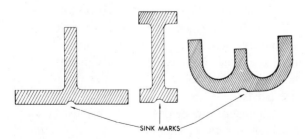

(C) Thickness of walls and ribs in themoplastic parts should be about 60 percent of the thickness of main walls. This will reduce the possibility of sink marks.

portant that all areas of the mold cavity be filled easily to uniformly minimize most of the stress in molding the part. A uniform wall thickness in the design is important to prevent uneven shrinkage of thin and thick sections. If wall thickness is not uniform, the molded part may distort, warp, and have internal stresses or cracks. From 0.24 to 0.51 in.

Table 21-2.   Plastics Processing Forms—Complexity of Part

Processing Form	Section Thickness, mm		Bosses	Undercuts	Inserts	Holes
	Max.	Min.				
Blow molding	> 6.35	0.254	Possible	Yes—but reduce production rate	Yes	Yes
Injection moldings	>25.4; normally 6.35	0.381	Yes	Possible—but undesirable; reduce production speed and increase cost	Yes—variety of threaded and nonthreaded	Yes—both through and blind
Cut extrusions	12.7	0.254	Yes	Yes—no difficulty	Yes—no difficulty	Yes—in direction of extrusion only; 0.50–1.0 mm min.
Sheet moldings (Thermoforming)	76.2	0.00635	Yes	Yes—but reduce production rate	Yes	No
Slush moldings		0.508	Yes	Yes—flexibility of vinyl allows drastic undercuts	Yes	Yes
Compression moldings		0.889–3.175	Possible	Possible—but not recommended	Yes—but avoid long, slender, delicate inserts	Yes—both through and blind; but should be round, large, and at right angles to surface of part
Transfer moldings		0.889–3.175	Possible	Possible—but should be avoided; reduce production rate	Yes—delicate inserts may be used	Yes—should be round, large, and at right angles to surface of part
Reinforced plastics moldings	Bag: 25.4 matched die: 6.35	Bag: 2.54 matched die: 0.762	Possible	Bag: yes; matched die: no	Bag: yes; matched die: possible	Bag: only large holes; matched die: yes
Castings		3.175–4.762	Yes	Yes—but only with split and cored molds	Yes	Yes

Source: *Materials Selector,* Materials Engineering, Reinhold Publishing Corp., Subsidiary of Litton Publications, Inc., Division of Litton Industries.

[6 to 13 mm] may be considered heavy wall thicknesses in molded parts.

In general, the ribs should have a width at the base equal to one-half the thickness of the adjacent wall. They should be no higher than three times the wall thickness. Boss designs should have an outside diameter equal to twice the inside diameter of the hole. They should be no higher than twice their diameter.

Plastics parts should have liberal fillets and rounds to increase strength, assist molten material flow, and reduce points of stress concentration. All radii should be generous, and the recommended minimum radius is 0.020 in. [0.50 mm]. Optimum design is obtained with a radius-to-thickness ratio of 1:6.

Undercuts (internal or external) in parts should be avoided if possible. Undercuts usually increase tooling costs by requiring techniques for molding, part removal (which usually requires movable parts in the mold), and cooling jigs. A slight undercut could be tolerated in some products when using tough, elastic materials. The molded part can be successfully snapped or stripped out of the cavity while hot. Undercut dimensions should be less than 5 percent of the part diameter.

Decorating may be considered an important functional factor in plastics design. The product may include textures, instructions, labels, or letters, and it must be decorated in a manner that will not complicate removal from the mold (Fig. 21-3). Decorations should provide durable service to the consumer. Letters are commonly engraved, hobbed, or electrochemically etched into mold cavities (Fig. 21-4).

Design Limitations

Next to material selection, tooling and processing have a marked effect on the properties and quality of all plastics products.

Closely related to production is the design of the product, and ultimately, the design of the mold to produce the product. Output rates, parting lines, dimensional tolerances, undercuts, finish, and ma-

Fig. 21-3. Familiar example of texture on plastics, which helps to hide imperfections. (Mold-Tech. Roehlen Industries)

Fig. 21-4. The electrical part was molded in one of two cavities of the injection mold. Note engraved letters in the mold cavities.

terial shrinkage are among factors that must be kept in mind by the mold maker or tool designer. For example, undercuts and inserts slow output rates and require a more costly mold.

The problem of material shrinkage is of equal importance to both the designer of molds and the designer of molded products. The loss of solvents, plasticizers, or moisture during molding, together with the chemical reaction of polymerization in some materials, results in shrinkage.

In the injection molding of crystalline materials, shrinkage is affected by the rate of cooling.

Thicker sections which take longer to cool will experience greater shrinkage than adjacent thinner sections which cool quickly.

The thermal contraction of the material must also be considered. The thermal expansion values for most plastics are relatively large. This is an asset in removal of molded products from mold cavities. If close tolerances are needed, material shrinkage and dimensional stability must be considered. Material shrinkage is sometimes used to ensure snug, or shrink, fits of metal inserts.

By using a computer model in a CAD system, it is possible to show the stress response of the part with a specific geometry, reinforcement content, reinforcement orientation, and molding orientation (flow). The computer model may require the use of ribs, contours, or other configurations to produce isotropic or anisotropic properties.

After the preliminary design is done, a physical prototype is often produced. This allows the design engineer and others to see and test a working prototype mold. Simulated performance and service tests may be performed with the prototype. If design or material errors were made, specifications for redesign can be made.

Both material and production considerations are important when designing products of plastics. The problem met in producing plastics products often demand selection of the production techniques before the material considerations are discussed.

18-3 PRODUCTION CONSIDERATIONS

In any product design, the behavior of the material and cost are often reflected in molding, fabrication, and assembly techniques. The tooling design must consider material shrinkage, dimensional tolerance, mold design, inserts, decorations, knockout pins, parting lines, production rates, and other post-processing operations (Table 21-3).

Trimming, cutting, boring holes, or other fabricating or assembly techniques may slow production, lower performance properties, and increase costs.

The part shape, size, matrix formulation, and polymer form will often limit the means of production to one or two possibilities.

Processes of Manufacture

With new technology and materials, processing is often the decisive competitive factor. Today, there are fewer limitations in processing thermo-

Table 21-3. Design Considerations

Plastics	Approximate Cost, ¢/in³	Approximate Cost, ¢/cm³	Linear Mold Shrinkage, mm/mm	Practical Dimensional Tolerances mm/mm (Single Cavity)			Taper Required, degrees		
				Fine	Standard	Coarse	Fine	Standard	Coarse
ABS	3–4	0.41	0.127–0.203	0.051	0.102	0.152	0.25	0.50	1.00
Acetal	7–8	0.46	0.508–0.635	0.102	0.152	0.229	0.50	0.75	1.00
Acrylic	4–5	0.27	0.025–0.102	0.076	0.127	0.178	0.25	0.75	1.25
Alkyd (filled)	5–6	0.33	0.102–0.203	0.051	0.102	0.127	0.25	0.50	1.00
Amino (filled)	3.5–5	0.27	0.279–0.305	0.051	0.076	0.102	0.125	0.5	1.00
Cellulosic	5–6	0.52	0.076–0.254	0.076	0.127	0.178	0.125	0.5	1.00
Chlorinated polyether	5–6	0.33	0.102–0.152	0.102	0.152	0.229	0.25	0.50	1.00
Epoxy			0.025–0.102	0.051	0.102	0.152	0.25	0.50	1.00
Fluoroplastic (CTFE)	240	14.7	0.254–0.381	0.051	0.076	0.127	0.25	0.50	1.00
Ionomer	4–5	0.27	0.076–0.508	0.076	0.102	0.152	0.50	1.00	2.00
Polyamide (6,6)	6–7	0.39	0.203–0.381	0.127	0.178	0.279	0.125	0.25	0.50
Phenolic (filled)	2.5–3	0.18	0.102–0.229	0.038	0.051	0.064	0.125	0.50	1.00
Phenylene oxide.	4.5–6	0.30	0.025–0.152	0.051	0.102	0.152	0.25	0.50	1.00
Polyallomer	2–2.5	0.14	0.254–0.508	0.051	0.102	0.152	0.25	0.50	1.00
Polycarbonate	7–8	0.46	0.127–0.178	0.076	0.152	0.203	0.25	0.50	1.00
Polyester (thermoplastic)	9–10	0.58	0.076–0.457	0.051	0.102	0.152	0.25	0.50	1.00
Polyethylene (high density)	1.5–2	0.15	0.508–1.270	0.076	0.127	0.178	0.50	0.75	1.50
Polypropylene	1–2	0.93	0.254–0.635	0.076	0.127	0.178	1.00	1.50	2.00
Polystyrene	2–2.5	0.14	0.025–0.152	0.051	0.102	0.152	0.25	0.50	1.00
Polysulfone	18–19	1.14	0.152–0.178	0.102	0.127	0.152	0.25	0.50	1.00
Polyurethane	7–8	0.46	0.254–0.508	0.051	0.102	0.152	0.25	0.50	1.00
Polyvinyl (PVC) (rigid)	35–40	2.26	0.025–0.127	0.051	0.102	0.152	0.25	0.50	1.00
Silicone (cast)	38–45	2.57	0.127–0.152	0.051	0.102	0.152	0.125	0.25	0.50

plastics and thermosetting materials than in years past. Processes that were unthinkable a few years ago have now become routine. Many thermosets are now injection molded or extruded. Some materials are processed in thermoplastic equipment and later cured. Polyethylene may be crosslinked after extrusion by chemical or radiation methods. Both thermoplastics and thermosets may be made cellular. Because of moldability, output rates, and other material properties, seemingly costly materials become inexpensive products.

In Table 21-4, a comparison of processing and economic factors is shown.

Material Shrinkage

Irregularities in wall thickness may create internal stresses in the molded part. Thick sections cool more slowly than thin ones and may create *sink marks* as well as differential shrinkage in crystalline plastics. As a general rule, injection molded crystalline plastics have high shrinkage while amorphous plastics shrink less.

Much pressure must be exerted to force the material through thin wall sections in the mold creating further problems because of material shrinkage. Polyethylenes, polyacetals, polyamides, polypropylenes, and some polyvinyls shrink between 0.020 and 0.030 in. [0.50 and 0.76 mm] after molding. Molds for these crystalline and other amorphous plastics must allow for material shrinkage.

Normally unfilled injection molded plastics shrink more in the direction of flow as opposed to the axis transverse to flow. This is mainly caused by the orientation pattern developed by the flow direction from the gate or gates. The differential shrinkage results because oriented plastics normally have a higher shrinkage than non-oriented plastics. An exception is fiber reinforced polymers.

Fiber reinforced polymers will shrink more along the axis transverse to flow than along the axis of material flow. Typical shrinkage of fiber reinforced polymers is about one-third to one-half that of nonreinforced polymers. The reason is that the fibers that are oriented in the direction of flow prevent the normal free shrinkage of the plastics or polymer.

Tolerances

Closely related to shrinkage is maintaining dimensional tolerances. Molding items with precision tolerances requires careful materials selection, and tooling costs are greater for precision molding. Di-

Table 21-4. Economic Factors Associated with Different Processes

Production Method	Economic Minimum	Production Rates	Equipment Cost	Tooling Cost
Autoclave	1–100	Low	Low	Low
Blow molding	1 000–10 000	High	Low	Low
Calendering (metres)	1 000–10 000	High	High	High
Casting processes	100–1 000	Low-high	Low	Low
Coating processes	1–1 000	High	Low-high	Low
Compression molding	1 000–10 000	High	Low	Low
Expanding processes	1 000–10 000	High	Low-high	Low-high
Extrusion (metres)	1 000–10 000	High	High	Low
Filament winding	1–100	Low	Low-high	Low
Injection molding	10 000–100 000	High	High	High
Laminating (continuous)	1 000–10 000	High	Low	High
Lay-up	1–100	Low	Low	Low
Machining	1–100	Low	Low	Low
Matched die	1 000–10 000	High	High	High
Mechanical forming	1–100	Low-high	Low	Low
Pressure-bag	1–100	Low	Low	Low
Pulforming (metres)	1 000–10 000	Low-high	Low	High
Pultrusion (metres)	1 000–10 000	Low-high	High	Low-high
Rotational casting	100–1 000	Low	Low	Low
Spray-up	1–100	Low	Low	Low
Thermoforming	100–1 000	High	Low	Low
Transfer molding	1 000–10 000	High	Low	High
Vacuum-bag	1–100	Low	Low	Low

mensional tolerances of single-cavity molded articles may be held to ±0.002 in./in. [±0.05 mm/mm] or less with selected plastics. Errors in tooling, variations in shrinkage between multicavity pieces, and differences in temperature, loading, and pressure from cavity to cavity all increase the critical dimensional tolerances of multicavity molds. If, for example, the number of cavities is increased to 50, the closest practical tolerance may then be ±0.010 in./in. [±0.25 mm/mm].

Tolerance standards have been established by technical custom molders and by the *Standards Committee* of the *Society of the Plastics Industry, Inc.* These standards are to be used only as a guide since the individual plastics material and the design must be considered in determining dimensions.

There are three classes of dimensional tolerances for molded plastics parts. They are expressed as plus and minus allowable variations in inches per inch (in./in.) or millimetres per millimetre (mm/mm). *Fine* tolerance is the narrowest possible limit of variation possible under controlled production. *Standard* tolerance is the dimensional control that can be maintained under average conditions of manufacture. *Coarse* tolerance is acceptable on parts where accurate dimensions are not important or critical.

So that the part may be removed with ease from the molding cavity, draft should be provided (both inside and out). The degree of draft may vary according to molding process, depth of part, type of material, and wall thickness. A draft of 0.25 degree is sufficient for all shallow molded parts. For textured designs and cores, the draft angles should be increased.

If a part has a depth of 10 in. [250 mm] and 0.125 degree draft for a fine dimensional tolerance of 0.002 2 in./in. [0.056 mm/mm], the total draft of the piece will be 0.22 in. [0.559 mm] per side (Fig. 21-5).

Mold Design

Mold design is an important factor in determining molding output. Because mold design is a complex subject, only a broad discussion is possible here. (See Chapters 10 and 22). A design for a typical two-piece, two-cavity injection mold is shown in Figure 21-6. A three-plate mold is shown in Figure 21-7.

As the hot, molten material is forced from the nozzle into the mold, it flows through channels or passageways. The terms *sprue, runner,* and *gate* are used to designate these channels (Fig. 21-8).

The heavy tapered channel that connects the nozzle with the runners is called the *sprue.* In a single-cavity mold, the sprue feeds material directly through a gate into the mold cavity (Fig. 21-9). If the sprue feeds directly, it eliminates the need for a separate runner and gate. In most single cavity molds, there is no need for a

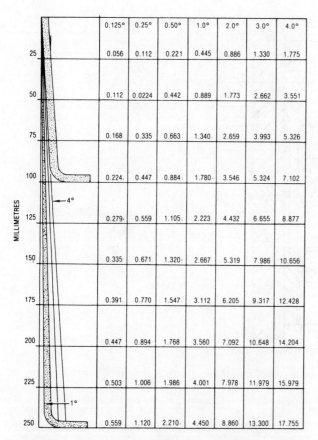

MILLIMETRES	0.125°	0.25°	0.50°	1.0°	2.0°	3.0°	4.0°
25	0.056	0.112	0.221	0.445	0.886	1.330	1.775
50	0.112	0.0224	0.442	0.889	1.773	2.662	3.551
75	0.168	0.335	0.663	1.340	2.659	3.993	5.326
100	0.224	0.447	0.884	1.780	3.546	5.324	7.102
125	0.279	0.559	1.105	2.223	4.432	6.655	8.877
150	0.335	0.671	1.320	2.667	5.319	7.986	10.656
175	0.391	0.770	1.547	3.112	6.205	9.317	12.428
200	0.447	0.894	1.768	3.560	7.092	10.648	14.204
225	0.503	1.006	1.986	4.001	7.978	11.979	15.979
250	0.559	1.120	2.210	4.450	8.860	13.300	17.755

Fig. 21-5. Degree of draft per side, in millimeters. (SPI)

runner unless the part is being gated in more than one place.

Runners are narrow channels that convey the molten plastics from the sprue to each cavity. In multicavity molds, the runner system should be designed so that all materials move the same distance from the sprue to each cavity (Fig. 21-10).

When a series of runners and gates is used, trapezoidal, half-round or full-round runners are machined into the mold. With trapezoidal and half-round runners and gates, only one half of the mold or die plate is machined (Fig. 21-11). Trapezoidal and half-round runners are easy to machine, but generally require more molding pressure. Round runners are advised for transfer molding if extruder-type plasticizers are used. Pinpoint (submarine) gating, as shown in Figure 21-12, may be used; however, this system also requires greater molding pressures.

A pinpoint gate is an opening of 0.030 in. [0.08 mm] or less through which the melt flows into a mold cavity. The submarine is a type of edge gate where the opening from the runner into the mold is located below the parting line or mold surface. The part is broken from the runner system or ejection from the mold.

The cost of elaborate mold designs and high waste from culls, sprues, and flash are two major limitations of transfer molding.

In some molds, there is only one cavity; others have many. Regardless of the number of cavities the *gate* is the point of entry into each mold cavity. In multicavity molds, there is a gate entering each mold cavity. The gates may be any shape or size, however, they are usually small so as to leave as small a blemish as possible. Gates must allow a smooth flow of molten material into the

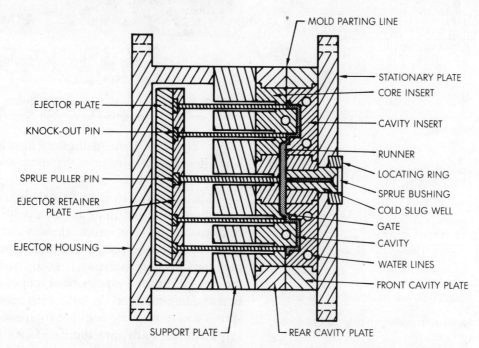

Fig. 21-6. Two-plate injection mold. (*Gulf Oil Chemicals Co.*)

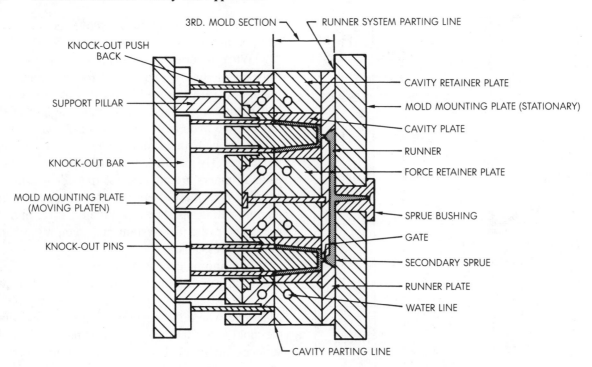

Fig. 21-7. Three-plate injection mold. (*Gulf Oil Chemicals Co.*)

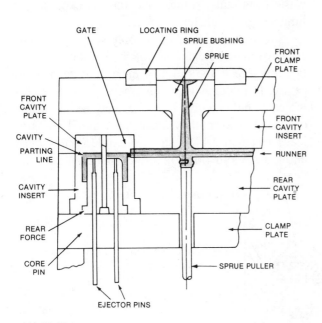

(A) Mold.

(B) Molded part, runner, gate, and sprue; and typical part.

Fig. 21-8. Construction of an injection mold, and a molded part.

Fig. 21-9. Injection mold, with molded product. Note the flash, or fins, on runner and parts. (Hull Corp.)

cavity (Fig. 21-13). A small gate will help the finished item break away from the sprue and runners cleanly (Fig. 21-14).

The sprues, gates, and runners are usually cooled and removed from the mold with the parts from each cycle. The sprues, runners and gates are removed from the parts, and then reground for molding. This reprocessing is costly and restricts the mass of molded articles per cycle of the injection molding machine. In *hot runner molding,* the sprues and runners are kept hot by means of heating elements built into the mold. As the mold

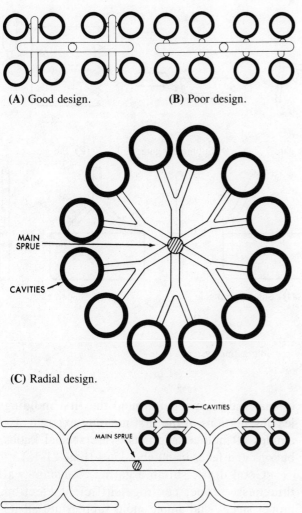

(A) Good design.   (B) Poor design.

(C) Radial design.

(D) Sweeping-curve design.

(E) H design.

Fig. 21-10. Some typical runner designs. (Du Pont)

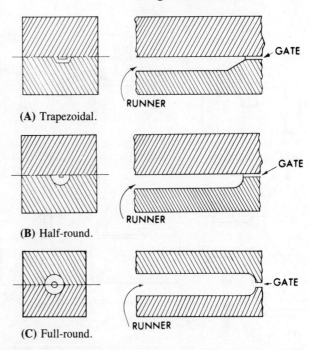

(A) Trapezoidal.

(B) Half-round.

(C) Full-round.

Fig. 21-11. Three basic runner systems used for transfer and injection molding.

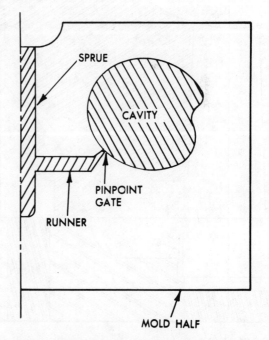

Fig. 21-12. Pinpoint (submarine) gate system, which eliminates gate breakoff problems. It also reduces or eliminates finishing problems caused by large gate marks.

opens, the cured part pulls free from the still-molten hot runner system. On the next cycle, the hot material remaining in the sprue and runner is forced into the cavity (Fig. 21-15).

A similar system called *insulated runner molding* is used in molding polyethylene or other materials with low thermal transfer (Fig. 21-16). This system is also called *runnerless molding*. In this design, large runners are used. As the molten material is forced through these runners, it begins

to solidify, forming a plastics lining that serves as insulation for the inner core of molten material. The hot inner core continues to flow through the tunnel-like runner to the mold cavity. A heated torpedo or probe may be inserted in each gate. This helps control freeze-up and drool.

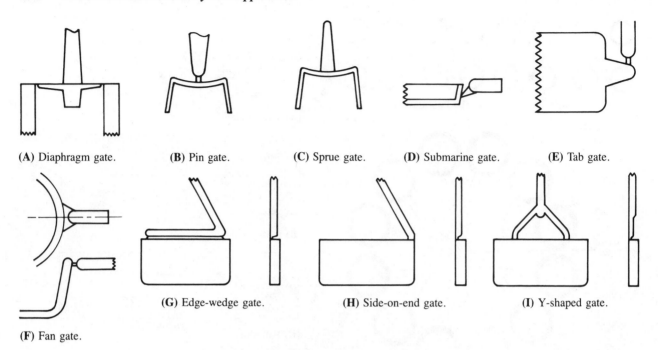

(A) Diaphragm gate.   (B) Pin gate.   (C) Sprue gate.   (D) Submarine gate.   (E) Tab gate.

(G) Edge-wedge gate.   (H) Side-on-end gate.   (I) Y-shaped gate.

(F) Fan gate.

Fig. 21-13. Some of the many possible gating systems.

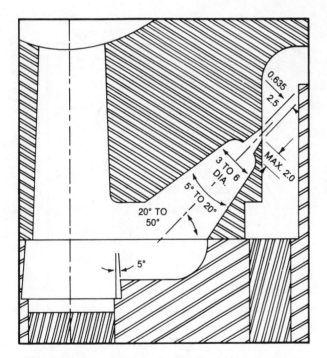

Fig. 21-14. A typical tunnel gate, which may be considered a variation of the pinpoint gate design. (Mobay Chemical Co.)

There are many other mold designs that may include valve-gated molds, single or multiple molds (Fig. 21-17), unscrewable molds (for internal or external threads), cam and pin molds (for undercuts and cores), and multicolor, or multimaterial, molds. In multicolor molds, a second color or material is injected around the first molding, leaving portions of the first molding exposed. Examples of multicolor products are special knobs, buttons and letter or number keys (Fig. 21-18).

Good draft, suitable gating, constant wall thickness, proper cooling, sufficient ejection, proper steels, and ample mold support are all important factors in mold design. (See Chapter 22, Tooling and Moldmaking).

Compression-Mold Design. There are three different types of compression-mold designs. Compression molds are usually produced of hardened steel that can withstand the great pressure and abrasive action of the hot plastics compound as it liquifies and flows into all parts of the mold cavity.

The *flash mold* is the least complex and most economical from the standpoint of original mold cost (Fig. 21-19). In this mold, excess material is forced out of the molding cavity to form a flash that becomes waste and must be removed from the molded part.

In positive mold design, vertical or horizontal flash provision is made for excess material in the cavity (Fig. 21-20). Vertical flash is easier to remove. The preforms must be measured with care if all parts are to be the same density and thickness. If too much material is loaded in the cavity, the mold may not fully close. In positive molds, little or no flash occurs. This design is used for molding laminated, heavily filled, or high-bulk materials.

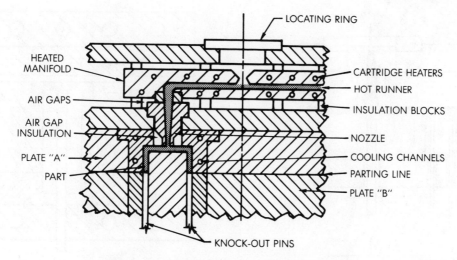

Fig. 21-15. Schematic drawing of hot runner mold.

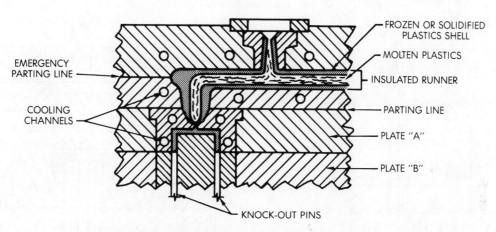

Fig. 21-16. Principle of insulated runner molding.

With fully positive molds, gases freed during the chemical curing of thermosets may be trapped in the mold cavity. The mold may be opened briefly to allow gases to escape. This operation is sometimes called breathing.

Semipositive molds have horizontal and vertical flash waste (Fig. 21-21). This design is costly to make and maintain, but it is the most practical where many parts or long runs are needed. The design allows for some inaccuracy of charge by allowing flash thus giving a dense, uniform molded part. As the mold charge is compressed in the cavity, any excess material escapes. As the mold body continues to close, very little material is allowed to flash. When the mold fully closes, the telescoping male half is stopped by the *land*.

Blow-Mold Design. Construction of a *blow mold* is less costly. Aluminum, beryllium-copper, or steel are used as basic materials. Aluminum is one of the lowest cost blow-mold materials

(Fig. 21-22). It is light and transfers heat rapidly. Beryllium-copper is harder and more wear resistant, but it is also more expensive. Steel is used at pinch-off points. If ferrous molds are used, they are plated to prevent rusting or pitting. (See Chapters 10 and 22).

Parting Lines. Parting lines are usually placed at the greatest radius of the molded part (Fig. 21-23). If the parting line isn't at the plane of greatest dimension, the mold must have movable parts or flexible molds or materials must be used. If the parting line cannot be placed on an inconspicuous edge or concealed, finishing is generally needed.

Ejector or Knockout Pins. Knockout or ejector pins push the hardened parts from the mold. They must touch the part in hidden or inconspicuous areas and should avoid contact with a flat surface, unless decorative designs can help conceal the marks. Pins should be made as large as possible for longest tool

2 PLATE MOLD

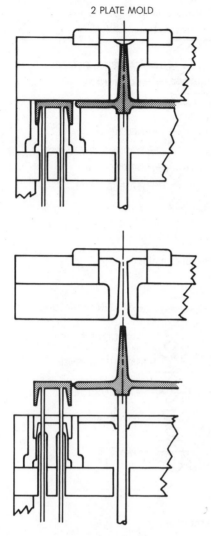

Fig. 21-17. Multiplate mold design for injection molding. (Du Pont Co.)

Fig. 21-18. Calculator keys are an example of a multicolored injection-molded product.

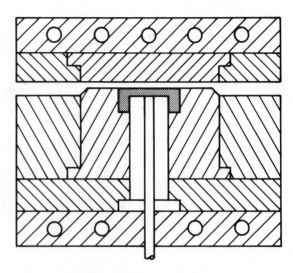

Fig. 21-19. Flash mold design.

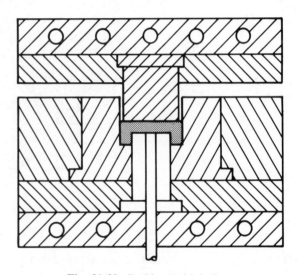

Fig. 21-20. Positive mold design.

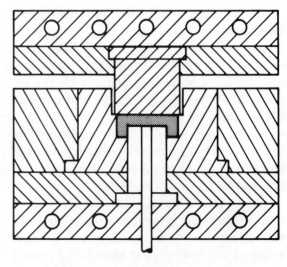

Fig. 21-21. Semipositive mold design.

life. Knockout pins, blades or bushing shapes should not press against thin part areas.

The pins are usually attached to a master bar or pin plate. They are drawn flush with the surface

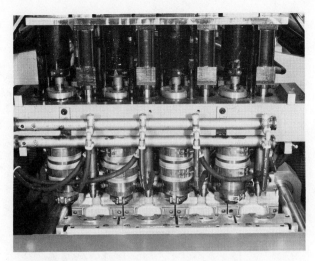

Fig. 21-22. Four containers are produced at one time from four parisons. Note aluminum tooling and cavity texture. (Uniloy Blowmolding Machinery Division, Hoover Universal)

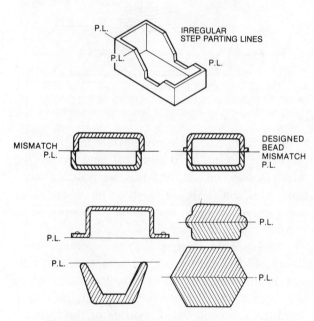

Fig. 21-23. Various locations for parting lines.

of the mold by spring action. When the die is opened, a rod attached to the master bar or pin plate strikes a stationary stop, pushing all pins forward forcing the part from the cavity.

All molds need venting. Ejector pins may be altered to provide venting.

Inserts. Inserts and holes must be designed and placed with care in the molding cavity or part. A liberal draft should be given long pins and plugs. A general rule is to never have a hole depth more than four times the diameter of the pin or plug. Long pins are often made to meet halfway through a hole allowing twice the hole-depth limitation. Long pins are easily broken and bent by pressures of the flowing plastics.

The placement of molding gates in relation to holes is important. As the molten material is forced into the mold, it must flow around the pins which protrude into the mold cavity. These pins are withdrawn when the mold opens. If parts are to be assembled on these holes, the material should be made thicker by producing a boss (Fig. 21-24). The boss adds strength, preventing the material from cracking. The pins often restrict the flow of material and may cause flow marks, weld lines or possible cracking because of molding stress (Fig. 21-25). Flow lines and patterns are shown in Figure 21-26.

When metal inserts are molded in place, the molten plastics material is forced around the insert. As the plastics cools, it shrinks around the metal insert, substantially contributing to the holding power of the insert. Inserts may be placed in thermoplastic parts by ultrasonic techniques. Before molding, inserts may be positioned automatically or by hand on small locating pins in the mold cavity. Enough material must be provided around all inserts to avoid cracking.

Pieces produced with pinpoint gating in the mold may not need any gate or sprue cutting. Many molds are designed so that gates are automatically sheared from the part as the press is opened.

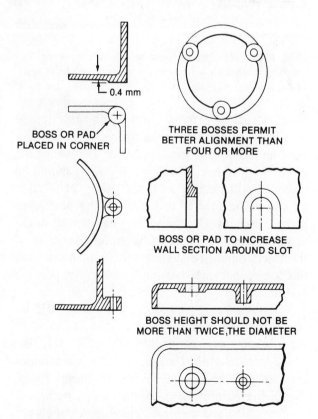

Fig. 21-24. Importance of boss design. (SPI)

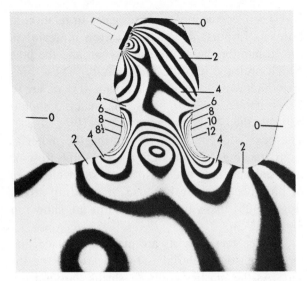

Fig. 21-25. Molding stress lines in this injection-molded part are visible under polarized light.

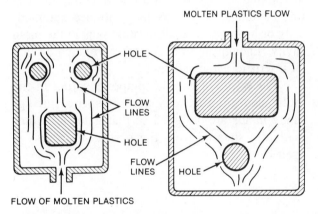

Fig. 21-26. Flow patterns must be kept in mind when designing molds. Flow lines may appear around holes and ribs, and opposite gates. (Vishay Intertechnology, Inc.)

Internal or external threads may be molded into plastics parts. Internal threaded parts may need an unscrewing mechanism for removal of the mold. A clearance of 0.03 in. [0.8 mm] should be provided at the end of all threads (Fig. 21-27A).

Metallic inserts molded into the part have greater strength. Generally, the ratio of wall thickness around the insert to the outer diameter of the insert should be slightly greater than one. Do not forget that materials have different expansion coefficients.

For blind core pin holes, a minimum 0.015 in. [0.04 mm] of clearance for screws, inserts, or other holding devices should be left (Fig. 21-27B).

Matched-Mold Design. The design parameters are similar to compression molding. Large composite parts such as sanitary tubs, bathroom shower stalls, and numerous automobile panels are

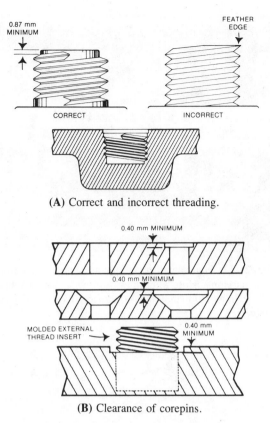

(A) Correct and incorrect threading.

(B) Clearance of corepins.

Fig. 21-27. Putting threads and corepins in plastics.

molded from SMC. Many SMC operations are precut, require no mold pinch-off, and produce no flash. Bosses, inserts, and ribs are possible when using SMC and TMC. If layers or pieces are used, make the overlapping bond as large as possible to prevent stress cracking.

Open-Mold Design. Layup, sprayup, autoclave, and bag techniques are similar. Careful attention must be given to the matrix formulation and orientation of reinforcements. This may have more influence on the properties of the finished composite than the design. Reinforcements should be overlapped two inches and all joints staggered. Bosses and ribs are used for added strength but must be liberally tapered. Simple, integrated part designs with gradual changes in thickness are desired. To aid in part removal, blow-out holes (pneumatic) may be located in the bottom of the mold.

Pultrusion Design. Bosses, holes, raised numbers, or textured surfaces are not possible with this continuous reinforcing process. Sharp corners or thickness transitions may result in resin rich zones with broken fibers.

Filament Winding Design. In this open-mold, continuous reinforcing process, the fibers are oriented to match the direction and magnitude of stresses. Computer-controlled placement of filaments is designed to compensate for angle, contour reinforcing, band width, equipment backlash, and other design considerations. Designs may call for permanent or removable mandrels (molds). (See Tooling in Chapter 22).

Laminar Design. The principle design criterion is concerned with reinforcement orientation in each layer. A design close to optimum to resist all loads may be a laminate consisting of plies at 0°, ±45°, and 90°. There must be a −45° ply for every +45° ply to avoid distortion of the laminate. See Figure 21-28. The lamina should be oriented in the principle direction of anticipated stresses. See Figure 21-29. Fibers arranged in a random

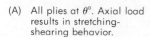

(A) All plies at θ°. Axial load results in stretching-shearing behavior.

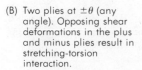

(B) Two plies at ±θ (any angle). Opposing shear deformations in the plus and minus plies result in stretching-torsion interaction.

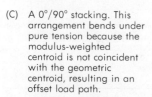

(C) A 0°/90° stacking. This arrangement bends under pure tension because the modulus-weighted centroid is not coincident with the geometric centroid, resulting in an offset load path.

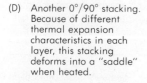

(D) Another 0°/90° stacking. Because of different thermal expansion characteristics in each layer, this stacking deforms into a "saddle" when heated.

Fig. 21-28. Symmetry effects on deflection of composites. (*Machine Design*)

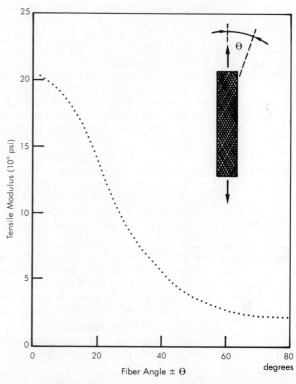

Fig. 21-29. Tensile modulus of carbon/epoxy composites drops steeply as the angle between the fibers and the direction of tensile load is increased. (*Machine Design*)

manner (isotropic), will have equal strength in all directions. The failure modes and deflection of composites are shown in Figure 21-30.

For sandwich laminates, the high density facings must resist most of the applied and bending forces. The lightweight core must resist transverse tension, compression, shear, and buckling.

Table 21-5 shows advantages and limitations of various processes.

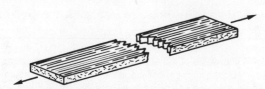

(A) FIBER TENSILE FAILURE FOR ALL FIBERS IN THE DIRECTION OF LOAD (0°).

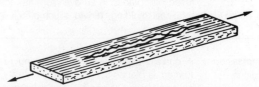

(C) RESIN SHEAR FAILURE THROUGH THE THICKNESS, BETWEEN FIBERS AT 0°. USUALLY CAUSED BY POOR FIBER-TO-RESIN ADHESION.

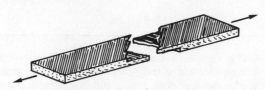

(B) RESIN SHEAR FAILURE FOR FIBERS AT ±45°.

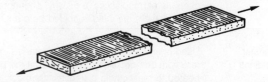

(D) RESIN TENSILE FAILURE BETWEEN FIBERS AT 90° TO LOAD.

Fig. 21-30. Failure modes of composites in tension. Tensional strength of a carbon/epoxy structural composite is always related to fiber direction. A simple tensile test shows strikingly different failure behavior in composites having different fiber orientation. In multidirectional composite, single plies can fail without overall structural failure. Recognition of the various failure modes and knowing how composites fail are prerequisites to determining a fix. (*Machine Design*)

Table 21-5. Plastics Processing Method — Processes, Advantages, Limitations

Processing Method	Process	Advantages	Limitations
Injection molding	Similar to die casting of metals. A thermoplastic molding compound is heated to plasticity in a cylinder at a controlled temperature and then forced under pressure through sprues, runners, and gates into a cool mold; the resin solidifies rapidly, the mold is opened, and the parts ejected; with certain modifications, thermosetting materials can be used for small parts	Extremely rapid production rate and hence low cost per part; little finishing required; excellent surface finish; good dimensional accuracy; ability to produce variety of relatively complex and intricate shapes	High tool and die costs; high scrap loss; limited to relatively small parts; not practical for small runs
Extrusion	Thermoplastic molding powder is fed through a hopper to a chamber where it is heated to plasticity and then driven, usually by a rotating screw, through a die having the desired cross section; extruded lengths are either used as is or cut into sections; with modifications, thermosetting materials can be used	Very low tool cost; material can be placed where needed; great variety of complex shapes possible; rapid production rate	Close tolerances difficult to achieve; openings must be in direction of extrusion; limited to shapes of uniform cross section (along length)
Thermoforming	VACUUM FORMING — Heat-softened sheet is placed over a male or female mold; air is evacuated from between sheet and mold, causing sheet to conform to contour of mold. There are many modifications, including vacuum snapback forming, plug-assist, drape forming, etc.	Simple procedure; inexpensive; good dimensional accuracy; ability to produce large parts with thin sections	Limited to parts of low profile
	BLOW OR PRESSURE FORMING — The reverse of vacuum forming in that positive air pressure rather than vacuum is applied to form sheet to mold contour	Ability to produce deep drawn parts; ability to use sheet too thick for vacuum forming; good dimensional accuracy; rapid production rate	Relatively expensive; molds must be highly polished
	MECHANICAL FORMING — Sheet metal equipment (presses, benders, rollers, creasers, etc.) forms heated sheet by mechanical means. Localized heating is used to bend angles; where several bends are required, heating elements are arranged in series	Ability to form heavy and/or tough materials; simple; inexpensive; rapid production rate	Limited to relatively simple shapes
Blow moldings	An extruded tube (parison) of heated plastics within the two halves of a female mold is expanded against the sides of the mold by air pressure; the most common method uses injection molding equipment with a special mold	Low tool and die cost; rapid production rate; ability to produce relatively complex hollow shapes in one piece	Limited to hollow or tubular parts; wall thickness difficult to control
Slush, rotational, dip castings	Powder (polyethylene) or liquid material (usually vinyl plastisol or organosol) is poured into a closed mold, the mold is heated to fuse a specified thickness of material adjacent to mold surface, excess material is poured out, and the semifused part placed in an oven for final curing. A variation, rotational molding, provides completely enclosed hollow parts	Low cost molds; relatively high degree of complexity; little shrinkage	Relatively slow production rate; choice of materials limited
Compression moldings	A partially polymerized thermosetting resin, usually preformed, is placed in a heated mold cavity; mold is closed, heat and pressure applied, and the material flows and fills mold cavity; heat completes polymerization and mold is opened to remove hardened part. Method is sometimes used for thermoplastics, e.g., vinyl phonograph records; in this operation, the mold is cooled before it is opened.	Little waste of material and reduced finishing costs due to absence of sprues, runners, gates, etc.; large, bulk parts possible	Extremely intricate parts involving undercuts, side draws, small holes, delicate inserts, etc., not practical; extremely close tolerances difficult to achieve
Transfer moldings	Used primarily for thermosetting materials, this method differs from compression molding in that the plastic is 1) first heated to plasticity in a transfer chamber, and 2) fed, by means of a plunger, through sprues, runners and gates into a closed mold	Thin sections and delicate inserts are easily used; flow of material is more easily controlled than in compression molding; good dimensional accuracy; rapid production rate	Molds are more elaborate than compression molds, and hence more expensive; loss of material in cull and sprue; size of parts somewhat limited

Table 21-5. Plastics Processing Method — Processes, Advantages, Limitations (Continued)

Processing Method	Process	Advantages	Limitations
Open-mold processing	CONTACT — The lay-up, which consists of a mixture of reinforcement (usually glass cloth or fibers) and resin (usually thermosetting), is placed in mold by hand and allowed to harden without heat or pressure	Low cost; no limitations on size or shape of part	Parts are sometimes erratic in performance and appearance; limited to polyesters, epoxies and some phenolics
	AUTOCLAVE — The vacuum-bag setup is simply placed in an autoclave with hot air at pressures up to 1.38 MPa	Better quality moldings	Slow rate of production
	FILAMENT WOUND — Glass filaments, usually in the form of rovings, are saturated with resin and machine wound onto mandrels having the shape of desired finished part; finished part is cured at either room temperature or in an oven, depending on resin used and size of part	Provides precisely oriented reinforcing filaments; excellent strength-to-mass ratio; good uniformity	Limited to shapes of positive curvature; drilling or cutting reduces strength
	SPRAY MOLDING — Resin systems and chopped fibers are sprayed simultaneously from two guns against a mold; after spraying, layer is rolled flat with a hand roller. Either room temperature or oven cure	Low cost; relatively high production rate; high degree of complexity possible	Requires skilled workers; lack of reproducibility
Castings	Plastics material (usually thermosetting except for the acrylics) is heated to a fluid mass, poured into mold (without pressure), cured, and removed from mold	Low mold cost; ability to produce large parts with thick sections; little finishing required; good surface finish	Limited to relatively simple shapes
Cold moldings	Method is similar to compression molding in that material is charged into a split, or open, mold; it differs in that it uses no heat — only pressure. After the part is removed from mold, it is placed in an oven to cure to final state	Because of special materials used, parts have excellent electrical insulating properties and resistance to moisture and heat; low cost; rapid production rate	Poor surface finish; poor dimensional accuracy; molds wear rapidly; relatively expensive finishing; materials must be mixed and used immediately
Bag molding	VACUUM BAG — Similar to contact except a flexible polyvinyl alcohol film is placed over layup and a vacuum drawn between film and mold (about 82 kPa)	Greater densification allows higher glass contents, resulting in higher strengths	Limited to polyesters, epoxies and some phenolics
	PRESSURE BAG — A variation of vacuum bag in which a rubber blanket (or bag) is placed against film and inflated to apply about 350 kPa	Allows greater glass contents	Limited to polyesters, epoxies and some phenolics
Matched-die molding	MATCHED DIE — A variation of conventional compression molding, this process uses two metal molds which have a close-fitting, telescoping area to seal in the resin and trim the reinforcement; the reinforcement, usually mat or preform, is positioned in the mold, a premeasured quantity of resin is poured in, and the mold is closed and heated; pressures generally vary between 1.04 and 2.75 MPa	Rapid production rates; good quality and excellent reproducibility; excellent surface finish on both sides; elimination of trimming operations; high strength due to very high glass content	High mold and equipment costs; complexity of part is restricted; size of part limited

Source: Modified from Materials Selector, *Materials Engineering*, Penton/IPC a subsidiary of Pittway Corp.

Performance Testing

The true test of any product is performance under actual service conditions. Tests can be used as indicators for product design, redesign, and reliable product service. The term *testing* implies that methods or procedures are employed to determine if parts meet the required or specified properties. A pressure tank, rocket engine case, or pole vaulter's pole may be subjected to a critical pass/fail (proof) test. *Quality control* procedures must be used to determine if a product is being manufactured to specifications. It is primarily a technique used by management to achieve quality. *Inspection* ensures that manufacturing personnel check technique procedures, gauge readings, and detect flaws in processing or materials. Inspection is part of quality control. See Appendix G for additional sources of information.

VOCABULARY

The following vocabulary words are found in this chapter. Look up the definition of any of these words you do not understand as they apply to plastics in the glossary, Appendix A.

Apparent density
Boss
Bulk factor
CAD
CAE
CAM
CAMM
CIM
Fillet
Flow line
Flow mark
Gate
Inspection
Orientation
Parameter
Parting lines
Quality control
Ribs
Runners
Safety factor
Specification
Sprue
Standards
Testing
Undercut

21-1. In ___?___ the sprues and runners are kept hot by means of heating elements built into the mold.

21-2. The ___?___ is the point of entry into the mold cavity.

21-3. Narrow channels that convey the molten plastics from the sprue to each cavity are called ___?___ .

21-4. Parting lines are normally placed at the ___?___ of the molded part.

21-5. The ___?___ is the opening in the mold where the product is formed.

21-6. The economic minimum number of pieces produced by hand lay-up is ___?___ .

21-7. Closely related to shrinkage is dimensional ___?___ .

21-8. The tapered channel connecting the nozzle and runners is called ___?___ .

21-9. In ___?___ compression mold designs no provision is made for placing excess material in the cavity.

21-10. Molded parts are pushed from the mold by ___?___ or ___?___ .

21-11. With ___?___ gating, pieces may not require any gate or sprue cutting. The dies are designed so that gates are sheared off automatically.

21-12. The three overriding requirements in plastics designing are:

21-13. If a mold costs $5,000 and 10 000 parts are to be made, the mold cost would be ___?___ per part.

21-14. In nearly every design, a compromise must be made between highest performance, attractive appearance, efficient production and ___?___ .

21-15. List four favorable properties possessed by most plastics:

21-16. List four unfavorable properties possessed by most plastics.

21-17. In addition to electrical, chemical, mechanical and economic considerations, name four additional requirements to consider before a product is made.

21-18. The most meaningful comparison in estimating the cost of a plastics product is cost per ___?___ .

21-19. Plastics are lighter per ___?___ than most materials.

21-20. Problems encountered in producing plastics products often require the selection of the ___?___ before material or ___?___ considerations.

21-21. There were many problems associated with early applications of plastics because designers forgot that the finished product must ___?___ as designed and desired.

21-22. Undercuts in parts usually increase ___?___ costs.

21-23. With relatively few exceptions, the ___?___ of plastics developed in the past have been by trial and error.

21-24. Plastics have replaced metals for many applications because of energy and ___?___ savings.

21-25. If wall thickness is not ___?___ the molded part may distort, warp or have internal stresses.

21-26. The mark on a molded piece resulting from the meeting of two or more flow fronts during the molding operation is called a ___?___ .

21-27. Wavey surface appearances caused by improper flow of the hot plastics into the mold cavity are known as ___?___ .

21-28. To help prevent flow marks, ___?___ or change gate location.

21-29. Name three methods of producing an internal thread in a plastics part:

Chapter 22

Tooling and Moldmaking

In this chapter, tool-making processes, equipment, and methods for shaping plastics will be discussed. Not all of the information will apply to all molding processes, since some processes are very specialized.

Computer-assisted moldmaking (CAMM) is a result of the microprocessor technology and may improve productivity by more than 100 percent. Computer assisted design and manufacturing (CAD/CAM) systems are used to design and aid in machining molds. This equipment automatically adjusts cavity dimensions for different resins or special contours. Designing and machining information is stored in the system's memory for making multicavity molds or for the replacement of cores or cavities.

CAMM could allow the plastics industry to produce high quality configurations, close tolerances, and reliable designs at a competitive price. The industry of moldmaking and tooling is highly labor intensive. Here, computer systems are sophisticated tools that increase productivity. The drafting and design departments can utilize CAD systems to automatically dimension the drawing and allow for material shrinkage. The data base in CAMM programs can select the mold base and recommend the placement of ejector pins, sleeves, locating ring, return pins, pullers, or other details. While drafting time and changing tooling parameters represent a significant percentage of the total cost of the part, CAMM systems greatly reduce this time and facilitate modifications or changes in tooling parameters.

Most of the moldmaking industry is composed of custom shops that specialize in moldmaking or in offering a special service such as plating, polishing, heat-treatment, engraving, or machining molds.

Molding information is given in the discussion of each process, however, Chapter 21 should be reviewed for basic design considerations. You should also review the descriptions of the plastics families, because they contain information about properties and design that affect moldability.

22-1 PLANNING

Most mold designs begin with sketches that allow the moldmaker to make decisions about layout and visualize how parts will be made. Final CAD drawings will contain notes, dimensions, and tolerances. Some critical dimensions may need tool tolerances as low as 9.8×10^{-7} in. [+0.0025 mm] for some parts. The design will also show any special requirements. Shrinkage tolerances, finish, engraving, plating, special materials, or other dimensioning factors will be noted.

CAM systems allow the user to determine cooling systems, finish, tool path, part geometry, feeds, and inherent equipment limitations (back lash, tool wear) before machining. After the program is verified it is stored for later use.

CAM systems can utilize the stored information to cut and form the molds. Nearly 80% of the moldmakers' time is devoted to setting up the machine tool. Only 20% is actually spent cutting the material. CAD/CAM systems substantially reduce setup, lead, and machining time. See Fig. 22-1.

Nearly every phase of the product development process, from concept to completion, can utilize CAMM to save time and reduce costs.

Fig. 22-1. With this sophisticated CAD-CAM installation, engineers can design, test, and refine even the most complex molds directly on the computer. The need for building costly prototypes is eliminated. (AGIE USA LTD)

22-2 TOOLING

Collectively, jigs, fixtures, molds, dies, gauges, clamping devices and inspection equipment will be referred to as **tooling**. The terms **jig** and **fixture** are often used interchangeably; both are devices used to locate and hold a workpiece in the correct position during machining, inspection, or assembly. A **jig** guides the tool during a manufacturing operation such as boring. A **fixture** does not have built-in tool guides. It is used primarily to hold the work securely during machining, cooling, and drying.

Part of the manufacturing cost of plastics is the cost of special fixtures. These are tools used to help measure or load a plastics charge in a molding machine, remove flash, remove molded parts, or hold parts for cooling. Some are used to aid machining. These include holding blocks, drill jigs, and punch dies.

Tooling Costs

Many factors affect tooling costs: the size of the production run, production technique, reinforcements, additives, fiber orientation, matrix, the complexity of design, the tolerance needed, the amount of mold maintenance, and machining.

Multiple-cavity molds, or those with inserts or special surface finishes, add to tooling costs. Designing a mold set to have interchangeable cavities may lower tooling costs because new cavities may be inserted to form other parts, thus extending the use of the original mold. For a small number

of parts, a multiple-cavity mold may not be economical because tooling tolerances are much harder to maintain, and mold maintenance and machining are more costly for multiple-cavity molds.

Two remarks often quoted in the moldmaking industry are, "There is no such thing as a simple plastics part," and "The part is only as good as the mold that makes it." Each of these remarks show the importance of tooling and mold design. See Figure 22-2.

There are four broad types of tooling: 1) prototype 2) temporary 3) short run and 4) production. These are shown in Table 22-1.

Types of materials used in tooling include gypsum plasters, plastics, wood, and metals.

Gypsum Plasters. The United States Gypsum Company has developed a number of high-strength plaster materials. These materials have enough strength to produce prototype models, die models,

Fig. 22-2. The importance of the tool and die personnel can not be understated. This tool and die maker puts finishing touches on a steel injection mold. (Bethlehem Steel)

Table 22-1.   General Types of Tooling

Tool Classification	Number of Parts	Tooling Materials
Prototype	1–10	Plaster, wood, reinforced plasters
Temporary	10–100	Faced plasters, reinforced plasters; faced plasters, backed metal-deposition, cast & machined soft metals
Short Run	100–1000	Soft metals steel
Production	>1000	Steel, soft metals for some processes

transfer (takeoff) tools, patterns, and die molds for forming plastics (Fig. 22-3).

Fibers, expanded metal, or other materials are often used to further strengthen the plaster tooling. Metal bases and frames provide secure mountings. In some techniques templates (loft or templates) help shape the wet plaster. Even a pile of rocks or an expanded bladder (balloon) can be used to help form the general contour. Plaster is easily shaped by hand. Models may be produced by placing plaster over clay, wax, wooden, or wire frame shapes. Rocks, wax, and other model forms are generally removed and replaced by reinforced plastics support. Some plaster molds are designed to be used only once. Hollow or some of the break-away designs require that the plaster mold

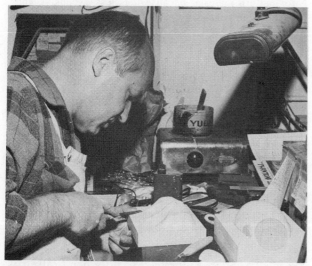

(A) This worker is making a gypsum plaster pattern for forming plastics. (Revell, Inc.)

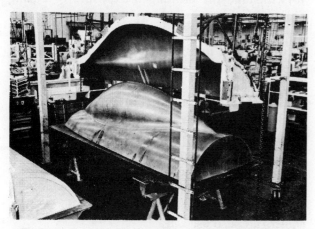

(B) An epoxy resin master model of an aircraft pilot's enclosure after removal from the plaster mold. (U. S. Gypsum Co.)

Fig. 22-3. Gypsum plaster pattern and plaster mold.

be broken and washed away. Typical plaster molds are faced with metal deposition, polymer coating, or composite facing, to provide a durable surface for part removal. Metal coatings also improve thermal conductivity necessary in some molding techniques. Cooling coils may be cast in the tooling.

The trade names Ultracal, Hydrocal, and Hydro-stone are found on plasters used for tooling. Vacuum-formed pattern molds are often made of low-cost plasters. Hydrostone has an average compressive strength of nearly 11 000 psi [76 MPa].

Plaster is an important master pattern material from which metal or polymer skin master molds are produced (Fig. 22-4).

Plastics. Polymer tooling (plastics and elastomers) are used to make master patterns, transfer tools, cores, boxes, templates, draw dies, jigs, fixtures, inspection tools, and prototypes. They are replacing wooden and plaster tooling (Fig. 22-5). Laminated, reinforced, and filled plastics are used mainly for making dies, jigs, and foundry patterns (Fig. 22-6).

Polymer molds are generally divided into two groups: those that are backed and those that are not. Foams and honeycomb sheets are often used to provide strong, lightweight support of tools.

The use of plastics in die fabrication is growing rapidly. Metal-filled and glass-reinforced phenolics, ureas, melamines, polyesters, epoxies, silicones, and polyurethanes are strong, light, and easy to machine. Alumina and steel are common fillers which offer improved thermal conductivity, machinability, strength, and extended service temperature and life. Many may be foam-filled. These materials are used in tooling in both the plastics and metals industries. Plastics tools have been

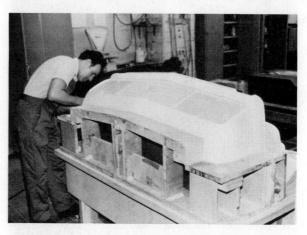

Fig. 22-4. Tooling for female RIM mold. (Mobay Chemical Co.)

(A) Basic shape and form (loft frame).

(B) Filling with syntatic foam.

(C) Machining to final shape and size.

Fig. 22-5. Light, strong tooling may be fabricated with honeycomb structures. Bonding and fill is done with syntactic foams of extrudable epoxy. (Ren Plastics, Inc.)

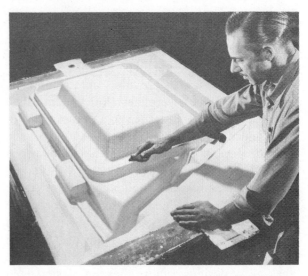

Fig. 22-6. A large foundry pattern being made by laminating glass cloth and epoxy resins. Plastics is replacing metal for pattern material because it is less expensive and because plastics patterns may be made faster. (U.S. Gypsum Co.)

used as bending dies, stretch-forming dies, and drop-hammer dies. Acetals, polycarbonates, high-density polyethylene, fluoroplastics, and polyamides are also used as tooling. These materials find use as matched stamping dies, jigs, and fixtures. The acceptability of plastics tools is verified by their wide use in the aerospace, aircraft, and automotive industries.

Large epoxy-reinforced molds are popular for layup and sprayup techniques. These toolings are backed and supported by metal frames and bases (Figs. 22-7). In Figure 22-8, polymer tooling is used to form SMC. Polymer tooling must be able to withstand the prolonged exposure of curing temperatures.

Plastics tooling has several advantages over metal or wood tools. Plastics tools may be cast in inexpensive molds, they duplicate easily, and they allow frequent changes in design. Plastics tooling is also light in mass and corrosive resistant. Hot-melt compounds are replacing wood and steel in dies, hammers, mockups, prototypes, and other fixtures used by industry (Fig. 22-9).

Complex furniture parts are sometimes made from flexible polymer molds. Silicones are best known, but polysulfide and polyurethane elastomers are also used.

Unbacked, flexible molds are used in the furniture industry to faithfully reproduce wood grain designs (Fig. 22-10). The basic concept of flexible-plunger or elastomer press molding is shown in Figure 22-11.

Fig. 22-7. Two layers of 10 oz. glass cloth and eight layers of 20 oz. cloth are alternately laid with a matrix of epoxy to produce this spa mold with intricate designs. (Ren Plastics)

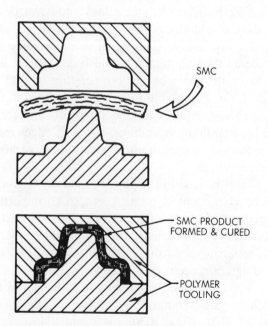

Fig. 22-8. Matched-die molding of SMC using polymer tooling.

SMC

SMC PRODUCT FORMED & CURED

POLYMER TOOLING

Wood. Wooden tooling is used for prototypes and some short-run work (Fig. 22-12). It is also used for some thermoforming dies and for pattern work.

(A) Low cost plastics pattern development.

(B) Duplicate milling of mold halves from plastics pattern on right.

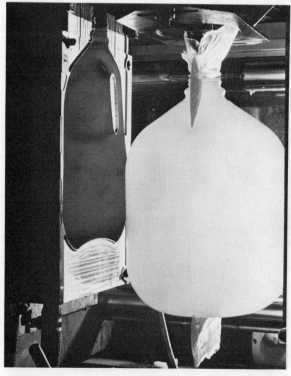

(C) Plastics product from metal-filled plastics mold. Note pinch-off and parting line. (Chemplex Co.)

Fig. 22-9. Low cost molds made of metal-filled plastics.

Fig. 22-10. Woodgrain detail and miter joints are faithfully reproduced in this polyurethane furniture component. (The Upjohn Co.)

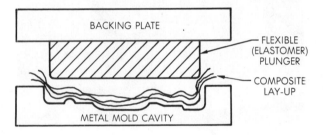

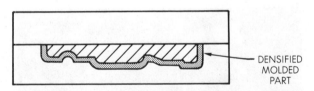

Fig. 22-11. Flexible-plunger or elastomeric press molding.

Fig. 22-12. A wooden pattern is made from photographs and drawings. It is double the size of the finished part. (Bethlehem Steel Corp.)

Metals. Although plastics may become the dominant material in the future, we cannot have plastics products without metals.

For prototype or short-run work, low-melting-point metals may be used. Zinc, lead, tin, bismuth, cadmium, and aluminum are used in thermoforming dies, casting patterns, and duplicate models. Aluminum is a popular metal for many molding processes because it is light, easy to machine, and is a good thermal conductor.

Aluminum 7075 is a popular, high-grade, heat-treated alloy. It is sometimes annodized and plated to improve surface hardness and prolong tooling life. Aluminum (7075-T652) is used extensively in blow molding, bag molding, thermoforming, and prototypes. It is soft and lacks sufficient strength and hardness to prevent material wash in injection molds or pinchoff areas in blow molds.

Berylium-copper (C17200) is used in some injection molds and blow-molding processes (Fig. 22-13). This material can be cast, wrought, and drawn. It can also be pressure cast using a master hob. Many cavities can be produced with one hob. The BeCu is generally supplied to the moldmaker tempered to 38-42 Rockwell C. Berylium-copper will reproduce fine detail. Wood grain patterns are examples of the detail possible with BeCu molds.

Kirksite (Zamak, zinc alloy) molds, made of an alloy of aluminum and zinc, are used because they are low in cost. This metal will reproduce detail better than aluminum and also last longer. Because of the low pouring temperature of 800 °F [425 °C], it is possible to cast cooling lines in the mold. These alloys are used for short- and long-run blow-molding operations. Pinchoff edges must be protected by steel inserts against excessive stress concentrations.

Steel is vital to the plastics industry. Carbon, the basic element in plastics, is also an important ingredient of steel, but steel must also have additional alloying elements for moldmaking. Mold hardness can vary from 35 to 65 Rockwell C, depending upon needs (Fig. 22-14).

A popular, non-alloyed carbon steel used for machine bases, frames, and structural components is AISI 1020, 1025, 1030, 1040, and 1045.

For long production runs, various steel alloys may be used. In uses where high compressive strength and wear resistance at elevated temperatures are required, an AISI Type H21 steel may be used. The symbol H indicates hot-work tool steels.

(A) Steps in producing a mold for simulated wood furniture components. From left: hand-carved wooden model, silicone impression, gypsum model, and beryllium-copper injection molds.

(B) The finished product.

Fig. 22-13. Beryllium-copper injection molds were used to produce this simulated wood furniture.

Type H21 steel analysis:
Carbon 0.35%
Manganese 0.25%
Silicon 0.50%
Chromium 3.25%
Tungsten 9.00%
Vanadium 0.40%

AISI Type W1 is a high-quality, straight-carbon, water-hardened tool steel used in various tooling applications.

Type W1 steel analysis:
Carbon 1.05%
Manganese 0.20%

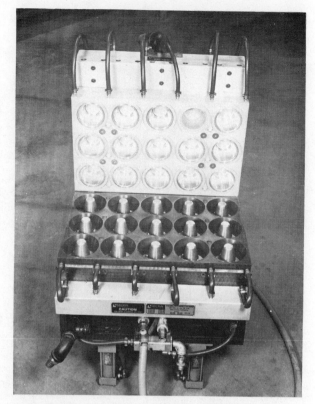

Fig. 22-14. A 15-cavity thermoforming mold. This is long-run, high production metal tooling. (Brown Machine Co.)

Silicon 0.20%
Alloys None

Toolmakers have long wished to have a nondeforming die steel. One that combines the deep-hardening characteristics of air-hardened steels with the simplicity of low-temperature heat treatments possible in many oil-hardened steels is AISI Type A6. The symbol A is used for air-hardened steels.

Type A6 steel analysis:
Carbon 0.70%
Manganese 2.25%
Silicon 0.30%
Chromium 1.00%
Molybdenum 1.35%
Plus alloy sulfides

Popular, alloyed steels, including tool steels, are AISI type A2 and A6 for injection, transfer, compression, and master hob molds. Fully hardened steels such as D2 and D3 are also popular. The high-carbon and chromium D3 steels have good wear resistance. Mold cavities, back plates, knives, and die are made from the chromium-molybdenum steel 4140.

The symbol P means a precipitation-hardened steel. AISI Type P20 is usable for all types of injection mold cavities. It may be hardened to a core hardness of 38 Rockwell C.

Type P20 steel analysis:
Carbon 0.30%
Molybdenum 0.25%
Chromium 0.75%

Mold cavities, holding blocks, dies, and other tooling are made of P20 and P21 steels.

Stainless steel Type 420 and 440C may be used where corrosion resistance is needed or adverse atmospheric conditions exist. It can develop hardness of 45 to 50 Rockwell C.

Oil-hardened steels are sometimes selected for slides and bushings. Type 02 is most common.

Tungsten carbide is widely used for cutting taps because it has good wear and abrasion resistance. Mold cavities may be made from this ceramic type of material but it is brittle and harder to shape than tool steels.

Miscellaneous Tooling. There are numerous miscellaneous and innovative tooling techniques. Wax and soluble salts are sometimes used. Mandrels, bladders, or other mold shapes have been made of air inflated elastomers. Even glass shapes are used as molds. After casting, forming, or filament winding, the glass is sometimes broken and removed. Concrete and ice have been used to produce unique tooling in some applications. The advantages and disadvantages of selected tooling materials are shown in Table 22-2.

Table 22-2.   Advantages and Disadvantages of Selected Tooling Materials.

Tool Material	Advantages	Disadvantages
Aluminum	Low cost; good heat transfer; easily machined corrosion resistant; lightweight; doesn't rust	Porosity; softness; galling; thermal expansion; easily damaged; limited runs
Copper alloyed (brass, bronzes, berylium)	Easily machined; good surface detail; high thermal conductivity; doesn't rust	Softness; copper may inhibit cure; attacked by some acids; easily damaged; limited runs
Miscellaneous (salts, inflatables, wax, ceramics)	Some are low cost; reusable; designed with undercuts; easily fabricated; light; hard; thermal conductors; ceramics are high temperature materials	Some are soft; easily damaged; dimensionally unstable; damaged by high temperature or chemicals; poor thermal conductors
Plaster	Low cost; easily shaped; good dimensional stability; doesn't rust	Porosity; softness; poor thermal conductivity; easily damaged; limited runs; limited thermal and strength range
Polymers (laminated, reinforced, filled)	Low cost; easily fabricated; thermal expansion similar to many composites; lightweight; doesn't rust; large designs economical; fewer parts	Limited design; poor thermal conductivity; limited dimensional stability; limited runs; limited thermal range
Steel	Most durable; high thermal resistance; strong and wear resistant; thermal conductivity	Most expensive tooling; machining more difficult; size limitations; many parts; rusts; heavy
Wood	Low cost; easily machined; lightweight; doesn't rust	Porosity; poor dimensional stability; soft; limited runs; poor thermal conductivity and resistance
Zinc alloyed (lead, tin)	Low cost; easily machined; good thermal conductivity; good detail; doesn't rust	Soft; easily damaged; limited runs; limited thermal and strength range

22-3 MACHINE PROCESSING

There are a number of processes used in making tools and dies from steels. Milling, turning, drilling, boring, grinding, hobbing, casting, planing, etching, electroforming, electrical-discharge machining, plating, welding, and heat-treating are only a few.

Toolmaking should be viewed as a tightly controlled process rather than a sequence of discrete tasks. Control of the toolmaking process may be done with the aid of the computer. CIM systems may control individual computer numerically controlled (CNC) machine tools. The manufacturing of tooling begins with planning and balancing the workload, specifications, and techniques with machine capabilities. Mistakes in the planning process can be costly and result in excessive production time and even tool failure. Machining of molds and tooling is sometimes separated into two broad areas: 1) initial machining which involves rough turning and milling and 2) final machining which involves grinding, EDM, and polishing. These processes usually take place after hardening.

Milling, turning, drilling, boring, and grinding are cutting-tool processes. Shapers, planers, lathes, drilling machines, grinding machines, milling machines, and various pantograph duplicating machines are often used to cut metal in making molds and dies. Figure 22-15A shows steel being removed by a cutting tool in a vertical milling machine.

In Figure 22-15B, a specially modified vertical milling machine is used to cut and duplicate molds in steel from the master pattern. The tracing head on the right controls movements of the work table and cutter spindle. This pattern is made of metal and the duplicator setup is machining a cavity for a blow mold. The workpiece ratio is 1:1.

(B) Duplicating a steel mold from a pattern, at right. (Cincinnati Milacron)

(C) Pantograph machine with a plaster master model used to machine a metal piece. (U. S. Gypsum Co.)

(A) Making a steel mold on a milling machine. (Revell, Inc.)

Fig. 22-15. Making tools and dies from steel and plastics.

Pantograph machines are similar to duplicating machines, except that they operate on variable ratios up to 20:1. In Figure 22-15C, a plaster master is shown much larger than the steel workpiece. The large ratio reduction allows the steel to be machined with very delicate detail by coordinated movements of the table and cutting tool.

Hobbing, etching, electroforming, and electrical discharge machining are *metal-displacement* processes. They are used in making molds where no cutting tools are involved.

Cold hobbing involves pushing a piece of very hard steel into a blank of unhardened steel (Fig. 22-16). The process is performed at room temperatures. Pressures vary from 200 to 400 × 10^3 psi [1 380 MPa to 2 760 MPa] depending on the hobbing metals and blanking material. Hobbing machines may need press capacity as high as 3 000 tons [2 722 tonnes].

In a proprietary procedure called CAVAFORM, a master hob can supply an unlimited number of impressions distinguished by a remarkable fidelity to the size and finish of the hob. This cold-forming swaging procedure is accomplished in most steel in an annealed state. A cavity steel can be chosen to satisfy molding requirements with low heat treat distortion. It may also be vacuum-heat treated to minimize polishing after heat treatment. A major disadvantage is that this process generally requires a through hole at the bottom of the impression.

Hobs are often made from oil-hardened tool steels containing a high percentage of chromium. It may be economical to hob single die cavities,

but hobbing is usually used for making large numbers of impressions for multicavity molds. Multicavity molds frequently are numbered to permit instant location of any molding troubles.

A slight draft must be provided to allow removal of the hob from the forming blank. The hob must be clean because even a pencil mark on the hob may be transferred to the cavity during the hobbing operation.

After hobbing, the blank is machined and hardened before placement in mold bases. In Figure 22-17A a finished hob (right) has formed the cavity (center) in the blank of steel. At left is the *force* or male portion of the compression mold. A finished, compression-molded part is shown in Figure 22-17B. It was molded in the finished mold cavity.

Electrical erosion or electrical-discharge machining (EDM) is a fairly slow method of removing metal, compared with mechanical methods. Steel is removed at about 0.016 in³/min [4.37 × 10^{-4} mm³/s]. The workpiece may be hardened be-

(A) Hob, at right, was pressed into steel block, at center.

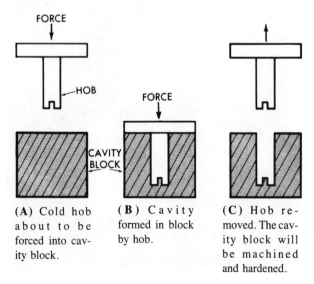

(A) Cold hob about to be forced into cavity block.

(B) Cavity formed in block by hob.

(C) Hob removed. The cavity block will be machined and hardened.

Fig. 22-16. Diagram of metal displacement (hobbing) process.

Fig. 22-17. Hob, compression mold, and finished part.

fore the cavity is formed, which removes any problems from heat treatment after machining or forming. In the machining process, a master pattern is made of copper, zinc, or graphite. The pattern is then placed about 0.025 mm [0.000 98 in.] from the workpiece and both the workpiece and the master are submerged in a poor dielectric fluid, such as kerosene or light oil. Current is forced across the gap between the master and the workpiece, and each discharge removes minute amounts of substance from both. The loss of material from the tool master must be compensated for to obtain accurate cavities in the workpiece.

Most modern EDM machines now incorporate a multi-axis orbital movement in their spindle head that enables one electrode to be used for roughing, final sizing, and finishing.

For inexpensive tool masters made of carbon or zinc, the ratio of material removed from the workpiece to that removed from the tool may be more than 20:1. Accuracy may be within ±0.001 in. [±0.025 mm] with a finish cut of less than 30 microinches [0.007 62 mm].

Both wirecut and diesinking EDM techniques often eliminate the need for secondary finishing operations. Wire EDM will produce accurate, intricate cavities that are difficult or impossible to produce with conventional techniques. In Figure 22-18 a CNC controlled wire EDM cuts the final shape of these molding dies. The EDM principle is shown in Figure 22-19.

Chemical Erosion

In chemical erosion (etching) an acid or alkaline solution is used to create a depression or cavity. The process usually involves the use of chemically resistant maskants such as wax, plastics-based paints, or films. The maskant is removed from those areas where the metal is to be chemically removed. Shallow cavities or designs are often reproduced with textures duplicating fabrics and leather. Photosensitive resistant materials are commonly used in the printing industry.

Allowance must be made to compensate for the effects of *etching radius,* or *etch factor.* As the etchant acts on the workpiece, it tends to undercut the maskant pattern. In deep cuts, the undercut may be serious. Figure 22-20 depicts the effects of the etch factor in chemical erosion.

Casting and *electroforming* are sometimes called *metal-deposition* processes. They involve

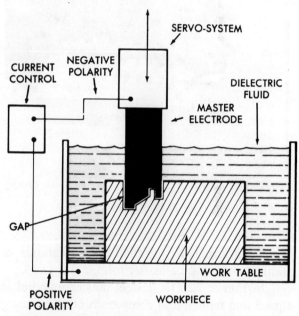

(A) Note the gap between the workpiece and master tool is uniform.

Fig. 22-18.

(B) A production model EDM

Fig. 22-19. Electrical erosion (EDM) machining of a die.

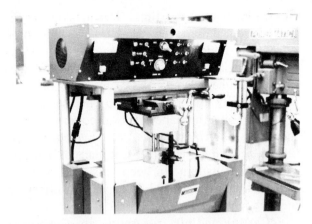

(C) A laboratory size EDM

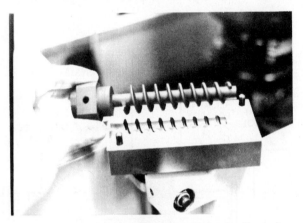

(D) Carbon master electrode used to make the die cavity.

Fig. 22-19. (continued)

depositing a metallic (or sometimes a ceramic or plastics) coating on a master form.

In Figure 22-21A a steel master mandrel is dipped into molten lead compounds until a coating is formed over it. The mandrel may then be removed and used again. Casting resins may be poured into the shell and removed when they are polymerized.

The hot casting of metals by the lost wax or sand processes, or by permanent metal molds may be used to produce precision molds. Molten metal may be poured over a hardened steel master to form a cavity, as shown in Figure 22-21B. This process is sometimes called *hot hobbing*. Molten metal is cast over a hob. Pressure is applied during the cooling.

Electroforming is an electroplating process. An accurate mandrel of plastics, glass, wax, or various metals is used as a master to electrically deposit the metallic ions from a chemical solution (Fig. 22-22). The molds are thin shelled and may have severe undercuts, but they usually have a

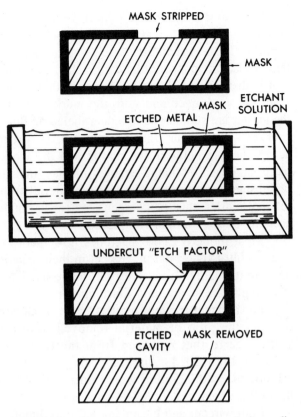

Fig. 22-20. Chemical erosion method of producing a die cavity.

highly polished finish (Fig. 22-23). The cavity may be strengthened by copper plating the back of the shell. Further strength may be provided by placing the die cavity into filled epoxy. The cavities may then be used for thermoforming, blow molding, or injection molding (Fig. 22-24).

The major advantages of electroformed molds are their accurate reproduction of detail, zero porosity, zero shrinkage, and lower cost. The ma-

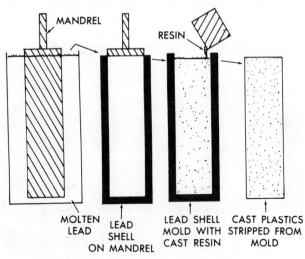

(A) Steps in casting plastics using cast molds.

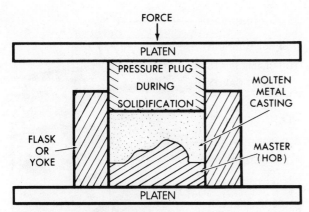

FORCE

PLATEN

PRESSURE PLUG
DURING
SOLIDIFICATION

MOLTEN
METAL
CASTING

FLASK
OR
YOKE

MASTER
(HOB)

PLATEN

(B) Molten metal is cast on master. Pressure from plug results in a dense, sound casting.

Fig. 22-21. Casting of plastics and metal.

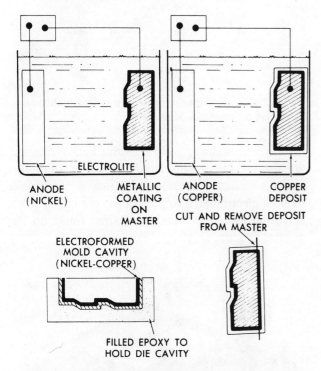

ANODE
(NICKEL)

METALLIC
COATING
ON
MASTER

ELECTROLITE

ANODE
(COPPER)

COPPER
DEPOSIT

CUT AND REMOVE DEPOSIT
FROM MASTER

ELECTROFORMED
MOLD CAVITY
(NICKEL-COPPER)

FILLED EPOXY TO
HOLD DIE CAVITY

Fig. 22-22. Electroformed mold cavity. Copper coating and filled epoxy backing strengthen the nickel cavity.

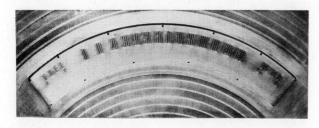

Fig. 22-23. An electroformed cavity showing raised detail from a polished background. The master is an engraved brass plate. (Electromold Corp.)

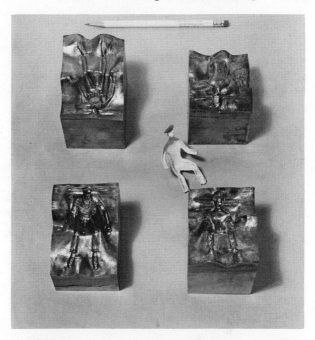

Fig. 22-24. A pair of cavities used to mold two halves of a hollow figure with a very elaborate match line. The cavities were filled with wax and the force plugs were electroformed against the cavity, producing perfect matched lines. Pencil shows relative scale. (Electromold Corp.)

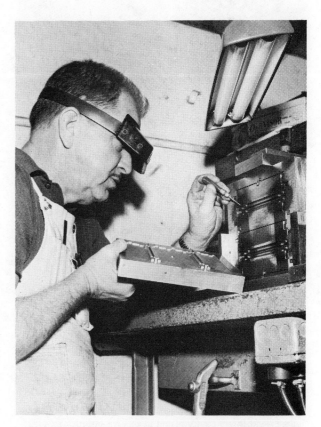

Fig. 22-25. Extreme care must be taken to prevent marring molds. (Revell, Inc.)

jor disadvantages include design limitations, relative softness and difficulty with multiple cavities.

Sometimes it is desirable to have the molded piece cling slightly to one half of the mold, depending on the knockout mechanism. Excessive sticking may be caused by dents or undercuts in the mold or by a dirty cavity surface. In cleaning the dies and cavities a wax, lubricant, or silicone spray is often used. For stubborn spots, a wooden scraper or brass brush may be used. Never use steel scrapers when cleaning cavities because they may scratch or damage the polished finish of the cavity surfaces. (Fig. 22-25).

There are a number of other metal-depositing methods that may be used for mold making. Flame spraying of metals and vacuum metallizing are two such methods. (See Chapter 17 Coating Processes.)

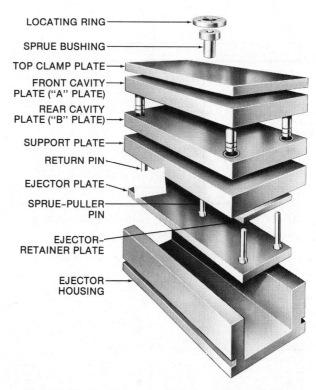

Fig. 22-27. Exploded view of a standard mold base, showing the parts. (D–M–E Co.)

Electroplating, welding, and heat treatment are also used in making molds. Many finishing operations of molds are done by hand. A steel mold may be electroplated to protect the die cavity from corrosion and to provide the desired finish on the plastics product (Fig. 22-26).

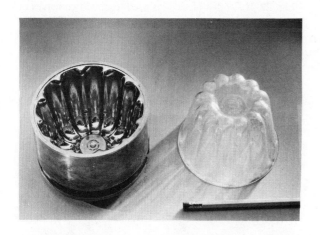

Fig. 22-26. This thermoforming cavity has an electroplated surface. (Electromold Corp.)

(A) Standard cavity insert rounds showing bored holes in upper and lower cavity plated to receive inserts.

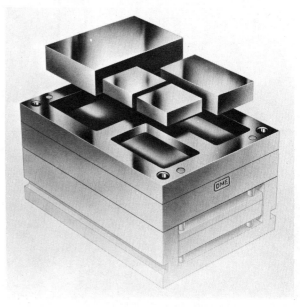

(B) Standard rectangular cavity insert blocks shown with pockets machined in mold base.

Fig. 22-28. Standard cavity insert blocks. (D–M–E Co.)

Mold bases are important to the tool-maker. Bases hold the cavities in place and are made with enough thickness to provide heating and cooling for the cavities. They are made from steel, and are available in standard sizes. A standard mold base is shown in Figure 22-27. Bases may be purchased to accept most custom or proprietary molds (Fig. 22-28).

Alignment pins ensure proper matching of cavities when the mold base is brought together. If the assembly is not properly aligned and the parting lines are not in register, the mold cavities may have to be repositioned. If the molded part sticks in the die cavity, last-minute stoning, hand grinding, and polishing of the mold may be needed.

VOCABULARY

The following vocabulary words are found in this chapter. Look up the definition of any of these words you do not understand as they apply to plastics in the glossary, Appendix A.

Air-hardening
Alignment pins
Chemical erosion
Deep-hardening
Die
EDM (Electrical-discharge machining)
Electroplating
Fixtures
Gypsum
Hobbing
Jig
Kirksite
Mold
Oil-hardened steel
Plastics tooling
Tools

22-1. Special fixtures, molds, dies etc., which enable a manufacturer to produce parts are called ___?___ .

22-2. Name an appliance for accurately guiding and locating tools during the operation involved in producing interchangeable parts.

22-3. Most mold designs begin from preliminary ___?___ .

22-4. A fundamental element in both plastics and steel is ___?___ .

22-5. Hobbing, etching, electroforming and electrical-discharge machining are generally classed as ___?___ processes.

22-6. An alloy of aluminum and zinc used for molds that has high thermal conductivity is known as ___?___ .

22-7. Much of the mold-making industry is composed of ___?___ shops.

22-8. Tooling ___?___ are more difficult to maintain in multiple cavity molds.

22-9. Tools for long run production work are usually made of ___?___ .

22-10. The trade names Ultracal, Hydrocal, and Hydrostone refer to ___?___ used for tooling.

22-11. Name the tooling material that has several advantages, including lightness, corrosion resistance, and low cost.

22-12. Devices that maintain proper alignment of the cavity as the mold closes are called ___?___ .

22-13. The act of shaping plastics or resins into finished products by heat and/or pressure is called ___?___ .

22-14. The tooling used to hold the cavities in place is called the ___?___ .

22-15. List four methods that may be used to produce a mold where no cutting tools are involved.

22-16. Because ___?___ are relatively inexpensive and have sufficient strength to produce some types of molds, they are used in prototype or experimental tooling.

22-17. Many metal tools and dies are being replaced by ___?___ because they are strong, light, and easy to machine.

22-18. A popular material for thermoforming and blow-molding molds is ___?___ because it is light, easy to machine, and has good thermal conductivity.

22-19. If a blow-molding mold requires fine mold detail, and if it must be easily machined but stronger than aluminum, ___?___ may be used.

22-20. For high strength, wear resistance, and high production runs, molds should be made of ___?___ .

22-21. A ___?___ machine is similar to and functions like a duplicator machine except that it is normally adjusted to operate at ratios as high as 20 to 1.

22-22. No cutting tools are used in machining operations classed as ___?___ processes.

22-23. The process where a piece of very hard steel is pushed into a blank of unhardened, unheated steel to form a mold cavity is called ___?___ .

22-24. An electrical means of removing metal by electrical erosion is called ___?___ .

22-25. In ___?___ or etching, an acid or alkaline solution is used to create a cavity.

22-26. Part of the manufacturing cost of plastics products is due to special ___?___ or tooling.

22-27. Gang or ___?___ molds and those with inserts require additional tooling costs.

22-28. Steel for molds may vary between 35 and 65 Rockwell C in ___?___ depending upon requirements.

22-29. Name the type of tooling you would select to produce:

(a) ten thermoformed serving trays

(b) ten thousand cast woodcut moldings for furniture doors

(c) ten thousand molded polypropylene chair seats

(d) ten million molded electrical switch-plate covers

22-30. Where corrosion resistance is needed or adverse atmospheric conditions exist, ___?___ steel may be used for molds.

Chapter 23

Commercial Considerations

In this chapter, you will learn that production personnel and management must work together for a successful plastics business. Financing, equipment, price quotations, plant locality, and other factors also must be considered.

Producing plastics parts is a competitive business. Material selection, processing techniques, production rates, and other variables must be considered in the selling price. (See Chapters 21 and 22.)

Resin manufacturers and custom molders are the best sources of information about the performance of a plastics material. The quantity and the compounding of the resin ingredients are important variables. When estimating or planning new items, confer with resin manufacturers and indicate all specifications. You should be able to answer the following questions:

1. How is the part to be used?
2. What grade of resin is to be used?
3. What physical requirements must the finished part meet?
4. What processing techniques will be used?
5. How many parts are to be made?
6. Will new capital investment for equipment be needed?
7. What are the specifications for reliability and quality of each part?
8. Will we be able to produce what the customer wants at a profit?
9. How and when will the customer pay for our services?

Price quotations may vary a great deal depending upon quantity. The price for polyethylene may exceed $0.75/lb in one pound bags, $0.48/lb in fifty pound bags, and $0.35/lb in a truck or train carload.

23-1 FINANCING

Strong interest and great prospects are not the only prerequisites to starting an industrial enterprise because without enough financing, no business firm can succeed and grow.

One of the major functions of management is to plan the capital structure of the business with care. In a proprietorship private capital or loans are used for financing, while in a corporation the sale of stock is used for capitalization.

Some equipment firms provide financing for purchases of their machinery through delayed or deferred payments, lease purchase plans, or direct financing. Insurance companies, commercial banks, mortgages, private lenders, and others are also sources of capital. The Small Business Investment Act of 1958 has helped many small firms. The Small Business Administration (a federal agency) has helped thousands of business to secure loans.

23-2 MANAGEMENT AND PERSONNEL

It has often been said, "a business is as strong or successful as its management operations." Many enterprises fail each year while others continue to struggle, barely surviving. Many of the problems of struggling and failing businesses may be associated with poor management. Management must coordinate the enterprise, regulating assets, personnel, and time to make a profit.

A major concern in the plastics industry is that the labor supply is not keeping pace with its growth rate. Employment opportunities for women are great since they make up nearly half the plastics industry workforce. The area of research in particular needs men and women to work with polymers, processes, and fabrication. Most plastics companies have need for professional personnel including executives, engineers, and supervisors; for technical personnel, including technicians and para-engineers; for skilled workers such as machinists, assistant technicians, and machinery set-up people; and for semiskilled and unskilled workers, who include material handlers, equipment operators, and packers.

Most of the unskilled or semiskilled personnel may be trained by the company; however, professional, technical, and skilled personnel must have college or technical school training.

The plastics industry will continue to compete for a limited supply of skilled designers, engineers, and moldmakers. The use of CAD, CAM, CAE, CAMM, and CIM systems might prove to be one way to meet the challenge of skilled personnel shortages and increase productivity.

Management must maintain good labor relations, which may include collective bargaining with labor unions. Successful relations between labor and management are part of the successful enterprise.

23-3 PLASTICS MOLDING

Much has been written about the general properties and forming processes of plastics. Molding plastics is difficult and often demands considerable experience to solve production problems. Technology in the plastics industry is constantly changing. Only basic information and precautions about molding plastics may be given here. (See also Chapters 10 and 21.)

The molding capacity of equipment may limit production output. Limitations include the available pressures of the press, amount of material it will mold, and physical size. Compression presses, for example, may vary in capacity from less than 5.5 to more than 1 653 tons [5 to 1 500 metric tonnes] of pressure. Extruder machines may plasticize less than 17 lb to more than 5.5 tons [8 to 5 000 kilograms] per hour. Injection machines may range from less than 0.70 oz to more than 44 lb [20 g to 20 kg] per cycle. The clamping pressures vary from less than 2.2 to more than 1 653 tons [2 to 1 500 metric tonnes]. It is common to run most equipment at 75 percent capacity, rather than maximum capacity (Table 23-1).

Many open-mold composite techniques are accomplished by hand. Placing layers of composite tape over specialized tools is slow. Handwork has the added disadvantages of possible incorrect tape orientation, process-induced voids,

Table 23-1. Advantages and Disadvantages of Selected Manufacturing Methods

Manufacturing Method	Injection-molding	Filament Winding	Blow-molding	Pultrusion	Rotational Casting	Bag	Extrusion	Sprayup	Thermoforming	Layup
Capital machine cost	high	low-high	low	high	low	low	high	low	low	low
Tool/mold costs	high	low-high	low	low-high	low	low	low	low	low	low
Material costs	high	high	high	high	low	high	high	high	high	high
Cycle times	low-high	high	low-high	high	high	high	low	high	high	high
Output rate	high	low	high	low	low	low	high	low	high	low
Dimensional accuracy	good	fair	fair	fair	fair	fair	fair	fair	poor	fair
Finishing stages	none	some	some	yes	some	some	yes	some	yes	some
Thickness variation	low	fair	fair	fair	low	low	low	low	high	low
Stress in molding	some	some	some	some	none	some	some	some	some	some
Can mold threads	yes	no	yes	no	yes	no	no	no	no	no
Can mold holes	yes	yes	yes	no	yes	yes	no	yes	no	yes
Open-ended components molded	yes	yes	yes	yes	yes	yes	yes	yes	yes	yes
Inserts molded-in	yes	yes	no	no	yes	yes	no	yes	no	yes
Waste material	none	some	some	none	none	some	some	some	some	some

and/or porosity. One solution to reduce manufacturing costs and assure consistent part quality is to utilize automated tape-laying equipment as part of the molding process. In Figure 23-1, course after course, layer upon layer, the part is formed by laminating the graphite/epoxy tape in a computer-designed, cross-ply pattern. It is then cured in an autoclave under controlled heat and pressure. The resulting aircraft part has superb structural strength plus a weight advantage unmatched by any metal (Fig. 23-2).

23-4 AUXILIARY EQUIPMENT

Plastics materials are poor conductors of heat. Some plastics are *hygroscopic*, that is, moisture-absorbing. For these reasons preheating auxiliary equipment may be needed to reduce moisture content and the polymerization or forming time. Often, hopper dryers are used on injection and extruders to remove moisture from the molding compounds and help assure consistent molding. Preheating of thermosets may be done by various thermal heating methods such as with infrared, sonic, or radio-frequency energy. Preheating may

Fig. 23-2. Significant portions of the F-16, including the vertical fin and horizontal stabilizer, are being fabricated from composite graphite/epoxy materials. Future aircraft will utilize even larger amounts of composite materials. (General Dyamics)

reduce cure and cycle time and prevent streaking, color segregation, molding stress, and part shrinkage. It also may allow for a more even flow of heavily filled molding compounds.

Molders often must compound their own additives into resins or other compounds. Both hot and dry mixers may be needed to blend plasticizers, colorants, or other additives. Material silos or other conveyor systems may be required.

Microprocessors (CIM systems) can control many materials handling operations. Through centralized monitoring and control, all settings can be made and the status of all stations can be checked. These systems store molding parameters and actual batch data (formulas) for future use or to prevent the wrong materials from entering a process. Hoppers, loaders, and blenders are controlled to accurately weigh and meter ingredients for each machine. Most dryer suppliers agree that microprocessor-control is the key in their technology. Microprocessors can accurately control energy use to create high performance, dust-free drying systems.

Preform equipment and loaders are important in compression molding and many composite systems.

Injection molding, blow molding, thermoforming, and other molding techniques may re-

Fig. 23-1. This automated tape-laying machine for laying exact courses of graphite/epoxy tape is an important step in the production of F-16 aircraft components. (General Dynamics)

quire the use of regrinders or granulators to grind sprues, runners, or other cutoff scrap pieces into usable molding materials. Some of these operations are done in-line. Processors look for noise and floorspace reduction, as well as energy efficiency and ease of maintenance when selecting granulators.

Annealing tanks are used with thermoplastic products to reduce sink marks and distortion. Large housings or similar large pieces are often placed over shrink blocks, or in jigs or dies, to help maintain correct dimensions and minimize warpage during cooling. Many automobile steering wheels once cracked after a length of time because of latent shrinkage of the molded part. Proper annealing and selection of materials have solved such problems.

Closed-loop processing with computer-controlled chillers and cooling towers is helping to reduce energy consumption and increase productivity.

The growth of the use of robotics has been phenomenal. The main reason for this growth is increased productivity. In addition, robots may remove human operators from hot, boring, and highly fatiguing jobs. The use of efficient and flexible robots can release humans for the creative and problem-solving tasks of the company (Fig. 23-3).

Although parts removal is the main application for robots, they could perform a variety of secondary operations that improve part quality and reduce labor costs. Robots may be used as parts-handling devices, transporting parts to secondary stations and packaging. This could eliminate non-value-added functions in the plant, such as assigning employees to package parts. Post-molding operations by robots may include assembling, gluing, sonic welding, decorating, or sprue snipping.

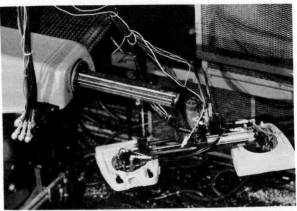

(B) Robot turns molded housings and places the parts on cooling conveyor.

(C) Robots, shown above removing a car bumper from an injection molding press, do practically all of the material handling in the production process. Station-to-station movement is by automatically guided vehicles or an automated overhead electric monorail system. Painting also is done by robots. The bumpers, manufactured from a polycarbonate polyester alloy called Xenoy, are as strong as steel but lighter, less expensive, and easier to paint. They meet five-mile-an-hour crash standards and will not rust. (Ford Motor Co.)

Fig. 23-3. Use of robots in plastics products manufacturing. (Prab Robots, Inc.)

(A) Robot completes broaching operation.

Some manufacturers utilize bar coding to keep track of material flow, inventory, and part production. Bar coding is used in flexible manufacturing systems (FMS). FMS consists of manufacturing cells that are equipped with a number of machining or molding operations or other automated equipment, all controlled by computer. As different coded parts come down the line, FMS or

CIM systems determine which operations (assembly, decorate, finish) are to be performed.

23-5 MOLDING TEMPERATURE CONTROL

One of the most important factors in efficient molding is the control of temperature. The temperature control system may consist of four basic parts including the thermocouple, temperature controller, power output device, and heaters. A control system using these four parts is shown in Figure 23-4.

The thermocouple is a device made of two unlike metals. Combinations of iron with constantan or copper with constantan are often used. When heat is applied to the junction of the two metals, electrons are freed, producing an electric current that is measured on a meter calibrated in degrees. Over 98 percent of all temperature sensors used by the plastics industry are thermocouples although resistance temperature detectors (resistance bulbs) have been used for a few installations. The West thermocouple system, used to minimize thermal variation, is shown in Fig. 23-5.

The millivolt and the potentiometric controllers are two types of basic temperature controllers used widely. The potentiometric controller differs from the millivoltmeter in that the signal from the thermocouple is electronically compared to a setpoint temperature. Millivolt and potentiometric controllers may be designed to control the power to the heaters and hold them at a set temperature (stepless control), or they may be designed to turn off the power to the heaters when a

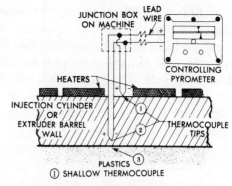

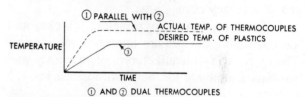

(A) Physical arrangement.

① SHALLOW THERMOCOUPLE

② DEEP THERMOCOUPLE

③ AREA IN PLASTICS WHERE UNIFORM TEMPERATURE IS DESIRED.

(B) Graph of temperatures.

Fig. 23-5. The physical and electrical arrangements of dual thermocouples used to minimize temperature variations. (West Instruments, Gulton MCS Division)

set temperature is reached (proportional control). The power input device controls the power to the heaters. It usually consists of a mechanical relay or a solid-state control circuit. Instrumentation may also be used to control cooling cycles. Figure 23-6 shows an instrument panel for a complete operation.

Most modern machines are operated by hydraulic power and electrical energy. Vacuum, compressed air, hot water, and chilled water supplies may also be needed. For the most part, cold

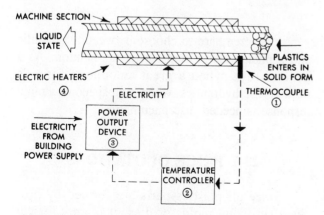

Fig. 23-4. The four basic parts of a temperature control system used in the plastics industry. (West Instruments, Gulton MCS Division)

Fig. 23-6. This instrument panel is used to control a large blow-molding operation. (Chemplex Co.)

water is used to cool mold dies in order to reduce the molding cycle time.

It is important in injection molding that the cooling system is able to remove the total heatload generated in each cycle. The system must also be designed to insure that all molded sections, thick or thin, will cool at the same rate to minimize potential differential shrinkage.

The microprocessor has made major changes in all areas of plastics processing, including chillers and mold-temperature controllers. Microprocessor-based mold temperature controllers may keep molds or chillers within ±1 °F. Electromechanical and solid-state controllers usually have ±3 °F accuracy ratings.

Balances, scales, pyrometers, clocks, and various timing devices are important accessories. There are many manufacturers of machines and auxiliary equipment for the plastics industry.

23-6 PNEUMATICS AND HYDRAULICS

Pneumatic- and hydraulic-actuated accessories and equipment are important in plastics processing. (Fig. 23-7).

Pneumatics are used to activate air-cylinders and provide compact, light, vibrationless power. Filters, air dryers, regulators, and lubricators are needed accessories for pneumatic systems.

Improper oil, or the presence of air or moisture, may cause noise in hydraulic lines. Chattering valves, pump wear, and high oil temperature may also cause noise problems. (See Chapter 3.)

Hydraulic power systems may be divided into four basic components:

1. *Pumps* that force the fluid through the system
2. *Motors or cylinders* that convert the fluid pressure into rotation or extension of a shaft
3. *Control valves* that regulate pressure and direction of fluid flow
4. *Auxiliary components,* which include piping, fittings, reservoirs, filters, heat exchangers, manifolds, lubricators, and instrumentation

Graphic symbols are used to show information about the fluid power system schematically. These symbols do not represent pressures, flow, or compound settings. Some devices have schematic diagrams printed on their faces (Fig. 23-8).

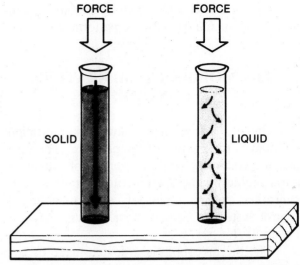

(A) Comparison of transmitting a force through a solid and a liquid.

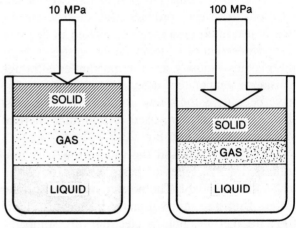

(B) Gas can be compressed, but a liquid resists compression.

Fig. 23-7. Force is transmitted differently by solids, liquids, and gases. A solid transmits force only in the direction of the applied force. Liquids (hydraulic systems) and gases (pneumatic systems) transmit force in all directions.

It is often hard to choose between a hydraulic and a pneumatic system for an application. As a general rule, when a great amount of force is needed, use hydraulics; when high speed or a rapid response is needed, use pneumatics.

23-7 PRICE QUOTATIONS

Since all plastics parts are formed by processes utilizing molds or dies, it is only logical that much consideration be given to them. (See Chapter 19.)

The cornerstone of any plastics business is tooling. In a tough, competitive market, a business must produce molds faster, cheaper and better than the competition can if the business is to be profitable (Fig. 23-9).

(C) Directional control valve.

Fig. 23-8. Fluid power control devices. (Sperry Vickers)

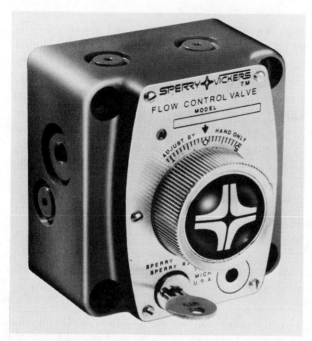

(A) Flow control valve compensates for fluid pressure and temperature variations.

(A) Male mold with part.

(B) Control panel with schematic on face.

(B) Female mold cavities with part.

Fig. 23-9. Example of male and female molds.

Computer numerically controlled (CNC) machine tools and computer assisted designs and manufacturing (CAD/CAM) equipment help produce better, more accurate molds for a substantial savings in cost and time.

All price quotations for molded products should be based on the design of the mold and its general condition. In the long run, the less costly die is usually not the best buy. Compression, transfer, and injection molds are costly.

It is common for price quotations to be based on *custom molds*. These are molds owned by the customer and turned over to the molder to produce parts. The greatest danger in quoting prices is failure to take into consideration the condition of the custom mold. All quotations should be based on approval of the custom mold because repair or alteration of the mold may be necessary.

If the die is owned or made by the molder, it is called a *proprietary mold*. In quotations for products to be made with proprietary molds, the molder must make enough profit to amortize, or pay for, the mold. Often a part of the cost of making the mold is calculated into the price of each piece or each thousand pieces.

With custom or proprietary molds, corrosion is an enemy. Molds should be given a moisture-resistant, noncorrosive coating and water, air, or steam holes should be dried and given a coating of oil. The mold should then be placed in storage with all its accessories.

Large tooling (panels, boat hulls, sinks) require careful consideration of production space and adequate storage facilities. Both add to the cost of the product.

23-8 PLANT SITE

The plant site should be determined in relation to the nearness of raw material and potential market. Freight must be reflected in the cost of each plastics product. A quotation must reflect tax rates, labor conditions, and wages. If there is an anticipated increase in taxes or wages, the price quotation must include such an increase in production costs.

Many states and communities encourage new enterprises. They offer reduced taxes and adequate labor force. Labor relations and tax incentives may be major concerns in plant location.

Other considerations in site location are an available skilled-labor pool and proximity to educational institutions.

23-9 SHIPPING

Shipping plastics, resins, and chemicals falls under special governmental regulations. The United States Postal Service has ruled that any liquid that gives off flammable vapors at or below a temperature of 20 °F [−7 °C] is not mailable (Section 124.2d, Postal Manual). Poisonous materials are generally regarded as nonmailable by the provisions of 124.2d. Caustic or corrosive substances are prohibited in 124.22. The prohibitions mentioned are specified by the law in Section 1716 of Title 18, U. S. Code. Acids, alkalies, oxidizing materials, or highly flammable solids, highly flammable liquids, radioactive materials, or articles emitting objectionable odors are considered to be nonmailable as well.

Postmasters and other employees at post offices will not give opinions with regard to mailability of materials. To determine if material is properly mailable, the mailer should write to the Mailability Division, Office of the General Counsel, Washington, D. C., for instructions.

The Department of Transportation (DOT) regulates the interstate transportation of cellulose nitrate plastics by rail, highway, or water, and special packaging must be used when shipping this material.

The Interstate Commerce Commission (ICC) regulates the packing, marking, labeling, and transportation of dangerous or restricted materials. Standards for many shipping containers are specified and regulations also specify the use of labels. For example, flammable liquids require a red label, flammable solids a yellow label, and corrosive liquids a white label. There are also other labels for poisons and shipments of radioactive material. It is the responsibility of the shipping agency to check the label to make certain it is correct and that it is filled in by the shipper.

If there is any doubt about the shipment of flammable or plastics materials, it is best to check local ICC offices. Certain cities have ordinances that prohibit vehicles containing flammable items from operating through tunnels or via bridges. Such Interstate Commerce Commission regulations must always be observed.

VOCABULARY

The following vocabulary words are found in this chapter. Look up the definition of any of these words you do not understand as they apply to plastics in the glossary, Appendix A.

Amortization
Annealing
Auxiliary equipment
Custom molds
Estimating
Flexible manufacturing systems (FMS)
Hydraulics
Hygroscopic
Pneumatics
Proprietary molds
Pyrometer

23-1. The greatest danger in quoting custom mold prices, is not taking into consideration the ___?___ of the custom mold.

23-2. Often a portion of the cost of making a mold is calculated into the production of each piece. This is called ___?___ .

23-3. Molds made by the customer and used by the molder are called ___?___ molds.

23-4. The transmission of power through a controlled flow of liquids is known as ___?___ .

23-5. Molds made and owned by the molder are called ___?___ molds.

23-6. The ___?___ is a device made of two dissimilar metals. Combinations of iron and constantan or copper and constantan are commonly used.

23-7. A ___?___ material tends to absorb moisture.

23-8. The ___?___ of the molding equipment will limit product size.

23-9. Name four pieces of auxiliary equipment that may be required for injection molding operations.

23-10. Name four factors that may influence plant locality.

23-11. Name four factors that may influence the cost of each plastics product when estimating the planning new items.

23-12. The Interstate Commerce Commission regulates the packing, marking, labeling and transportion of ___?___ or ___?___ materials.

23-13. Would you select a hydraulic or a pneumatic power system if great force is required?

23-14. When high speed and rapid response are required, would you select a hydraulic or a pneumatic power system?

23-15. The best sources of information about the performance of a plastics material are ___?___ manufacturers and ___?___ molders.

23-16. In a ___?___ enterprise, private capital or loans are used for financing.

23-17. The sale of stock shares is used for capitalization in a ___?___ enterprise.

23-18. Planning the capital structure of an enterprise is the major function of ___?___ .

23-19. Successful relations between ___?___ and ___?___ are part of the successful enterprise.

23-20. Name the four basic components of hydraulic power systems.

23-21. The cornerstone of any plastics business is ___?___ .

23-22. It is common to run or operate most molding equipment at ___?___ percent of maximum capacity.

23-23. Name the four general classifications of plastics personnel.

23-24. Approximately ___?___ percent of the plastics industry work force is female.

23-25. If ___?___ and CNC equipment is used, better, more accurate molds may be produced with a savings in cost and time.

23-26. In addition to increased productivity, ___?___ may be used to relieve human operators from hot, boring, and highly fatiguing jobs.

GLOSSARY

The numbers in parentheses following each glossary entry refer to the chapter in which the word is found.

A-stage — An early stage in the reaction of a thermosetting resin in which the material is fusible and is still soluble in certain liquids. (8)

Ablative plastics — Plastics composites used as heat shields on aerospace vehicles. The intense heat encountered erodes and chars the top layers. The charred layer and some cooling effects from evaporation insulates the inner areas against further penetration of the intense heat. (4)

Acetals (poly) — A polymer having the molecular structure of a linear acetal (polyformaldehyde), consisting of unbranched polyoxymethylene chains. (7)

Acrylic — A synthetic resin prepared from acrylic acid or from a derivative of acrylic acid. (7)

Acrylonitrile — A monomer with the structure $CH_2 = CHON$. It is most useful in copolymers. Its copolymer with butadiene is nitrile rubber, and several copolymers with styrene exist that are tougher than polystyrene. (7)

Acrylonitrile-butadiene-styrene (ABS) — Acrylonitrile and styrene liquids and butadiene gas are polymerized together in a variety of ratios to produce the family of ABS resins. (7)

Acute Toxicity — The adverse effect on a human or animal body, with severe symptoms developing rapidly and quickly coming to a crisis. Examples include dizziness, nausea, skin rashes, inflammation, tearing of the eyes, unconsciousness, and even death. (3)

Additional polymerization — Polymers formed by the combination of monomer molecules without the splitting-off of low molecular mass by-products, such as water. (2)

Adhesion — The state in which two surfaces are held together by interfacial forces which may consist of interlocking action (mechanical means). (3, 18)

Adhesive — A substance capable of holding materials together by surface attachment. (3)

Air-assist forming — A method of thermoforming in which air flow or air pressure is used to partially preform the sheet just before the final pull-down into the mold using vacuum. (15)

Air-hardening — Refers to steel that is cooled in air. (22)

Air pollution — Filling the atmosphere with undesirable particles, some of which may be toxic. (5)

Air slip forming — A thermoforming process in which air pressure is used to form a bubble and a vacuum is then used to form the hot plastics against the mold. (12)

Alignment pins — Devices that maintain proper cavity alignment as a mold closes. (22)

Aliphatic molecules — Organic compounds whose molecules do not have their carbon atoms arranged in a ring structure. (2)

Alkanes — Hydrocarbons with the general formula C_nH_{2n+2}. (2)

Alkenes — Hydrocarbons that have the general formula C_nH_{2n} and possess double covalent bonds. (2)

Alkyd — Polyester resins made with some fatty acid as a modifier. (8)

Alkynes — Hydrocarbons that have the general formula C_nH_{2n-2} and possess a triple bond between two carbon atoms. (2)

Allowances — The intentional differences in the dimensions of two parts. (18)

Allyl — A synthetic resin formed by the polymerization of chemical compounds containing the group $CH_2 = CH - CH_2$. The principal commercial allyl resin is a casting material that yields allyl carbonate polymer. (7)

Alpha particle — A subatomic particle composed of two protons and two neutrons; hence it is the same as the nucleus of a helium atom. (20)

Amids—Organic compounds containing a —$CONH_2$ group, derived from organic acids. (7)

Amines—Organic derivatives of ammonia (NH_3) obtained by substituting hydrocarbon radicals for one or more hydrogen atoms. (7)

Amino—Chemical name showing the presence of an NH_2 or NH group. Also, materials with these groups. (8)

Amortization—The gradual repayment of the cost of equipment such as molds. It may be done by contribution to a sinking fund at the time of each periodic interest payment. (23)

Anisotropic—Exhibiting different properties when tested along axes in different directions. (1)

Annealing—A process of holding a material at a temperature near, but below, its melting point, for a period of time to relieve internal stress without shape distortion. (9, 10)

ANSI—American National Standards Institute; a privately funded, voluntary membership organization that identifies industrial and public needs for national consensus standards and coordinates development of such standards. Many ANSI standards relate to safe design/performance of equipment, such as safety shoes, eyeglasses, smoke detectors, firepumps and household appliances; and safe practices or procedures, such as noise measurement, testing of fire extinguishers and flame arrestors, industrial lighting practices, and the use of abrasive wheels. (3, 4)

Antioxidant—A stabilizer that retards breakdown of the plastics by oxidation. (6)

Antistatic—An additive that reduces static charges on a plastics surface. (6)

Apparent density—The mass per unit volume of a material; includes the voids inherent in the material. (21)

Aromatic hydrocarbons—Hydrocarbons derived from or characterized by the presence of unsaturated ring structures. (2)

Ashing—The use of wet abrasives on wheels to sand and polish plastics. (9)

ASTM—American Society for Testing Materials; voluntary membership organization with members from broad spectrum of individuals, agencies, and industries concerned with materials. As the world's largest source of voluntary consensus standards for materials, products, systems, and services, ASTM is a resource for sampling and testing methods, health and safety aspects of materials, safe performance guidelines, and effects of physical and biological agents and chemicals. (3, A/G)

Atactic stereoisomerism—A random arrangement of molecular chains in a polymer. (2)

Atomic mass (gram-atom)—Relative mass of an atom of any element, as compared with that of one atom of carbon taken as 12 g. (2)

Atomic number (atomic mass number)—A number equal to the number of protons in the nucleus of an atom of the element. (2)

Atoms—The smallest particles of an element that can combine with particles of other elements to produce the molecules of compounds. Atoms consist of a complex arrangement of electrons revolving about a positively charged nucleus containing particles called protons and neutrons. (2)

Autoclave—Pressure vessel that can maintain temperature and pressure of a desired air or gas for the curing of organic-matrix composite materials. (12, 13)

Autoclave molding—A molding method in which, after final layup, an entire assembly is put into a steam or electrically heated autoclave at elevated pressure. Additional pressure achieves higher reinforcement loadings and improved removal of air. (12, 13)

Automatic cycle—A machine operation that will perform repetitive cycles. (10)

Auxiliary equipment—Equipment needed to help control or form the product. Filters, vents, ovens, and takeup reels are examples. (23)

Avogadro's number—The number of atoms in a gram-atom mass of an element: $6.024\,86 \times 10^{23}$. (2)

Azo group—The group —N=N—, generally combined with two aromatic radicals. A whole class of dyestuffs is characterized by the presence of this group. (7)

B-stage—An intermediate stage in the reaction of a thermosetting resin. In this stage, the material softens when heated and swells in contact with certain liquids, but it does not entirely fuse or dissolve. Resins in thermosetting molding compounds are usually in this stage. (8)

Back pressure—The viscous resistance of a material to continued flow when a mold is closed. In extrusion, resistance to the forward flow of molten material. (11)

Baffle—A device used to restrict or divert the passage of fluid or gases through a pipeline or channel. (10)

Banbury—A machine for compounding materials. The machine contains a pair of contrarotating rotors that masticate and blend the materials. (10)

Barrel—The cylindrical housing in which the extruder screw rotates. (11)

Barrel vent—An opening in a barrel wall to permit the escape of air and volatile matter from the material being processed. (10, 11)

Benzene—A clear, flammable liquid, C_6H_6. It is the most important aromatic chemical. (2)

Beta particle—An elementary particle (electron) having 1.602×10^{-19} coulomb of negative charge, with the rest mass of an electron, which is emitted by an atomic nucleus. (20)

Biaxial winding—A type of winding, in filament winding, in which the helical band is laid in sequence, side by side, with no crossover fibers. (13)

Bifunctional—A molecule with two active functional groups. (2)

Binder—Resin that holds together filler or carriers in a pre-preg or molding compound. **Carrier** is the fabric portion of a composite. Fiberglass is the most commonly used fabric for composites. (5, 13)

Blanket—Plies laid up in a complete assembly and placed on or in the mold all at one time (flexible-bag process); also, the form of bag in which the edges are sealed against the mold. (13)

Blanking—The cutting of flat sheet stock to shape by striking it sharply with a punch while it is supported on a mating die. Punch presses are used. (9)

Bleeder cloth—A nonstructural layer of material used in manufacture of composite parts to allow the escape of excess gas and resin during curing.

Block polymer—A polymer molecule that is made up of comparatively long sections that are of one chemical composition, those sections being separated from one another by segments of different chemical character. (2)

Blow molding—A method of fabrication in which a parison is forced into the shape of the mold cavity by internal air pressure. (11)

Blow systems—Any method or process used to cause polymers to expand or become cellular. (13)

Blowup ratio—In blow molding, the ratio of the mold-cavity diameter to the parison diameter. In blown film, the ratio of the final tube diameter to the original die diameter. (11)

Boss—A protuberance on a part designed to add strength or to facilitate assembly. (21)

Branching—Side chains attached to the main chain of the polymer. Side chains may be long or short. (2)

Breaker plate—A perforated metal plate located between the end of the screw and the die head. (11)

Breather—Porous material, such as a fabric or cilitate, for removal of air, moisture, and volatiles during curing. (13)

Breathing—The opening and closing of a mold to allow gases to escape early in the molding cycle. Also called degassing. (10)

Brittleness temperature—The temperature at which plastics and elastomers rupture by impact under certain conditions. (4)

Buffing—An operation to provide a high luster to a surface. The operation, which is not intended to remove much material, usually follows polishing. (18)

Bulk density—The mass per unit volume of a molding powder as determined in a reasonably large volume.

Bulk factor—The ratio of the volume of any given weight of loose plastics to the volume of the same weight of the material after molding or forming. (21)

Bulk polymerization—The polymerization of a monomer without added solvents or water. (2)

C-stage—The final stage in the reaction of a thermosetting resin, in which the material is relatively insoluble and infusible. Thermosetting resins, if fully cured plastics, are in this stage. (8)

CAD—Computer-aided design; a computer system that aids or assists in the creation, modification, and display of a design. It is used to produce three dimensional designs and illustrations of the proposed part. The term **CAD** is also used to refer to computer-aided drafting. (22, 23)

CAE—Computer-aided engineering; a computer system that assists in the engineering or design cycle. It analyzes the design and calculates the performance predictions of service life and safety design factors. (22, 23)

Calendering—Process of forming a continuous sheet by squeezing the material between two or more parallel rolls to impart the desired finish or to ensure uniform thickness. (11)

CAM—Computer-aided manufacturing; the utilization of computer systems in the management, control, and operations of the manufacturing facility through either direct or indirect computer interface with the physical and human resources of the company. (22, 23)

CAMM—Computer-assisted mold making; a computer system used to analyze mold temperatures and flows through finite element analysis or boundary element analysis. Orientation and placement of reinforcements in composites are also shown in some data base systems. This information is then used to compensate for matrix shortage in the design and machining of the mold. (22, 23)

Casein—A protein material precipitated from skimmed milk by the action of either rennin or dilute acid. Rennet casein is made into plastics. (8)

Casting—The process of pouring a heated plastics or other fluid resin into a mold to solidify and take the shape of the mold by cooling, loss of solvent, or completing polymerization. No pressure is used. *Casting* should not be used as a synonym for *molding*. (10, 14)

Catalyst—A chemical substance added in minor quantity (as compared to the amounts of primary reactants) that markedly speed up the cure or polymerization of a compound. See also Initiator. (2, 6)

Carcinogen — A substance or agent that can cause a growth of abnormal tissue or tumors in humans or animals. A material identified as an animal carcinogen does not necessarily cause cancer in humans. Examples of human carcinogens include coal tar, which can cause skin cancer, and vinyl chloride, which can cause liver cancer. (3)

Caul plate — A smooth metal plate used in contact with the layup during curing to transmit normal pressure and to provide a smooth surface to the finished part. (13)

Cellular or foamed — A sponge form. The sponge may be flexible or rigid, the cells closed or interconnected. The density of a cellular plastics may be anything from that of the solid parent resin down to 32 kg/m^3 [2 lbs/ft^3]. The terms *cellular, expanded,* and *foamed* plastics are used synonymously; however, *cellular* is most descriptive of the product. (3)

Celluloid — A strong, elastic plastics made from nitrocellulose, camphor, and alcohol. Celluloid is used as a trade name for some plastics. (1)

Cellulosics — A family of plastics with the polymeric carbohydrate cellulose as the main constituent. (7)

Cement — To bond together, as to adhere with a liquid adhesive employing a solvent base of the synthetic elastomer or resin variety. (3)

Centipoise — One one-hundredth of a poise, a unit of viscosity. Water at room temperature has a viscosity of about one centipoise. (4)

CERCLA — Comprehensive Environmental Response, Compensation, and Liability Act (3)

CFR — Code of Federal Regulations (3)

Charging — In molding, the measurement or placing of the material in a mold. In polishing, the deposit of abrasive on a revolving wheel. (9)

Chemical erosion — A chemical method of removing metal. (22)

Chlorinated polyether — The polymer obtained from pentaerythritol by preparing a chlorinated oxethane and polymerizing it to a polyether by means of opening the ring structure. (7)

Chronic Toxicity — Adverse (chronic) effects resulting from repeated doses of or exposures to a substance over a relatively prolonged period of time. Ordinarily used to denote effects in experimental animals. (3)

CIM — Computer-integrated manufacturing; the logical organization of individual engineering, production, and marketing, or other support functions into a computer-integrated system. Functional areas such as design, inventory control, physical distribution, cost accounting, planning, purchasing, etc. are integrated with direct materials management and shop floor management. (22, 23)

Closed-cell — Describing the condition of individual cells that make up cellular or foamed plastics when cells are not interconnected. (3)

Coating — Placing a permanent layer of material on a substrate. (17)

Cohesion — The propensity of a substance to adhere to itself; the internal attraction of molecular particles toward each other; the ability to resist partition from the mass. (3, 18)

Cold flow — See *Creep*.

Cold molding — A procedure in which a composition is shaped at room temperature and cured by subsequent baking. (10)

Colorants — Dyes or pigments that impart color to plastics. (6)

Composite — A combination of two or more materials (generally a polymer matrix with reinforcements). The structural components of the composites are sometimes subdivided into: fibrous, flake, laminar, particulate, and skeletal. (1, 5)

Composting — Grinding trash into small pieces and mixing it with soil. (5)

Compound — A substance composed of two or more elements joined together in definite proportions. (2)

Compression mold — A mold that is open when the substance is introduced and that shapes the material by heat and by the pressure of closing. (10)

Compression molding — A technique in which the molding compound is placed in an open mold cavity, the mold closed, and heat and pressure applied until the material has cured or cooled. (10)

Compressive strength — The highest load sustained by a test sample in a compressive test, divided by the original area of the sample. (4)

Condensation — A chemical reaction in which two or more molecules combine, with the separation of water or some other simple substance. If a polymer is formed, the condensation process is called *polycondensation*. (8)

Condensation polymerization — Polymerization by chemical reaction that also produces a by-product. (2)

Copolymerization — Addition polymerization involving more than one type of mer. (2)

Corona discharge — A method of oxidizing a film of plastics to make it printable. Achieved by passing the film between electrodes and subjecting it to a high-voltage discharge. (19)

Coumarone — A compound (C_8H_6O) found in coal tar and polymerized with indene to form thermoplastic resins that are used in coatings and printing inks. (7)

Covalent bonding — Atomic bonding by sharing electrons. (2)

CPSC — Consumer Products Safety Commision; Federal agency with responsibility for regulating hazardous materials when they appear in consumer goods. For CPSC purposes, hazards are defined in the Hazardous Substances Act and the Poison Prevention Packaging Act of 1970. (3)

Cracking — Thermal or catalytic decomposition of organic compounds to break down the high-boiling compounds into lower-boiling fractions. (7)

Creep — The permanent deformation of a material resulting from prolonged application of a stress below the elastic limit. A plastics subjected to a load for a period of time tends to deform more than it would from the same load released immediately after application. The degree of the deformation is dependent on the load duration. Creep at room temperature is sometimes called *cold flow.* (4)

Crosslinking — The tying together of adjacent polymer chains. (2)

Crystallization — The process or state of molecular structure in some plastics that denotes uniformity and compactness of the molecular chains forming the polymer; usually attributed to the formation of solid crystals having a definite geometric form. (2)

Custom molds — Molds owned by the customer and used by the molder. (23)

Cyclic hydrocarbons — Cyclic or ring compounds. Benzene (C_6H_6) is one of the most important cyclic hydrocarbons. (2)

Damping — Variations in properties resulting from dynamic loading conditions (vibrations). Damping provides a mechanism for dissipating energy without excessive temperature rise, prevents premature brittle fracture, and is important to fatigue performance. (4)

Daylight opening — The clearance between platens when the press is fully opened. The opening must be large enough to allow a part to be ejected when the mold is in a fully opened position. (10)

Debond — Area of separation with or between plies in a laminate, or within a bonded joint, caused by contamination, improper adhesion during processing, or damaging interlaminar stresses. (12, 13)

Deep-hardening — Refers to the depth of hardening possible in a piece of steel. (22)

Degree of polymerization (DP) — Average number of structural units per average molecular mass. In most plastics, the DP must reach several thousand to achieve worthwhile physical properties. (2)

Delamination — Debonding process primarily resulting from unfavorable interlaminar stresses; edge delamination, however, can be effectively prevented by a wrap-around reinforcement. (12, 13)

Denier — The mass in grams of 9 000 m (29 527 ft) of synthetic fiber in the form of a single continuous filament. (3)

Density — Mass per unit volume of a substance, expressed in grams per cubic centimeter, or kilograms per cubic meter. (4)

Diamines — Compounds containing two amino groups. (7)

Dibasic acid — An acid that has two replaceable hydrogen atoms. (7)

Die — A forming piece used in shaping parts for quantity production. (19)

Die cutting — Blanking or cutting shapes from sheet stock by striking sharply with a steel-rule die, which is a knife edge made in the shape of the item to be cut. (9)

Difunctional — See Bifunctional

Dimensional stability — The ability of a plastics part to keep the precise shape in which it was molded, fabricated, or cast. (4)

Dip casting — The process of submerging a hot mold into a resin. After cooling, the product is removed from the mold. (14)

Dip coating — Applying a coating by dipping an item into a tank of melted resin or plastisol, then chilling. The object may be heated and powders used for the coating; the powders melt as they strike the hot object. (17)

Drape forming — Method of forming thermoplastic sheet in a movable frame. The sheet is heated and draped over high points of a male mold. Vacuum is then pulled to complete the forming. (15)

Drawing — The process of stretching a thermoplastic sheet, rod, film, or filaments to reduce cross-sectional area and change physical properties. (11)

Dwell — A pause in the application of pressure to a mold, made just before the mold is completely closed. The pause allows gas to escape from the molding material. (10)

EDM — See Electrical-discharge machining.

Elastic limit — The extent to which a material can be stretched or deformed before taking on a **permanent set.** Permanent set occurs when a material that has been stressed does not recover its original dimensions, as when a 12-inch piece of rubber that has been stretched becomes 13 inches long when relaxed. (4)

EMI — *Electromagnetic Interference;* refers to the use of conductive materials in composites to make the composite conductive; thus, capable of protecting (shielding) electronic devices from unwanted electrical interference (radiation, static electricity, lightning). (5)

EMP — *Electromagnetic Pulse;* see EMI. (5)

EPA — U.S. Environmental Protection Agency; federal agency with environmental protection regulatory and enforcement authority. Administers Clean Air Act, Clean Water Act, FIFRA, RCRA, TSCA, other federal environmental laws. (3)

Elastomer — A rubberlike substance that can be stretched to several times its original length, and which, on release of the stress, returns rapidly to almost its original length. (1)

Electric dipole — A nonuniform charge distribution, in which one end of a molecule or ion is positive and the other negative. (2)

Electrical-discharge machining (EDM) — A process in which a high-frequency intermittent electric spark is used to erode the workpiece. (22)

Electromagnetic bonding — See Induction bonding.

Electron — A negatively charged particle that is present in every atom. (2)

Electroplating — A method of applying metallic coatings to a substrate. (22)

Electrostatic printing — The deposit of ink on a plastics surface where electrostatic potential is used to attract the dry ink through an open area defined by opaquing. (19)

Embedment — Enclosing an object in an envelope of transparent plastics by immersing it in a casting resin and allowing the resin to polymerize. (3)

Emulsion polymerization — A process where monomers are polymerized by a water-soluble initiator while dispersed in a concentrated soap solution. (2)

Encapsulating — Enclosing an item, usually an electronic component, in an envelope of plastics by immersing it in a casting resin and allowing the resin to solidify by polymerizing or cooling. (3, 14)

Engineering plastics — Those with good physical and mechanical properties, designed to meet a special need or use. (7, 8)

Engraving — The act of cutting figures, letters, or symbols into a surface. A plastics web is often printed or decorated by interposing a resilient offset roll between an engraved roll and the web. (19)

Epoxy — Material based on ethylene oxide, its derivatives or homologs. Epoxy resins form straight-chain thermoplastics and thermosetting resins. (8)

Ester — A compound formed by the replacement of the acidic hydrogen of an organic acid by a hydrocarbon radical; a compound of an organic acid and an alcohol formed with the elimination of water. (7)

Esterification — The process of producing an ester by reaction of an acid with an alcohol and the elimination of water. (7)

Estimating — The act of determining from statistical examples, experience, or other parameters, the cost of a product or service. (23)

Exothermic — Evolving heat during reaction (cure). (6)

Expanded (foamed) plastics — Plastics that are cellular or spongelike. (16)

Expand-in-place — A process in which resin, catalyst, expanding agents and other ingredients are mixed and poured at the site where needed. Expansion takes place at room temperature. (16)

Extrudate — The product delivered by an extruder. (11)

Extrusion — The compacting and forcing of a plastics material through an orifice in more-or-less continuous fashion. (11)

Extrusion coating — The resin coating placed on a substrate by extruding a thin film of molten resin and pressing it into or onto the substrate (or both), without use of adhesives. (11, 17)

FDA — The U.S. Food and Drug Administration; under the provisions of the Federal Food, Drug and Cosmetic Act, the FDA establishes requirements for the labeling of foods and drugs to protect consumers from misbranded, unwholesome, ineffective, and hazardous products. FDA also regulates materials for food contact service and the conditions under which such materials are approved. (3)

Fatigue strength — The highest cyclic stress a material can withstand for a given number of cycles before failure occurs. (4)

Feed — The distance the cutting tool moves into the work with each revolution. (9)

Feedstocks — Refers to sources of raw material from which we make polymers. (1)

Female mold — The indented half of a mold designed to receive the male half. (10)

Fiber orientation — Fiber alignment in a nonwoven or a mat laminate where the majority of fibers are in the same direction, resulting in a higher strength in that direction. (5, 13)

Fibers — This terms usually refers to relatively short lengths of various materials with very small cross sections. Fibers can be made by chopping filaments. (3)

Filament — A fiber characterized by extreme length, with little or no twist. A filament is usually produced without the spinning operation required for fibers. (3)

Filament winding — A composite fabrication process that consists of winding a continuous reinforcing fiber (impregnated with resin) around a rotating and removable form (mandrel). (13)

Filler — An inert substance added to a plastic to make it

less costly. Fillers may improve physical properties. The percentage of filler used is usually small in contrast to reinforcements. (6)

Fillet — A rounded filling of the internal angle between two surfaces of a plastic molding. (21)

Fire resistant — A term for a substance that does not easily burn. (5)

Fixture — A device used to support the work during processing or manufacture. (22)

Flame spraying — Method of applying a plastics coating in which finely powdered plastics and suitable fluxes are projected through a cone of flame onto a surface. (17)

Flash — Extra plastics attached to a molding along the parting line. It must be removed to make a finished part. (10)

Flash point — The temperature at which it gives off vapors sufficient to form an ignitable mixture with the air near the surface of the liquid. (3)

Flexible manufacturing systems — A series of machines and associated workstations linked by a hierarchical common control, and providing for automatic production of a family of workpieces. A transportation system, for both the workpiece and the tooling, is as integral to an FMS as is computerized control. (22)

Flexural modulus — The ratio, within the elastic limit, of the applied stress on a test sample in flexure to the matching strain in the outermost fibers of the sample. (4)

Flexural strength (modulus of rupture) — The highest stress in the outer fiber of a sample in flexure at the moment of crack or break. In the case of plastics, this value is usually higher than tensile strength. (4)

Flow line — Sometimes called a weld line. A mark on a molded piece made by the meeting of two flow fronts during molding. (21)

Flow mark — Wavy surface appearance caused by improper flow of hot plastics into the mold cavity. (21)

Fluidized bed — A method of coating heated items by immersion in a dense-phase fluidized bed of powdered resin. The objects are usually heated in an oven to provide a smooth coating. (17)

Fluorescence — A property of a substance that causes it to produce light while it is being acted on by radiant energy such as ultraviolet light or X-rays. (6)

Fluoroplastics — A group of plastics materials containing the element fluorene (F). (7)

Foamed — See Cellular.

Foaming agents — Chemicals that generate inert gases on heating or chemical reaction, causing the resin to assume a cellular structure. (16)

Formalin — A commercial 40 percent solution of formaldehyde in water. (8)

Free forming — Air pressure is used to blow a heated sheet of plastics, the edges of which are being held in a frame, until the desired shape or height is attained. (15)

Frost line — In extrusion, a ring-shaped zone located at the point where the film reaches its final diameter. (11)

Gamma ray — Electromagnetic radiation originating in an atomic nucleus. (20)

Gate — In injection and transfer molding, the orifice through which the melt enters the cavity. Sometimes the gate has the same cross section as the runner leading to it. (10)

Gel coat — A thin layer of resin that serves as the surface of the product. Reinforced layers may then be build up. (13)

Glass transition — The change in amorphous or partially crystalline polymers from a viscous or rubbery condition to a hard and relatively brittle one (from hard to viscous condition). (4)

Glass transition temperature (Tg) — A characteristic temperature at which glassy amorphous polymers become flexible or rubberlike because of the motion of molecular segments. (4)

Glass, Type C — Chemically resistant glass fibers. (6)

Glass, Type E — Electrical grade glass fibers. (6)

Glue — Formerly, an adhesive prepared from animal hides, tendons, and other by-products by heating with water. In general use, the term is now synonymous with the term "adhesive." (3)

Gradient-density column — A means of conveniently measuring small plastics samples. The column is a glass gradient tube filled with a heterogeneous mixture of two or more liquids. The density of the mixture varies linearly or in other known fashion with the height. The specimen is placed in the gradient tube and falls to a position of equilibrium that shows its density by comparison with positions of known standard samples. (4)

Graft copolymer — A combination of two or more chains of constitutionally or configurationally different features, one of which serves as a backbone main chain, and at least one of which is bonded at some point(s) along the backbone and constitutes a side chain. (2, 4, 20)

Gray (Gy) — The unit of measurement of the absorbed dose of ionizing radiation, defined as one joule per kilogram (1 Gy = 1 J/kg). (20)

Gutta percha — A rubberlike product obtained from certain tropical trees. (1)

Gypsum — Crystalline hydrated sulfate of calcium ($CaSO_4 \cdot 2H_2O$), used for making plaster of Paris and Portland cement. (22)

Halogens — The elements fluorine, chlorine, bromine, iodine, and astatine. (4)

Hand layup — Method of positioning successive layers of reinforcement mat or web (which may or may not be preimpregnated with resin) on a mold or by hand. Resin is used to impregnate or coat the reinforcement, followed by curing of the resin to permanently fix the formed shape. (13)

Hardness — The resistance of a material to compression, indentation, and scratching. (4)

Haze — The cloudy or turbid appearance of an otherwise transparent sample, caused by the light scattered from within the sample or from its surfaces. (4)

Heat transfer decorating — A process in which the image is transferred from the carrier film to the product by stamping with rigid or flexible shapes, using heat and pressure. (19)

Heated-tool bonding — A method of joining plastics by simultaneous application of heat and pressure to areas in contact. Heat may be applied by conduction or dielectrically. (18)

High-pressure laminates — Laminates formed and cured at pressures higher than 7 000 kPa (1 015 psi). (12)

Hobbing — Forming cavities for multiple molds by forcing a hardened-steel shape, called a hob, into soft steel or beryllium-copper cavity blanks. (22)

Hollow ground — A saw blade that has been specially ground so that the cutting teeth are the thickest portion to prevent binding in the kerf. (9)

Homopolymer — A polymer consisting of like monomer structures. (2, 1)

Honeycomb — A manufactured product of metal, paper, or other materials that is resin-impregnated and has been formed into hexagonal-shaped cells. Used as a core material for sandwich or laminated construction. (3)

Hopper dryer — A combination feeding and drying device for extrusion and injection molding of thermoplastics. (10)

Hot-gas welding — A technique of joining thermoplastics materials in which the materials are softened by a jet of hot air from a welding torch and joined together at the softened points. Generally, a thin rod of the same material is used to fill and consolidate the gap. (18)

Hot-leaf stamping — Decorating operation for marking plastics in which a metal leaf or paint is stamped with heated metal dies onto the face of the plastics. Ink compounds can also be used. (19)

Hot melt — A general term referring to thermoplastic synthetic resins composed of 100 percent solids and used as adhesives at temperatures between 248 and 392 °F [120 °C and 200 °C]. (3)

Hydraulics — Referring to the branch of science that deals with liquids in motion; the transmission, control, or flow of energy by liquids. (23)

Hydrocarbon — An organic compound containing only carbon and hydrogen and often occurring in petroleum, natural gas, coal, and bitumens. (1)

Hydrocarbon plastics — Plastics based on resins made by polymerizing monomers composed solely of carbon and hydrogen. (2)

Hygroscopic — Tending to absorb and retain moisture. (23)

Impact strength — The ability of a material to withstand shock loading. (4)

Impingement — A method of mixing in which two or more materials collide. (10)

Impregnate — To provide liquid penetration into a porous or fibrous material; the dipping or immersion of a fibrous substrate into a liquid resin. Generally, the porous material serves as a reinforcement for the plastic binder after curing. (12, 13)

Impregnation — The process of thoroughly soaking a material such as wood, paper, or fabric with synthetic resin, so that the resin gets within the body of the material. (3)

Incineration — Burning of waste in a specially designed enclosed chamber. (5)

Index of refraction (refractive index) — The ratio of the velocity of light in a vacuum to its velocity in a transparent sample. It is expressed as the ratio of the sine of the angle of incidence to the sine of the angle of refraction. The index of refraction of a substance usually varies with the wavelength of the refracted light. (4)

Induction bonding — High-frequency electromagnetic fields are used to excite the molecules of metallic inserts placed in the plastics or in the interfaces, thus fusing the plastics. The inserts remain in the joint. (18)

Industrial plastics — A plastics waste generated by various industrial sectors. (3)

Inert (rare) gases — Gases that do not combine with other elements; helium, argon, neon, krypton, xenon, and radon (Group 8 of the periodic table). (1, 2)

Inhalation — The breathing in of a substance in the form of a gas, vapor, fume, mist, or dust. (3)

Inhibitor — A substance that slows down a chemical reaction. Inhibitors sometimes used in certain monomers and resins to prolong storage life. (6)

Initiation phase — The first of three steps in addition polymerization. Refers to producing a reactive state of the molecules, usually be some high-energy source catalysts, or radiation. (2)

Initiator — An agent necessary to cause polymerization, especially in emulsion polymerization processes. (6)

Injection blow molding — A blow-molding process in which the parison to be blown is formed by injection molding. (11)

Injection molding — A molding procedure in which a heat-softened plastics is forced from a cylinder into a relatively cool cavity which gives the item the desired shape. (10)

In-mold decorating — Making decorations or patterns on molded products by placing the pattern or image in the mold cavity before the actual molding cycle. The pattern becomes part of the plastics item as it is fused by heat and pressure. (19)

In-situ foaming — The technique of depositing a foamable plastics into the place where foaming will take place. (16)

Inspection — A term used to indicate that during manufacture of a part, personnel will conduct visual examinations of materials, placement of plies, gauge readings, etc. (4)

Interference — The negative allowance used to assure a tight shrink or press fit. (18)

Interlaminar shear — The shear strength at rupture in which the plane of fracture is located between the layers of reinforcement of a laminate. (12)

Intermolecular forces — Secondary valence, or van der Waals, forces between different molecules. (2)

Interpenetrating polymner network (IPN) — An entangled combination of two cross-linked polymers that are not bonded to each other. (2)

Ion — An atom or group of atoms with a positive or negative electrical charge. (2)

Ionic bonding — Atomic bonding by electrical attraction of unlike ions. (2)

Ionomer — A polymer with ethylene as its major component, but containing both covalent and ionic bonds. The polymer exhibits very strong interchain ionic forces. These resins have high transparency, resilience, tenacity, and many of the characteristics of polyethylene. (4, 7)

Irradiation — As applied to plastics, refers to bombardment with a variety of kinds of ionizing and nonionizing radiation. Irradiation has been used to begin polymerization and copolymerization of plastics, and in some cases, to bring about changes in the physical properties of a plastics. (20)

Isocyanate resins — Resins synthesized from isocyanates and alcohols. Most uses are based on their combination with polyols. (8)

Isomers — Molecules with the same chemical composition but different structures. (2)

Isotatic stereoisomerism — A sequence of regularly spaced asymmetric atoms arranged in like configuration in a polymer chain. (2)

Isotope — One of a group of nuclides having the same atomic number but differing atomic mass. (20)

Isotropic — Properties of a material are equal in all directions. (1)

Jig — An appliance for accurately guiding and locating tools during the making of interchangeable parts. (22)

Kerf — The slit or notch made by a saw or cutting tool. (9)

Kirksite — An alloy of aluminum and zinc used for molds. It has high thermal conductivity. (22)

Knife coating — A method of coating a substrate by an adjustable knife or bar set at a suitable angle to the substrate. (17)

Lac — A dark-red resinous substance deposited by scale insects on the twigs of trees; used in making shellac. (1)

Lamellar — Sheetlike or platelike in shape. Referring to the aligned, looped molecular structure of crystalline polymers. (2)

Laminar composite — Referring to a composite composed of layers of materials held together by the polymer matrix. They are divided into two classes: laminates and sandwiches. (5)

Laminated plastics — A dense, tough solid produced by bonding together layers of sheet materials impregnated with a resin and curing them by application of heat, or heat and pressure. (12)

Laminated vapor plating — The process of vacuum-metalizing alternate layers of metal coatings on a polymer substrate. (12, 19)

Laminates — Two or more layers of material bonded together. The term usually applies to preformed layers joined by adhesives or by heat and pressure. The term also applies to composites of plastics films with other films, foil, and paper, even though they have been made by spread coating or by extrusion coating. A reinforced laminate usually refers to superimposed layers of resin-impregnated or resin-coated fabrics or fibrous reinforcements that have been bonded, especially by heat and pressure. When the bonding pressure is at least 7 000 kPa [1 015 psi] the product is called a high-pressure laminate. Products pressed at pressures under 7 000 kPa are called low-pressure laminates. Products produced with little or no pressure, such as hand lay-ups, filament-wound structures, and spray-ups, are sometimes called contact-pressure laminates. (3, 12)

Laminating — The process of producing a composite laminate. (5)

Lamination—The process of preparing a laminate; also, any layer in a laminate. (12)

Laser cutting—A means of cutting materials by laser energy. (9)

Latex—An emulsion of natural or synthetic resin particles dispersed in a watery medium. (1)

LD—Lethal dose; a concentration of a substance being tested which will kill a test animal. (3)

Leaching—Removing a soluble component from a polymer mix with solvents. (16)

Limits—The maximum and minimum dimensions that define the tolerance. (18)

Linear—Refers to a long straight-chain molecule, as contrasted with one having many side chains or branches. (2)

Liquid Reaction Molding (LRM)—Same as reaction injection molding. (10)

Low-pressure laminates—In general, laminates molded and cured at pressures ranging from 0.4 ksi (2.8 MPa) to contact pressure. (12)

Lubrication bloom—An irregular, cloudy, greasy film on a plastics surface caused by excess lubricants. (6)

Luminescence—Light emission by the radiation of photons after initial activation. Luminescent pigments are activated by ultraviolet radiation, producing very strong luminescence. (6)

Macromolecules—The large (giant) molecules that make up the high polymers. (2)

MM/RIM—Mat molding reaction injection molding. (10)

Mandrel—A form around which a filament-wound and pultruded composite structures are shaped. (13)

Mass—The quantity of matter; the physical amount of matter. When gravity acts on a mass of matter we say it has weight. See also weight. (2)

Matched-mold forming—Forming hot sheets between matched male and female molds. (13)

Matrix—The polymer material used to bind the reinforcements together in a composite. (5)

Mechanical fastening—Mechanical means of joining plastics with machine screws, self-tapping screws, drive screws, rivets, spring clips, clips, dowels, catches, or other devices. (18)

Mechanical forming—Heated sheets of plastics are shaped or formed by hand or with the aid of jigs and fixtures. No mold is used. (15)

Mer—The smallest repetitive unit in a polymer. (2)

Methyl methacrylate—A colorless, volatile liquid derived from acetone cyanohydrin, methanol, and dilute sulphuric acid and used in production of acrylic resins. (7)

m/s—metres per second. (9)

Microballoons—Hollow glass spheres. (6)

Mold—The cavity or matrix in which plastics are formed. Also, to shape plastics or resins into finished items by heat or heat and pressure. (22)

Molding compounds—Plastics or resin materials in varying stages of formulation (powder, granular, or preform) comprising resin, filler, pigments, plasticizers, or other ingredients ready for use in the molding operation. (3)

Molecular mass—The sum of the atomic mass of all atoms in a molecule. In high polymers, the molecular masses of individual molecules vary widely, so that they must be expressed as averages. Average molecular mass of polymers may be expressed as number-average molecular mass (Mn) or mass-average molecular weight (Mw). Molecular mass measurement methods include osmotic pressure, light scattering, solution pressure, solution viscosity, and sedimentation equilibrium. (2)

Molecule—The smallest particle of a substance that can exist independently while retaining the chemical identity of the substance. (2)

Monohydric—Containing one hydroxyl (OH) group in the molecule. (7)

Monomers—A simple molecule capable of reacting with like or unlike molecules to form a polymer; the smallest repeating structure of a polymer, also called a mer. (1)

NIOSH—National Institute for Occupational Safety and Health of the Public Health Service, U.S. Department of Health and Human Services (DHHS); federal agency that recommends occupational exposure limits for various substances and assists OSHA and MSHA in occupational safety and health investigations and research. (3)

NFPA—National Fire Prevention Association. (5)

Negative rake—The angle of the face of a cutting tool ground so that the cutting end of a tool with positive rake is more blunt than that of a tool with no rake (rake angle equals zero). (9)

Nitrocellulose (cellulose nitrate)—Material formed by the action of a mixture of sulphuric acid and nitric acid on cellulose. The cellulose nitrate used for Celluloid manufacture usually contains 10.8 to 11.1 percent nitrogen. (1)

Novolac—A phenolic-aldehyde resin that remains permanently thermoplastic unless a source of methylene groups is added. (8)

Nuisance plastics—Waste plastics that cannot be reprocessed under the existing technoeconomic conditions. (3)

OSHA — Occupational Safety and Health Administration of the U.S. Department of Labor; federal agency with safety and health regulatory and enforcement authorities for most U.S. industry and business. (3, 5)

Offset printing — A printing technique in which ink is transferred from the printing plate to a roller. Subsequently, the roller transfers the ink to the object to be printed. (19)

Oil-hardened steel — Steel that is cooled by an oil bath. (22)

Open-celled — Referring to the interconnecting of cells in cellular or foamed plastics. (3)

Open dumping — Placing trash or waste in open, uncontrolled areas on the land. (5)

Open-grit sandpaper — Coarse sandpaper (number 80 or less). (9)

Organosol — A dispersion usually of vinyl or polyamide, in a liquid phase containing one or more organic solvents. (3)

Oxygen index — A test for the minimum oxygen concentration in a mixture of oxygen and nitrogen that will support a flame of a burning polymer. (3)

Parameters — A term used loosely to denote a specified range of variables, characteristics, or properties relating to the subject being discussed; also, an arbitrary constant. (4, 21)

Parison — The hollow plastics tube from which a product is blow-molded. (11)

Particulate — Small particles with various shapes and sizes used to reinforce a polymer matrix. (5)

Parting lines — Marks on a molding or casting where halves of the mold meet in closing. (21)

Pellets — One of the many formulations of molding compounds. (3)

Periodic table — An arrangement of the elements in order of increasing atomic number, forming groups the members of which show similar physical and chemical properties. (2)

Phenolic — A synthetic resin produced by the condensation of an aromatic alcohol with an aldehyde, particularly of phenol with formaldehyde. (8)

Phenoxy — A high-molecular-mass thermoplastic polyester resin based on bisphenol A and epichlorohydrin. (7)

Phosphorescence — Luminescence that lasts for a period after excitation. (6)

Photon — The least amount of electromagnetic energy that can exist at a given wavelength. A quantum of light energy is analogous to the electron. (6, 20)

Photosynthesis — Refers to the synthesis of chemicals with the aid of radiant energy from the light of the sun. (1)

Pinch-off — A raised edge around the cavity in the mold that seals off the part and separates the excess material as the mold closes around the parison. (11)

Plastic — An adjective, meaning pliable and capable of being shaped by pressure. Plastic is often incorrectly used as the generic word for the plastics industry and its products. (1)

Plasticizer — Chemical agent added to plastics to make them softer and more flexible. (6)

Plastics — A noun; an organic substance, usually synthetic or semisynthetic, that can be formed into various shapes by heat and pressure and retain those shapes after heat and pressure have been removed. In its finished state, it is a rigid or flexible (but not elastic) solid containing a polymer of high molecular mass (weight). (1)

Plastics alloy — Plastics made by physically mixing two or more polymers during the melt. (2)

Plastics tooling — Tools, dies, jigs, or fixtures primarily for the metal-forming trades, constructed of plastics. (Usually laminates or casting materials). (22)

Plastic strain — The strain permanently given to a material by stresses that exceed the elastic limit. (4)

Platens — The mounting plates of a press to which the mode assembly is bolted. (10)

Polar winding — A winding in which the filament path passes tangent to the polar opening at the other end. A one-circuit pattern is inherent in the system. (13)

Polyacrylate — A thermoplastic resin made by the polymerization of an acrylic compound. (7)

Polyallomer — Crystalline polymers produced from two or more olefin monomers. (7)

Polyamide — A polymer in which the structural units are linked by amide or thioamide groupings. (7)

Polyblend — Plastics that have been modified by the addition of an elastomer. (2)

Polycarbonate — Polymers derived from the direct reaction between aromatic and aliphatic dihydroxy compounds with phosgene or by the ester exchange reaction with appropriate phosgene-derived precursors. (7)

Polyester — A resin formed by the reaction between a dibasic acid and a dihydroxy alcohol, both organic. Modification with multifunctional acids or acids and bases and some unsaturated reactants permits crosslinking to thermosetting resins. Polyesters modified with fatty acids are called alkyds. (8)

Polyethylene — A thermoplastic material composed of polymers of ethylene. One of polyolefin family. (7)

Polyimide — A group of resins made by reacting pyromellitic dianhydride with aromatic diamines. The polymer is characterized by rings of four carbon atoms tightly bound together. (7)

Polymer — A compound with high molecular mass (weight), either natural or synthetic, whose structure can be represented by a repeated small unit (the mer). Some polymers are elastic and some are plastics. (1)

Polymerization — The process of growing large molecules from small ones. (2)

Polymerization Reaction — A chemical reaction in which the molecules of a monomer are linked together to form large molecules. (1)

Polymethyl methacrylate — See Methyl methacrylate.

Polymethylpentene — An isotactically arranged aliphatic polyolefin of 4-methyl-pentene-1. (7)

Polyolefin — A term used to indicate a family of polymers produced from hydrocarbons with double carbon-to-carbon bonds. Includes polyethylene, polypropylene, polymethylpentene. (7)

Polyphenylene oxide — Currently made as a polyether of 2,6-dimethyl-phenol by an oxidative coupling process involving air or pure oxygen in the presence of a copper-amine complex catalyst. (7)

Polypropylene — A plastics material made by the polymerization of high-purity propylene gas in the presence of an organometallic catalyst at relatively low pressures and temperatures. One of the polyolefin family. (7)

Polystyrene — A thermoplastic material produced by the polymerization of styrene (vinyl benzene). (7)

Polysulfone — A thermoplastic consisting of benzene rings connected by a sulfone group (SO_2), an isopropylidene group, and an ether linkage. (7)

Polyurethane — A family of resins produced by reacting diisocyanate with organic compounds containing two or more active hydrogens to form polymers having free isocyanate groups. These groups, under the influence of heat or certain catalysts, will react with each other, or with water, glycols, or other materials to form a thermoset. (8)

Polyvinyls — A broad family of plastics derived from the vinyl group ($CH_2 = H —$). (7)

Pneumatics — A branch of science dealing with the mechanical properties of gases. (23)

Postconsumer plastics waste — A plastics waste generated by a consumer. (3)

Potting — An embedding process for parts, similar to encapsulating except that the object may be simply covered and not surrounded by an envelope of plastics. Normally considered a coating process. (8, 14)

ppm — Parts per million; a way of expressing tiny concentrations. In air, ppm is usually a volume/volume ratio; in water, a weight/volume ratio. (3)

Pre-expanded — When polymer beads or granules are partially expanded prior to molding into cellular parts. (16)

Preplasticator — The act of softening material before forcing it into the mold, another molding machine, or accumulator. (10)

Preplasticizer — The act of adding a softening agent (plasticizer) before molding. (6)

Prepuffs — The pre-expanded pieces of polymers used to make cellular polymer parts. (16)

Pressure-bag molding — A process for molding reinforced plastics, in which a tailored flexible bag is placed over the contact layup on the mold, sealed, and clamped in place. Compressed air forces the bag against the part to apply pressure while the part cures. (13)

Primary bonds — A strong association (interatomic attraction) between atoms. (2)

Primary colors — The basic colors from which all others are made. (6)

Primary recycling — The processing of scrap plastics into the same or similar types of product from which it has been generated, using standard plastics processing methods. (3)

Promoter — A chemical, itself a feeble catalyst, that greatly speeds up the activity of a given catalyst. (2, 6)

Propagation phase — The second step in addition polymerization. Refers to rapid growth or addition of monomer units to the molecular chain. (2)

Proportional limit — The greatest stress that a material can sustain without deviation from proportionality of stress and strain (Hooke's law); the point at which elastic strain becomes plastic strain. It is expressed in force per unit area (Pa). (4)

Proprietary molds — Molds made and owned by the molder. (23)

Proton — A positively charged particle in the nucleus of an atom. Its charge is equal to but the opposite of that of the electron. (2)

Pultrusion — A continuous process for manufacturing composites with a constant cross-sectional shape. The process consists of pulling a fiber reinforcing material through a resin-impregnation bath and into a shaping die where the resin is subsequently cured. (13)

Purging — Cleaning one color or type of material from the cylinder of a molding machine. (10)

Pyrometer — A device used to measure thermal radiation. (23)

Pyrolysis — Chemical decomposition of a substance by heat, and pressure used to change waste into usable compounds. (5)

Quality control — A procedure to determine if a product is being manufactured to specifications; a technique of management for achieving quality. Inspection is part of that technique. (21, 23)

Quarternary recycling—The recovery of energy from waste plastics. (3)

Radiation—See Irradiation.

Radical—A group of atoms of different elements that behave as a single atom in chemical reactions. (7)

Rare gases—See inert gases.

Reaction injection molding (RIM)—The molding process in which two or more liquid polymers are mixed by impingement-atomizing in a mixing chamber, then injected into a closed mold. (10)

Recycling—Collection and reprocessing of waste materials. (5)

Reinforced molding compound—A material reinforced with special fillers, fibers, or other materials to meet design needs. (13)

Reinforced plastics—Plastics with strength increased by addition of filler and reinforcing fibers, fabrics, or mats to the base resin. (13)

Relative density—The density of any material divided by that of water at a standard temperature, usually 68 or 73 °F [20° or 23 °C]. Since water's density is nearly 1.00 g/cm^3, density in grams per cubic centimeter and relative density are numerically equal. (4)

Rennin—An enzyme of gastric juice that causes the coagulation of milk. (8)

Resin—Gum-like solid or semisolid substances that may be obtained from certain plants and trees or made from synthetic materials. (1)

Resin-rich area—Localized area filled with resin and lacking reinforcing material. (13)

Resin transfer molding (RTM)—The transfer of catalyzed resin into an enclosed mold in which the fiber reinforcing has been placed. Also called *resin-injection molding* and *liquid resin molding (LRM)*. (10)

rpm—revolutions per minute. (9)

r/s—revolutions per second. (9)

Rib—A reinforcing member of a fabricated or molded part. (21)

Rotational casting—A method used to make hollow objects from plastisols or powders. The mold is charged and rotated in one or more planes. The hot mold fuses the substance into a gel during rotation, covering all surfaces. The mold is then chilled and the product removed. (14)

Roving—A bundle of untwisted strands, usually of fibrous glass. (6)

Runners—Channels through which plastics flow from the sprue to the gates of mold cavities. (10, 21)

Safety factor—The ratio of the ultimate strength of the material to the allowable working stress. (21)

Sandwich—A class of laminar composites composed of a lightweight core material (honeycomb, foamed plastics, etc.) to which two thin, dense, high-strength faces or skins are adhered. (5, 12)

Sandwich construction—A structure consisting of relatively dense, high-strength facings bonded to a less dense, lower-strength intermediate material or core. (12)

Sandwich panel—Panel consisting of two thin face sheets bonded to a thick, lightweight, honeycomb or foam core. (12)

Sanitary landfill—The controlled filling of lowlands or trenches with solid waste. (3)

Saturated compounds—Organic compounds that do not contain double or triple bonds and thus cannot add elements or compounds. (8)

Scleroscope—An instrument for measuring impact resilience by dropping a ram with a flattened-cone tip from a given height onto the sample and then noting the height of rebound. (4)

Scrap plastics—Waste plastics that are capable of being reprocessed into commercially acceptable plastics products. (3)

Secondary bonds—Forces of attraction, other than primary bonds, that cause many molecules to join. (2)

Secondary color—A color obtained by mixing two or more primary colors. (6)

Secondary recycling—The processing of scrap plastics into plastics products with less demanding properties. (3)

Self-extinguishing—A somewhat loosely used term describing the ability of a material to cease burning once the source of flame has been removed. (5)

Shell—A region about the nucleus of an atom in which electrons move; each electron shell corresponds to a definite energy level. (2)

Shellac—A natural polymer; refined lac, a resin usually produced in thin, flaky layers or shells and used in varnish and insulating materials. (1)

Shrink fit—A joining method in which an insert is put into a plastics part while that part is hot. Shrink fitting takes advantage of the fact that plastics expand when heated and shrink when cooled. The plastics is normally heated and the insert placed in an undersized hole. On cooling, the plastics shrinks around the insert. (18)

Shrink wrapping—A technique of packaging in which the strains in a plastics film are released by raising the temperature of the film. This causes it to shrink over the package. (15)

Silicosis—A disease of the lungs caused by the inhalation of silica dust. (13)

Sintering—Forming items from fusible powders. The

process of holding the pressed powder at a temperature just below its melting point. (10)

Slush casting — Resin in liquid or powder form is poured into a hot mold, where a viscous skin forms. The excess slush is drained off, the mold is cooled, and the casting removed. (14)

Snap-back forming — A technique in which a plastics sheet is stretched to a bubble shape by vacuum or air pressure, a male mold is inserted into a bubble and the vacuum or air pressure is released allowing the plastics to snap-back over the mold. (15)

Solid waste — Refuse that does not rot or decay, such as dirt, concrete, bricks, and many plastics. (5)

Solution polymerization — A process where inert solvents are used to cause monomer solutions to polymerize. (2)

Solvent — A substance, usually a liquid, in which other substances are dissolved. The most common solvent is water. (6, 13)

Solvent resistance — The ability of a plastics material to withstand exposure to a solvent. (4)

Specification — A statement of a set of requirements to be satisfied by a product, material, process, or system indicating (when appropriate) the procedure by which it may be determined whether the requirements are satisfied. Specifications may cite standards, be expressed in numerical terms, and include contractual agreements or requirements between the buyer and seller. (21, 22, 23)

Spherulite — A rounded aggregate of radiating crystals with fibrous appearance. Spherulites are present in most crystalline plastics and may range in diameter from a few tenths of a micron to several millimeters. (2)

Spinneret — A type of extrusion die with many tiny holes. A plastics melt is forced through the holes to make fine fibers and filaments. (11)

Spin bonding or welding — A process of fusing two objects by forcing them together while one or both are spinning, until frictional heat melts the interface. Spinning is then stopped and pressure held until the parts are frozen together. (18)

Sprayup — A general term covering several processes using a spray gun. In reinforced plastics, the term applies to the simultaneous spraying of resin and chopped reinforcing fibers onto the mold or mandrel. (13)

Sprue — In the mold, the channel or channels through which the plastics is led to the mold cavity. (21)

Stabilizer — An ingredient used in the formulation of some plastics (especially elastomers) to assist in holding the physical and chemical properties of the compounded materials at their initial values throughout the processing and service life of the material. (6)

Standard — A document or an object for physical comparison to define nomenclature, concepts, processes, materials, dimensions, relationships, interfaces, or test methods. (21, 23)

Stereoisomerism — The arrangement of molecular chains in a polymer. *Atactic* pertains to an arrangement that is more or less random. *Isotactic* pertains to a structure containing a sequence of regularly spaced asymmetric atoms arranged in like configuration in a polymer chain. *Syndiotactic* pertains to a polymer molecule in which groups of atoms that are not part of the primary backbone structure alternate regularly on opposite sides of the chain. (2)

Stiffness — The capacity of a material to resist a bending force. (4)

Strain — The ratio of the elongation to the gauge length of a test sample; that is, the change in length per unit of original length. (4)

Strand — A bundle of filaments. (6)

Strength-to-mass ratio — Pertaining to materials that are strong for their mass (weight). The strength/density value of a material. (4)

Stress — The force producing, or tending to produce, deformation of a substance. Expressed as the ratio of applied load to the original cross-sectional area. (4)

Structural foam — Cellular plastics with integral skin. (16)

Substrate — A material onto which an adhesive or similar substance is applied. (3)

Suspension polymerization — A process in which liquid monomers are polymerized as liquid droplets suspended in water. (2)

Syndiotatic stereoisomer — A polymer molecule in which atoms that are not part of the primary structure alternate regularly on opposite sides of the chain. (2)

Syntactic foam — Cellular resins or plastics with low-density fillers. (19)

Synthetic — Materials produced by chemical means, rather than natural origin. (1)

TLV — Threshold limit value; a term used by ACGIH to express the airborne concentration of a material to which nearly all persons can be exposed day after day, without adverse effects. ACGIH expresses TLVs in three ways: 1) *TLV-TWA:* the allowable time weighted average concentration for a normal 8-hour work day or 40-hour work week. 2) *TLV-STEL:* the short-term exposure limit, or maximum concentration for a continuous 15-minute exposure period (maximum of four such periods per day, with at least 60 minutes between exposure periods, and provided that the daily TLV-TWA is not exceeded). 3) *TLV-C:*

the ceiling limit—the concentration that should not be exceeded even instantaneously. (3)

Tampo-Print—A process of transferring ink from an engraved ink-filled surface to a product surface by the use of a flexible printing (transfer) pad. (19)

Technology—The science of efficient application of scientific knowledge. (1)

Termination phase—The last of three steps in addition polymerization. Refers to ending molecular growth of polymers by adding chemicals. (2)

Tertiary recycling—The recovery of chemicals from waste plastics. (3)

Testing—A term that implies that methods or procedures are used to determine physical, mechanical, chemical, optical, electrical, or other properties of a part. (4)

Tex—An ISO standard unit of linear density used as a measure of yarn count. One tex is the linear density of a fabric that has a mass of 1 g and a length of 1 km and is equal to 10^{-6} kg/m. (6)

Thermal Expansion Resin Transfer Molding (TERTM)—A variation of the RTM process in which, after the resin is injected, heat causes the cellular core material to expand forcing reinforcements and matrix against the mold walls. (10)

Thermoforming—Any process of forming thermoplastic sheet that consists of heating the sheet and pulling it down onto a mold surface. (15)

Thermoplastic—(adj.) Capable of being repeatedly softened by heat and hardened by cooling. (n) A linear polymer that will repeatedly soften when heated and harden when cooled. (1)

Thermoset or *Thermosetting*—A network polymer that will undergo or has undergone a chemical reaction by the action of heat, catalysts, ultraviolet light, etc., leading to a relatively infusible state. (1)

Thixotropy—State of materials that are gel-like at rest but fluid when agitated. Liquids containing suspended solids are apt to be thixotropic. (4, 6)

Tools—Special fixtures, molds, dies, or other devices that enable a manufacturer to produce parts. (22)

Tool steels—Steels used to make cutting tools and dies. Many of these steels have considerable quantities of alloying elements such as chromium, carbon, tungsten, molybdenum, and other elements. They form hard carbides that provide good wearing qualities but at the same time decrease machinability. Tool steels in the trade are classified for the most part by their applications, such as hot die, cold work die, high speed, shock resisting, mold, and special purpose steels. (22)

Toughness—A term with a wide variety of meanings, no single mechanical definition being generally recognized. Represented by the energy required to break a material, equal to the area under the stress-strain curve. (4)

Toxicity—The degree of danger posed by a substance to animal or plant life. (3)

Toxic Substance—Any substance which can cause acute or chronic injury to the human body, or which is suspected of being able to cause diseases or injury under some conditions. (3)

Trade name—A name given to a product to make it easy to recognize, spell, and pronounce. In the plastics industry, a trade name is used by the manufacturer to identify a particular resin or product. (1)

Transfer molding—A method of molding plastics where the material is softened by heat and pressure in a transfer chamber, then forced by high pressure through the sprues, runners, and gates in a closed mold for final curing. (10)

Tripoli—A silica abrasive. (9)

Tumbling—Finishing operation for small plastics articles. Gates, flash, and fins are removed and surfaces are polished by rotating the parts in a barrel or on a belt together with wooden pegs, sawdust, and some polishing compounds. (9)

USDA—U.S. Department of Agriculture; prior to 1971, USDA performed tests and issued approvals on respirators for use with pesticides. In 1971, the Bureau of Mines took over the pesticide respirator testing/approval functions/procedures later delegated to the Testing and Certification Branch (TCB) of NIOSH. (3)

Ultrasonic bonding—A method of joining using vibratory mechanical pressure at ultrasonic frequencies. Electrical energy is changed to ultrasonic vibrations through the use of either a magetostrictive or piezoelectric transducer. The ultrasonic vibrations generate frictional heat to melt the plastics allowing them to join. (18)

Undercut—Having a protuberance or indention that impedes withdrawal from a two-piece rigid mold. Flexible materials can be ejected intact with slight undercuts. (21)

Vacuum forming—Method of sheet forming in which the edges of the plastics sheet are clamped in a stationary frame, and the plastics is heated and drawn down by a vacuum into a mold. (15)

Vacuum metalizing—A process in which surfaces are thinly coated by exposing them to a metal vapor under vacuum. (17)

Valence electrons—The electrons in the outermost shell of an atom. (2)

Van der Waals' forces — Weak secondary interatomic attraction arising from internal dipole effects. (2)

Vibrational microlamination (VIM) — A casting process where heated molds are vibrated in a bed of polymer pellets or powder. (14)

Vicat softening point — The temperature at which a flat-ended needle of 1 mm^2 circular or square cross section will penetrate a thermoplastic specimen to a depth of 1 mm under a specified load using a uniform rate of temperature rise (Definition from ASTM D-1525.) (4)

Viscosity — A measure of the internal friction resulting when one layer of fluid is caused to move in relationship to another layer. (4)

Warp — The lengthwise direction of the weave in cloth or roving; also the dimensional distortion of a plastic object. See Weft. (6)

Waste plastics — A plastics resin or product that must be reprocessed or disposed of. (3)

Water pollution — Waste products discharged into rivers and waterways. (5)

Weft — The transverse threads or fibers in a woven fabric; those fibers running perpendicular to the warp; also called *fill, filler, yarn, woof, pick*. (6)

Weight — Weight is the force exerted by a mass as a result of gravity. In common usage, weight is used as a synonym for mass. The word "weight" should be avoided in technical practice, and replaced by gravitational force acting on the object, measured in newtons. (2)

Whiskers — Single crystals used as reinforcements. (6)

Whiting — Calcium carbonate powder abrasive. (9)

Yarn — Bundle of twisted strands. (6)

ABBREVIATIONS FOR
SELECTED MATERIALS

Abbreviation	Polymer Term or Generic Name
ABS	Acrylonitrile-butadiene-styrene
ACS	Acrylonitrile-chlorinated polyethylene-styrene
AES	Acrylonitrile-ethylpropylene-styrene
AI	Amide-imide polymers
AMMA	Acrylonitrile-methyl-methacrylate
AN	Acrylonitrile
AP	Ethylene propylene
ASA	Acrylic-styrene-acrylonitrile
AU	Polyester polyurethane
BFK	Boron fiber reinforced plastic
BMC	Bulk molding compounds
CA	Cellulose acetate
CAB	Cellulose acetate-butyrate
CAP	Cellulose acetate propionate
CAR	Carbon fiber
CF	Cresol-formaldehyde
CFRP	Carbon fiber reinforced plastics
CMC	Carboxymethyl cellulose
CN	Cellulose nitrate
CP	Cellulose propionate
CPE	Chlorinated polyethylene
CPET	Crystallized PET
CPVC	Chlorinated polyvinyl chloride
CS	Casein
CTFE	(Poly) Chlorotrifluoro-ethylene
DAIP	Diallyl isophthalate resin
DAP	Diallyl phthalate resin
DCPD	(Poly) Dicyclopentadiene
DMC	Dough molding compound
DP	Degree of polymerization
EC	Ethyl cellulose
ECTFE	Ethylene-chlorotrifluoroethylene
EEA	Ethylene-ethyl acrylate
EMA	Ethylene-methyl acrylate
EP	Epoxy
EPDM	Ethylene propylene diene rubber
EPE	Epoxy resin ester
EP-G-G	Prepreg of epoxy resin and glass fabric
EPM	Ethylene propylene copolymer
EPR	Ethylene and propylene copolymer
EPS	Expanded polystyrene
ETFE	Ethylene-tetrafluoroethylene
EU	Polyether polyurethane
EVA	Ethylene vinyl acetate
EVOH	Ethylene vinyl alcohol
FEP	Fluorinated ethylene-propylene (Also PFEP)
FRP	Glass fiber reinforced polyester
FRTP	Fiberglass reinforced thermoplastics
GF	Glass fiber reinforced
GF-EP	Glass fiber reinforced epoxy resin
GR	Glass fiber reinforced
GRP	Glass reinforced plastics

Abbreviation	Polymer Term or Generic Name
HIPS	High-impact polystyrene
HMW-HDPE	High molecular weight-high density polyethylene
IPN	Interpenetrating polymer network
LCP	Liquid crystal polymers
LDPE	Low-density polyethylene
LIM	Liquid impingement molding
LLDPE	Linear low-density polyethylene
LRM	Liquid reaction molding
MF	Melamine formaldehyde
OPP	Oriented polypropylene
OPVC	Oriented polyvinylchloride
OSA	Olefin-modified styrene-acrylonitrile
PA	Polyamide
PAA	Polyacrylic acid
PAI	Polyamide-imide
PAN	Polyacrylonitrile
PAPI	Polymethylene polyphenyl isocyanate
PBAN	Polybutadiene-acrylonitrile
PBS	Polybutadiene-styrene
PBT	Polybutylene terephthalate
PC	Polycarbonate
PCTFE	Polymonochlorotrifluoroethylene
PDAP	Polydiallyl phthalate
PE	Polyethylene
PEEK	Polyetheretherketone
PEI	Polyetherimide
PES	Polyether sulfone
PET	Polyethylene terephthalate
PF	Phenol-formaldehyde resin
PFA	Perfluoroalkoxy
PFEP	Polyfluoroethylenepropylene
PI	Polyimide
PMCA	Polymethylchloroacrylate
PMMA	Polymethyl methacrylate
POM	Polyoxymethylene
PP	Polypropylene
PPC	Polyphthalate carbonate
PPE	Polyphenylene ether
PPO	Polyphenylene oxide
PPSO	Polyphenylsulfone
PPS	Polystyrene
PSO	Polysulfone
PTFE	Polytetrafluoroethylene
PTMT	Polytetramethylene terephthalate
PU	Polyurethane
PUR	Polyurethane rubber
PVAc	Polyvinyl acetate
PVAl	Polyvinyl alcohol
PVB	Polyvinyl butyral
PVC	Polyvinyl chloride
PVDC	Polyvinylidene chloride
PVDF	Polyvinylidene fluoride

ABBREVIATIONS FOR
SELECTED MATERIALS (continued)

Abbreviation	Polymer Term or Generic Name	Abbreviation	Polymer Term or Generic Name
PVF	Polyvinyl fluoride	TPX	Polymethylpentene
SAN	Styrene-acrylonitrile	UF	Urea-formaldehyde
SBP	Styrene-butadiene plastics	UHMWPE	Ultra-high molecular weight polyethylene
SI	Silicone	UP	Urethane plastics
SMA	Styrene-maleic anhydride	VCP	Vinyl chloride-propylene
SMC	Sheet molding compounds	VDC	Vinylidene chloride
SRP	Styrene-rubber plastics	VLDPE	Very low density polyethylene
TPE	Thermoplastic elastomer		

Appendix C

TRADE NAMES AND MANUFACTURERS

Trade Name	Polymer	Manufacturer
Abasfil	Reinforced ABS	Dart Industries, Inc.
Absinol	ABS	Allied Resinous Products, Inc.
Abson	ABS resins and compounds	BF Goodrich Chemical Co.
Acelon	Cellulose acetate film	May & Baker, Ltd.
Acetophane	Cellulose acetate film	UCB-Sidac
Aclar	CTFE fluorohalo-carbon films	Allied Chemical Corp.
Acralen	Ethylene-vinyl acetate polymer	Verona Dyestuffs Div. Verona Corp.
Acrilan	Acrylic (acrylontrile-vinyl chloride)	Monsanto Co.
Acroleaf	Hot stamping foil	Acromark Co.
Acrylaglas	Fiber-glass rein-forced styrene-acrylonitrile	Dart Industries, Inc.
Acrylicomb	Acrylic-sheet-faced honeycomb	Dimensional Plastics Corp.
Acrylite	Acrylic molding compounds; cast arylic sheets	American Cyanamid Co.
Acryloid	Acrylic modifiers for PVC; coating resins	Rohm & Haas Co.
Acrylux	Acrylic	Westlake Plastics Co.
Aeroflex	Polyethylene extrusions	Anchor Plastics Co.
Aeron	Plastic-coated nylon	Flexfilm Products, Inc.
Aerotuf	Polypropylene extrusions	Anchor Plastics Co.
Afcolene	Polystyrene and SAN copolymers	Pechiney-Saint-Gobain
Afcoryl	ABS copolymers	Pechiney-Saint-Gobain
Alathon	Polyethylene resins	E. I. du Pont de Nemours & Co.
Alfane	Thermosetting epoxy resin cement	Atlas Minerals & Chemicals Div., of ESB Inc.
Alpha-Clan	Reactive monomer	Marvon Div., Borg-Warner Corp.
Alphalux	PPO	Marbon Chemical Co.

TRADE NAMES AND MANUFACTURERS

Trade Name	Polymer	Manufacturer
Alsynite	Reinforced plastic panels	Reichhold Chemicals, Inc.
Amberlac	Modified alkyd resins	Rohm & Haas Co.
Amberol	Phenolic and maleic resins	Rohm & Haas Co.
Amer-Plate	PVC sheet material	Ameron Corrosion Control Div.
Ampol	Cellulose acetates	American Polymers, Inc.
Amres	Thermosetting liquid resins	Pacific Resins & Chemicals, Inc.
Ancorex	ABS extrusions	Anchor Plastics Co.
Anvyl	Vinyl extrusions	Anchor Plastics Co.
Apogen	Epoxy resin series	Apogee Chemical, Inc.
Araclor	Polychlorinated polyphenyls	Monsanto Co.
Araldite	Epoxy resins and hardeners	CIBA Products Co.
Armorite	Vinyl coating	John L. Armitage & Co.
Arnel	Cellulose triacetate fiber	Celanese Corp.
Arochem	Modified phenolic resins	Ashland Chemical Co.
Arodure	Urea resins	Ashland Chemical Co.
Arofene	Phenolic resins	Ashland Chemical Co.
Aroplaz	Alkyd resins	Ashland Chemical Co.
Aroset	Acrylic resins	Ashland Chemical Co.
Arothane	Polyester resin	Ashland Chemical Co.
Artfoam	Rigid urethane foam	Strux Corp.
Arylon	Polyarl ether compounds	Uniroyal, Inc.
Arylon T	Polyaryl ether	Uniroyal, Inc.
Ascot	Coated spunbonded polyolefin sheet	Appleton Coated Paper Co.
Astralit	Vinyl copolymer sheets	Dynamit Nobel of America, Inc.
Astroturf	Nylon, polyethylene	Monsanto Co.
Atlac	Polyester resin	Atlas Chemical Industries, Inc.
Averam	Inorganic	FMC Corp.
Avisco	PVC films	FMC Corp.
Avistar	Polyester film	FMC Corp.
Avisun	Polypropylene	Avisun Corp.

Trade Name	Polymer	Manufacturer
Bakelite	Polyethylene, ethylene copolymers, epoxy, phenolic, polystyrene, phenoxy, ABS and vinyl resins and compounds	Union Carbide Corp.
Beetle	Urea molding compounds	American Cyanamid Co.
Betalux	TFE-filled acetal	Westlake Plastics Co.
Blanex	Crosslinked polyethylene compounds	Reichhold Chemicals, Inc.
Blapol	Polyethylene compounds and color concentrates	Reichhold Chemicals, Inc.
Blapol	Polyethylene molding and extrusion compounds	Blane Chemical Div., Reichhold Chemicals, Inc.
Blendex	ABS resin	Marbon Div., Borg-Warner Corp.
Bolta Flex	Vinyl sheeting and film	General Tire & Rubber Co., Chemical/Plastics Div.
Bolta Thene	Rigid olefin sheets	General Tire & Rubber Co., Chemical/Plastics Div.
Boltaron	ABS or PVC rigid plastic sheets	General Tire & Rubber Co., Chemical/Plastics Div.
Boronal	Polyolefins with boron	Allied Resinous Products
Bostik	Epoxy and polyurethene adhesives	Bostik-Finch, Inc.
Bronco	Supported vinyl or pyroxylin	General Tire & Rubber Co., Chemical/Plastics Div.
Budene	Polybutadiene	Goodyear Tire & Rubber Co., Chemical/Plastics Div.
Butaprene	Styrene-butadiene latexes	Firestone Plastics Co. Div., Firestone Tire & Rubber
Cadco	Plastics rod, sheet, tubing, and film	Cadillac Plastic & Chemical Co.
Capran	Nylon 6 film	Allied Chemical Corp.
Capran	Nylon films and sheet	Allied Chemical Corp.
Carbaglas	Fiber-glass reinforced polycarbonate	Fiberfil Div., Dart Industries, Inc.
Carolux	Filled urethane foam, flexible	North Carolina Foam Industries, Inc.
Carstan	Urethane foam catalysts	Cincinnati Milacron Chemicals, Inc.
Castcar	Cast polyolefin films	Mobil Chemical Co.

Trade Name	Polymer	Manufacturer
Castethane	Castable molding urethane elastomer system	Upjohn Co., CPR Div.
Castomer	Urethane elastomer system	Baxenden Chemical Co.
Castomer	Urethane elastomer and coatings	Isocyanate Products Div., Witco Chemical Corp.
Celanar	Polyester film	Celanese Plastics. Co.
Celanex	Thermoplastic polyester	Celanese Plastics Co.
Celcon	Acetal copolymer resins	Celanese Plastics Co.
Cellasto	Microcellular urethane elastomer parts	North American Urethanes, Inc.
Cellofoam	Polystyrene foam	US Mineral Products Co.
Cellonex	Cellulose acetate	Dynamit Nobel of America, Inc.
Celluliner	Resilient expanded polystyrene foam	Gilman Brothers Co.
Cellulite	Expanded polystyrene foam	Gilman Brothers Co.
Celpak	Rigid polyurethane foam	Dacar Chemical Products Co.
Celthane	Rigid polyurethane foam	Decar Chemical Products
Chem-o-sol	PVC plastisol	Chemical Products Corp.
Chem-o-thane	Polyurethane elastomer casting compounds	Chemical Products Corp.
Chemfluor	Fluorocarbon plastics	Chemplast, Inc.
Chemglaze	Polyurethane-based coating materials	Hughson Chemical Co., Div., Lord Corp.
Chemgrip	Epoxy adhesives for TFE	Chemplast, Inc.
Cimglas	Fiber glass reinforced polyester moldings	Cincinnati Milacron, Molded Plastics Div.
Clocel	Rigid urethane foam system	Baxenden Chemical Co.
Clopane	PVC film and tubing	Clopay Corp.
Cloudfoam	Polyurethane foam	International Foam Div., Holiday Inns of America
Co-Rexyn	Polyester resins and gel coats; pigment pastes	Interplastic Corp., Commercial Resins Div.
Cobocell	Cellulose acetate butyrate tubing	Cobon Plastics Corp.
Coboflon	Teflon tubing	Cobon Plastics Corp.
Cobothane	Ethylene-vinyl acetate tubing	Cobon Plastics Corp.
Colorail	Polyvinyl chloride handrails	Blum, Julius & Co.
Colovin	Calendered vinyl sheeting	Columbus Coated Fabrics

TRADE NAMES AND MANUFACTURERS

Trade Name	Polymer	Manufacturer
Conathane	Polyurethane casting, potting, tooling and adhesive compounds	Conap, Inc.
Conolite	Polyester laminate	Woodall Industries, Inc.
Cordo	PVC foam and films	Ferro Corp., Composites Div.
Cordoflex	Polyvinylidene fluoride solutions, etc.	Ferro Corp., Composites Div.
Corlite	Reinforced foam	Snark Products, Inc.
Coror-Foam	Urethane foam systems	Cook Paint & Varnish
Coverlight HTV	Vinyl-coated nylon fabric	Reeves Brothers, Inc.
Creslan	Acrylic	American Cyanamid Co.
Crystic	Unsaturated polyester resins	Scott Bader Co.
Cumar	Coumarone-indene resins	Neville Chemical Co.
Curithane (Series)	Polyaniline polyamine; organo-mercury catalyst	Upjohn Co., Polymer Chemicals Div.
Curon	Polyurethane foam	Reeves Brothers, Inc.
Cycolac	ABS resins	Marbon Div., Borg-Warner Corp.
Cycolon	Synthetic resinous compositions	Marbon Div., Borg-Warner Corp.
Cycoloy	Alloys of synthetic polymers w/ABS resins	Marbon Div., Borg-Warner Corp.
Cycopac	ABS and nitrile barrier	Borg-Warner Chemicals
Cyovin	Self-extinguishing ABS graft-polymer blends	Marbon Div., Borg-Warner Corp.
Cyglas	Glass-filled polyester molding compound	American Cyanamid Co.
Cymel	Melamine molding compound	American Cyanamid Co.
Dacovin	PVC compounds	Diamond Shamrock Chemical Co.
Dacron	Polyester	E. I. du Pont de Nemours & Co.
Dapon	Diallyl phthalate resin	FMC Corp., Organic Chemicals Div.
Daran	Polyvinylidene chloride emulsion coatings	W. R. Grace & Co., Polymers & Chemicals Div.
Daratak	Polyvinyl acetate homopolymer emulsions	W. R. Grace & Co., Polymers & Chemicals Div.
Darex	Styrene-butadiene latexes	W. R. Grace & Co., Polymers & Chemicals Div.
Davon	TFE resins & reinforced compounds	Davies Nitrate Co.

TRADE NAMES AND MANUFACTURERS

Trade Name	Polymer	Manufacturer
Delrin	Acetal resin	E. I. du Pont de Nemours & Co.
Densite	Molded flexible urethane foam	General Foam Div., Tenneco Chemical, Inc.
Derakane	Vinyl ester resins	Dow Chemical Co.
Dexon	Propylene-acrylic	Exxon Chemical USA
Diaron	Melamine resins	Reichhold Chemicals, Inc.
Dielux	Acetal	Westlake Plastics Co.
Dion-Iso	Isophthalic polyesters	Diamond Shamrock Chemical Co.
Dolphon	Epoxy resin & compounds polyester resins	John C. Dolph Co.
Dorvon	Molded polystyrene foam	Dow Chemical Co.
Dow Corning	Silicones	Dow Corning Corp.
Dri-Lite	Expanded polystyrene	Poly Foam, Inc.
Duco	Lacquers	E. I. du Pont de Nemours & Co.
Duracel	Lacquers for cellulose acetate & other plastics	Maas & Waldstein Co.
Duracon	Acetal copolymer	Polyplastics Co.
Dural	Acrylic modified semirigid PVC	Alpha Chemical & Plastics Corp.
Duramac	Oil modified alkyds	Commercial Solvents Corp.
Durane	Polyurethane	Raffi & Swanson, Inc.
Duraplex	Alkyd resins	Rohm & Haas Co.
Durelene	PVC flexible tubing	Plastic Warehousing Corp.
Durethene	Polyethylene film	Sinclair-Koppers Co.
Durez	Phenolic resins	Hooker Chemical & Plastics Corp.
Duron	Phenolic resins & molding compounds	Firestone Foam Products Co.
Dyal	Alkyd and styrenated-alkyd resins	Sherwin Williams Chemicals
Dyalon	Urethane elastomer material	Thombert, Inc.
Dyfoam	Expanded polystyrene	W. R. Grace & Co.
Dylan	Polyethylene	ARCO/Polymers, Inc.
Dylel	ABS plastics	Sinclair-Koppers Co.
Dylene	Polystyrene resin and oriented sheet	Sinclair-Koppers Co.
Dylite	Expandable polystyrene bead, extruded sheets, etc.	Sinclair-Koppers Co.
E-Form	Epoxy molding compounds	Allied Products Corp.
Easy-Kote	Fluorocarbon release compound	Borco Chemicals, Inc.

TRADE NAMES AND MANUFACTURERS

Trade Name	Polymer	Manufacturer
Easypoxy	Epoxy adhesive kits	Conap, Inc.
Ebolan	TFE compounds	Chicago Gasket Co.
Eccosil	Silicone resins	Emerson & Cuming, Inc.
El Rexene	Polyethylene, polypropylene, polystyrene and ABS resins	Dart Industries, Inc.
Elastolit	Urethane engineering thermoplastic	North American Urethanes, Inc.
Elastollyx	Urethane engineering thermoplastic	North American Urethanes, Inc.
Elastolur	Urethane coatings	BASF
Elastonate	Urethane isocyanate prepolymers	BASF
Elastonol	Urethane polyester polyols	North American Urethanes, Inc.
Elastopel	Urethane engineering thermoplastics	North American Urethanes, Inc.
Electroglas	Cast arylic	Glasflex Corp.
Elvace	Acetate-ethylene copolymers	E. I. du Pont de Nemours & Co.
Elvacet	Polyvinyl acetate emulsions	E. I. du Pont de Nemours & Co.
Elvacite	Acrylic resins	E. I. du Pont de Nemours & Co.
Elvamide	Nylon resins	E. I. du Pont de Nemours & Co.
Elvanol	Polyvinyl alcohols	E. I. du Pont de Nemours & Co.
Elvax	Vinyl resins; acid terpolymer resins	E. I. du Pont de Nemours & Co.
Ensocote	PVC lacquer coating	Uniroyal, Inc.
Ensolex	Cellular plastic sheet material	Uniroyal, Inc.
Ensolite	Cellular plastic sheet material	Uniroyal, Inc.
Epi-Rez	Basic epoxy resins	Celanese Coatings Co.
Epi-Tex	Epoxy ester resins	Celanese Coatings Co.
Epikote	Epoxy resin	Shell Chemical Co.
Epocap	Two-part epoxy compounds	Hardman, Inc.
Epocast	Epoxies	Furane Plastics, Inc.
Epocrete	Two-part epoxy materials	Hardman, Inc.
Epocryl	Epoxy acrylate resin	Shell Chemical Co.
Epocure	Epoxy curing agents	Hardman, Inc.
Epolast	Two-part epoxy compounds	Hardman, Inc.
Epolite	Epoxy compounds	Hexcel Corp., Rezolin Div.
Epomarine	Two-part epoxy compounds	Hardman, Inc.
Epon	Epoxy resin; hardener	Shell Chemical Co.
Eponol	Linear polyether resin	Shell Chemical Co.

TRADE NAMES AND MANUFACTURERS

Trade Name	Polymer	Manufacturer
Eposet	Two-part epoxy compounds	Hardman, Inc.
Epotuf	Epoxy resins	Reichold Chemicals, Inc.
Estane	Polyurethane resins and compounds	BF Goodrich Chemical Co.
Estron	Acetate	Eastman Kodak Co.
Ethafoam	Polyethylene foam	Dow Chemical Co.
Ethocel	Ethyl cellulose resin	Dow Chemical Co.
Ethofil	Fiber-glass reinforced polyethylene	Fiberfil Div., Dart Industries, Inc.
Ethoglas	Fiber-glass reinforced polyethylene	Fiberfil Div., Dart Industries, Inc.
Ethosar	Fiber-glass reinforced polyethylene	Fiberfil Div., Dart Industries, Inc.
Ethylux	Polyethylene	Westlake Plastics Co.
Evenglo	Polystyrene resins	Sinclair-Koppers Co.
Everflex	Polyvinyl acetate co-polymer emulsion	W. R. Grace & Co., Polymers & Chemicals Div.
Everlon	Urethane foam	Stauffer Chemical Co.
Excelite	Polyethylene tubing	Thermoplastic Processes
Exon	PVC resins, compounds and latexes	Firestone Tire & Rubber Co.
Extane	Polyurethane tubing	Pipe Line Service Co.
Extrel	Polyethylene & polypropylene films	Exxon Chemical, USA
Extren	Fiber-glass reinforced polyester shapes	Morrison Molded Fiber Glass Co.
Fabrikoid	Pyroxylin-coated fabric	Stauffer Chemical Co.
Facilon	Reinforced PVC fabrics	Sun Chemical Corp.
Fassgard	Vinyl coating on nylon	M. J. Fassler & Co.
Fasslon	Vinyl coating	M. J. Fassler & Co.
Felor	Nylon filaments	E. I. du Pont de Nemours & Co.
Fiber foam	Polyester-reinforced foam	Weeks Engineered Plastics
Fibro	Rayon	Courtaulds NA, Inc.
Flexane	Urethanes	Devcon Corp.
Flexocel	Urethane foam systems	Baxenden Chem Co.
Floranier	Cellulose for esters	ITT Rayonier, Inc.
Fluokem	Teflon spray	Bel-Art Products
Fluon	TFE resin	ICI American, Inc.
Fluorglas	PTFE coated and impregnated woven glass fabric, laminates, belting	Dodge Industries, Inc.
Fluorocord	Fluorocarbon material	Raybestos Manhattan
Fluorofilm	Cast Teflon films	Dilectrix Corp.

TRADE NAMES AND MANUFACTURERS

Trade Name	Polymer	Manufacturer
Fluoroglide	Dry-film lubricant of TFE	Chemplast, Inc.
Fluororay	Filled fluorocarbon	Raybestos Manhattan
Fluorored	Compounds of TFE	John L. Dore Co.
Fluorosint	TFE-fluorocarbon base composition	Polymer Corp.
Foamthane	Rigid polyurethane foam	Pittsburgh Corning Corp.
Formadall	Polyester premix compound	Woodall Industries, Inc.
Formaldafil	Fiber-glass reinforced acetal	Fiberfil Div., Dart Industries, Inc.
Formaldaglas	Fiber-glass reinforced acetal	Fiberfil Div., Dart Industries, Inc.
Formaldasar	Fiber-glass reinforced acetal	Fiberfil Div., Dart Industries, Inc.
Formica	High-pressure laminate	American Cyanamid Co.
Formrez	Urethane elastomer chemicals	Witco Chemical Corp., Organics Div.
Formvar	Polyvinyl formal resins	Monsanto Co.
Forticel	Cellulose propionate flake, resins	Celanese Plastics Co.
Fortiflex	Polyethylene resins	Celanese Plastics Co.
Fortrel	Polyester	Fiber Industries, Inc.
Fosta-Net	Polystyrene foam extruded mesh	Foster Grant Co.
Fosta Tuf-Flex	High-impact polystyrene	Foster Grant Co.
Fostacryl	Thermoplastic polystyrene resins	Foster Grant Co.
Fostafoam	Expandable polystyrene beads	Foster Grant Co.
Fostalite	Light-stable polystyrene molding powder	Foster Grant Co.
Fostarene	Polystyrene molding powder	Foster Grant Co.
Futron	Polyethylene powder	Fusion Rubbermaid Co.
Gelva	Polyvinyl acetate	Monsanto Co.
Genal	Phenolic compounds	General Electric Co.
Genthane	Polyurethane rubber	General Tire & Rubber Co.
Gentro	Styrene butadiene rubber	General Tire & Rubber Co.
Geon	Vinyl resins, compounds latexes	BF Goodrich Chemical Co.
Gil-Fold	Polyethylene sheet	Gilman Brothers Co.
Glaskyd	Alkyd molding compound	American Cyanamid Co.
Glyptal	Alkyd resins	General Electric Co.
Gordon Superdense	Polystyrene in pellet form	Hammond Plastics, Inc.

TRADE NAMES AND MANUFACTURERS

Trade Name	Polymer	Manufacturer
Gordon Superflow	Polystyrene in granular or pellet form	Hammond Plastics, Inc.
Gracon	PVC compounds	W. R. Grace & Co.
GravoFLEX	ABS sheets	Hermes Plastics, Inc.
GravoPLY	Acrylic sheets	Hermes Plastics, Inc.
Halon	TFE molding compounds	Allied Chemical Corp.
Haylar	CTFE	Allied Chemical Corp.
Haysite	Polyester laminates	Synthane-Taylor Corp.
Herculon	Olefin	Hercules, Inc.
Herox	Nylon filaments	E. I. du Pont de Nemours & Co.
Hetrofoam	Fire retardant urethane foam systems	Durez Div., Hooker Chemical Corp.
Hetron	Fire retardant polyester resins	Durez Div., Hooker Chemical Corp.
Hex-One	High-density polyethylene	Gulf Oil Co.
Hi-fax	Polyethylene	Hercules, Inc.
Hi-Styrolux	High-impact polystyrene	Westlake Plastics
Hydrepoxy	Water-based epoxies	Acme Chemicals Div., Allied Products Corp.
Hydro Foam	Expanded phenolformaldehyde	Smithers Co.
Implex	Acrylic molding powder	Rohm & Haas Co.
Intamix	Rigid PVC compounds	Diamond Shamrock Chemical Co.
Interpol	Copolymeric resinous systems	Freeman Chemical Corp.
Irvinil	PVC resins and compounds	Great American Chem.
Isoderm	Urethane rigid and flexible intergral-skinning foam	Upjohn Co., CPR Div.
Isofoam	Urethane foam systems	Witco Chemical Corp.
Isonate	Diisocyanates and urethane systems	Upjohn Co., CPR Div.
Isoteraglas	Isocyanate elastomer-coated Dacron-glass fabric	Natvar Corp.
Isothane	Flexible polyurethane foams	Bernel Foam Products Co.
Jetfoam	Polyurethane foam	International Foam
K-Prene	Urethane cast material	Di-Acro Kaufman
Kalex	Two-part polyurethane elastomers	Hardman, Inc.
Kalspray	Rigid urethane foam system	Baxenden Chemical Co.

TRADE NAMES AND MANUFACTURERS

Trade Name	Polymer	Manufacturer
Kapton	Polyimide	E. I. du Pont de Nemours & Co.
Keltrol	Vinyl toluene copolymer	Spencer Kellogg
Ken-U-Thane	Polyurethanes; urethane foam ingredients	Kenrich Petrochemicals, Inc.
Kencolor	Silicone/pigments dispersion	Kenrich Petrochemicals, Inc.
Kodacel	Cellulosic film and sheeting	Eastman Chemical Products, Inc.
Kodar	Copolyester thermoplastics	Eastman Chemical Products, Inc.
Kodel	Polyester	Eastman Kodak Co.
Kohinor	Vinyl resins and compounds	Pantasote Co.
Korad	Acrylic film	Rohm & Haas Co.
Koroseal	Vinyl films	BF Goodrich Chemical Co.
Kralastic	ABS high-impact resin	Uniroyal, Inc.
Kralon	High-impact styrene and ABS resins	Uniroyal, Inc.
Kraton	Styrene-butadiene polymers	Shell Chemical Co.
Krene	Plastic film and sheeting	Union Carbide Corp.
Krystal	PVC sheet	Allied Chemical Corp.
Krystaltite	PVC shrink films	Allied Chemical Corp.
Kydene	Acrylic/PVC powder	Rohm & Haas Co.
Kydex	Acrylic/PVC sheets	Rohm & Haas Co.
Kynar (Series)	Polyvinylidene fluoride	Pennwalt Corp.
Lamabond	Reinforced polyethylene	Lamex, Columbian Carbon Co.
Lamar	Mylar vinyl laminate	Morgan Adhesives Co.
Laminac	Polyester resins	American Cyanamid Co.
Last-A-foam	Plastic foam	General Plastics Mfg.
Lexan	Polycarbonate resins, film, sheet	General Electric Co., Plastics Dept.
Lucite	Acrylic resins	E. I. du Pont de Nemours & Co.
Lumasite	Acrylic sheet	American Acrylic Corp.
Lustran	SAN and ABS molding and extrusion resins	Monsanto Co.
Lustrex	Polystyrene molding and extrusion resins	Monsanto Co.
Lycra	Spandex	E. I. du Pont de Nemours & Co.
Macal	Cast vinyl film	Morgan Adhesives Co.
Marafoam	Polyurethane foam resin	Marblette Co.

TRADE NAMES AND MANUFACTURERS

Trade Name	Polymer	Manufacturer
Maraglas	Epoxy casting resin	Marblette Co.
Maraset	Epoxy resin	Marblette Co.
Marathane	Urethane Compounds	Allied Products Corp.
Maraweld	Epoxy resin	Marblette Co.
Marlex	Polyethylenes, polypropylenes, other polyolefin plastics	Phillips Petroleum
Marvinol	Vinyl resins and compounds	Uniroyal, Inc.
Meldin	Polyimide and reinforced polyimide	Dixon Corp.
Merlon	Polycarbonate	Mobay Chemical Co.
Metallex	Cast acrylic sheets	Hermes Plastics, Inc.
Meticone	Silicone rubber dies and sheets	Hermes Plastics, Inc.
Metre-Set	Epoxy adhesives	Metachem Resins Corp.
Micarta	Thermosetting laminates	Westinghouse Electric Corp.
Micro-Matte	Extruded acrylic sheet with matte finish	Extrudaline, Inc.
Micropel	Nylon powders	Nypel, Inc.
Microsol	Vinyl plastisol	Michigan Chrome & Chemical Co.
Microthene	Powdered polyolefins	U.S. Industrial Chemicals Co.
Milmar	Polyester	Morgan Adhesives Co.
Mini-Vaps	Expanded polyethylene	Malge Co., Agile Div.
Minit Grip	Epoxy adhesives	High-Strength Plastics Corp.
Minit Man	Epoxy adhesive	Kristal Draft, Inc.
Mipoplast	Flexible PVC sheets	Dynamit Nobel of America, Inc.
Mirasol	Alkyd resins: epoxy ester	C. J. Osborn Chemicals, Inc.
Mirbane	Amino resin	Showa Highpolymer Co.
Mirrex	Calendered rigid PVC	Tenneco Chemicals, Inc., Tenneco Plastics Div.
Mista Foam	Urethane foam systems	M. R. Plastics & Coatings, Inc.
Mod-Epox	Epoxy resin modifier	Monsanto Co.
Molycor	Glass-fiber reinforced epoxy tubing	A. O. Smith, Inland, Inc.
Mondur	Isocyanates	Mobay Chemical Co.
Monocast	Direct Polymerized nylon	Polymer Corp.
Moplen	Isotactic Polypropylene	Montecatini Edison S.p.A.
Multrathane	Urethane elastomer chemical	Mobay Chemical Co.
Multron	Polyesters	Mobay Chemical Co.
Mylar	Polyester film	E. I. du Pont de Nemours & Co.

TRADE NAMES AND MANUFACTURERS

Trade Name	Polymer	Manufacturer
Napryl	Polypropylene	Pechiney-Saint-Gobain
Natene	High-density polyethylene	Pechiney-Saint-Gobain
Naugahyde	Vinyl coated fabrics	Uniroyal, Inc.
NeoCryl	Acrylic resins and resin emulsions	Polyvinyl Chemicals, Inc.
NeoRez	Styrene emulsions and urethane solutions	Polyvinyl Chemicals, Inc.
NeoVac	PVA emulsions	Polyvinyl Chemicals, Inc.
Nestorite	Phenolic and urea-formaldehyde	James Ferguson & Sons
Nevillac	Modified coumarone-indene resin	Neville Chemical Co.
Nimbus	Polyurethane foam	General Tire & Rubber Co.
Nitrocol	Nitrocellulose base pigment dispersion	C. J. Osborn Chemicals, Inc.
Nob-Lock	PVC sheet material	Ameron Corrosion Control Div.
Nopcofoam	Urethane foam systems	Diamond Shamrock Chemical Co., Resinous Products Div.
Norchem	Low-density polyethylene resin	Northern Petrochemical Co.
Noryl	Modified polyphenylene oxide	General Electric Co., Plastics Dept.
Nupol	Thermosetting acrylic resins	Freeman Chemical Corp.
Nyglathane	Glass-filled polyurethane	Nypel, Inc.
Nylafil	Fiber-glass reinforced nylon	Fiberfil Div., Dart Industries, Inc.
Nylaglas	Fiber-glass reinforced nylon	Fiberfil Div., Dart Industries, Inc.
Nylasar	Fiber-glass reinforced nylon	Fiberfil Div., Dart Industries, Inc.
Nylasint	Sintered nylon parts	Polymer Corp.
Nylatron	Filled nylons	Polymer Corp.
Nylo-Seal	Nylon 11 tubing	Imperial-Eastman Corp.
Nylux	Nylon	Westlake Plastics Co.
Nypelube	TFE-filled Nylons	Nypel, Inc.
Nyreg	Glass-reinforced Nylon molding compounds	Nypel, Inc.
Oasis	Expanded phenol-formaldehyde	Smithers Co.
Oilon Pv 80	Acetal-based resin sheets rods, tubing, profiles	Cadillac Plastic & Chemical Co.
Olefane	Polypropylene film	Amoco Chemicals Corp.
Olefil	Filled polypropylene resin	Amoco Chemicals Corp.
Oleflo	Polypropylene resin	Amoco Chemicals Corp.
Olemer	Copolymer polypropylene	Amoco Chemicals Corp.
Oletac	Amorphous polypropylene	Amoco Chemicals Corp.
Opalon	Flexible PVC materials	Monsanto Co.
Oppanol	Polyisobutylene	BASF Wyandotte Corp.
Orgalacqe	Epoxy and PVC powders	Aquitaine-Organico
Orgamide R	Nylon 6	Aquitaine-Organico
Orlon	Acrylic fiber	E.I. du Pont de Nemours & Co.
Panda	Vinyl and urethane coated fabric	Pandel-Bradford Inc.
Papi	Polymethylene polyphenylisocyanate	Upjohn Co., Polymer Chemicals Div.
Paradene	Dark coumarone-indene resins	Neville Chemical Co.
Paraplex	Polyester resins and plasticizers	Rohm & Haas Co.
Pelaspan	Expandable polystyrene	Dow Chemical Co.
Pelaspan-Pac	Expandable polystyrene	Dow Chemical Co.
Pellethane	Thermoplastic urethane	Upjohn Co., Polymer Chemicals Div.
Pellon Aire	Nonwoven textile	Pellon Corp.
Penton	Chlorinated polyether	Hercules, Inc.
PermaRex	Cast epoxy	Permali, Inc.
Permelite	Melamine molding compound	Melamine Plastics, Inc.
Petra	Polyester sheet	Allied Chemical Corp.
Pethrothene	Low-, medium-, and high-, density polyethylene	U.S. Industrial Chemical Co.
Petrothene XL	Crosslinkable polyethylene	U.S. Industrial Chemical Co.
Phenoweld	Phenolic adhesive	Hardman, Inc.
Philjo	Polyolefin films	Phillips-Joana Co.
Philprene	Styrene-butadiene	Phillips Chemical Corp.
Piccoflex	Acrylontrile-styrene resins	Pennsylvania Industrial Chemical Corp.
Piccolastic	Polystryrene resins	Pennsylvania Industrial Chemical Corp.
Piccotex	Vinyl-toluene copolymer	Pennsylvania Industrial Chemical Corp.
Piccoumaron	Coumarone-indene resins	Pennsylvania Industrial Chemical Corp.
Piccovar	Alkyl-aromatic resins	Pennsylvania Industrial Chemical Corp.
Pienco	Polyester resins	Mol-Rex Div., American Petrochemical Corp.
Pinpoly	Reinforced polyurethane foam	Holiday Inns of America, Inc.
Plaskon	Plastic molding compounds	Allied Chemical Corp.

TRADE NAMES AND MANUFACTURERS

Trade Name	Polymer	Manufacturer
Plastic Steel	Epoxy tooling and repair	Devcon Corp.
Pleogen	Polyester resins and gel coats; polyurethane systems	Mol-Rex Div., Whittaker Corp.
Plexiglas	Acrylic sheets and molding powders	Rohm & Haas Co.
Plicose	Polyethylene film, sheeting, tubing, bags	Diamond Shamrock Corp.
Pliobond	Adhesive	Goodyear Tire & Rubber Co.
Pliolite	Styrene-butadiene resins	Goodyear Tire & Rubber Co.
Pliothene	Polyethylene-rubber blends	Ametek/Westchester Plastics
Pliovic	PVC resins	Goodyear Tire & Rubber Co.
Pluracol	Polyethers	BASF Wyandotte Corp.
Pluragard	Urethane foams	BASF Wyandotte Corp.
Pluronic	Polyethers	BASF Wyandotte Corp.
Plyocite	Phenolic-impregnated overlays	Reichhold Chemicals, Inc.
Plyophen	Phenolic resins	Reichhold Chemicals, Inc.
Polex	Oriented acrylic	Southwestern Plastics, Inc.
Pollopas	Urea-formaldehyde compounds	Dynamit Nobel of America, Inc.
Polvonite	Cellular plastic material in sheet form	Voplex Corp.
Poly-Dap	Diallyl phthalate electrical molding compounds	U.S. Polymeric, Inc.
Poly-Eth	Low-density polyethylene	Gulf Oil Corp.
Poly-Eth-Hi-D	High-density polyethylene	Gulf Oil Corp.
Polycarbafil	Fiber-glass reinforced polycarbonate	Fiberfil Div., Dart Industries, Inc.
Polycure	Crosslinked polyethylene compounds	Crooke Color & Chemical Co.
Polyfoam	Polyurethane foam	General Tire & Rubber Co.
Polyimidal	Thermoplastic polyimide	Raychem Corp.
Polylite	Polyester resins	Reichhold Chemicals, Inc.
Polymet	Plastic-filled sintered metal	Polymer Corp.
Polymul (series)	Polyethylene emulsions	Diamond Shamrock Chemical Co.
Polyteraglas	Polyester-coated Dacron-glass fabric	Natvar Corp.

TRADE NAMES AND MANUFACTURERS

Trade Name	Polymer	Manufacturer
Polywrap	Plastic film	Flex-O-Glass, Inc.
Poxy-Gard	Solventless epoxy compounds	Sterling, Div. Reichhold Chemicals, Inc.
PPO	Polyphenylene oxide	Reichhold Chemicals, Inc.
Pro-fax	Polypropylene	Hercules, Inc.
Profil	Fiber-glass reinforced polypropylene	Fiberfil Div., Dart Industries, Inc.
Proglas	Fiber-glass reinforced polypropylene	Fiberfil Div., Dart Industries, Inc.
Prohi	High-density polyethylene	Protective Lining Corp.
Propathene	Polypropylene polymers and compound	Imperial Chemical Ind., Ltd., Plastics Div.
Propylsar	Fiber-glass reinforced polypropylene	Fiberfil Div., Dart Industries, Inc.
Propylux	Polypropylene	Westlake Plastics Co.
Protectolite	Polyethylene film	Protective Lining Corp.
Protron	Ultrahigh-strength polyethylene	Protective Lining Corp.
Purilon	Rayon	FMC Corp.
Quelflam	Urethanes, low surface spread flame	Baxenden Chemical Co.
Rayflex	Rayon	FMC Corp.
Regalite	Press polished clear flexible PVC	Tenneco Advanced Materials, Inc.
REN-Shape	Epoxy material	Ren Plastics, Inc.
Ren-Thane	Urethane elastomers	Ren Plastics, Inc
Resiglas	Polyester resins, etc.	Kristal Draft, Inc.
Resimene	Melamine resins	Monsanto Co.
Resinol	Polyolefins	Allied Resinous Products, Inc.
Resinox	Phenolic resins	Monsanto Co.
Resorasabond	Resorcinol and phenol-resorcinol	Pacific Resins & Chemicals, Inc.
Restfoam	Urethane foam	Stauffer Chemical Co., Plastics Div.
Rexolene	Crosslinked polyolefin sheet	Brand-Rex Co.
Rexolite	Polystyrene rod and sheet stock	Brand-Rex Co.
Reynosol	Urethane, PVC	Hoover Ball & Bearing Co.
Rhodiod	Cellulose acetate sheet	M & B Plastics, Ltd.
Rhoplex	Acrylic emulsion	Rohm & Haas Co.
Richfoam	Urethane foam	E. R. Carpenter Co.
Rigidite	Modified acrylic and sheet polyester resins	American Cyanamid Co.
Rigidsol	Rigid plastisol	Watson-Standard Co.
Rolox	Two-part epoxy compounds	Hardman, Inc.

TRADE NAMES AND MANUFACTURERS

Trade Name	Polymer	Manufacturer
Royalex	Structural cellular thermoplastic sheet material	Uniroyal, Inc.
Royalite	Thermoplastic sheet material	Uniroyal Inc., Uniroyal Plastic Products
Roylar	Polyurethane elastoplastic	Uniroyal, Inc.
Rucoam	Vinyl film and sheeting	Hooker Chemical Corp.
Rucoblend	Vinyl compounds	Hooker Chemical Corp.
Rucon	Vinyl resins	Hooker Chemical Corp.
Rucothane	Polyurethanes	Hooker Chemical Corp.
Ryton	Polyphenylene Sulfide	Phillips Chemical Co.
Santolite	Aryl sulfonamide-formaldehyde resin	Monsanto Co.
Saran	Polyvinylidene chloride resin	Dow Chemical Co.
Satin Foam	Extruded polystyrene foam	Dow Chemical Co.
Scotchpak	Heat-sealable polyester film	3M Co.
Scotchpar	Polyester film	3M Co.
Selectrofoam	Urethane foam systems and polyols	PPG Industries, Inc.
Selectron	Polymerizable synthetic resins; polyesters	PPG Industries, Inc.
Shareen	Nylon	Courtaulds North America, Inc.
Shuvin	Vinyl molding compounds	Blane Chemical Div., Reichhold Chemicals, Inc.
Silastic	Silicone rubber	Dow Corning Corp.
Sipon	Alkyl and Aryl resin	Alcolac, Inc.
Siponate	Alkyl and aryl sulfonates	Alcolac, Inc.
Skinwich	Urethane rigid and flexible integral-skinning foam	Upjohn Co.
Softlite	Ionomer foam	Gilman Brothers Co.
Solarflex	Chlorinated polyethylene	Pantasote Co.
Solithane	Urethane prepolymers	Thiokol Chemical Corp.
Sonite	Epoxy resin compound	Smooth-On, Inc.
Spandal	Rigid urethane laminates	Baxenden Chemical Co.
Spandofoam	Rigid urethane foam board and slab	Baxenden Chemical Co.
Spandoplast	Expanded polystyrene board and slab	Baxenden Chemical Co.
Spectran	Polyester	Monsanto Textiles Co.
Spenkel	Polyurethane resins	Spencer Kellogg Div., Textron Inc.

TRADE NAMES AND MANUFACTURERS

Trade Name	Polymer	Manufacturer
Starez	Polyvinyl acetate resin	Standard Brands Chemical Ind., Inc.
Structoform	Sheet molding compounds	Fiberite Corp.
Stryton	Nylon	Phillips Fibers Corp.
Stylafoam	Coated polystyrene sheet	Gilman Brothers Co.
Stypol	Polyesters	Freeman Chemical Corp., Div., H. H. Robertson Co.
Styrafil	Fiber-glass reinforced polystyrene	Fiberfil Div., Dart Industries, Inc.
Styroflex	Biaxially oriented polystyrene film	Natvar Corp.
Styrofoam	Polystyrene foam	Dow Chemical Co.
Styrolux	Polystyrene	Westlake Plastics Co.
Styron	Polystyrene resin	Dow Chemical Co.
Styronol	Styrene	Allied Resinous Products, Inc.
Sulfasar	Fiber-glass reinforced polysulfone	Fiberfil Div., Dart Industries, Inc.
Sulfil	Fiber-glass reinforced polysulfone	Fiberfil Div., Dart Industries, Inc.
Sunlon	Polyamide resin	Sun Chemical Corp.
Super Aeroflex	Linear polyethylene	Anchor Plastic Co.
Super Coilife	Epoxy potting resin	Westinghouse Electric Corp.
Super Dylan	High-density polyethylene	Sinclair-Koppers Co.
Superflex	Grafted high-impact polystyrene	Gordon Chemical Co.
Superflow	Polystyrene	Gordon Chemical Co.
Sur-Flex	Ionomer film	Flex-O-Glass, Inc.
Surlyn	Ionomer resin	E. I. du Pont de Nemours & Co.
Syn-U-Tex	Urea-formaldehyde and melamine-formaldehyde	Celanese Resins Div., Celanese Coatings Co.
Syntex	Alkyd and polyurethane ester resins	Celanese Resins Div., Celanese Coatings Co.
Syretex	Styrenated alkyd resins	Celanese Resins Div., Celanese Coatings Co.
TanClad	Spray or dip plastisol	Tamite Industries, Inc.
Tedlar	PVF film	E. I. du Pont de Nemours & Co.
Teflon	FEP and TFE fluorocarbon resins	E. I. du Pont de Nemours & Co.
Tenite	Cellulosic compounds	Eastman Chemical Products, Inc.
Tenn Foam	Polyurethane foam	Morristown Foam Corp.
Tere-Cast	Polyester casting compounds	Sterling Div., Reichhold Chemicals, Inc.

TRADE NAMES AND MANUFACTURERS

Trade Name	Polymer	Manufacturer
Terucello	Carboxymethyl cellulose	Showa Highpolymer Company
Tetra-Phen	Phenolic-type resins	Georgia-Pacific Corp. Chemical Div.
Tetra-Ria	Amino-type resins	Georgia-Pacific Corp., Chemical Div.
Tetraloy	Filled TFE molding compounds	Whitford Chemical Corp.
Tetran	Polytetrafluoroethylene	Pennwalt Corp.
Texin	Urethane elastomer molding compound	Mobay Chemical Co.
Textolite	Industrial laminates	General Electric Co., Laminated Products Dept.
Thermalux	Polysulfone	Westlake Plastics Co.
Thermasol	Vinyl plastisols and organosols	Lakeside Plastics International
Thermco	Expanded polystyrene	Holland Plastics Co.
Thorane	Rigid polyurethane foam	Dow Chemical Co.
T-Lock	PVC sheet material	Amercoat Corp.
TPX	Polymethyl pentene	Mitsue Petrochemical Industries
Tran-Stay	Flat polyester film	Transilwrap Co.
Transil GA	Precoated acetate sheets	Transilwrap Co.
Tri-Foil	TFE-coated aluminum foil	Tri-Point Industries, Inc.
Trilon	Polytetrafluoroethylene	Dynamit Nobel of America, Inc.
Triocel	Acetate	Celanese Fibers Marketing Co.
Trolen (series)	Polyethylene and polypropylene sheets	Dynamit Nobel of America, Inc.
Trolitan (series)	Phenol-formaldehyde compounds; boron	Dynamit Nobel of America, Inc.
Trolitrax	Industrial laminates	Dynamit Nobel of America, Inc.
Trosifol	Polyvinyl butyral film	Dynamit Nobel of America, Inc.
Tuffak	Polycarbonate	Rohm & Haas Co.
Tuftane	Polyurethane film and sheet	BF Goodrich Chemical Co.
Tybrene	Acrylonitrile-butadiene-styrene	Dow Chemical Co.
Tynex	Polyamide filaments	E. I. du Pont de Nemours & Co.
Tyril	Styrene-acrylonitrile resin	Dow Chemical Co.
Tyrilfoam	Styrene-acrylonitrile foam	Dow Chemical Co.
Tyrin	Chlorinated polyethylene	Dow Chemical Co.
U-Thane	Rigid insulation board stock urethane	Upjohn Co., CPR Div.

TRADE NAMES AND MANUFACTURERS

Trade Name	Polymer	Manufacturer
Uformite	Urea and melamine resins	Rohm and Haas Co.
Ultramid	Polyamide 6; 6,6; and 6,10	BASF Wyandotte Corp.
Ultrapas	Melamine-formaldehyde compounds	Dynamit Nobel of America, Inc.
Ultrathene	Ethylene-vinyl acetate resins and copolymers	U.S. Industrial Chemicals Co.
Ultron	PVC film and sheet	Monsanto Co.
Unifoam	Polyurethane foam	William T. Burnett & Co.
Unipoxy	Epoxy resins, adhesives	Kristal Kraft, Inc.
Urafil	Fiber-glass reinforced polyurethane	Fiberfil Div., Dart Industries, Inc.
Uraglas	Fiber-glass reinforced polyurethane	Fiberfil Div., Dart Industries, Inc.
Uralite	Urethane compounds	Rezolin Div., Hexcel Corp.
Uramol	Urea-formaldehyde molding compounds	Gordon Chemicals Co.
Urapac	Rigid urethane systems	North American Urethanes, Inc.
Urapol	Urethane elastomeric coating	Poly Resins
Uvex	Cellulose acetate butyrate sheet	Eastman Chemical Products, Inc.
Valox	Thermoplastic polyester	General Electric Co.
Valsof	Polyethylene emulsions	Valchem Div., United Merchants & Mfrs., Inc.
Varcum	Phenolic resins	Reichhold Chemicals, Inc.
Varex	Polyester resins	McCloskey Varnish Co.
Varkyd	Alkyd and modified alkyd resins	McCloskey Varnish Co.
Varkydane	Urethane vehicles	McCloskey Varnish Co.
Varsil	Silicone-coated fiber glass	New Jersey Wood Finishing Co.
V del	Polysulfone resins	Union Carbide Corp.
Vectra	Polypropylene fibers	Exxon Chemical USA
Velene	Styrene-foam laminate	Scott Paper Co., Foam Division
Velon	Film and sheeting	Firestone Plastics Co., Div., Firestone Tire & Rubber Co.
Versel	Thermoplastic polyester	Allied Chemical Corp.
Versi-Ply	Coextruded films	Pierson Industries, Inc.
Vibrathane	Polyurethane elastomer	Uniroyal, Inc.

TRADE NAMES AND MANUFACTURERS

Trade Name	Polymer	Manufacturer
Vibrin-Mat	Polyester-glass molding compound	Marco Chemical Div., W. R. Grace & Co.
Vibro-Flo	Epoxy and polyester coating powders	Armstrong Products Co.
Vinoflex	PVC resins	BASF Wyandotte Corp.
Vitel	Polyester resin	Goodyear Tire & Rubber Co., Chemical Div.
Vithane	Polyurethane resins	Goodyear Tire & Rubber Co., Chemical Div.
Vituf	Polyester resin	Goodyear Tire & Rubber Co., Chemical Div.
Volara	Closed-cell, low-density polyethylene foam	Voltek, Inc.
Volaron	Closed-cell, low density polyethylene foam	Voltek, Inc.
Volasta	Closed-cell, medium-density polyethylene foam	Voltek, Inc.
Voranol	Polyurethane resins	Dow Chemical Co.
Vult-Acet	Polyvinyl acetate latexes	General Latex & Chemical Corp.
Vultafoam	Urethane foam systems	General Latex & Chemical Corp.
Vultathane	Urethane coatings	General Latex & Chemical Corp.
Vycron	Polyester	Beaunit Corp.
Vygen	PVC resin	General Tire & Rubber Co., Chemical/Plastics Div.
Vynaclor	Vinyl chloride emulsion coatings and binders	National Starch & Chemical Corp.
Vynaloy	Vinyl sheet	BF Goodrich Chemical Co.

TRADE NAMES AND MANUFACTURERS

Trade Name	Polymer	Manufacturer
Vyram	Rigid PVC materials	Monsanto Co.
Weldfast	Epoxy and polyester adhesives	Fibercast Co.
Wellamid (series)	Polyamide 6 and 6,6 molding resins	Wellman, Inc., Plastics Div.
Well-A-Meld	Reinforced nylon resins	Wellman, Inc.
Westcoat	Strippable coatings	Western Coating Co.
Whirlclad	Plastic coatings	Polymer Corp.
Whitcon	Fluoroplastic lubricants	Whitford Chemical Corp.
Wicaloid	Styrene-butadiene emulsions	Wica Chemicals, Div. Ott Chemical Co.
Wicaset	Polyvinyl acetate emulsions	Wica Chemicals, Div. Ott Chemical Co.
Wilfex	Vinyl plastisols	Flexible Products Co.
Xylon	Polyamide 6 and 6,6	Fiberfil Div., Dart Industries, Inc.
Zantrel	Rayon	American Enka Co.
Zefran	Acrylic, nylon polyester	Dow Badische Co.
Zelux	Polyethylene films	Union Carbide Corp., Chemicals & Plastics Div.
Zendel	Polyethylene films	Union Carbide Corp., Chemicals & Plastics Div.
Zerlon	Copolymer of acrylic and styrene	Dow Chemical Co.
Zetafin	Ethylene copolymer resins	Dow Chemical Co.
Zytel	Nylon	E. I. du Pont de Nemours & Co.

EXPLOSION CHARACTERISTICS OF SELECTED DUSTS USED IN THE PLASTICS INDUSTRY*

Type of Dust	Ignition Temperature, °C [°F]	Explosibility	Ignition Sensitivity
Cornstarch	400 [752]	Severe	Strong
Wood flour, white pine	470 [878]	Strong	Strong
Acetal, linear	440 [824]	Severe	Severe
Methyl methacrylate polymer	480 [896]	Strong	Severe
Methyl methacrylate-ethyl acrylate-styrene copolymer	440 [824]	Severe	Severe
Methyl methacrylate-styrene-butadiene-acrylonitrile copolymer	480 [896]	Severe	Severe
Acrylonitrile polymer	500 [932]	Severe	Severe
Acrylonitrile-vinyl pyridine copolymer	510 [950]	Severe	Severe
Cellulose acetate	420 [788]	Severe	Severe
Cellulose triacetate	430 [806]	Strong	Strong
Cellulose acetate butyrate	410 [770]	Strong	Strong
Cellulose propionate	460 [860]	Strong	Strong
Chlorinated polyether alcohol	460 [860]	Moderate	Moderate
Tetrafluoroethylene polymer	670 [1238]	Moderate	Weak
Nylon polymer	500 [932]	Severe	Severe
Polycarbonate	710 [1310]	Strong	Strong
Polyethylene, high-pressure process	450 [842]	Severe	Severe
Carboxy polymethylene	520 [968]	Weak	Weak
Polypropylene	420 [788]	Severe	Severe
Polystyrene molding compound	560 [1040]	Severe	Severe
Styrene-acrylonitrile copolymer	500 [932]	Strong	Strong
Polyvinyl acetate	550 [1022]	Moderate	Moderate
Polyvinyl butyral	390 [734]	Severe	Severe
Polyvinyl chloride, fine	660 [1220]	Moderate	Weak
Vinylidene chloride polymer, molding compound	900 [1652]	Moderate	Weak
Alkyd molding compound	500 [932]	Weak	Moderate
Melamine-formaldehyde	810 [1490]	Weak	Weak
Urea-formaldehyde molding compound	460 [860]	Moderate	Moderate
Epoxy, no catalyst	540 [1004]	Severe	Severe
Phenol formaldehyde	580 [1076]	Severe	Severe
Polyethylene terephthalate	500 [932]	Strong	Strong
Styrene-modified polyester-glass-fiber mix	440 [824]	Strong	Strong
Polyurethane foam	510 [950]	Severe	Severe
Coumarone-indene, hard	550 [1022]	Severe	Severe
Shellac	400 [752]	Severe	Severe
Rubber, crude	350 [662]	Strong	Strong
Rubber, synthetic, hard	320 [608]	Severe	Severe
Rubber, chlorinated	940 [1724]	Moderate	Weak

Source: Compiled in part from *The Explosibility of Agricultural Dusts*, R1 5753, and *Explosibility of Dusts Used in the Plastics Industry*, R1 5971, U. S. Department of Interior.

ELASTOMERS

The word "rubber" was given to this elastic material by the English chemist Joseph Priestley. He observed that the material could "rub out" a pencil mark. Today, both natural and synthetic rubber products are known as elastomers. Since World War I, there has been a steady demand for and increased use of synthetic elastomers. Natural rubber has been supplemented but not replaced by synthetic sources.

The latex from the *Hevea brasiliensis* tree has been the main source of natural rubber. There are other potential sources, including guayule, dandelion, goldenrod, osage orange and numerous other plants. The natural rubber molecule can be duplicated by synthetic means.

Since natural and synthetic elastomers (rubbers) have properties not found in other materials, they are of tremendous economic importance. They are blended and copolymerized with other plastics and rubber monomers. The term *elastomer* is used to describe any rubberlike material, whether natural or synthetic.

The ASTM definition of an elastomer is "a polymeric material that at room temperature can be stretched to at least twice (200%) its original length and upon immediate release of the stress will return quickly to approximately its original length."

Elastomers are classified by chemical composition, common name and ASTM symbol, as shown in Table E-1.

ASTM 1418 has adopted the following class codes (last letter of symbol):

- M — Rubbers having a saturated chain of the poly methylene type
- N — Rubbers having nitrogen in the polymer chain
- O — Rubbers having oxygen in the polymer chain
- R — Rubbers having an unsaturated carbon chain (natural and synthetic)
- Q — Rubbers having silicone in the polymer chain
- T — Rubbers having sulfur in the polymer chain

TABLE E-1.
Common Elastomers

Monomer or Compound	Common Name	ASTM Symbol D-1418
Acrylate-butadiene		ABR
Bromoisobutene-isoprene		BIIR
Butadiene	Polybutadiene	BR
Chlorinated polyethylene		CM
Chloroisobutene-isoprene		CIIR
Chloroprene	Neoprene	CR
Chlorosulfonated polyethylene	Hypalon	CSM
Chlorotrifluoroethylene-vinylidene fluoride	Fluoroplastic	FKM
Diisocyanates (polyester and polyether)	Polyurethane or urethane	AV, EV
Epichlorohydrin (homo-polymer and copolymer)		CO, ECO
Ethylene-propylene copolymer	EPM and EP rubber	EPM
terpolymer	EPDM	EPDM
Fluorosilicone		FVMQ
Isobutene-isoprene	Butyl	IIR
Isoprene	Natural rubber or polyisoprene	IR
Natural rubber		PBR
Nitrile-butadiene		NBR
Nitrile-chloroprene		NCR
Nitrile-isoprene		NR
Polyacrylate	Polyacrylate or acrylic rubbers	ACM
Pyridine-butadiene		PBR
Pyridine-styrene-butadiene		PSBR
Silicone (methyl group; phenyl and methyl group; vinyl and mehtl group; phenyl, vinyl, and methyl group)	Silicone	MQ, PMQ, VMQ, PVMQ
Sodium tetrasulfide-ethylene-dichloride	Polysulfide (Thiokol)	T
Styrene-butadiene	SBR (GRS, old)	SBR
Styrene-chloroprene		SCR
Styrene-isoprene		SIR
Vinylidene fluoride-hexafluoropropylene	Fluoroplastics	FKM

- U — Rubbers having carbon, oxygen, and nitrogen in the polymer chain

Elastomers may be divided into two main types, thermosetting and thermoplastic elastomers. Sometimes it is difficult to distinguish elastomers from fully plasticized polyvinyls and other plastics.

Thermoplastic elastomers are generally more easily processed. Thermoplastic polyurethane, copolyester, olefins, and styrene copolymers are gaining popularity. They have similar properties to thermosetting (vulcanized) elastomers, but have shorter cycle times.

Appendix F

ENGLISH-METRIC CONVERSION

	If You Know	You Can Get	If You Multiply By*
LENGTH	Inches	Millimetres (mm)	25.4
	Millimetres	Inches	0.04
	Inches	Centimetres (cm)	2.54
	Centimetres	Inches	0.4
	Inches	Metres (m)	0.0254
	Metres	Inches	39.37
	Feet	Centimetres	30.5
	Centimetres	Feet	4.8
	Feet	Metres	0.305
	Metres	Feet	3.28
	Miles	Kilometres (km)	1.61
	Kilometre	Miles	0.62
AREA	Inches^2	Millimetres^2 (mm^2)	645.2
	Millimetres^2	Inches^2	0.0016
	Inches^2	Centimetres^2 (cm^2)	6.45
	Centimetres^2	Inches^2	0.16
	Foot^2	Metres^2 (m^2)	0.093
	Metres^2	Foot^2	10.76
CAPACITY-VOLUME	Ounces	Millilitres (ml)	30
	Millilitres	Ounces	0.034
	Pints	Litres (l)	0.47
	Litres	Pints	2.1
	Quarts	Litres	0.95
	Litres	Quarts	1.06
	Gallons	Litres	3.8
	Litres	Gallons	0.26
	Cubic Inches	Litres	0.0164
	Litres	Cubic Inches	61.03
	Cubic Inches	Cubic Centimetres (cc)	16.39
	Cubic Centimetres	Cubic Inches	0.061
WEIGHT (MASS)	Ounces	Grams	28.4
	Grams	Ounces	0.035
	Pounds	Kilograms	0.45
	Kilograms	Pounds	2.2
FORCE	Ounce	Newtons (N)	0.278
	Newtons	Ounces	35.98
	Pound	Newtons	4.448
	Newtons	Pound	0.225
	Newtons	Kilograms (kg)	0.102
	Kilograms	Newtons	9.807
ACCELERATION	Inch/Sec^2	Metre/Sec^2	.0254
	Metre/Sec^2	Inch/Sec^2	39.37
	Foot/Sec^2.	Metre/Sec^2 (m/s^2)	0.3048
	Metre/Sec^2	Foot/Sec^2	3.280
TORQUE	Pound-Inch (Inch-Pound)	Newton-Metres (N-M)	0.113
	Newton-Metres	Pound-Inch	8.857
	Pound-Foot (Foot-Pound)	Newton-Metres	1.356
	Newton-Metres	Pound-Foot	.737
PRESSURE	Pound/sq. in. (PSI)	Kilopascals (kPa)	6.895
	Kilopascals	Pound/sq. in.	0.145
	Inches of Mercury (Hg)	Kilopascals	3.377
	Kilopascals	Inches of Mercury (Hg)	0.296
FUEL PERFORMANCE	Miles/gal	Kilometres/litre (km/l)	0.425
	Kilometres/litre	Miles/gal	2.352
VELOCITY	Miles/hour	Kilometres/hr (km/h)	1.609
	Kilometres/hour	Miles/hour	0.621
TEMPERATURE	Fahrenheit Degrees	Celsius Degrees	5/9 (F° −32)
	Celsius Degrees	Fahrenheit Degrees	9/5 (C° +32) = F

*Approximate Conversion Factors to be used where precision calculations are *not* necessary

CONVERT CENTIGRADE TEMPERATURE
TO FAHRENHEIT AND VICE VERSA

Centigrade C° = 5/9 (F° − 32)   TEMPERATURE CONVERSION TABLES   To Fahrenheit F° = (9/5 X C°) + 32

C.		F.	C.		F.	C.		F.	C.		F.
−17.8	0	32	8.89	48	118.4	35.6	96	204.8	271	520	968
−17.2	1	33.8	9.44	49	120.2	36.1	97	206.6	277	530	986
−16.7	2	35.6	10.0	50	122.0	36.7	98	208.4	282	540	1004
−16.1	3	37.4	10.6	51	123.8	37.2	99	210.2	288	550	1022
−15.6	4	39.2	11.1	52	125.6	37.8	100	212.0	293	560	1040
−15.0	5	41.0	11.7	53	127.4	38	100	212	299	570	1058
−14.4	6	42.8	12.2	54	129.2	43	110	230	304	580	1076
−13.9	7	44.6	12.8	55	131.0	49	120	248	310	590	1094
−13.3	8	46.4	13.3	56	132.8	54	130	266	316	600	1112
−12.8	9	48.2	13.9	57	134.6	60	140	284	321	610	1130
−12.2	10	50.0	14.4	58	136.4	66	150	302	327	620	1148
−11.7	11	51.8	15.0	59	138.2	71	160	320	332	630	1166
−11.1	12	53.6	15.6	60	140.0	77	170	338	338	640	1184
−10.6	13	55.4	16.1	61	141.8	82	180	356	343	650	1202
−10.0	14	57.2	16.7	62	143.6	88	190	374	349	660	1220
− 9.44	15	59.0	17.2	63	145.4	93	200	392	354	670	1238
− 8.89	16	60.8	17.8	64	147.2	99	210	410	360	680	1256
− 8.33	17	62.6	18.3	65	149.0	100	212	413	366	690	1274
− 7.78	18	64.4	18.9	66	150.8	104	220	428	371	700	1292
− 7.22	19	66.2	19.4	67	152.6	110	230	446	377	710	1310
− 6.67	20	68.0	20.0	68	154.4	116	240	464	382	720	1328
− 6.11	21	69.8	20.6	69	156.2	121	250	482	388	730	1346
− 5.56	22	71.6	21.1	70	158.0	127	260	500	393	740	1364
− 5.00	23	73.4	21.7	71	159.8	132	270	518	399	750	1382
− 4.44	24	75.2	22.2	72	161.6	138	280	536	404	760	1400
− 3.89	25	77.0	22.8	73	163.4	143	290	554	410	770	1418
− 3.33	26	78.8	23.3	74	165.2	149	300	572	416	780	1436
− 2.78	27	80.6	23.9	75	167.0	154	310	590	421	790	1454
− 2.22	28	82.4	24.4	76	168.8	160	320	608	427	800	1472
− 1.67	29	84.2	25.0	77	170.6	166	330	626	432	810	1490
− 1.11	30	86.0	25.6	78	172.4	171	340	644	438	820	1508
− 0.56	31	87.8	26.1	79	174.2	177	350	662	443	830	1526
− 0	32	89.6	26.7	80	176.0	182	360	680	449	840	1544
0.56	33	91.4	27.2	81	177.8	188	370	698	454	850	1562
1.11	34	93.2	27.8	82	179.6	193	380	716	460	860	1580
1.67	35	95.0	28.3	83	181.4	199	390	734	466	870	1598
2.22	36	96.8	28.9	84	183.2	204	400	752	471	880	1616
2.78	37	98.6	29.4	85	185.0	210	410	770	477	890	1634
3.33	38	100.4	30.0	86	186.8	216	420	788	482	900	1652
3.89	39	102.2	30.6	87	188.6	221	430	806	488	910	1670
4.44	40	104.0	31.1	88	190.4	227	440	824	493	920	1688
5.00	41	105.8	31.7	89	192.2	232	450	842	499	930	1706
5.56	42	107.6	32.2	90	194.0	238	460	860	504	940	1724
6.11	43	109.4	32.8	91	195.8	243	470	878	510	950	1742
6.67	44	111.2	33.3	92	196.7	249	480	896	516	960	1760
7.22	45	113.0	33.9	93	199.4	254	490	914	521	970	1778
7.78	46	114.8	34.4	94	201.2	260	500	932	527	980	1796
8.33	47	116.6	35.0	95	203.0	266	510	950	532	990	1814

DECIMAL EQUIVALENTS OF FRACTIONS OF ONE INCH

1/64	.015625	1/4	.250000	31/64	.484375	3/4	.750000
1/32	.031250	17/64	.265625	1/2	.500000	49/64	.765625
3/64	.046875	9/32	.281250			25/32	.781250
1/16	.062500	19/64	.296875	33/64	.515625	51/64	.796875
5/64	.078125	5/16	.312500	17/32	.531250	13/16	.812500
				35/64	.546875		
3/32	.093750	21/64	.328125	9/16	.562500	53/64	.828125
7/64	.109375	11/32	.343750	37/64	.578125	27/32	.843750
1/8	.125000	23/64	.359375			55/64	.859375
9/64	.140625	3/8	.375000	19/32	.593750	7/8	.875000
5/32	.156250	25/64	.390625	39/64	.609375	57/64	.890625
				5/8	.625000		
11/64	.171875	13/32	.406250	41/64	.640625	29/32	.906250
3/16	.187500	27/64	.421875	21/32	.656250	59/64	.890625
13/64	.203125	7/16	.437500			15/16	.937500
7/32	.218750	29/64	.453125	43/64	.671875	61/64	.953125
15/64	.234375	15/32	.468750	11/16	.687500	31/32	.968750
				45/64	.703125		
				23/32	.718750	63/64	.984375
				47/64	.734375	1	1.000000

STANDARD DRAFT ANGLES

Depth	1/4°	1/2°	1°	1½°	2°	2½°	3°	5°	7°	8°	10°	12°	15°	Depth
1/32	.0001	.0003	.0005	.0008	.0011	.0014	.0016	.0027	.0038	.0044	.0055	.0066	.0084	1/32
1/16	.0003	.0006	.0011	.0016	.0022	.0027	.0033	.0055	.0077	.0088	.0110	.0133	.0168	1/16
3/32	.0004	.0008	.0016	.0025	.0033	.0041	.0049	.0082	.0115	.0132	.0165	.0199	.0251	3/32
1/8	.0005	.0010	.0022	.0033	.0044	.0055	.0066	.0109	.0153	.0176	.0220	.0266	.0335	1/8
3/16	.0008	.0016	.0033	.0049	.0065	.0082	.0098	.0164	.0230	.0263	.0331	.0399	.0502	3/16
1/4	.0011	.0022	.0044	.0066	.0087	.0109	.0131	.0219	.0307	.0351	.0441	.0531	.0670	1/4
5/16	.0014	.0027	.0055	.0082	.0109	.0137	.0164	.0273	.0384	.0439	.0551	.0664	.0837	5/16
3/8	.0016	.0033	.0065	.0098	.0131	.0164	.0197	.0328	.0460	.0527	.0661	.0797	.1005	3/8
7/16	.0019	.0038	.0076	.0115	.0153	.0191	.0229	.0383	.0537	.0615	.0771	.0930	.1172	7/16
1/2	.0022	.0044	.0087	.0131	.0175	.0218	.0262	.0438	.0614	.0703	.0882	.1063	.1340	1/2
5/8	.0027	.0054	.0109	.0164	.0218	.0273	.0328	.0547	.0767	.0878	.1102	.1329	.1675	5/8
3/4	.0033	.0065	.0131	.0196	.0262	.0328	.0393	.0656	.0921	.1054	.1322	.1595	.2010	3/4
7/8	.0038	.0076	.0153	.0229	.0306	.0382	.0459	.0766	.1074	.1230	.1543	.1860	.2345	7/8
1	.0044	.0087	.0175	.0262	.0349	.0437	.0524	.0875	.1228	.1405	.1763	.2126	.2680	1
1 1/4	.0055	.0109	.0218	.0327	.0437	.0546	.0655	.1094	.1535	.1756	.2204	.2657	.3349	1 1/4
1 1/2	.0064	.0131	.0262	.0393	.0524	.0655	.0786	.1312	.1842	.2108	.2645	.3188	.4019	1 1/2
1 3/4	.0076	.0153	.0305	.0458	.0611	.0764	.0917	.1531	.2149	.2460	.3085	.3720	.4689	1 3/4
2	.0087	.0175	.0349	.0524	.0698	.0873	.1048	.1750	.2456	.2810	.3527	.4251	.5359	2
Depth	1/4°	1/2°	1°	1½°	2°	2½°	3°	5°	7°	8°	10°	12°	15°	Depth

CONVERSION OF SPECIFIC GRAVITY TO GRAMS PER CUBIC INCH

16.39 x Specific Gravity = Grams/In.³

Specific Gravity	Grams/In.³	Specific Gravity	Grams/In.³
1.20	19.7	1.82	29.8
1.22	20.0	1.84	30.2
1.24	20.3	1.86	30.5
1.26	20.7	1.88	30.8
1.28	21.0	1.90	31.1
1.30	21.3	1.92	31.5
1.32	21.6	1.94	31.8
1.34	22.0	1.96	32.1
1.36	22.3	1.98	32.5
1.38	22.6	2.00	32.8
1.40	22.9	2.02	33.1
1.42	23.3	2.04	33.4
1.44	23.6	2.06	33.8
1.46	23.9	2.08	34.1
1.48	24.3	2.10	34.4
1.50	24.6	2.12	34.7
1.52	24.9	2.14	35.1
1.54	25.2	2.16	35.4
1.56	25.6	2.18	35.7
1.58	25.9	2.20	36.1
1.60	26.2	2.22	36.4
1.62	26.6	2.24	36.7
1.64	26.9	2.26	37.0
1.66	27.2	2.28	37.4
1.68	27.5	2.30	37.7
1.70	27.9	2.32	38.0
1.72	28.2	2.34	38.4
1.74	28.5	2.36	38.7
1.76	28.8	2.38	39.0
1.78	29.2	2.40	39.3
1.80	29.5		

To Determine the Cost/Cu./In.:
Price/Lb. x Sp. Gravity x .03163.
$1.32 x 1.76 x .03163 = $0.09/Cu./In.

DIAMETERS AND AREAS OF CIRCLES

Diam.	Area	Diam.	Area	Diam.	Area	Diam.	Area
1/64 "	.00019	7/8 "	2.7612	11/16 "	17.257	7/8 "	61.862
1/32	.00077	15/16	2.9483	3/4	17.721	9- "	63.617
3/64	.00173	2- "	3.1416	13/16	18.190	1/8	65.397
1/16	.00307	1/16	3.3410	7/8	18.665	1/4	67.201
3/32	.00690	1/8	3.5466	15/16	19.147	3/8	69.029
1/8	.01227	3/16	3.7583	5- "	19.635	1/2	70.882
5/32	.01917	1/4	3.9761	1/16	20.129	5/8	72.760
3/16	.02761	5/16	4.2000	1/8	20.629	3/4	74.662
7/32	.03758	3/8	4.4301	3/16	21.125	7/8	76.589
1/4	.04909	7/16	4.6664	1/4	21.648	10- "	78.540
9/32	.06213	1/2	4.9087	5/16	22.166	1/8	80.516
5/16	.07670	9/16	5.1572	3/8	22.691	1/4	82.516
11/32	.09281	5/8	5.4119	7/16	23.211	3/8	84.541
3/8	.11045	11/16	5.6727	1/2	23.758	1/2	86.590
13/32	.12962	3/4	5.9396	9/16	24.301	5/8	88.664
7/16	.15033	13/16	6.2126	5/8	24.850	3/4	90.763
15/32	.17257	7/8	6.4918	11/16	25.406	7/8	92.886
1/2	.19635	15/16	6.7771	3/4	25.967	11- "	95.033
17/32	.22165	3- "	7.0686	13/16	26.535	1/2	103.87
9/16	.24850	1/16	7.3662	7/8	27.109	12- "	113.10
19/32	.27688	1/8	7.6699	15/16	27.688	1/2	122.72
5/8	.30680	3/16	7.9798	6- "	28.274	13- "	132.73
21/32	.33824	1/4	8.2958	1/8	29.465	1/2	143.14
11/16	.37122	5/16	8.6179	1/4	30.680	14- "	153.94
23/32	.40574	3/8	8.9462	3/8	31.919	1/2	165.13
3/4	.44179	7/16	9.2806	1/2	33.183	15- "	176.71
25/32	.47937	1/2	9.6211	5/8	34.472	1/2	188.69
13/16	.51849	9/16	9.9678	3/4	35.785	16- "	201.06
27/32	.55914	5/8	10.321	7/8	37.122	1/2	213.82
7/8	.60132	11/16	10.680	7- "	38.485	17- "	226.98
29/32	.64504	3/4	11.045	1/8	39.871	1/2	240.53
15/16	.69029	13/16	11.416	1/4	41.282	18- "	254.47
31/32	.73708	7/8	11.793	3/8	42.718	1/2	268.80
1- "	.7854	15/16	12.177	1/2	44.179	19- "	283.53
1/16	.8866	4- "	12.566	5/8	45.664	1/2	298.65
1/8	.9940	1/16	12.962	3/4	47.173	20- "	314.16
3/16	1.1075	1/8	13.364	7/8	48.707	1/2	330.06
1/4	1.2272	3/16	13.772	8- "	50.265		
5/16	1.3530	1/4	14.186	1/8	51.849		
3/8	1.4849	5/16	14.607	1/4	53.456		
7/16	1.6230	3/8	15.033	3/8	55.088		
1/2	1.7671	7/16	15.466	1/2	56.745		
9/16	1.9175	1/2	15.904	5/8	58.426		
5/8	2.0739	9/16	16.349	3/4	60.132		
11/16	2.2465	5/8	16.800				
3/4	2.4053						
13/16	2.5802						

STEAM TEMPERATURE VERSUS GAUGE PRESSURE

Gauge Pressure Lbs.	Temp. Deg. F
50	297.5
55	302.4
60	307.1
65	311.5
70	315.8
75	319.8
80	323.6
85	327.4
90	331.1
95	334.3
100	337.7
105	341.0
110	344.0
115	347.0
120	350.0
125	353.0
130	356.0
135	358.0
140	361.0
145	363.0
150	365.6
155	368.0
160	370.3
165	372.7
170	374.9
175	377.2
180	379.3
185	381.4
190	383.5
195	385.7
200	387.5

WEIGHT OF 1000 PIECES IN POUNDS BASED ON WEIGHT OF ONE PIECE IN GRAMS

Weight Per Piece in Grams	Weight Per 1000 Pieces in Pounds	Weight Per Piece in Grams	Weight Per 1000 Pieces in Pounds
1	2.2	51	112.3
2	4.4	52	114.5
3	6.6	53	116.7
4	8.8	54	118.9
5	11.0	55	121.1
6	13.2	56	123.3
7	15.4	57	125.5
8	17.6	58	127.7
9	19.8	59	129.9
10	22.0	60	132.1
11	24.2	61	134.3
12	26.4	62	136.5
13	28.6	63	138.7
14	30.8	64	140.9
15	33.0	65	143.1
16	35.2	66	145.3
17	37.4	67	147.5
18	39.6	68	149.7
19	41.8	69	151.9
20	44.0	70	154.1
21	46.2	71	156.3
22	48.4	72	158.5
23	50.6	73	160.7
24	52.8	74	162.9
25	55.0	75	165.1
26	57.2	76	167.4
27	59.4	77	169.6
28	61.6	78	171.8
29	63.8	79	174.0
30	66.0	80	176.2
31	68.2	81	178.4
32	70.4	82	180.6
33	72.6	83	182.8
34	74.8	84	185.0
35	77.0	85	187.2
36	79.2	86	189.4
37	81.4	87	191.6
38	83.7	88	193.8
39	85.9	89	196.0
40	88.1	90	198.2
41	90.3	91	200.4
42	92.5	92	202.6
43	94.7	93	204.8
44	96.9	94	207.0
45	99.1	95	209.2
46	101.3	96	211.4
47	103.5	97	213.6
48	105.7	98	215.8
49	107.9	99	218.0
50	110.1	100	220.2

EQUIVALENT WEIGHTS
1 Gram = .0353 Oz.
.0625 Pounds = 1 Ounce = 28.3 Grams
454 Grams = 1 Pound

LENGTH EQUIVALENTS
Millimeters to Inches

Milli-meters	Inches	Milli-meters	Inches	Milli-meters	Inches
1	.03937	34	1.33860	67	2.63779
2	.07874	35	1.37795	68	2.67716
3	.11811	36	1.41732	69	2.71653
4	.15748	37	1.45669	70	2.75590
5	.19685	38	1.49606	71	2.79527
6	.23622	39	1.53543	72	2.83464
7	.27559	40	1.57480	73	2.87401
8	.31496	41	1.61417	74	2.91338
9	.35433	42	1.65354	75	2.95275
10	.39370	43	1.69291	76	2.99212
11	.43307	44	1.73228	77	3.03149
12	.47244	45	1.77165	78	3.07086
13	.51181	46	1.81102	79	3.11023
14	.55118	47	1.85039	80	3.14960
15	.59055	48	1.88976	81	3.18897
16	.62992	49	1.92913	82	3.22834
17	.66929	50	1.96850	83	3.26771
18	.70866	51	2.00787	84	3.30708
19	.74803	52	2.04724	85	3.34645
20	.78740	53	2.08661	86	3.38582
21	.82677	54	2.12598	87	3.42519
22	.86614	55	2.16535	88	3.46456
23	.90551	56	2.20472	89	3.50393
24	.94488	57	2.24409	90	3.54330
25	.98425	58	2.28346	91	3.58267
26	1.02362	59	2.32283	92	3.62204
27	1.06299	60	2.36220	93	3.66141
28	1.10236	61	2.40157	94	3.70078
29	1.14173	62	2.44094	95	3.74015
30	1.18110	63	2.48031	96	3.77952
31	1.22047	64	2.51968	97	3.81889
32	1.25984	65	2.55905	98	3.85826
33	1.29921	66	2.59842	99	3.89763
				100	3.93700

VOLUME EQUIVALENTS
1 c.c. = .061 cu. in.
1 cu. in. = 16.387 c.c.

APPENDIX G

SOURCES OF HELP AND BIBLIOGRAPHY

The following alphabetical list of service organizations, standards and specifications groups, trade associations, professional societies, reference, and U.S. governmental agencies may serve as sources for further information:

American Chemical Society
1155 16th Street, NW
Washington, DC 20036
(202) 872-4600

American Conference of Governmental Industrial
Hygienists (ACGIH)
6500 Glenway Avenue
Cincinnati, OH 45201
(513) 661-7881

American Industrial Hygiene Association (AIHA)
66 S Miller Road
Akron, OH 44130
(216) 762-7294

American Insurance Association (AIA)
85 John Street
New York, NY 10038
(212) 669-0400

American Medical Association (AMA)
535 N Dearborn Street
Chicago, IL 60610
(312) 645-5003

American National Standards Institute (ANSI)
1430 Broadway
New York, NY 10018
(212) 354-3300

American Petroleum Institute
1801 K Street, NW
Washington, DC 20006
(202) 682-8000

(The) American Society for Testing and Materials
(ASTM)
1916 Race Street
Philadelphia, PA 19103
(215) 299-5400

(The) American Society of Mechanical Engineers
(ASME)
United Engineering Center
345 E 47th Street
New York, NY 10017
(212) 705-7722

American Society of Safety Engineers
850 Busse Highway
Park Ridge, IL 60068
(312) 692-4121

Center for Plastics Recycling Research (CPRR)
PO Box 189
Kennett Square, PA 19348
(215) 444-0659

Chemical Manufacturers Association
2501 M Street, NW
Washington, DC 20037
(202) 887-1100

Defense Standardization Program Office (DSPO)
5203 Leesburg Pike, Suite 1403
Falls Church, VA 22041-3466

Department of Defense (DOD)
Office for Research and Engineering
Washington, DC 20301
(202) 545-6700

Department of Transportation (DOT)
Hazardous Materials Transportation
400 7th Street, SW
Washington, DC 20590
(202) 426-4000

Environmental Protection Agency (EPA)
401 M Street, SW
Washington, DC 20460
(202) 829-3535

Factory Mutual Engineering Corporation
1151 Providence Highway
Norwood, MA 02062
    (617) 762-4300

Federal Emergency Management Agency
PO Box 8181
Washington, DC 20024
    (202) 646-2500

Federal Register
U.S. Government Printing Office
Superintendent of Documents
Washington, DC 20402
    (202) 783-3238

Food and Drug Administration (FDA)
200 Independence Avenue
Washington, DC 20204
    (202) 245-6296

General Services Administration (GSA)
Federal Supply Service
18th and F Streets
Washington, DC 20406
    (202) 566-1212

Global Engineering Documentation Services, Inc.
3301 W MacArthur Boulevard
Santa Ana, CA 92704
    (714) 540-9870

Industrial Health Foundation Inc. (IHF)
34 Penn Circle
Pittsburgh, PA 15232
    (412) 363-6600

Instrument Society of America
400 Stanwix Street
Pittsburgh, PA 15222
    (412) 261-4300

International Organization for Standardization
    (ISO)
1 rue de Varembe,
CH 1211
Geneve 20 Switzerland/Suisse

Leidner, Jacob. *Plastics Waste Recovery of Economic Value*. New York: Marcel Dekker, Inc., 1981.

Manufacturing Chemists Association, Inc.
1825 Connecticut Avenue, NW
Washington, DC 20009
    (202) 887-1100

National Association of Manufacturers
1776 F Street, NW
Washington, DC 20006
    (202) 737-8551

National Bureau of Standards (NBS)
Standards Information & Analysis Section
Standards Information Service (SIS)
Building 225, Room B 162
Washington, DC 20234
    (301) 921-1000

National Conference on Weights and Measures
c/o National Bureau of Standards
Washington, DC 20234
    (301) 921-1000

National Fire Protection Association (NFPA)
470 Atlantic Avenue
Boston, MA 02210
    (617) 770-3000

National Institute for Occupational Safety and
    Health (NIOSH)
U.S. Department of Health, Education, and
    Welfare
Parklawn Building
5600 Fishers Lane
Rockville, MD 20852
    (301) 472-7134

National Safety Council
444 N Michigan Avenue
Chicago, IL 60611
    (312) 527-4800

Navy Publications and Printing Service Office
700 Robbins Avenue
Philadelphia, PA 19111
    (215) 697-2000

Occupational Safety and Health Administration
    (OSHA)
U.S. Department of Labor
Department of Labor Building
Connecticut Avenue, NW
Washington, DC 20210
    (202) 523-9361

Office of the Federal Register
1100 "L" Street NW, Rm 8401
Washington, DC 20408
    (202) 523-5240

Plastics Education Foundation
Society of Plastics Engineers, Inc.
14 Fairfield Drive
Brookfield Center, CT 06805
    (203) 775-0471

Safety Standards
U.S. Department of Labor
Government Printing Office (GPO)
Washington, DC 20402
    (202) 783-3238

Society of Plastics Engineers, Inc.
Plastics Education Foundation
14 Fairfield Drive
Brookfield Center, CT 06805
    (203) 775-0471

(The) Society of the Plastics Industry, Inc. (SPI)
1025 Connecticut Avenue, NW
Ste 409
Washington, DC 20036
    (202) 822-6700

Underwriters Laboratories (UL)
333 Pfingston Road
Northbrook, IL 60062
    (312) 272-8800

U.S. Government Printing Office
Superintendent of Documents
Washington, DC 20402
    (202) 783-3238

The following bibliography list may be useful for further study and more detailed discussion of selected topics presented:

Advanced Composites: Conference Proceedings, American Society for Metals, December 2–4, 1985.

Allegri, Theodore. *Handling and Management of Hazardous Materials and Waste.* New York: Chapman and Hall, 1986.

Berngardt, Ernest. *CAE Computer Aided Engineering for Injection Molding.* New York: Hanser Publishers, 1983.

Billmeyer, Fred W. *Textbook of Polymer Science.* 3rd ed. New York: Wiley, 1984.

Brooke, Lindsay. "Cars of 2000: Tomorrow Rides Again!" *Automotive Industries,* May 1986, pp 50–67.

Broutman, L., and R. Krock. *Composite Materials.* 6 vols. New York: Academic Press, 1985.

Budinski, Kenneth. *Engineering Materials: Properties and Selection.* 2nd ed. Reston: Reston Publishing Company, Inc., 1983.

Carraher, Charles E., Jr., and James Moore. *Modification of Polymers.* New York: Plenum Press, 1983.

"Chemical Emergency Preparedness Program Interim Guidance," Revision 1, #9223.01A. Washington, DC: United States Environmental Protection Agency, 1985.

Composite Materials Technology, Society of Automotive Engineers, 1986.

"Defense Standardization Manual: Defense Standardization and Specification Program Policies, Procedures and Instruction," DOD 4120.3-M, August 1978.

Dreger, Donald. "Design Guidelines of Joining Advanced Composites," *Machine Design,* May 8, 1980, pp 89–93.

Dym, Joseph. *Product Design with Plastics: A Practical Manual.* New York: Industrial Press, 1983.

Ehrenstein, G., and G. Erhard. *Designing with Plastics: A Report on the State of the Art.* New York: Hanser Publishers, 1984.

English, Lawrence. "Liquid-Crystal Polymers: In a Class of Their Own," *Manufacturing Engineering,* March 1986, pp 36–41.

English, Lawrence. "The Expanding World of Composites," *Manufacturing Engineering,* April 1986, pp 27–31.

Fitts, Bruce. "Fiber Orientation of Glass Fiber-Reinforced Phenolics," *Materials Engineering,* November 1984, pp. 18–22.

Grayson, Martin. *Encyclopedia of Composite Materials and Components.* New York: John Wiley and Sons Inc., 1984.

Johnson, Wayne, and R. Schwed. "Computer-Aided Design and Drafting," *Engineered Systems,* March/April 1986, pp 48–51.

Kliger, Howard. "Customizing Carbon-Fiber Composites: For Strong, Rigid, Lightweight Structures," *Machine Design,* December 6, 1979, pp 150–157.

Levy, Sidney, and J. Harry Dubois. *Plastics Product Design Engineering Handbook.* 2nd ed. New York: Chapman and Hall, 1984.

Lubin, George. *Handbook of Composites.* New York: Van Nostrand Reinhold Company, Inc., 1982.

Modern Plastics Encyclopedia. Vol 63 (10A), October 1986.

Mohr, G., and others. *SPI Handbook of Technology and Engineering of Reinforced Plastics/Composites.* 2nd ed. Malabar: Robert Krieger Publishing Company, 1984.

Moore, G. R., and D. E. Kline. *Properties and Processing of Polymers for Engineers.* Englewood Cliffs: Prentice-Hall, Inc., 1984.

Naik, Saurabh, and others. "Evaluating Coupling Agents for Mica/Glass Reinforcement of Engineering Thermoplastics," *Modern Plastics,* June 1985, pp 1979–1980.

Plunkett, E. R. *Handbook of Industrial Toxicology.* New York: Chemical Publishing Company, 1987.

Powell, Peter C. *Engineering with Polymers.* New York: Chapman and Hall, 1983.

Richardson, Terry. *Composites: A Design Guide.* New York: Industrial Press, 1987.

Schwartz, Mel. *Fabrication of Composite Materials: Source Book,* American Society for Metals, 1985.

Schwartz, M. M. *Composite Materials Handbook.* New York: McGraw-Hill Book Company, 1984.

Seymour, Ramold B., and Charles Carraher. *Polymer Chemistry.* New York: Marcel Dekker, Inc., 1981.

Shook, Gerald. *Reinforced Plastics for Commercial Composites: Source Book,* American Society for Metals, 1986.

"Standardization Case Studies: Defense Standardization and Specification Program," Department of Defense, March 17, 1986.

Stepek, J. and H. Daoust. *Additives for Plastics.* New York: Springer Verlag, 1983, p 260.

Von Hassell, Agostino. "Computer Integrated Manufacturing: Here's How to Plan for It," *Plastics Technology Productivity Series,* No. 1, 1986.

Wigotsky, Victor. "Plastics are Making Dream Cars Come True," *Plastics Engineering,* May 1986, pp 19–27.

Wigotsky, Victor. "U.S. Moldmakers Battle Foreign Prices for Survival," *Plastics Engineering,* November 1985, pp 22–23.

Wood, Stuart. "Patience: Key to Big Volume in Advanced Composites," *Modern Plastics,* March 1986, pp 44–48.

Index